WEB DEVELOPMENT FOR BUSINESS

*A Complete Approach to Front-End
and Server-Side Development*

George C. Philip

Jakob H. Iversen

The University of Wisconsin Oshkosh

Prospect
Press

Founded in 2014, Prospect Press serves the academic discipline of Information Systems by publishing innovative textbooks across the curriculum including introductory, core, emerging, and upper level courses. Prospect Press offers reasonable prices by selling directly to students. Prospect Press provides tight relationships between authors, publisher, and adopters that many larger publishers are unable to offer. Based in Burlington, Vermont, Prospect Press distributes titles worldwide. We welcome new authors to send proposals or inquiries to Beth.Golub@ProspectPressVT.com.

Editor: Beth Lang Golub
Production Management: Rachel Paul
ePub Conversion: Scribe Inc.
Cover Design: Annie Clark

Web Development for Business
eTextbook
- Edition 1.0
- ISBN: 978-1-943153-89-3
- Available from RedShelf.com and VitalSource.com

Printed Paperback
- Edition 1.0
- ISBN 978-1-943153-93-0
- Available from RedShelf.com

For more information, visit https://www.prospectpressvt.com/textbooks/philip-web-development

CONTENTS

Note that the following appendices are included in the e-textbook only. For those using the paperback version, PDFs of the appendices can be freely downloaded from the title's website at https://www.prospectpressvt.com/textbooks/philip-web-development. (Scroll to the red horizontal menu bar, then click on "Student Resources.")

PREFACE

Welcome to *Web Development for Business* with *HTML, CSS, Bootstrap, JavaScript, Ajax, ASP.NET Web Forms, C#, Razor Pages, and Core MVC.* This book provides a comprehensive presentation of the concepts, techniques, and tools for front-end and back-end development of web applications for business and other real-world examples.

Target Audience

This book is designed for introductory- and intermediate-level web development courses in IS, CIS, IT, and CS programs and in technical programs in web development at two-year and four-year institutions. The breadth and depth of coverage make this book suitable for a two-course sequence. However, curricula with a single web development course may choose from multiple possible tracks, depending on the area of focus, as described below in the section "Course Coverage Options."

Approach

A major challenge in learning web applications is understanding how the functionality of a web application is split between one or more servers and a web browser client, requiring both front-end and back-end development involving many different technologies. This book focuses on the use of both front-end and back-end technologies that have proven themselves enduring and relatively basic. Figure 0-1 shows an overview of the various technologies that are presented in this book for a full-stack treatment of web development.

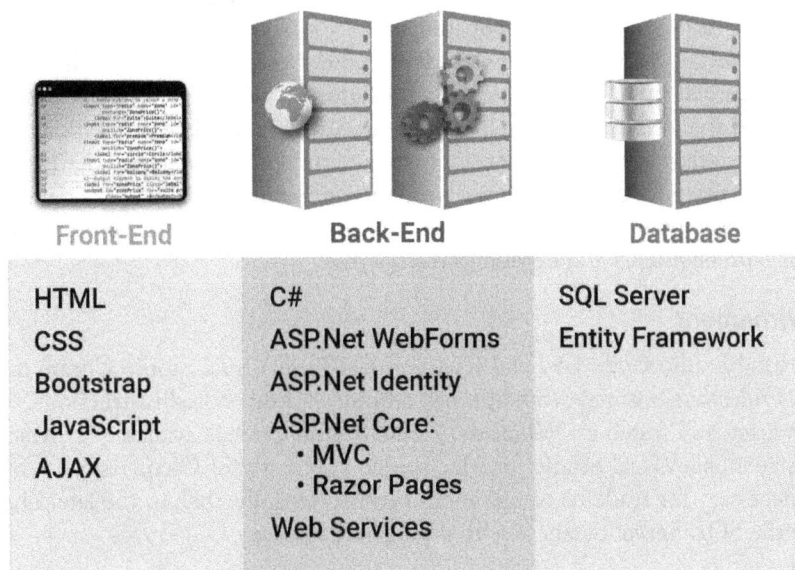

Front-End	Back-End	Database
HTML	C#	SQL Server
CSS	ASP.Net WebForms	Entity Framework
Bootstrap	ASP.Net Identity	
JavaScript	ASP.Net Core:	
AJAX	• MVC	
	• Razor Pages	
	Web Services	

Figure 0-1: Technologies used in the book

With the exception of C#, these technologies are covered in sufficient depth and detail to give students the skills and knowledge to build fully functioning and well-designed websites. Even though small chunks of C# code are used in some of the later chapters, C# programming per se is outside the scope of this book.

In order to integrate the different web development techniques and tools, tutorials guide students to apply them to develop a common application that runs across multiple chapters. Students also get a better understanding of the differences and similarities between different frameworks by developing the same application using multiple frameworks. For example, students will see both ASP.NET Web Forms and Core MVC as approaches to developing websites.

The chapters in this book are organized into four groups, which allows this book to be used in a wide range of scenarios—from a single introductory course on front-end development to a two-course sequence that covers all technologies and frameworks. Various possible tracks are discussed in the "Course Coverage Options." The below section "Overview of the Content and Organization" specifies the prerequisites.

Chapters 1–5 focus on front-end processing. This includes a presentation of information in web pages using HTML, CSS, and Bootstrap, accepting user input with HTML forms and client-side processing using JavaScript. This part could be used in an introductory web development class that is open to any major.

Chapters 6–10 describe the development of ASP.NET Web Forms to build web applications involving server-side processing, database access, and dynamic lists. Web forms, which is the traditional approach that uses drag-and-drop server controls, is particularly suitable for beginners. This part could be used to introduce back-end development to students who have had some exposure to C#.

Chapters 11–15 present user authentication, Ajax, API, and advanced topics in ASP.NET Web Forms.

Chapters 16–18 present Razor Pages and Core MVC frameworks, which are relatively new technologies. These frameworks are more complex, but they provide several advantages over ASP.NET Web Forms, particularly for larger and more complex applications.

In addition, **Appendices A–E** provide the background needed in several areas: data organization, SQL, creating SQL Server databases, object-oriented programming, and creating and running C# console applications in Visual Studio Code and Visual Studio Community. Note that these appendices are included in the e-textbook only. For those using the paperback version, PDFs of the appendices can be freely downloaded from the title's website at https://www.prospectpressvt.com/textbooks/philip-web-development. (Scroll to the red horizontal menu bar, then click on "Student Resources.")

Key Features

A key feature of this book is the breadth of web development tools and frameworks covered, without sacrificing depth: HTML, CSS, Bootstrap, JavaScript, Ajax, ASP.NET Web Forms, C#, Razor Pages, Core MVC, and Web Services. These technologies and frameworks are integrated into a common application to give students an understanding of their use in real-world applications.

This book enhances learning by introducing web development concepts in small chunks through examples and illustrations accompanied by hands-on tutorials. The tutorials, which are interspersed with the concepts, help students explore and apply what they learn immediately. Additionally, review questions and exercises within the chapters, along with comprehensive web development assignments, enhance student interest and learning.

Although the book is written for beginners, it is thorough, concise, and provides considerable depth on a range of more-advanced web development topics. Short code blocks and screenshots are used throughout the book to enhance learning for both beginners and experienced students.

Development Environment

Chapters 1–5 use Visual Studio Code 1.54 and Chapters 6–18 use Visual Studio Community 2019. Both are free for student use. Older versions may work but may require slightly revised instructions.

All tutorials are written and tested on Windows. Visual Studio Code is available for Mac and only requires slightly different instructions. Visual Studio for Mac should also work for Chapters 6–18, but there are differences that make it necessary for students to figure out certain steps. Further, in the later chapters, it would be necessary to replace the SQL Server database with a SQLite database.

Supplements

For Students: The Tutorial-Starts.zip file contains

- Starter projects for tutorials
- Code segments and graphics needed for tutorials
- Data files and databases used in projects

You may download the Tutorial-starts.zip file from the student resources site.

- Visit https://www.prospectpressvt.com/textbooks/philip-web-development.

- Scroll to the horizontal red menu bar.
- Click on "Student Resources."

For Instructors: Instructor resources include

- Completed tutorials
- PowerPoint slides for all chapters
- Test bank
- Starter projects for tutorials
- Code segments and graphics needed for tutorials
- Data files and databases used in projects
- Answers to Review Questions

To access instructor resources

- Visit https://www.prospectpressvt.com/textbooks/philip-web-development.
- Scroll to the red horizontal menu bar.
- Click on "Instructor Resources."
- Click on the yellow box with "Resources Login."
- Click on "Instructor Resources Access Request Form" and fill out the form. Only verified faculty will receive access information.

Installing Visual Studio

You may download Visual Studio Community 2019 and Visual Studio Code from the following website:
https://visualstudio.microsoft.com
You may use the default options for installation. If you need to add more components, you also can add them after installation.

Overview of the Content and Organization

This section contains brief descriptions of the chapters and their dependencies. A basic understanding of the C# language is required for certain chapters, as specified in the prerequisites. Those who are new to programming or new to C# may use online resources like https://www.learncs.org and https://www.w3schools.com/cs to become familiar with programming concepts and C#. You may also learn C# through our other book, available from Prospect Press, *Fundamentals of C# Programming for Information Systems*. You can develop and run C# programs in Visual Studio Code or Visual Studio Community, as described in Appendix E.

Chapter 1 presents the basic concepts of web application development, with a focus on using the HyperText Markup Language (HTML) to structure and give meaning to the contents of a static web page, in the Visual Studio Code environment. Sections 1.1–1.3 of this chapter are prerequisites for all other chapters, and the entire chapter is a prerequisite for Chapters 2–4, 12, and 16–18.

Chapter 2 introduces the use of Cascading Style Sheets (CSS) to control the layout and presentation of web pages. The chapter discusses the separation of HTML content from its presentation, as controlled by CSS rules.

Chapter 3 introduces students to creating responsive web pages using the Bootstrap framework. Students get an introduction to the basic Bootstrap classes, the grid model for controlling layout at different screen widths, and several standard layout controls. Since Bootstrap is based on CSS, Chapter 2 is a prerequisite for this chapter.

Chapter 4 discusses HTML web forms and how they are used to build the user interface for interactive web pages. Chapter 5 develops the JavaScript for client-side processing of input data. Chapter 1 is a prerequisite for this chapter. This chapter is a prerequisite for Chapters 5 and 16–18.

Chapter 5 describes the fundamentals of the JavaScript language. Students use JavaScript to develop client-side code that processes input data, dynamically changes the contents of a page, moves to another page, and shares data between pages. Students also learn how to use jQuery to simplify JavaScript code. Chapter 4 is a prerequisite for this chapter. This chapter is a prerequisite for Chapters 16–18.

Chapters 6–10 discuss the development of ASP.NET Web Forms to build web applications involving server-side processing, database access, and dynamic lists. Students use the ASP.NET framework to develop applications in Visual Studio. These chapters should be covered in sequence and require a basic understanding of C#. While Chapters 2–5 provide a helpful background before starting these chapters, it is possible to move directly from Chapter 1 to Chapter 6.

Chapters 11–15 cover several different advanced topics for building websites. Bootstrap is used in these chapters, so it may be helpful to at least cover Chapter 3 briefly before moving on to these chapters. These chapters can be covered independently. Even though they are designed to follow Chapters 6–10, they may be covered after a review of ASP.NET Web Forms and controls from Chapter 6. A basic understanding of C# is a prerequisite.

Chapter 11 introduces the concept of state and shows students how to store data when moving between pages in an application. The chapter covers several different strategies for managing state, including view state, session state, application state, as well as how to use cookies.

Chapter 12 shows students how to use Master Pages to control common content between pages. This chapter also covers how to build forms using the Bootstrap framework. Chapter 3 is a prerequisite for this chapter.

Chapter 13 discusses user authentication in web forms projects based on the ASP.NET Identity system. The chapter also introduces Entity Framework, since that is the basis for the Identity system. We also show how to control access to pages and folders on a website using role-based authentication.

Chapter 14 covers the Ajax approach to making websites interactive. The focus is primarily on using the five built-in Ajax controls in ASP.NET, followed by a brief discussion of the controls in the Ajax Control Toolkit.

Chapter 15 introduces students to WCF and Web API (REST) web services, and shows them how to build web services as well as how to consume them. The chapter also discusses common ways that web services are used in modern websites.

Chapters 16 presents Razor Pages and its use to create dynamic HTML pages that include server-side code and database access. Students develop applications in Visual Studio using the ASP.NET Core framework. Chapters 4 and 5, and a basic understanding of C#, are prerequisites for this chapter.

Chapters 17 and 18 present the Model View Controller (MVC) framework. Students develop ASP.NET Core MVC applications in Visual Studio. Basic C# and Chapter 17 are prerequisites for Chapter 18.

Here is a brief summary of Appendices A–E:

- Appendix A presents a list of HTML elements.
- Appendix B provides an introduction to SQL and data organization.
- Appendix C shows how to create an SQL Server database in Visual Studio.
- Appendix D describes the basic object-oriented concepts.
- Appendix E shows how to create and run C# console applications in Visual Studio Code and Visual Studio Community.

Note that these appendices are included in the e-textbook only. For those using the paperback version, PDFs of the appendices can be freely downloaded from the title's website at https://www.prospectpressvt.com/textbooks/philip-web-development. (Scroll to the red horizontal menu bar, then click on "Student Resources.")

Course Coverage Options

Here we show how the chapters in the book can be combined in different ways, in either a single course or a two-course sequence.

Single Course

Option 1: Focus on core front-end web technologies: Chapters 1–5 and, if time permits, selected topics from other chapters (e.g., Common Content, Razor Pages, MVC).

Option 2: Focus on ASP.NET Web Forms: Chapters 1 and 6–15. Chapters 2 and 3 will need to be covered briefly before some of the later chapters in this group.

Option 3: Front-end web technologies and ASP.NET Web Forms: Chapters 1–10.

Option 4: Focus on HTML forms and Razor Pages: Chapters 1–5 and 16, and selected topics from other chapters, if time permits (e.g., MVC, Authentication, and Web API).

Option 5: Focus on HTML forms and MVC: Chapters 1–5, 17, and 18, and selected topics from other chapters (e.g., Razor Pages, Authentication, and Web API).

Two-Course Sequence

Option 1

First course: Focus on ASP.NET Web Forms: Chapters 1 and 6–15. Chapters 2 and 3 will need to be covered briefly before some of the later chapters in this group.
Second course: Focus on HTML forms, Razor Pages, and MVC: Chapters 2–5 and 16–18.

Option 2

First course: Focus on HTML forms: Chapters 1–5 and 16.
Second course: Focus on MVC: An overview of ASP.NET Web Forms from Chapters 6–10, in-depth coverage of Chapters 17 and 18, and selected topics from Chapters 11–15.

Option 3

First course: Focus on HTML and ASP.NET forms: Chapters 1–10.
Second course: Advanced Web Technologies: Chapter 11–18.

Acknowledgments

We are thankful for the valuable assistance provided by many people in the preparation of this book.

We were fortunate to work with Beth Lang Golub, editor and president of Prospect Press, who was flexible and supportive of our goals to offer a high-quality comprehensive web development textbook at a reasonable price. Special thanks go to Rachel Paul for her painstaking attention to detail in editing this book, and to Andy Golub and Annie Clark for the cover design.

We wish to acknowledge the contributions of the following reviewers for their valuable guidance in improving the presentation and contents of this book:

Ian Barland, Radford University
Yurii Boreisha, Minnesota State University Moorhead
Kuan-Chou Chen, Purdue University Northwest
Susan Conrad, Marymount University
Silvana Faja, University of Central Missouri
Ross Foultz, Coastal Carolina University
Niki Giles, Augusta Technical College
Hector J. Hernandez, Department of Computer Science, St. Xavier University
Yousra Javed, Illinois State University
Jayne Klenner, PhD, King's College, Wilkes-Barre, PA
Huei Lee, PhD, Professor of CIS, Eastern Michigan University
Thom Luce, Ohio University
Katia Mayfield, Athens State University
Abhijit Kumar Nag, PhD, Department of Computer Information Systems, Texas A&M University-Central Texas
Dr. Chris Olson, Associate Professor, Dakota State University
Pete Partin, University of Saint Francis
Jacqueline C. Pike, Duquesne University
Josie Reynaud, Utah Valley University
Sherri L. Shade, Associate Professor, Kennesaw State University

About the Authors

Dr. George Philip is Professor Emeritus of Information Systems at the College of Business at the University of Wisconsin Oshkosh. He has more than twenty-five years of teaching and consulting experience in the information systems field. He has taught courses at both the undergraduate and graduate levels. He also served as chair of the Information Systems Team, and director of the MS in Information Systems program. He has published numerous articles in this field and a textbook on C# programming.

Dr. Jakob Iversen is a Professor of Information Systems with more than twenty years of experience in higher education teaching, research, and administration. He currently serves as Associate Dean for the College of Business at the University of Wisconsin Oshkosh. He earned his PhD at Aalborg University in Denmark, with a focus on software process improvement. He has been at UW Oshkosh since 2000, where he has taught a range of courses at both the undergraduate and graduate levels in information systems. Throughout his time at UW Oshkosh, he has worked on creating new degree programs in collaboration with other departments and universities in the University of Wisconsin System. As a scholar, he has published in leading academic journals, including *MIS Quarterly*, and is the coauthor of two textbooks on mobile app development and C# programming.

## Chapter 1

Introduction to Web Application Development

This chapter presents the basic concepts of web applications development, with a focus on the HyperText Markup Language (HTML). A distinguishing aspect of web application is that they are installed on web servers and run on browsers on client computers. The World Wide Web (the web) is fundamentally made up of two standards: Hypertext Transfer Protocol (HTTP), which is responsible for transferring data between web servers and clients, and HTML, which is the basic language that structures content and describes the web page to the browser.

This chapter briefly describes how HTTP functions, but primarily is devoted to HTML and demonstrating its use in building static web pages. Special emphasis is given to using HTML for structuring and giving meaning to the contents of a web page using semantic HTML elements, leaving the presentation of content to other tools like Cascading Style Sheets (CSS), which is introduced in this chapter and discussed in detail in the next two chapters. You will create a website and build a simple web page using HTML to display information on a web page.

Learning Objectives

After studying this chapter, you should be able to

- Describe how web applications differ from desktop applications.
- Describe the components of a web application.
- Describe the terms *HTTP request* and *HTTP response*.
- Explain the differences between static and dynamic web pages.
- Describe the structure of an HTML document.
- Explain the difference between block and inline elements, and semantic and non-semantic elements.
- Build static web pages using appropriate HTML elements.
- Apply simple inline CSS styling.

1.1 Introduction to Web Applications

Desktop (Windows Forms) applications like those you may have developed in an introductory programming class are installed and run on client computers running the Windows operating system. Hence, to access a desktop application from a computer, the application must be installed on that computer.

Web applications, on the other hand, are installed and run on a server(s) and accessed through a browser from clients like laptops, smartphones, and tablets over the Internet. Figure 1-1 shows the components of a simple web application that consists of clients, the Internet, and a web server that stores, maintains, and delivers web pages to the clients. To carry out these functions, the server is

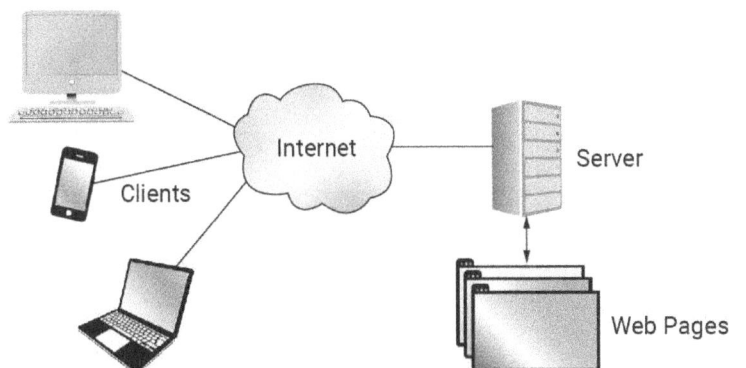

Figure 1-1: Components of a web application

configured with web server software like Microsoft Internet Information Server (IIS), Apache, or Nginx. The term *web server* is often used to refer to the server hardware and/or the software.

Because web applications are accessed through a browser, the clients don't have to be Windows machines.

1.1.1 How Does a Client Display a Web Page from the Server?

From a user's perspective, displaying a web page on a client is simple. You type the URL of the page into a browser, select a bookmark, or click a link. For instance, the following URL would display the ordering page on the **domain** *prospectpressvt* within the top-level domain .com:

> http://www.prospectpressvt.com/ordering

The first two components of the above URL are

`http` The web server protocol used by the client and the server to communicate with each other. Other types of protocols include SMTP (e-mail server protocol) and FTP (File Transfer Protocol).

`www` The default **host name**. If you don't specify a host name, it is assumed to be www. Thus yahoo.com is the same as www.yahoo.com. However, finance.yahoo.com finds the domain yahoo.com on the host *finance*.

Now let's look at what happens behind the screen when you display a web page. We will start with the simpler case of **static** pages and then look at the more general **dynamic** pages.

1.1.2 Static and Dynamic Web Pages

A **static** web page remains the same each time it is displayed unless it is manually altered by a developer. Thus a web page that displays product prices would be a static page if the prices are "hard-coded" into the contents of the page. However, it would be a dynamic page if the prices are read from a file/database and incorporated into the web page each time it is displayed. A dynamic web page typically is created on-the-fly when the page is requested. Dynamic pages generally are interactive, and interactive web pages often are dynamic. For example, a web page on an e-commerce site that lets the user look up the price of a product is both interactive and dynamic. We will use each of these terms to refer to pages with both attributes.

In addition to static and dynamic web pages written in HTML, a website might contain additional static files like .css files that contain formatting information, .js (JavaScript) files that contain client-side scripts, and .cs files containing server-side C# code.

1.1.3 Displaying Static Web Pages

To understand how static web pages are displayed, let's look at how a static page is stored on the server. Though the browser displays a web page with formatted text and graphics, a static web page is stored on the server as a plain text file written using the HTML markup language that specifies the contents of the web page and their structure. Note that HTML is referred to as a markup language rather than a programming language, because it is used primarily to provide structure and meaning to the contents of a web page. The files that contain the HTML markup generally have .htm or .html extensions. The HTML code for a static page typically includes the text displayed on the page. As an example, Figure 1-2 shows a segment of the text displayed on the Prospect Press website.

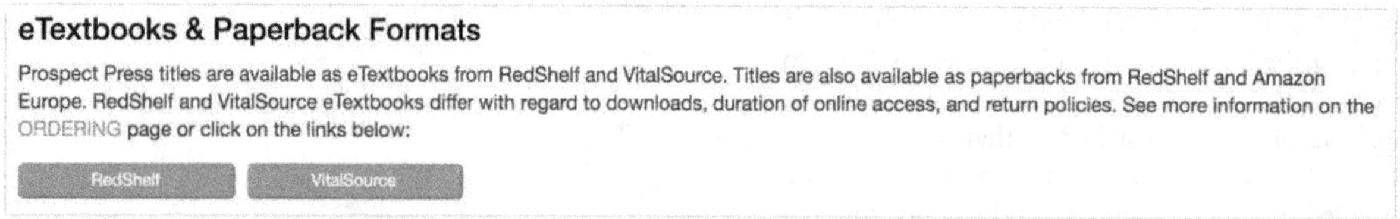

eTextbooks & Paperback Formats

Prospect Press titles are available as eTextbooks from RedShelf and VitalSource. Titles are also available as paperbacks from RedShelf and Amazon Europe. RedShelf and VitalSource eTextbooks differ with regard to downloads, duration of online access, and return policies. See more information on the ORDERING page or click on the links below:

RedShelf   VitalSource

Figure 1-2: A segment of a web page

The corresponding HTML markup is shown in Figure 1-3.

```
1   <h2>eTextbooks & Paperback Formats</h2>
2   <p>Prospect Press titles are available as eTextbooks from RedShelf and VitalSource. These
    are also available as paperbacks from RedShelf and Amazon Europe. RedShelf and VitalSource
    eTextbooks differ with regard to downloads, duration of online access, and return policies.
    See more information on the <a href="/ordering/direct-student-orders/" target=
    "_blank" rel="noopener noreferrer">ORDERING</a> page or click on the links below:</p>
3   <div class="button-section">
4      <a class="cta-small" href="http://redshelf.com" target="_blank"
5         rel="noopener noreferrer">RedShelf</a>
6      <a class="cta-small" href=" https://www.vitalsource.com/" target="_blank"
7         rel="noopener noreferrer">VitalSource</a>
8   </div>
```

Figure 1-3: A segment of HTML markup

We will discuss the structure of HTML later in this chapter.

What happens when you click on a link or type in a URL and press the Enter key to display a static web page? The browser on the client opens the connection to the server and sends a **request** (**HTTP request**, to be more precise) to the web server, which essentially tells the web server, "Hey, get me the file specified in this URL." The server retrieves the file that contains the HTML for the page and any related files, and it sends a **response** (**HTTP response**) that includes the files back to the browser. Thus a static page is delivered to the browser exactly as stored on the server.

When the browser receives the response, it closes the connection, creates the web page as specified in the HTML code, and displays it. The user won't be able to modify or interact with the static page displayed in the browser. The combination of request and response is called a *transaction*. Figure 1-4 shows this process.

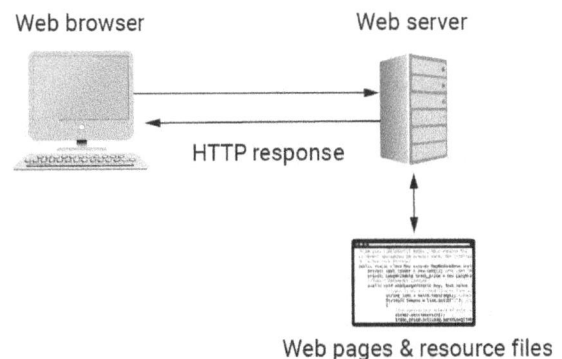

Figure 1-4: HTTP request and response

Dynamic web pages are described in Chapters 5 and 6. In addition to the web server that handles HTTP requests, a website that contains dynamic pages may use an **application server** and a **database server**. The application server provides an environment to execute scripts and create dynamic web pages, while the database server executes the commands that access the database.

In later chapters, you will learn about using **ASP.NET**, which is a platform for building web applications, and the C# language to develop dynamic web pages.

Review Questions

1. What is the difference between desktop (Windows Form) applications and web applications?

2. What is HTTP?

3. What are two differences between dynamic web pages and static web pages?

4. How is a static web page displayed when you click a link to the page? Describe the process of displaying the page.

5. What is the difference between an application server and a database server?

1.2 Creating Your First Website

Over the course of this chapter, you will learn the essentials of the HTML markup language and develop a static web page. Any lightweight text editor, including Notepad and TextEdit, can be used to develop the HTML code. However, there are better choices available beyond what comes standard with the operating system. Notepad++, Sublime Text, and Visual Studio Code are examples of free editors that have better features than Notepad. You also may use the Visual Studio environment, which has a powerful editor. However, for the sake of simplicity and platform independence, we will use Visual Studio Code editor for all examples and exercises in Chapters 1 to 5.

1.2.1 Characteristics of a Good Website

When you create a website, it is important that you are aware of what makes a good website. While a detailed discussion of the characteristics of a good website is outside the scope of this chapter, the following is a list of five areas and typical characteristics within each area that a web developer needs to pay attention to.

Usability
- Simple layout
- Easy to navigate
- Consistent between pages and within pages
- Easy to find information
- Works well in different screen resolutions and on a variety of devices

Appeal
- Attractive layout
- Appropriate color and graphics
- Text is easy to read

Content
- Valuable, accurate, and current
- Interesting and engaging
- Well-organized and easy to understand
- No spelling/grammar errors
- Respects copyright
- Contact information, if applicable

Functionality
- Quick response
- Works correctly and as expected
- Works well in different browsers
- Secure
- Track visitors

Search Engine Optimization
- Use important key words in links
- Use key words frequently
- Structure contents using semantic HTML elements

Keep in mind that the above is a general list, so certain characteristics may not apply to a specific website.

Tutorial: Developing a Simple Web Page: oatmeal-recipe

This tutorial develops a web page named *oatmeal-recipe*, which is a part of the P&I Recipes website. The tutorial introduces you to the basics of HTML and web development.

Figure 1-5 shows the web page to be developed, in two segments. As is typical, the web page includes the header for the website, which displays the site heading *P&I Recipes*, the site logo, and the site navigation links.

The focus of this tutorial is on structuring the contents of the page using HTML. You will significantly improve the appearance of the page using CSS in the next two chapters. Initially, you will develop the page without the site header and then add the site header. The complete page is developed in about twenty steps that are interspersed with the HTML topics to help you practice what you learned in one or more sections and to help you learn the development process.

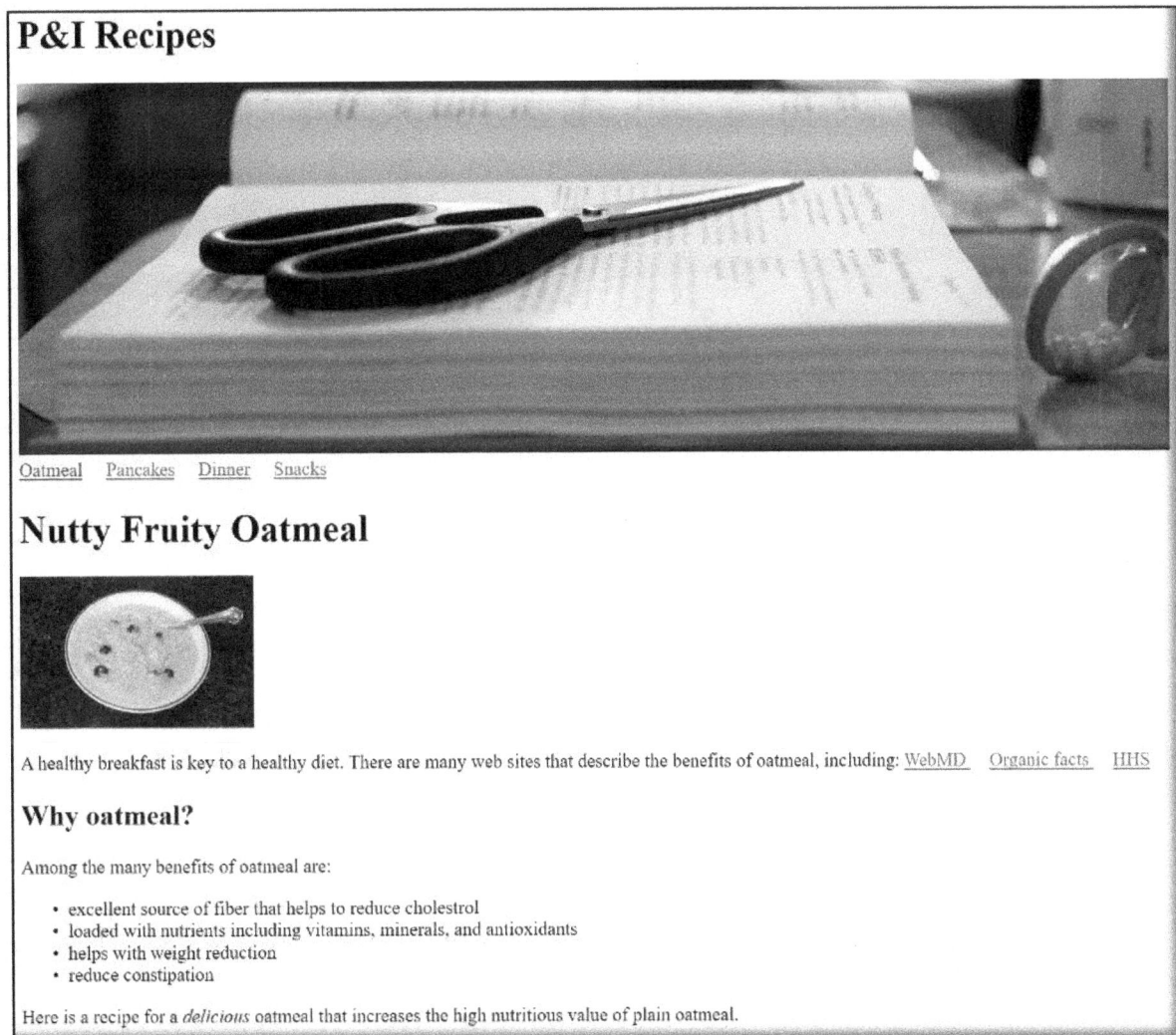

Figure 1-5: Oatmeal-recipe page. Cookbook photo by Scott Akerman/CC BY 2.0 (*continues*).

Ingredients

1/2 cup quick oats

      Quick oats have almost the same nutrition as old fashioned and steel cut oats; but take less time to cook

1/2 cup each of almond and vanila soy milk:

      You may substitute coconut milk or regular milk for soy milk

2 Tbsp chopped walnuts

1/2 cup each chopped banana and blueberries

A pinch each of cardamom and ginger powder

1/4 tsp cinnamon powder

Not sure what quick oats are? Quick oats are oat grains that are cut to smaller pieces and then steamed and rolled.

Directions

Mix all ingredients in a cooking pot, **add one cup of water**, bring to a boil and simmer for a couple of minutes. **Important:** Make sure you stir the mixture **frequently** while simmering.

Nutritional value

Nutritional value of 1/2 cup of
dry quick oats

Nutrition	Amount
Calories	150 cal (kcal)
Carbohydrates	27 g
Protein	5 g
Fat	3 g
Dietary fiber	4 g

You may find the nutritional value of almost any food from sources like Nutrition Value and What's In Food

Contacts: J. Iversen, iversen@uwosh.edu; G. Philip, philip@uwosh.edu

Figure 1-5: Oatmeal-recipe page. Cookbook photo by Scott Akerman/CC BY 2.0 (*continued*).

1.2.2 Working with Visual Studio Code

Visual Studio Code (VS Code) is a lightweight yet powerful source code editor, available for Windows, macOS, and Linux. VS Code provides a wide array of features, including syntax highlighting and IntelliSense, to support the development of HTML documents and JavaScript code. Support is also available for other languages, like Node.js, C#, Python, and PHP. You will be introduced to many of these languages as you develop the oatmeal-recipe website. Note that VS Code is available for free download at https://code.visualstudio.com/Download.

Step 1: Install Visual Studio Code.

You may use the default settings for installation.

Visual Studio Code opens, displaying the Release Notes.

Close VS Code.

Step 2: Open VS Code.

You will see the Welcome page, similar to the one shown in Figure 1-6.

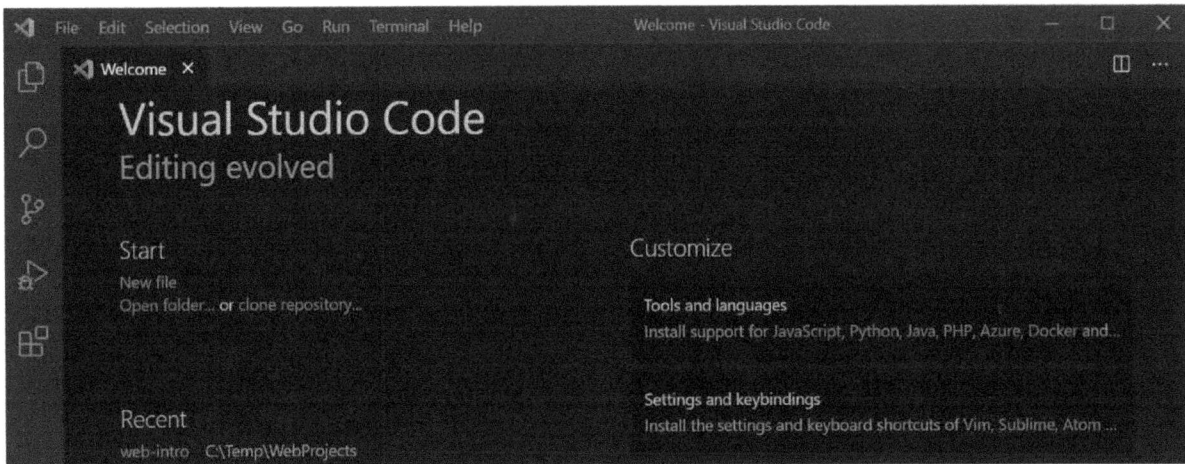

Figure 1-6: The Welcome page for VS Code

In VS Code, you may create one or more HTML files and save them to a common folder. If you open the folder in VS Code, you will have all files available so that you can work with them.

To work with more complex websites, VS Code also lets you create a **workspace**, which is essentially a collection of one or more project folders and related information like configuration data. When you save the workspace, VS Code creates a file named file ***workspacename*.code-workspace** that stores information about the workspace, including workspace settings. Opening this file opens the workspace with all its folders and settings.

Step 3: Create a project folder named *web-intro*.

Open Windows Explorer (or Finder on Mac) to create a folder named *WebProjects* on your local drive (e.g., C:) that will serve as the common folder for all your web projects, including the current one.

Create a project folder named *web-intro* within *WebProjects*.

Step 4: Add a workspace named *web-intro*.

Select *File > Add folder to Workspace from the* VS Code menu bar.

Select *web-intro*, and click *Add*.

Open the *Explorer* window of VS Code if it is not already open. (If Explorer is not displayed, select *View > Explorer*, or click the *Explorer* button on the toolbar.) It shows an untitled workspace that contains the single folder, *web-intro* (Figure 1-7).

From the VS Code menu, select *File > Save Workspace As*.

Because we will have only one project folder in the workspace, the workspace will be given the same name as the folder. Type in **web-intro** for *file name*, and click *Save*.

The Explorer should show the workspace name WEB-INTRO (Figure 1-8).

The web-intro folder now contains the workspace file, *web-intro.code-workspace*.

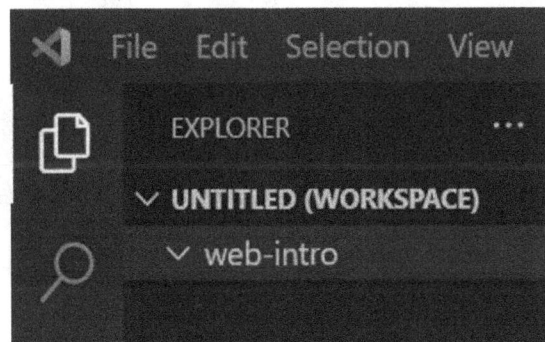

Figure 1-7: Untitled workspace

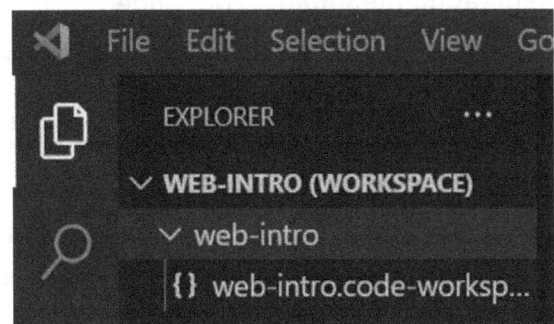

Figure 1-8: Named workspace

Close/Open Workspace

To close the workspace, select *File > Close Workspace*.
To open the closed workspace, select *File > Open Workspace*.
Select the workspace file, *web-intro.code-workspace*.
Instead of opening the workspace, you may open the single folder, *web-intro*, as follows:
Select *File > Open Folder*.
Select the folder, *web-intro*, and click *Open*.

Change Settings and Add an HTML File

You add HTML files to a folder as follows:
Right-click the folder name and select *New File*.
Type in the name of the file with the extension, .html.

Step 5: Change settings to automatically save changes.

Select, *File, Preferences, Settings*.

Under *User* tab, make sure *AutoSave* is set to *onFocusChange*. Close the *Settings* window.

Step 6: Add a .html file named *oatmeal-recipe.html* **to the folder** *web-intro*.

In VS Code Explorer, right-click the *web-intro* folder.

Select *New File*. Type in **oatmeal-recipe.html** for the file name. Press the *Enter* key.

The Explorer will display the newly added file. The window on the right of the screen shows the empty .html file (displaying line number 1; see Figure 1-9).

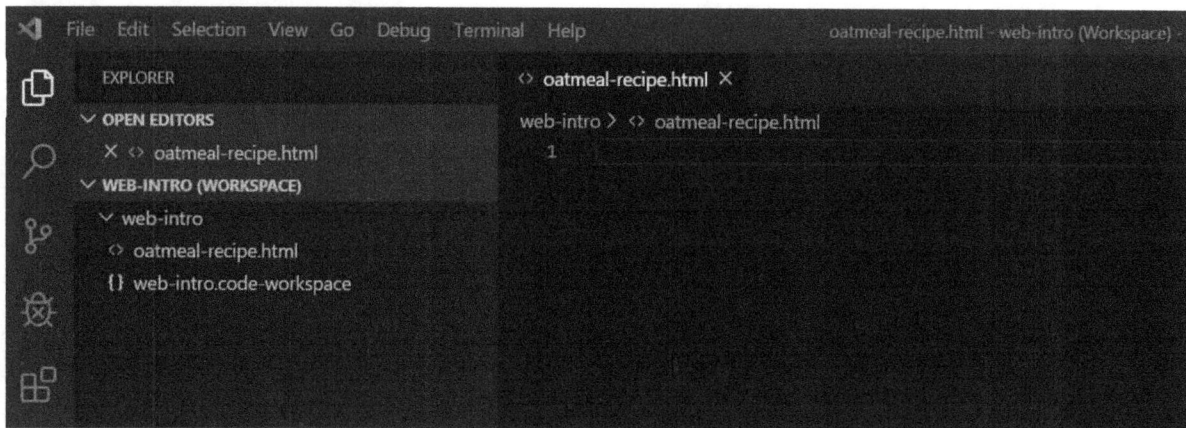

Figure 1-9: The Explorer window

Generating the Initial HTML Code

You may generate the basic elements of an HTML document by typing in a single exclamation symbol, "!", in the .html file.

Step 7: Generate the basic elements of the HTML document.

Type in a single exclamation symbol on line 1 and press the *Enter* key.

VS Code generates the following HTML code (Figure 1-10). If it doesn't, make sure you specified the file extension (.html). You may delete the file and add it with the extension.

```
<> oatmeal-recipe.html ●

web-intro > <> oatmeal-recipe.html > ⊘ html > ⊘ head > ⊘ meta
1    <!DOCTYPE html>
2    <html lang="en">
3    <head>
4        <meta charset="UTF-8">
5        <meta name="viewport" content="width=device-width, initial-scale=1.0">
6        <meta http-equiv="X-UA-Compatible" content="ie=edge">
7        <title>Document</title>
8    </head>
9    <body>
10
11   </body>
12   </html>
```

Figure 1-10: The initial HTML code

Essentially, the initial code provides the basic structure of an HTML document, discussed shortly.

For now, you will change the title and insert a heading into the document.

Step 8: Change the title and insert a heading.

In line 7, change the title ("Document") to "Oatmeal Recipe", as shown here.

```
<title>Oatmeal Recipe</title>
```

Add the heading "Nutty Fruity Oatmeal" into the body of the page:

```
<body>
    <h1>Nutty Fruity Oatmeal</h1>
</body>
```

Displaying an HTML file in the Browser

You can open an HTML file in the default browser by double-clicking the file in Windows Explorer (or in Finder on Mac computers). You may right-click the file in the VS Code Explorer and select *Reveal in Explorer* to open Windows Explorer.

However, there are extensions that you can install in VS Code so that you can display HTML files using shortcut keys.

Step 9: Install an extension to display HTML files in the browser.

Within VS Code, click the *Extensions* button (fourth button on the toolbar on the left).

Type in "open in browser" in the *Search* area, as shown in Figure 1-11.

Click the *Install* button to install the first app in the list. If the app doesn't show an Install button next to it, the app is already installed.

Click *View > Explorer* to display the Explorer window.

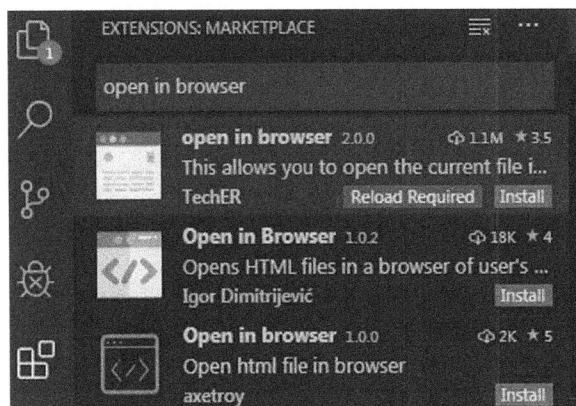

Figure 1-11: The Extensions Marketplace window

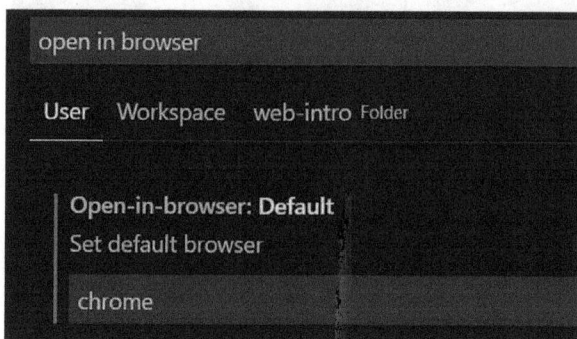

Figure 1-12: The Settings window

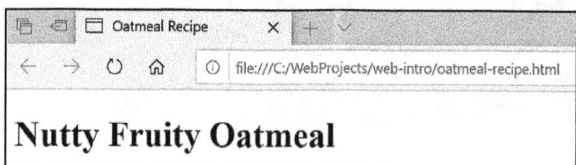

Nutty Fruity Oatmeal

Figure 1-13: The initial oatmeal-recipe page

Step 10: Change the default browser in VS Code.

Microsoft Edge is the default browser in VS Code. You may change the default browser as follows:

Select *File > Preferences > Settings*.

Type in "open in browser" in the search area, and type in the name of the browser (e.g., Chrome, Edge, or Firefox) that you want to set as default, as shown in Figure 1-12.

Close the Settings window.

Step 11: Display oatmeal-recipe.html in the browser.

Make sure the oatmeal-recipe.html file is open.

Press **Alt+b** to display the page in the default browser.

Figure 1-13 shows how the page is displayed in Edge.

1.3 Introduction to HTML

As stated earlier, **HTML** (Hypertext Markup Language) is a markup language that is used to structure and display the contents of a web page in a browser. The original intent of HTML was to add simple formatting tags to text—to *mark up* how the browser should display the text. For example, the **tag** <b> specifies bold and <i> specifies italics. These HTML tags were used in pairs to enclose some text that should be displayed in bold or italics, like this:

```
This is some <b>bold</b> and <i>italics</i> text.
```

In this example, bold would be in bold font and italics would be in italics font when displayed in the browser, as follows:

This is some **bold** and *italics* text.

Early versions of HTML primarily were concerned with specifying the appearance of the page, like boldface, italics, spacing, and color.

Over time, HTML has evolved to become more adept at structuring information in a consistent and standard way, to make it clear to humans and also to machines (e.g., web crawlers and screen readers) what the contents of a web page are. Thus HTML provides meaning (semantics) to the information presented. HTML elements that provide such meaning are called **semantic elements**.

The current recommendation is to use HTML primarily for the semantic structuring of information and to use Cascading Style Sheet (CSS) or other tools for the presentation of information. We will focus on using semantic elements in this chapter to structure the content and discuss CSS in the next two chapters to deal with the appearance of the page. To help understand these two aspects of HTML better, we will examine the basic structure of an HTML document.

1.3.1 Structure of an HTML Document

Elements

HTML is fairly easy to understand. It is made up of elements that start with an opening tag (start tag) and end with a closing tag (end tag), with contents, if any, in between. Here is an example of the title element of an HTML document, which specifies the title to be displayed on the browser tab and in search engine results:

```
<title>Oatmeal Recipe</title>
```

Here, `<title>` is the opening tag, "Oatmeal Recipe" is the content, and `</title>` is the ending tag.

- *Opening tag:* The opening tag consists of the name of the element and optional attributes, enclosed in a pair of angle brackets. In this example, *title* is the name of the element.
- *Closing tag:* The closing tag is similar to the opening tag, except that there is a forward slash before the name of the element.
- *Content:* The content, if any, is the content of the element.

In general, HTML is not case sensitive. Modern browsers try hard to render HTML as best they can and be forgiving of any mistakes made by the programmer. So, the element name *TITLE* is the same as *title*. However, while the original versions of HTML were not case sensitive, later versions like XHTML and HTML5 are. For a document to be fully compliant with the latest HTML versions, element and attribute names must be in lowercase letters.

It should be noted that the term *tag* is often used in place of *element*, which could be confusing. We will use the term *tag* to mean opening and closing tags, not *element*.

Attributes of an Element

An opening tag of an element may contain one or more attributes that provide information pertaining to the element. For example, the following anchor element (`<a>`), which creates a hyperlink to another page, includes an attribute named `href`, which specifies the link to another web page.

```
<a href="https://www.webmd.com/diet/oatmeal-benefits#1"> WebMD </a>
```

Here, "WebMD" is the content of the element, and `href=https://www.webmd.com/diet/oatmeal-benefits#1` is the attribute. The attribute name (`href`) is followed by an equal sign and the value of the attribute `https://www.webmd.com/diet/oatmeal-benefits#1` within quotes. The anchor element is described later in more detail.

Elements without an End Tag

Some elements, like `img`, may not have any content or end tag:

```
<img src="Nutty-fruity-oatmeal.jpg" width="250" alt="Nutty & Fruity Oatmeal" />
```

The `img` element defines an image to be displayed on the page. The `src` attribute specifies the filename /URL of the image, the `width` attribute specifies the width, and the `alt` attribute describes what the image is. The browser retrieves the image and displays it on the page when the page is loaded, while the value of the alt attribute is commonly used by web crawlers and screen readers. As you can see, this element doesn't have any content or an end tag, with everything contained in the opening tag with a slash at the end to close the tag.

Before you can use elements to build a web page, you need to understand the flow of content in an HTML page.

Line Breaks and White Spaces

An important characteristic of HTML is that **HTML ignores any line breaks and extra white spaces in the code**. Thus, it doesn't matter whether there are spaces between two elements, or they are placed in the same line or different lines. Similarly, extra spaces between words and line breaks in a block of text are ignored.

To become familiar with the elements, their syntax, and how a block of text is displayed in a browser, let's type in some HTML code with two elements and a block of text, and display the page.

Step 12: Enter the initial HTML code for the web page.

Insert the code shown in Figure 1-14 below the heading element you already typed into the oatmeal-recipe page.

```
1   <body>
2       <h1>Nutty & Fruity Oatmeal</h1>
3       <img src="img/Nutty-fruity-
         oatmeal.jpg" width="250" alt="Nutty & Fruity Oatmeal" />
4       A healthy breakfast is key to a helthy diet.There are many web sites
5       that describe the benefits of oatmeal, including:
6       <a href="https://www.webmd.com/diet/oatmeal-benefits#1"> WebMD </a>
7   </body>
```

Figure 1-14: Initial HTML code for the oatmeal-recipe page

Create a new folder named *img* to the project folder *WebProjects/web-intro* as follows:

Right-click the project name *web-intro* within the Windows Solution Explorer.

Select *Add > New Folder*.

Change the name of the new folder to *img*.

Copy the image file *Nutty-fruity-oatmeal.jpg* from the *Tutorial-starts\Data-files\img* folder to the *img* folder (within *WebProjects\web-intro*).

Display the web page in the browser by clicking the browser name in the toolbar.

Depending on the width of the browser window, the text that follows the image may be in line with the image shown in Figure 1-15.

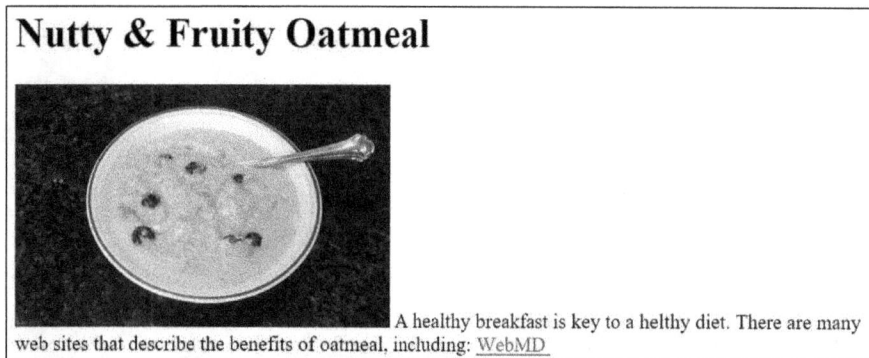

Figure 1-15: Text in line with the image

Note that the line break between lines 4 and 5 does not affect how the text is displayed. Further, the block of text is displayed on the same line as the image, and the link is displayed on the same line as the text. Thus all line breaks are ignored, and the extra spaces between the two sentences within the block of text also are ignored. The number of lines displayed in a browser is determined by the size of the browser window. For now, don't worry about the appearance of the contents.

Click the *Close* button of the browser to close it, and click the *Stop Debugging* button to stop the application.

So, how can we add line breaks? Later, you will use *block elements* like paragraph (<p>), lists (<ul>, <ol>, and <dl>), and table (<table>) to control the layout of content. You also will learn to use the line break (<br>) element for special cases like an address where the name, street, city, and zip code are displayed on separate lines.

Let's look at how additional spaces are displayed on a web page using the special HTML character called a *non-breaking space* ( ).

Non-breaking Space ( )

You can use the special HTML character   (**non-breaking space**) to add an extra space between words or elements when the page is displayed in the browser. Thus, to add three extra spaces, you would add three   characters in sequence, as in   . The space added by   is non-breaking, meaning that words or elements separated by   will stick together in the same line when the page is displayed and will not break into a new line.

Next, you will add links to two more websites and use the   character to provide some space between the links.

Step 13: Add non-breaking spaces.

Insert links to two more websites as shown in Figure 1-16, lines 7 and 8.

```
 1  <body>
 2      <h1>Nutty & Fruity Oatmeal</h1>
 3      <img src="img/Nutty-fruity-
           oatmeal.jpg" width="250" alt="Nutty & Fruity Oatmeal" />
 4      A healthy breakfast is key to a helthy diet.There are many web sites that
 5      describe the benefits of oatmeal, including:
 6      <a href="https://www.webmd.com/diet/oatmeal-benefits#1"> WebMD </a>
           
 7      <a href="https://www.organicfacts.net/oatmeal.html"> Organic facts </a>
           
 8      <a href="https://www.hhs.gov/fitness/eat-healthy/
 9          dietary-guidelines-for-americans/index.html"> HHS </a>
10  </body>
```

Figure 1-16: Non-breaking spaces

Make sure you add the   characters at the end of lines 6 and 7 to provide extra space between the links and to keep them together in the same line.

Display the page. Note the spaces between links, as shown in Figure 1-17.

Close the browser.

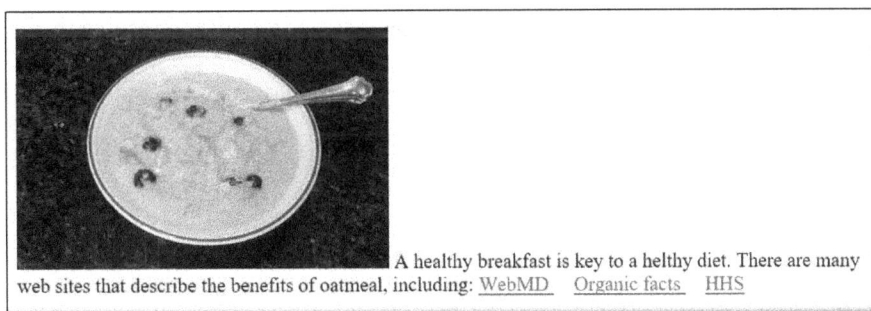

A healthy breakfast is key to a helthy diet. There are many web sites that describe the benefits of oatmeal, including: WebMD   Organic facts   HHS

Figure 1-17: Non-breaking spaces between links

HTML Document

The general structure of an HTML document has the pattern shown in Figure 1-18.

```
1  <!DOCTYPE html>
2  <html>
3     <head>
4            <title>Title of the page</title>
5            <meta attributes>
6     </head>
7     <body>
8
9     </body>
10 </html>
```

Figure 1-18: Structure of an HTML document

Here we will take a brief look at the key elements of this document.

DOCTYPE Declaration

The HTML document starts with the DOCTYPE declaration, which tells the browser that the version of HTML used is *html*, which means the current version, HTML5. The DOCTYPE declaration looks as follows:

```
<!DOCTYPE html>
```

html (<html>) Element

The entire HTML code is wrapped in the html element:

```
<html>
      (head element)
      (body element)
</html>
```

Note that the head and body elements are nested within the html element. Similarly, the title and meta elements are nested within the head element, as shown in Figure 1-18.

Head Element

The head element contains information about the page that is not displayed on the page for the user. The head must contain the title element, which specifies the title that is displayed on the browser tab and in search engine results.

The head may also contain CSS styles and additional meta elements as shown to specify metadata, which is information about the document, like the name of the author, key words, and the description that appears in search engine results. The meta element includes the *name* and *content* attributes.

The head element in Figure 1-19 shows the runat="server" attribute, the title element, and examples of some common meta elements.

```
1  <head runat="server">
2         <title>Oatmeal Recipe</title>
3         <meta name="description" content="Nutty and Fruity Oatmeal recipe">
4         <meta name="keywords" content="oatmeal, recipe, oats, wholegrain">
5  </head>
```

Figure 1-19: The runat="server" attribute

Body Element

The body contains everything you see on the web page. A wide variety of elements are used to define the structure and meaning of the contents of the body element. Next, we will turn to look at some more commonly used HTML elements.

1.4 HTML Elements

As stated earlier, this chapter focuses on structuring the contents of a page using HTML semantic elements. The presentation of the contents using CSS is discussed in detail in the next two chapters. In the rest of this chapter, you will learn about the more common use of elements and gain a background to explore additional features from online resources. You will also get a brief peek into CSS.

1.4.1 Block and Inline Elements

Block-level elements display the contents on a new line and cause a line break at the end so that any material that follows is displayed on a new line. They occupy the entire width of their parent element. The headings (`<h1>` ... `<h6>`), paragraph (`<p>`), and lists (`<ul>`, `<ol>`, `<li>`, and `<dl>`) described in this section are block-level elements.

Note that the anchor (`<a>`) and image (`<img>`), presented earlier, are not block elements. They are **inline elements**, which do not cause their contents to be displayed on a new line. They occupy only the space needed by their contents. Thus they allow contents of other elements to be in the same line on the left and/or right.

We start with a few commonly used `block` elements.

Headings <h1> to <h6>

HTML provides six different elements, `<h1>`, `<h2>`, `<h3>`, `<h4>`, `<h5>`, and `<h6>`, to specify six levels of headings, with `<h1>` for the most important heading and `<h6>` for the least important. By default, each heading element has a specific font size and appearance. The effect of specifying heading elements `<h1>`, `<h3>`, and `<h5>` is shown in Figure 1-20.

```
<h1>Nutty & Fruity Oatmeal</h1>
<h3>Nutty & Fruity Oatmeal</h3>
<h5>Nutty & Fruity Oatmeal</h5>
```

Nutty & Fruity Oatmeal

Nutty & Fruity Oatmeal

Nutty & Fruity Oatmeal

Figure 1-20: Effect of applying `<h1>`, `<h3>`, and `<h5>`

Paragraph <p>

The paragraph element `<p>` specifies that the content is a separate paragraph, which is a block of text or elements separated from other blocks by white spaces before and after.

As discussed earlier, the browser ignores any line breaks and extra white spaces between words or elements. The number of lines displayed in a browser depends on the size of the browser window.

Let's use the paragraph element to display text that immediately follows an image and links to three websites as a separate paragraph, as shown in Step 14.

Step 14: Insert paragraph elements.

Insert the paragraph elements into the oatmeal-recipe page, as shown in Figure 1-21.

```
1   <body>
2       <h1>Nutty & Fruity Oatmeal</h1>
3       <img src="Nutty-fruity-oatmeal.jpg" width="250" alt="Nutty & Fruity Oatmeal"/>
4       <p>
5           A healthy breakfast is key to a helthy diet. There are many web sites that
6           describe the benefits of oatmeal, including:
7           <a href="https://www.webmd.com/diet/oatmeal-benefits#1">WebMD</a>
8           <a href="https://www.organicfacts.net/oatmeal.html"> Organic facts </a>
9           <a href="https://www.hhs.gov/fitness/eat-healthy/
10              dietary-guidelines-for-americans/index.html"> HHS </a>
11      </p>
12  </body>
```

Figure 1-21: The paragraph element

Test your code by displaying the page in a browser. Note that the block of text that is marked up as a paragraph starts on a new line. Any content that is added after the paragraph will be displayed on a new line. As before, the line breaks within the text and between the text and the links are ignored.

Close the browser and stop the application.

It's time to practice! Do Exercise 1-1 (see the end of the chapter).

Unordered List <ul>

HTML provides three different elements to display a list of items:

- Unordered List <ul>
- Ordered List <ol>
- Description List <dl>

Each item within unordered and ordered lists is marked by the list element <li>.

Figure 1-22 shows how you would mark up an unordered list of ingredients for an oatmeal recipe.

```
1   <ul>
2       <li> 1/2 cup quick oats </li>
3       <li> 1/2 cup each of almond and vanilla soy milk:  </li>
4       <li> 2 Tbsp chopped walnuts </li>
5       <li> 1/2 cup each of chopped banana and blueberries </li>
6       <li> A pinch each of cardamom and ginger powder </li>
7   </ul>
```

Figure 1-22: HTML for an unordered list

By default, the list is displayed indented, with bullets for each item (Figure 1-23).

The item markers can be changed to circles, squares, and more (or removed entirely) using tools like CSS.

- 1/2 cup quick oats
- 1/2 cup each of almond and vanilla soy milk:
- 2 Tbsp chopped walnuts
- 1/2 cup each of chopped banana and blueberries
- A pinch each of cardamom and ginger powder

Figure 1-23: An unordered list in the browser

Ordered List <ol>

Ordered lists are marked by the element <ol>, as shown in Figure 1-24.

```
1   <ol>
2       <li> 1/2 cup quick oats </li>
3       <li> 1/2 cup each of almond and vanilla soy milk </li>
4       <li> 2 Tbsp chopped walnuts </li>
5       <li> 1/2 cup each chopped banana and blueberries </li>
6       <li> A pinch each of cardamom and ginger powder </li>
7   </ol>
```

Figure 1-24: HTML for an ordered list

By default, the items are marked by numbers, as shown in Figure 1-25.

You also may display items with uppercase or lowercase letters or with Roman numeral markers by specifying the item type. This is how you would specify lowercase Roman numerals:

```
<ol type="i">
```

1. 1/2 cup quick oats
2. 1/2 cup each of almond and vanilla soy milk
3. 2 Tbsp chopped walnuts
4. 1/2 cup each chopped banana and blueberries
5. A pinch each of cardamom and ginger powder

Figure 1-25: An ordered list in the browser

Other options are `type="I"` for uppercase Roman numerals; `type="A"` for uppercase letters; or `type="a"` for lowercase letters.

Let's use an unordered list to display a few benefits of oatmeal on your web page.

Step 15: Display information on the health benefits of oatmeal.

Add HTML code to display the information shown in Figure 1-26.

Note that lists are block elements. Thus spaces will be automatically added before and after the list.

Display the page to test your code. You may change the list to an ordered list and display the list with different markers, and then change it back to an unordered list.

Close the browser and stop the application.

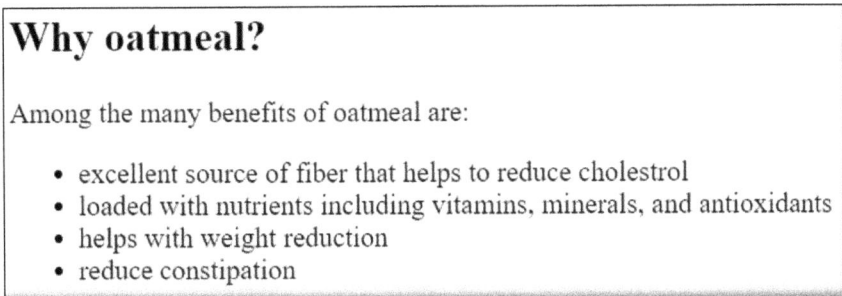

Why oatmeal?

Among the many benefits of oatmeal are:

- excellent source of fiber that helps to reduce cholestrol
- loaded with nutrients including vitamins, minerals, and antioxidants
- helps with weight reduction
- reduce constipation

Figure 1-26: A segment of a web page

Description List <dl>

The description list <dl> allows you to display a list of items with an optional description for each item. Each item (term) within the list is marked by the element <dt>, and each description is marked by the element <dd>, as shown in Figure 1-27.

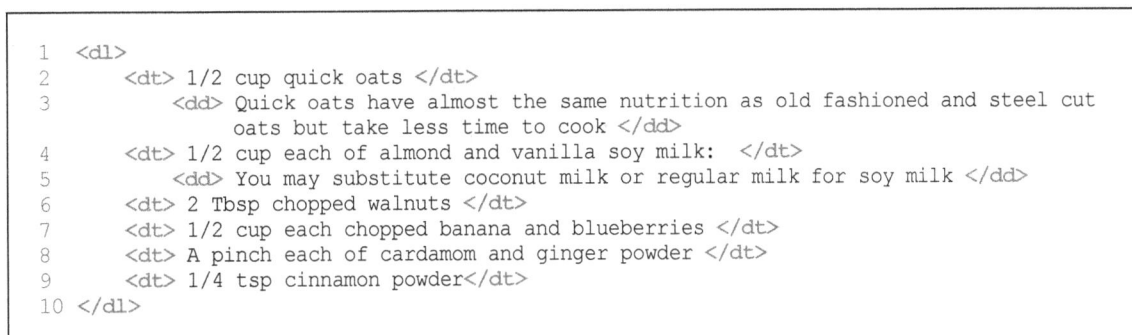

```
1   <dl>
2       <dt> 1/2 cup quick oats </dt>
3           <dd> Quick oats have almost the same nutrition as old fashioned and steel cut
                oats but take less time to cook </dd>
4       <dt> 1/2 cup each of almond and vanilla soy milk:  </dt>
5           <dd> You may substitute coconut milk or regular milk for soy milk </dd>
6       <dt> 2 Tbsp chopped walnuts </dt>
7       <dt> 1/2 cup each chopped banana and blueberries </dt>
8       <dt> A pinch each of cardamom and ginger powder </dt>
9       <dt> 1/4 tsp cinnamon powder</dt>
10  </dl>
```

Figure 1-27: The HTML for a description list

Description lists are displayed with no item markers (bullets, numbers, etc.), as shown in Figure 1-28.

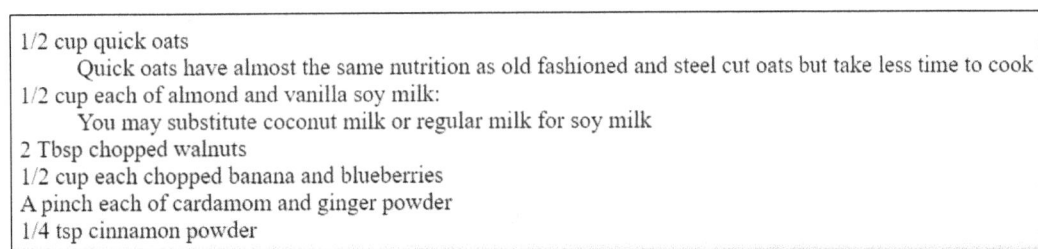

1/2 cup quick oats
        Quick oats have almost the same nutrition as old fashioned and steel cut oats but take less time to cook
1/2 cup each of almond and vanilla soy milk:
        You may substitute coconut milk or regular milk for soy milk
2 Tbsp chopped walnuts
1/2 cup each chopped banana and blueberries
A pinch each of cardamom and ginger powder
1/4 tsp cinnamon powder

Figure 1-28: A description list in the browser

You may nest other HTML elements, including another list, image, or link, within a list. For example, you may want to create a list that looks like a description list, but the items need to be numbered. You can do this by creating an ordered list with an unordered list nested within each list element `<li>` that needs a description.

Step 16: Add a Level 2 heading and definition list.

Add HTML code to your web page to display the introductory text, the level 2 heading *Ingredients,* and a description list of ingredients as shown in Figure 1-29.

Display the page to test your code. Close the browser window and stop the application.

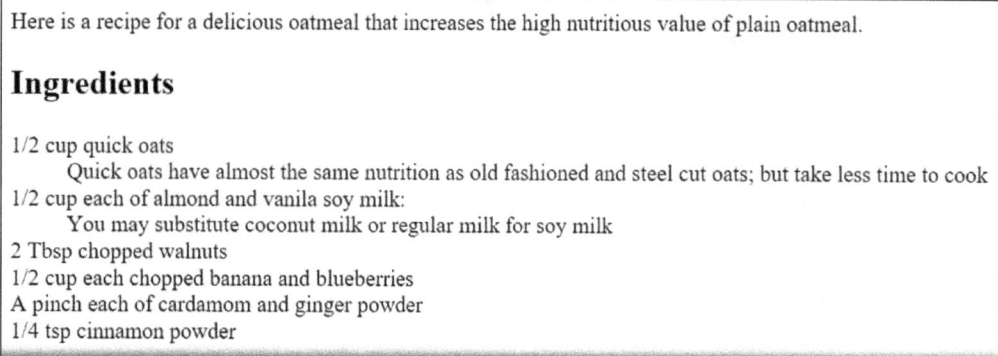

Here is a recipe for a delicious oatmeal that increases the high nutritious value of plain oatmeal.

Ingredients

1/2 cup quick oats
    Quick oats have almost the same nutrition as old fashioned and steel cut oats; but take less time to cook
1/2 cup each of almond and vanila soy milk:
    You may substitute coconut milk or regular milk for soy milk
2 Tbsp chopped walnuts
1/2 cup each chopped banana and blueberries
A pinch each of cardamom and ginger powder
1/4 tsp cinnamon powder

Figure 1-29: A web page with a description list

Table (<table>) Element

The table element is designed to present two-dimensional tabular data, consisting of rows and columns of cells. Though the table element is often conveniently used to control layout (e.g., spacing and alignment), it is not recommended for layout because of certain drawbacks: (1) tables don't automatically adjust to the width of their container windows, (2) using tables for layout makes it harder for screen readers to interpret the data, and (3) the code is more difficult to maintain. You should use CSS or similar tools for layout.

An HTML table is defined by the `<table>` element, along with one or more of the elements shown in Table 1-1.

The example shown in Figure 1-30 demonstrates the use of the elements in Table 1-1 to display the nutritional value of oats.

Table 1-1: Table elements

`<caption>`	defines the caption
`<tr>`	defines a row
`<th>`	defines a column heading (table heading).
`<td>`	defines a column

```
1   <table border="1">
2       <caption> Nutritional value of 1/2 cup of dry quick oats</caption>
3       <tr>
4           <th>Nutrition</th>
5           <th>Amount</th>
6       </tr>
7       <tr>
8           <td>Calories</td>
9           <td>150 cal (kcal)</td>
10      </tr>
11      <tr>
12          <td>Carbohydrates</td>
13          <td>27 g</td>
14      </tr>
15      <tr>
16          <td>Protein</td>
17          <td>5 g</td>
18      </tr>
19  </table>
```

Figure 1-30: HTML for a table

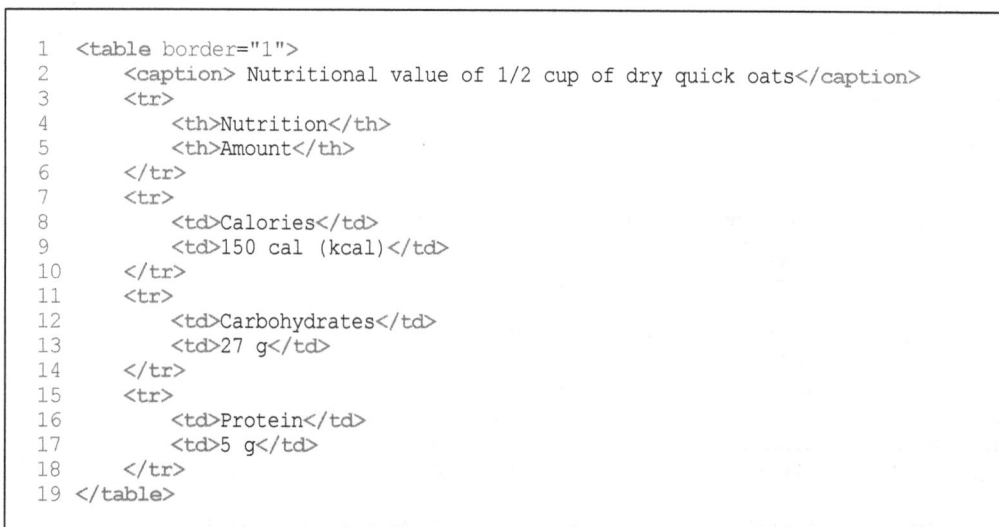

The **border attribute** is used here as a quick and easy way to specify a border for the table and cells by determining its thickness using a number. A value of "1" specifies a thinner border. This attribute is not supported in HTML5. Later, you will use CSS to add borders and for additional styling.

Figure 1-31 shows how the code in Figure 1-30 would display the caption, column headings, and other contents of the table.

Additional elements that are used with a table include <thead>, <tfoot>, and <tbody>, which specify the header, body, and footer of a table to provide meaning to those sections.

Nutritional value of 1/2 cup of dry quick oats	
Nutrition	**Amount**
Calories	150 cal (kcal)
Carbohydrates	27 g
Protein	5 g

Figure 1-31: A table on a web page

Attributes of the Table Element Not Supported in HTML5

In addition to the border attribute, the following attributes of the table element, though commonly used in existing code, are not supported in HTML5:

"align", "bgcolor", "cellpadding", "cellspacing", "frame", "rules", "summary", and "width"

Step 17: Add a table.

Add HTML code to your web page to display the nutritional values of oats in a table as shown in Figure 1-32. Use the style attribute to specify a color for the table, as described in the next section on CSS.

Display the page and test your code. Close the browser window.

Nutritional value of 1/2 cup of dry quick oats	
Nutrition	**Amount**
Calories	150 cal (kcal)
Carbohydrates	27 g
Protein	5 g
Fat	3 g
Dietary fiber	4 g

Figure 1-32: A table on nutritional values

1.4.2 A Peek at CSS

CSS (Cascade Styling Sheets) is a language that provides extensive features to specify the appearance of web pages. CSS stores the rules that specify the look and feel of a web page in a separate style sheet(s). There are three simple ways to apply CSS:

1. Use the style attribute of HTML elements (inline styling).
2. Specify styles in an internal style sheet in the head section.
3. Use an external style sheet.

You will learn these methods in the next chapter. Because we use inline styling in a few examples in this chapter, we will briefly describe it here.

Using the Style Attribute of HTML Elements

HTML provides a **style attribute** for every element, which enables you to apply a segment of CSS styling to the content of any individual element. The following two examples show how to specify a color and font size for a paragraph and an ordered list, respectively:

```
<p style="color: darkblue; font-size: 14px">
<ol type="1" style="color:brown; font-size: 20px">
```

Similarly, you may specify the width of an image to be 100% of the width of the browser window, as follows:

```
<img src="img/Nutty-fruity-oatmeal.jpg" style="width: 100%" alt="Oatmeal" />
```

This method of specifying styling within a tag is called *inline CSS* or *inline styling*.

1.4.3 Semantic versus Non-semantic (Generic) Elements

All elements that were presented so far, like headings (<h1> to <h6>), paragraph, lists, and table, are called semantic elements because they convey information about their contents to browsers, screen readers, and search engines. For example, a search engine understands the relative importance of the headings <h1>, <h2>, and so forth, and the structure of the contents in a table element. That is, a semantic element provides meaning to its content. A benefit of using semantic elements is that the information they covey could help search engine optimization (SEO). Similarly, screen readers could use the information to enhance the listener experience.

Non-semantic (Generic) Elements

Non-semantic elements, also called *generic elements*, do not convey the importance of content or what it is. That is, they don't convey any meaning. Thus it is recommended that you use non-semantic HTML elements like <span>, <div>, <b> (bold), and <i> (italics) only when no other semantic element is appropriate. In general, you should use CSS or similar tools to specify the appearance of the contents.

We will look at two elements, span and div, that are not semantic elements but are commonly used in HTML code. The div element is also widely used to apply CSS to a group of elements. Thus it is important for you to be familiar with these two non-semantic elements.

Span (<span>): A Generic Element

The span element is a generic (non-semantic) inline element, which serves as a container for a segment of text. It is commonly used to apply CSS style to a word or group of words using its style attribute.

The following example shows how the span element is used to apply bold to the segment of text **benefits of oatmeal**.

```
<p> Among the many <span style="font-weight: bold"> benefits of oatmeal</span> are: </p>
```

The corresponding browser display appears as follows:

Among the many **benefits of oatmeal** are:

As stated earlier, non-semantic elements like <span> do not convey any meaning. For example, you can make a text look like a heading by specifying the text as the content of the span element and using its style attribute to specify the font size of the text:

```
<span style="font-size: 200%">Nutty & Fruity Oatmeal</span>
```

The corresponding display would look similar to an <h1> heading:

Nutty & Fruity Oatmeal

However, unlike the heading elements, the span element doesn't convey the importance of its content or what it is. Hence the use of span element when a semantic alternative like <h1> is available should be avoided.

Step 18: Add headings and span.

Insert HTML code immediately after the list of ingredients (before the table of nutritional values) to display the following heading and the block of text.

> ## Directions
>
> Mix all ingredients in a cooking pot, **add one cup of water**, bring to a boil and simmer for a couple of minutes. Important: Make sure you stir the mixture frequently while simmering.

Make sure that the segment of the text, **add one cup of water**, is in bold. You may use the style attribute of the `<span>` element to display the text in bold. You may also use a level-2 heading.

Display the page to test the code. Close the browser window and stop the application.

Next, you will learn the use of another non-semantic element, `<div>`, that has been commonly used in HTML.

Division (<div>): A Generic Element

The division element is a generic (non-semantic) block element that serves as a container for one or more elements. A common use of `<div>` is to specify CSS styles for the contents of multiple elements to avoid repeating the styles within each element. For example, the following `<div>` element shown in Figure 1-33 uses the style attribute to apply a common color and font size to a paragraph and an unordered list contained within it.

```
1   <div style="color: darkblue; font-size: 14pt">
2       <p> Among the many benefits of oatmeal are: </p>
3       <ul>
4           <li>excellent source of fiber that helps to reduce cholestrol</li>
5           <li>loaded with nutrients including vitamins, minerals, and antioxidants</
li>
6           <li>helps with weight reduction</li>
7           <li>reduce constipation</li>
8       </ul>
9   </div>
```

Figure 1-33: The division element

The browser displays both the paragraph and the list in dark blue with a larger font size (Figure 1-34).

> Among the many benefits of oatmeal are:
>
> - excellent source of fiber that helps to reduce cholestrol
> - loaded with nutrients including vitamins, minerals, and antioxidants
> - helps with weight reduction
> - reduce constipation

Figure 1-34: The division element in the browser

Because the `<div>` element is **not** a semantic element, it doesn't have any special meaning. So it should not be used in place of semantic elements like `<header>`, `<footer>`, `<nav>`, and `<section>`, which convey the meaning and importance of those sections and hence are better suited to structure HTML documents. You will learn about these elements shortly.

Next, we will look at some commonly used inline semantic elements.

1.4.4 Inline Elements

Recall that inline elements do not cause their contents to be displayed on a new line, and they allow the contents of other elements to be on the same line on the left and/or the right.

Strong (<strong>) Element

The <strong> element is used where the text conveys strong emphasis/importance. By default, the strong element, which is relatively new with HTML5, displays the text in bold. You would use <strong> where you want to hear the text in a strong voice if it is read. Here is an HTML markup that shows the use of <strong>, along with the corresponding display in a browser:

```
<strong>Important:</strong> Make sure you stir the mixture <strong>frequently</strong> while
simmering.
```

> **Important:** Make sure you stir the mixture **frequently** while simmering.

Here, <strong> conveys the importance of the words *Important* and *frequently* to benefit both humans and machines. By using <strong> to mark up the words, it is implied that if you were to read them, you would use a strong voice. You may change the style for <strong> from the default bold to something else like underline using CSS.

You should not use non-semantic elements like <span> with CSS styling in place of <strong> because that doesn't say anything about the importance of the words.

What if you want to change only the appearance of text to bold so it stands out from the rest of the text to draw attention to it without conveying any special importance? Bold is often used for key words and terms. For example, say you want to display the names of ingredients in bold as shown here:

> 1. 1/2 cup **quick oats**
> 2. 1/2 cup each of **almond and vanilla soy milk**
> 3. 2 Tbsp **chopped walnuts**

In this case, you would use CSS to format the styling.

The old method of specifying the bold style was to use the non-semantic HTML element bold (<b>). But, as explained earlier, non-semantic HTML elements like <b> and <i> (italics) should be used only as a last resort. Using bold and italic elements may also create problems when the page is displayed in different languages because the reasons for using bold and italics may be different in different languages. Even within the same language, it may become necessary to change some bold text to a different style and not others. The use of CSS with an external style sheet would make it easier to change the appearance consistently across web pages.

Emphasis (<em>)

The emphasis (<em>) element conveys gentle emphasis of the text to both humans and machines. Typically, <em> displays the text in italics. You would use <em> in instances where you want to emphasize pronunciation if the text is spoken. The following example shows the use of <em> and the corresponding output in the browser:

```
I would <em>really</em> love to get a job as a web developer.
```

> I would *really* love to get a job as a web developer.

Here, it is assumed that the reader would emphasize pronunciation of the word *really* if the text is read. Another example would be

```
Here is a recipe for a <em>delicious</em> oatmeal.
```

Italics are typically used for new terms when they are introduced the first time, such as technical terms, titles of books and articles, foreign words, and so forth. To change just the appearance of text to italics in such instances, you would use CSS.

Break (<br>)

The break element inserts a line break. It is intended to insert line breaks that are part of the content, as in addresses and poems, and should not be used for other structures like lists. The break element is a **void element**—that is, it doesn't have an end tag. It is coded as a self-closing tag by adding a forward slash, as in `<br />`. The following example shows HTML source code that uses the break element, along with the corresponding browser display:

```
Prospect Press <br />
47 Prospect Parkway <br />
Burlington, VT 05401 USA
```

```
Prospect Press
47 Prospect Parkway
Burlington. VT 05401 USA
```

The break element may be used within other elements, like paragraphs, headings, and lists.

Let's put what you learned into practice. Develop the HTML code to mark up the text shown in the next two steps.

Step 19: Mark up the word *delicious* **to emphasize pronunciation when it is spoken in the following text.**

Here is a recipe for a *delicious* oatmeal that increases the high nutritious value of plain oatmeal.

Step 20: Mark up the words *Important* **and** *frequently* **to convey strong emphasis.**

Directions

Mix all ingredients in a cooking pot, **add one cup of water**, bring to a boil and simmer for a couple of minutes. **Important:** Make sure you stir the mixture **frequently** while simmering.

1.4.5 Additional Semantic Elements

The later versions of HTML introduce several additional semantic elements that are important in structuring the contents of web pages.

Header (<header>) and Footer (<footer>)

The `header` element for a page typically contains the heading and introductory content, like a logo, table of contents, search form, and navigational links. Much of the information on the page header may repeat on each page. In addition to the page, the sections within a page also may have headers that contain introductory information about the section.

Because the recipe page was developed as a stand-alone page without regard for other pages within the website, it doesn't contain any information that is repeated across multiple pages. Let's consider an expanded version of the website that includes three additional pages named **pancakes.html**, **dinner.html**, and **snacks.html**. In such a multi-page website, you would add a header to each page containing content that is common to all pages, like the heading for the website, the logo, and site navigation links, as shown in Figure 1-35.

The corresponding HTML code is marked up with the header element (Figure 1-36).

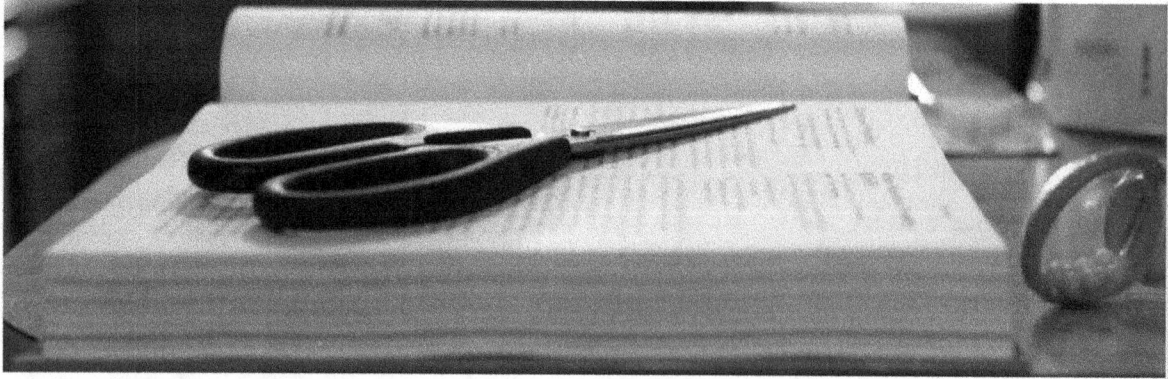

Figure 1-35: The page header. Cookbook photo by Scott Akerman/CC BY 2.0.

```
1   <header>
2       <h1>P&I Recipes</h1>
3       <img src="img/cookbook-header.png" width="100%" />
4           <a href="oatmeal-recipe.html">Oatmeal</a>    
5           <a href="pancakes.html">Pancakes</a>    
6           <a href="dinner.html">Dinner</a>    
7           <a href="snacks.html">Snacks</a>
8   </header>
```

Figure 1-36: The header element

Here, *P&I Recipes* is the website heading and *cookbook-header.png* is the file containing the logo, which is stored within the *img* folder of the project.

Note that marking up a group of elements as a header won't affect how the contents are displayed. The purpose of the header element is to convey to machines and humans that the contents form the header of the page.

The **footer** element for a page may contain copyright information, information about the author, or additional links that are not appropriate in the header. It is recommended that information about the author be provided in an address block element within the footer, as shown in Figure 1-37.

```
1   <footer>
2       Copyright @ 2022 P&I Traders. All rights reserved.
3       <address>
4           Contacts: J. Iversen, iversen@uwosh.edu; G. Philip, philip@uwosh.edu
5       </address>
6   </footer>
```

Figure 1-37: The footer element

The contents of the footer generally repeat on each page. In addition to the page footer, sections within a page also may have footers.

Main (<main>)

The main element represents content that is unique to the page (doesn't repeat across pages) and is related to the central topic of the page. Thus the main element excludes the site header and the footer. A page must not have more than one main element. Figure 1-38 shows how the header, footer, and main elements for the recipe page would appear.

```
1  <header>
2      <h1>P&I Recipes </h1>
3      <img src="img\cookbook-header.png" width="100%" />
4          <a href="oatmeal-recipe.html">Oatmeal</a>
5          <a href="pancakes.html">Pancakes</a>
6          <a href="dinner.html">Dinner</a>
7          <a href="snacks.html">Snacks</a>
8  </header>
9  <main>
10     <h1>Nutty & Fruity Oatmeal</h1>
11     <img src="img/Nutty-fruity-
              oatmeal.jpg" width="200" alt="Nutty & Fruity Oatmeal"/>
12     <p>
13         A healthy breakfast is key to a helthy diet. There are many web sites that
14         describe the benefits of oatmeal, including:
15         ...
16         ...
17 </main>
18 <footer>
19     Copyright @ 2019 P&I Traders. All rights reserved.
20     <address>
21         Contacts: J. Iversen, iversen@uwosh.edu; G. Philip, philip@uwosh.edu
22     </address>
23 </footer>
```

Figure 1-38: The header, footer, and main elements

Step 21: Add the header, main, and footer elements.

Add the necessary HTML code to your page to add the header element, mark up the main element, and add a footer element, as shown in Figure 1-38, except the footer shows your name and email.

Copy the file *cookbook-header.png* to the *img* folder within the project directory.

Display the page to test the code.

Navigation (<nav>)

The navigation element is intended to mark up a group of major navigation links that link to other pages or parts within the same page. It is a semantic element that conveys to other applications that its contents are navigation links. You may have multiple nav elements on a page. In Figure 1-39, the nav element describes the set of four site navigation links that you created within the header.

```
1  <nav>
2  <a href="oatmeal-recipe.html">Oatmeal</a>    
3  <a href="pancakes.html">Pancakes</a>    
4  <a href="dinner.html">Dinner</a>    
5  <a href="snacks.html">Snacks</a>
6  </nav>
```

Figure 1-39: The navigation element

The nav element is a block element, so it displays the group of links in a separate block, as shown here:

| Oatmeal | Pancakes | Dinner | Snacks |

Step 22: Insert the navigation element to mark up the group of four links that you created in the header as navigation links.

Aside <aside>

Aside is a semantic block-level element that is typically used to mark up a block of text that is only tangentially related to the content around it. Removal of the aside element should not affect the readability or flow of the surrounding text. Examples include comments, excerpts from articles, related links, and related content. Figure 1-40 shows an aside that describes XHTML within a section that discusses HTML.

```
1  <p>
2      HTML (Hypertext Markup Language) is used to display information in web pages.
3      Essentially, HTML is used to tell the browser how to structure and present
4      content in a web page.
5      <aside style="margin-left: 30px; font-size: 14px">
6          A language closely related to HTML is XHTML (EXtensible HTML), which
7          is almost identical to HTML with stricter rules to conform to
8          XML (EXtensible Markup Language).
9      </aside>
10 </p>
```

Figure 1-40: The aside element

Because an aside is typically indented with a smaller font size, the code shown in Figure 1-40 uses the optional style attribute to specify the left margin and font size. Figure 1-41 shows how the aside in the code is displayed.

HTML (Hypertext Markup Language) is used to display information in web pages. Essentially, HTML is used to tell the browser how to structure and present content in a web page.

A language closely related to HTML is XHTML (EXtensible HTML), which is almost identical to HTML with stricter rules to conform to XML (EXtensible Markup Language).

Figure 1-41: The aside element in the browser

Aside may also be used at the page (or site) level, where it is relevant to the whole page or site. This could be additional navigation, the author's biography, the glossary, and so forth.

You should not use a paragraph element in place of the aside element because the paragraph doesn't convey the information that the content is an aside.

Step 23: Add the necessary HTML code to your page to display the following information in a smaller font size with indentation, as an aside right below the table of nutritional values.

You may find the nutritional value of almost any food from sources like Nutrition Value and What's In Food

The URL for the two links are https://www.nutritionvalue.org and https://www.nutrition.gov/subject/whats-in-food, respectively.

Definition (<dfn>)

The definition element is a semantic inline element that marks a newly defined term. The term should be defined within the same element (e.g., the paragraph element) that contains the <dfn> element. That is, the term is defined in the surrounding text. Figure 1-42 shows a paragraph element that contains the <def> element to mark the term *semantic elements*.

```
1  <p>
2      Elements like headings, paragraph, lists and table, are called
3      <dfn>semantic elements</dfn>, which are elements that convey information about
   their contents to browsers, screen readers and search engines.
4  </p>
```

Figure 1-42: The definition element

There are different variations of how you would use the <dfn> element. For example, you may use an abbreviation of a term as the content and specify the expanded term using the title attribute of the <dfn> element.

Let's use the <dfn> element in your recipe page where you define quick oats.

> **Step 24: Add the necessary HTML code to your page to display the following paragraph that contains a definition of *Quick oats* immediately after the description list of ingredients.**

> Not sure what quick oats are? Quick oats are oat grains that are cut to smaller pieces and then steamed and rolled.

Make sure you mark the term *Quick oats* using the <dfn> element to convey that this term is defined in the surrounding text.

Section (<section>)

The section element is used to group related contents together. A primary benefit of such grouping is that it helps create machine-generated outlines. Typically, you would use section to group a heading and related information presented in elements like paragraphs, tables, and lists. For example, you would mark up the heading *Ingredients* and the related description list and paragraph as a section (Figure 1-43).

```
1  <section>
2      <h2>Ingredients</h2>
3      <dl>
4          <dt> 1/2 cup quick oats </dt>
5          <dd>
6              Quick oats have almost the same nutrition as old fashioned and steel cut
7              oats; but take less time to cook
8          </dd>
9          <dt> 1/2 cup each of almond and vanilla soy milk:  </dt>
10         <dd> You may substitute coconut milk or regular milk for soy milk </dd>
11         ...
12     </dl>
13     <p>
14         Not sure what quick oats are?
15         <def>Quick oats</def> are oat grains that are cut to smaller pieces and then
               steamed and rolled.
16     </p>
17 </section>
```

Figure 1-43: The section element

Similarly, each of the other level 2 headings ("Why oatmeal," "Directions," and "Nutritional value") and related contents may be marked as a separate section.

A `section` may be nested inside another `section`, just like topics in an outline contain subtopics, and a level 1 heading may contain level 2 headings.

Step 25: Mark each level 2 heading as a separate section.

Add the necessary HTML code to your page to mark each level 2 heading ("Why oatmeal," "Ingredients," "Directions," and "Nutritional value") and related contents as a separate section.

Figure 1-46 (displayed at the end of this chapter) provides the complete HTML code specified in previous steps for the web page displayed earlier in Figure 1-5.

The rest of the elements presented in this chapter are not appropriate for the current contents of the page. However, you may find them useful in other applications.

Article (<article>)

The `article` is similar to a `section`, except that the content of the article must make sense by itself even when it is taken out of context. A major goal of the `article` element is to make it easy for other computer programs to understand, extract, and share (syndicate) the contents independent of the rest of the document. An `article`, rather than a `section`, should be used when a group of contents is to be shared. So, an `article` can be viewed as a special type of `section` that is distributable.

Some examples of `article` are magazine articles, blog posts with associated user comments, self-contained user feedback, and web pages that can stand on their own, including *About* and *Contact Us* pages.

Code (<code>)

The `code` element is a semantic element that defines a segment of computer code within a document. The browser would display the contents of the `code` element using its default monospace font (a fixed-width font like Courier). You may use the `class` attribute to specify the language name prefixed with `"language-"` as follows:

```
<code class="language-csharp">
     (code segment)
</code>
```

The `code` element, however, does not maintain the typographic format of the code (e.g., white spaces, tabs, and line breaks) when the code is displayed in the browser. The purpose of the `code` element is to indicate that its content is computer code. You may use the style attribute to provide CSS styling to maintain the format as shown in Figure 1-44.

```
1  <code class="language-csharp" style="white-space: pre-wrap; font-family: monospace">
2      switch (rblZones.SelectedValue.ToString())
3      {
4          case "Zone2":
5              zonePrice = 10;
6              break;
7          case "Zone3":
8              zonePrice = 20;
9              break;
10     }
11 </code>
```

Figure 1-44: The code element

For now, it is enough for you to know that the `white-space` property determines how white space inside an element is handled. The value `pre-wrap` specifies that white spaces are maintained and that text will wrap on line breaks and as necessary. The code will be displayed in the browser in monospace font, maintaining all spacing, as shown in Figure 1-45.

```
switch (rblZones.SelectedValue.ToString())
{
    case "Zone2":
        zonePrice = 10;
        break;
    case "Zone3":
        zonePrice = 20;
        break;
}
```

Figure 1-45: The code element in the browser

Preformatted text other than computer code, like poems and transcripts of conversations, may be displayed using the style attribute to provide CSS styling (without using the `<code>` element) as shown here:

```
<p style="white-space: pre-wrap; font-family: monospace">
    (preformatted text)
</p>
```

A less desirable option is to use the `<pre>` element, which is a non-semantic element commonly found in older HTML code. The `pre` element maintains the typographic format when the text is displayed in the browser.

Escaping HTML Characters

Certain characters like "<" and ">" have special meaning in HTML. They are called *reserved characters*. What if such characters are a part of the text you want to display in the browser? For example, to describe the HTML break element on a web page, you may want to display the following sentence that contains the code `"<br/>"`:

> The break element <br/> is intended to insert line breaks that are part of the content, as in an address

Consider using the following HTML code to display the above sentence:

```
<p> The break element <br/> is intended to insert line breaks
that are part of the content, as in an address </p>
```

The code as shown here **won't** work, because the browser will render `<br/>` as a line break, rather than displaying the code, and display the rest of the sentence on a new line.

To display an HTML character like "<" and ">", it is necessary to *escape* it by replacing it with the corresponding HTML **character entity**. For example, the HTML entity for "<" is "&lt" and the entity for ">" is "&gt." If you escape the opening bracket, you don't have to escape the closing bracket. So, the provided HTML code should be changed as follows:

```
<p> The break element &ltbr/> is intended to insert line breaks
that are part of the content, as in an address </p>
```

1.4.6 Less Frequently Used Elements

There are several additional semantic and non-semantic elements that you are less likely to use when coding static HTML pages. Appendix A* provides a partial list of such elements and those already described in this chapter.

Review Questions

6. Using examples, describe the difference between block and inline HTML elements.
7. Using examples, describe the difference between semantic and non-semantic (generic) HTML elements.

_____

* Appendix can be freely downloaded from https://www.prospectpressvt.com/textbooks/philip-web-development at "Student Resources."

1.5 The Complete HTML Code

Figure 1-46 shows the complete code for the oatmeal-recipe page.

```
1   <!DOCTYPE html>
2   <html>
3   <head>
4       <meta charset="utf-8" />
5       <title>Oatmeal Recipe</title>
6   </head>
7   <body>
8       <header>
9           <h1>P&I Recipes</h1>
10          <img src="img/cookbook-header.png" width="100%" />
11          <nav>
12              <a href="oatmeal-recipe.html">Oatmeal</a>
13              <a href="pancakes.html">Pancakes</a>
14              <a href="dinner.html">Dinner</a>
15              <a href="snacks.html">Snacks</a>
16          </nav>
17      </header>
18      <main>
19          <h1>Nutty & Fruity Oatmeal</h1>
20          <img src="img/Nutty-fruity-
21              oatmeal.jpg" width="200" alt="Nutty & Fruity Oatmeal" />
22          <p>
23              A healthy breakfast is key to a helthy diet.There are many web sites that
                describe the benefits of oatmeal, including:
24              <a href="https://www.webmd.com/diet/oatmeal-benefits#1"> WebMD </a>
               
25              <a href="https://www.organicfacts.net/oatmeal.html"> Organic facts </a>
               
26              <a href="https://www.hhs.gov/fitness/eat-healthy/dietary-guidelines-
                for-americans/index.html"> HHS </a>
27          </p>
28          <section>
29              <h2>Why oatmeal?</h2>
30              <p> Among the many benefits of oatmeal are: </p>
31              <ul>
32                  <li>excellent source of fiber that helps to reduce cholestrol</li>
33                  <li>loaded with nutrients including vitamins, minerals, and
34                      antioxidants</li>
35                  <li>helps with weight reduction</li>
36                  <li>reduce constipation</li>
37              </ul>
38              <p>
39                  Here is a recipe for a <em>delicious</em> oatmeal that increases the
40                  high nutritious value of plain oatmeal.
41              </p>
42          </section>
43          <section>
44              <h2>Ingredients</h2>
45              <dl>
46                  <dt> 1/2 cup quick oats </dt>
47                      <dd> Quick oats have almost the same nutrition as old fashioned
48                          and steel cut oats; but take less time to cook. </dd>
49                  <dt> 1/2 cup each of almond and vanila soy milk:  </dt>
50                      <dd>You may substitute coconut milk or regular milk for soy</dd>
51                  <dt> 2 Tbsp chopped walnuts </dt>
52                  <dt> 1/2 cup each chopped banana and blueberries </dt>
53                  <dt> A pinch each of cardamom and ginger powder </dt>
54                  <dt> 1/4 tsp cinnamon powder</dt>
55              </dl>
56              <p>
57                  Not sure what quick oats are?
```

Figure 1-46: The complete HTML code for the oatmeal-recipe page (*continues*)

```
58              <dfn>Quick oats</dfn> are oat grains that are cut to smaller pieces
                and then steamed and rolled.
59          </p>
60      </section>
61      <section>
62          <h2>Directions</h2>
63          <p>
64              Mix all ingredients in a cooking pot,
                <span style="font-weight: bold"> add one cup of water</span>,
                bring to a boil and simmer for a couple of minutes.
65              <strong>Important:</strong> Make sure you stir the mixture
66              <strong>frequently</strong> while simmering.
67          </p>
68      </section>
69      <section>
70          <h2>Nutritional value</h2>
71          <table border="1" style="color: chocolate; font-size: 14px">
72              <caption> Nutritional value of 1/2 cup of dry quick oats</caption>
73              <tr>
74                  <th>Nutrition</th>
75                  <th>Amount</th>
76              </tr>
77              <tr>
78                  <td>Calories</td>
79                  <td>150 cal (kcal)</td>
80              </tr>
81              <tr>
82                  <td>Carbohydrates</td>
83                  <td>27 g</td>
84              </tr>
85              <tr>
86                  <td>Protein</td>
87                  <td>5 g</td>
88              </tr>
89              <tr>
90                  <td>Fat</td>
91                  <td>3 g</td>
92              </tr>
93              <tr>
94                  <td>Dietary fiber</td>
95                  <td>4 g</td>
96              </tr>
97          </table>
98          <p>
99          <aside style="margin-left: 30px; font-size: 14px">
100            You may find the nutritional value of almost any food from sources like
               <a href="https://www.nutritionvalue.org"> Nutrition Value </a> and
101            <a href="https://www.nutrition.gov/subject/ whats-in-food">
102                What's In Food</a>
103         </aside>
104         </p>
105     </section>
106 </main>
107 <footer>
108     Copyright @ 2019 P&I Traders. All rights reserved.
109     <address>
110         Contacts: J. Iversen, iversen@uwosh.edu; G. Philip, philip@uwosh.edu
111     </address>
112 </footer>
113</body>
114</html>
```

Figure 1-46: The complete HTML code for the oatmeal-recipe page (*continued*)

Exercises

Exercise 1-1: The Grand Intro

This exercise develops a small segment of the home page for the Grand Oshkosh theater. Create a folder named Assignments within the WebProjects folder. Add a new HTML page named Default.html to the Assignments folder. Develop the following web page using appropriate HTML elements in the Default.html page. The page tab should show the title "The Grand Oshkosh" when it is displayed in the browser.

The Grand Oshkosh

The Grand is a historic theater locatted in <u>Oshkosh</u>, Wisconsin. It has a seating capacity of 660 that re-creates the warm, intimate atmosphere of a European-style theater, while show-casing some of the best talent today.

Photo courtesy of the Grand Oshkosh.

The link for Oshkosh should point to the Wikipedia page for Oshkosh, Wisconsin (https://en.wikipedia.org/wiki/Oshkosh,_Wisconsin).

To display the image, add a folder named *img* to your project folder, *Theater*, and copy the file *InsideGrand.jpg* from the **Tutorial-starts\img** folder (in **StudentResources**) to the **Theater\img** folder.

Next, use an *img* element that specifies `src="img/InsideGrand.jpg"`.

Exercise 1-2: The Grand Finale

This exercise makes significant additions to the web page *Default.html* that you created in Exercise 1-1. (If you haven't already done Exercise 1-1, you need to do it before proceeding to this exercise.) In this exercise, you should create the complete web page shown in the three screenshots below. This exercise focuses on structuring the contents of the page using HTML code. You will significantly improve the appearance of the page using CSS and Bootstrap later in the next two chapters. You should be able to create the web page with the elements that were used in this chapter's tutorial.

The images for the logo ("The GRAND OSHKOSH"), inside view (InsideGrand.jpg), and the seat map (seatmap.jpg) can be found in the folder **Tutorial-starts\img** (in **StudentResources**). The text displayed in the logo image doesn't have a background color, so you may use the style attribute of the *img* element to set the *background-color* to **darkmagenta**. Set the width to 100% so that the browser adjusts the width of the image to fit the browser width.

Create a second page named *about.html* that displays just the logo and the copyright information. Provide navigation links to *default.html*, *about.html reservation.html*, and *events.html*. You don't have to create *reservation.html* and *events.html*.

You must structure the HTML code for the page with appropriate semantic elements like header, main, footer, section, heading, strong, emphasis, address, aside, and nav, if applicable. You may use the style attribute of the span element to change the appearance of a segment of text without conveying any emphasis. Minimize the use of non-semantic elements.

Make sure you verify that the HTML code works correctly and that it displays the web page as shown in the screenshots. Note that the actual number of lines displayed on the page will depend on the width of the browser window.

The
GRAND
OSHKOSH

Home   About   Reservation   Events

The Grand is a historic theater locatted in Oshkosh Wisconsin. It has a seating capacity of 550 that re-creates the warm, intimate atmosphere of a European-style theater, while show-casing some of the best talent today.

A Variety of Events

The Grand hosts nearly 100 public performances each year. This includes performances by:

National touring artists
Oshkosh Community Players (OCP)
    - the resident community theatre at the Grand Opera House
Oshkosh Symphony Orchestra (OSO)
    - Wisconsin's most exciting and innovative professional orchestra
Hysterical Productions (HP)
    - brings stories to life on the stage, in the park, at historical locations and even in your own home!
Area high schools
Regional arts groups

In addition, The Grand hosts a series of educational events, called **Student Discovery Series** .

The Grant and its *magnificent* facilities also may be **rented** for events like weddings, seminars,photo shoots and paranormal investigations.

    The Grand is said to be *haunted*, and it was the subject of investigation in a 2014 documentary film "Haunted State: Theatre of Shadows"

Admission Tickets

The theater provides you a wide range of seating choices at affordable prices:

STANDARD

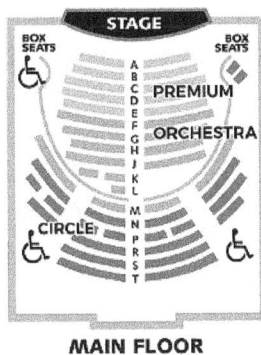

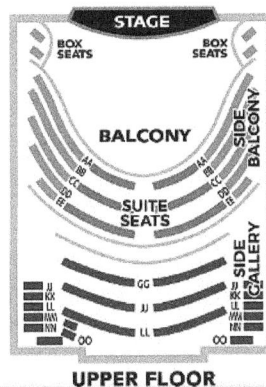

Current Performances

There are seven prices for each performance corresonding to the seven sections:

Suite A, Suite B, Premium, Orchestra, Circle, Side balcony, Side gallery

Performances for this Month

Performance	Days/times	Prices
The Young Irelanders	Mon, Wed: 7:30 p.m.	$65 | 60 | 50 | 48 | 45 | 35 / 25
Justin Hayward "Voice of The Moody Blues"	Tue: 7:30 p.m.	$70 | 65 | 60 | 55 | 50 | 40 |30
TEDxOshkosh	Thu: 7:30 p.m.	$99 | 99 | 89 | 89 | 89 | 79 | 79
OSO Orchestral Presents	Fri: 7:30 p.m.	$45 | 40 | 30 | 30 | 25 | 20 | 20
The Quebe Sisters	Sat: 2 p.m., 7:30 p.m.	$52 | 46 | 40 |36 | 32 | 28 | 25

Box Office
(920) 424-2350
Monday-Friday 11:30 a.m.-5:00 p.m.
Saturday 11:00 a.m.-2:00 p.m.
The Box Office is open one hour before each performance.

Policies

Ticket Returns & Exchanges:

All sales are final unless a performance is canceled, with the following exceptions: Exchanges are allowed within a performance, subject to a fee, providing that tickets are returned to the Box Office at least 24 hours prior to the initial performance.

Donors and package purchasers may exchange tickets without a fee from one season event to another providing that tickets are returned at least 24 hours prior to the initial event.

Photography, Recording and Electronic Devices:

Use of distracting devices like cameras, cellular phones, laser pointers and recording devices are **not allowed**.

The Grand Oshkosh
100 High Ave
Oshkosh, WI 54901
(920) 424-2350

## Chapter 2

Cascading Style Sheets (CSS)

When HTML was created, it was designed to do simple formatting of text documents—to "mark up" the text and format it for graphical display. Websites were small and simple, so formatting could be applied relatively simply by specifying how each HTML tag was to be displayed. However, over time websites have become far larger and more complex, and expectations have become higher for graphically intense websites. This made HTML unsuitable for modern websites. The solution has been the introduction and adoption of Cascading Style Sheets (CSS), which is a cornerstone web standard alongside HTML and JavaScript. Just like HTML, CSS is maintained by W3C. CSS doesn't replace HTML; it augments it by allowing a developer to place presentation instructions in a separate stylesheet and let HTML only describe the content and structure of the page.

This separation of concerns allows for targeting different output devices for the same content. For instance, you may have different style sheets for mobile users, printing, or even for braille devices. By placing styles in a single external style sheet, you can also provide consistent formatting for multiple pages in the same site—and change the formatting across the entire site in a single place.

Learning Objectives

After studying this chapter, you should be able to
- Describe the difference between content and presentation in web pages.
- Describe how styles are used to apply presentation to HTML tags.
- Describe how the specificity of CSS rules affects how a page is rendered by a web browser.
- Use the CSS box model to determine the size of HTML elements on the page.
- Use common CSS formatting rules for web pages.
- Apply CSS rules to affect the flow of content on a page—positioning.
- Use CSS for the layout of an entire web page using flex box and grid layout.

2.1 Introduction to Styles

CSS is a relatively simple language that specifies the look and feel of a web page. A style sheet specifies a list of rules, with each rule specifying a *selector* and a *declaration block*. The selector specifies which part of the HTML is going to be styled, and the declaration block specifies what the styling is.

Figure 2-1 shows an example of a few CSS rules. This example has two selectors: h1 in line 1 and p in line 7. The rules in the declaration blocks will only apply to the specific HTML tag. So, any h1 tag in the document will be red, with a font size of 16 pixels, and a 1-pixel solid blue border. Text in any paragraphs marked by p will be green, on the other hand. Together, the combination of a selector and a declaration block is called a *ruleset* or just a *rule*.

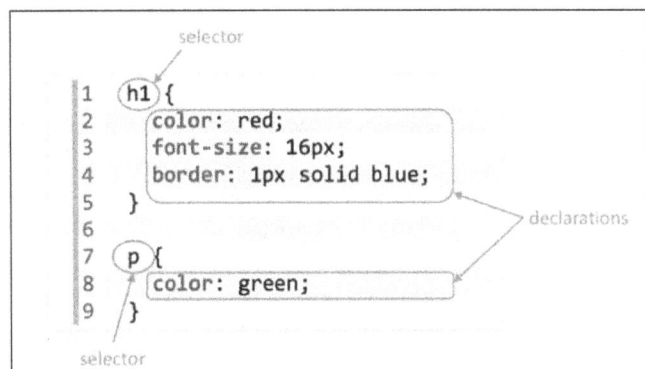

```
1  h1 {
2      color: red;
3      font-size: 16px;
4      border: 1px solid blue;
5  }
6
7  p {
8      color: green;
9  }
```

Figure 2-1: CSS example

2.1.1 Declarations

Declarations are the workhorses in CSS and are made up of two elements: *Properties* and *Values*. The two are separated by a colon. Multiple declarations in a declaration block are separated by a semicolon. For example, in line 2 of Figure 2-1, the property is `color` and the value is `red`. This declaration will set the text color of any `h1` tag to red. Line 4 shows multiple values for the same property; this is called *Shorthand* because it is shortened from multiple separate properties for the border. In this case, a solid blue border of 1 pixel is added to the `h1` tag. The declaration in line 4 is the same as these three lines:

```
1   border-width: 1px;
2   border-style: solid;
3   border-color: blue;
```

CSS is very flexible, so you can mix and match shorthand and longhand notation. The order of the values you use isn't even important when using shorthand. The browser will detect the value and assign it to the proper property. You can also omit values for some of the possible shorthand properties, in which case they will be set to their default values.

For example, the declaration in line 4 in Figure 2-1 could also be written like any of the following with the exact same result:

```
1   border: 1px solid blue;
2   border: solid blue 1px;
3   border: blue 1px solid;
```

On the other hand, if you omitted the color (blue), the border would turn black; if you omitted the width (1px), the border would be thicker at 3px; but if you omitted the style (solid), the box would disappear as the default style is no border.

Should you make a mistake in a declaration, CSS will try to recover as quickly as possible by simply ignoring the problematic rule. This can be both good and bad in that you will get output in the browser when the page is rendered, but it can also be difficult to troubleshoot where a problem occurred.

CSS predefines a large number of properties, so while the syntax is easy to learn, CSS has a lot of depth and mastering it takes a lot of practice exploring all the possible ways to use it. In this chapter, you will see fairly simple ways to apply CSS that will show you the principles and some examples of how to work with CSS. To truly master it, you will need to consult additional references and practice what you learn. We recommend W3 Schools (w3schools.com/css) and Mozilla Developer Network (developer.mozilla.org/css) as good starting places.

The properties are predefined in CSS. Table 2-1 contains a list of commonly used properties.

Table 2-1: List of common CSS properties

Property	Description
background	Shorthand for setting a number of properties related to the background of an element (e.g., -color, -image, -size, -repeat, ...). Often used to place images in the background of an element.
border	Used to set the border for an element. Allows shorthand for -width, -style, and -color.
border-radius	Allows for specifying a rounded border around an element.
bottom	Specifies an element's vertical position. The effect depends on how the element is positioned.
color	Sets the color of an element. Can be applied using named colors, hex values, RGB, or HSL values.
column-count	Divides an element into multiple columns.
display	Specifies the way content is rendered in an element. Allows for specifying many different types of rendering, such as inline (default), block, flow, and contents. By specifying none, the element is not displayed at all.
filter	Allows for applying graphical filters, such as blur, contrast, drop-shadow, invert, and saturation to an image.
font	Used to specify the font used for a text element. Can specify a list of increasingly available fonts, for example: `font-family: Helvetica, Arial, sans-serif` You can also specify `font-style`, which can set italics, `font-weight` to set bold, and `font-size`, which can set the size in pixels, em, or percent.

(continues)

`height/width`	Sets the height or width of an element in percent or pixels.
`hyphens`	Allows you to control how hyphens are added to text at line breaks. You can choose `manual`, `none`, or `auto`. In the `auto` mode, the browser uses the language you specify for the text to use its built-in rules to hyphenate words.
`left/right/ top/bottom`	These two properties help specify the horizontal and vertical positioning of an element. The functioning depends on which positioning is used (e.g., absolute, relative, etc.). For example, if an element's `position` is set to `absolute` or `fixed`, the `left/right/top/bottom` properties specify the distance between the edge of the element and the corresponding edge of its containing block.
`list-style`	This allows you to specify how a list is rendered. You can control the shape of bullets (circles, squares, images) or the type of numbering used.
`margin`	Using this property, you can control the space on all four sides of an element.
`opacity`	With this property, you can set how transparent an element is. This is often used with images.
`overflow`	You can decide what to do with content that doesn't fit in a containing element. Options include adding scrollbars, clipping, or displaying content overflowing from the container.
`padding`	Sets the padding area around an element. It uses shorthand to set each of the four sides of an element (-top, -right, -bottom, and -left).
`position`	Specifies the type of positioning used to place an element (static, relative, absolute, fixed, or sticky).
`text-align`	Specifies how text is aligned (justified).
`top/bottom`	These properties work in concert with a specific positioning property to specify where the top or bottom of an element is rendered. The specific effect depends on which positioning is used. For example, if the position is absolute or fixed, `top` sets the top edge of the element above/below the top edge of its nearest positioned ancestor, whereas if the position is relative, `top` specifies how far the top edge is moved from its normal position.
`verti- cal-align`	Vertical alignment of an element (baseline, top, middle, bottom, …).
`visibility`	Specifies whether an element is hidden or visible.
`z-index`	Sets the stack order of overlapping elements. The higher the z-index value, the further to the top an element is placed.

2.1.2 Selectors

While most of the action in CSS, is in the declarations, the selectors are also important to understand. You already saw how CSS rules can be applied to an HTML tag by specifying an HTML tag as the selector in the CSS rule. However, you can also use the *class* selector, which allows you to select your own name for a selector and then apply that to whichever HTML tag you choose. Figure 2-2 shows a brief example of how classes are used in CSS.

```
1    .ingredient {
2        color: blue;
3    }
4
5    .measurement {
6        font-weight: bold;
7        background-color: #dff5d6;
8    }
```

- **1/2 tsp** cinnamon
- **1 cup** sugar

```
1    <ul>
2        <li class="ingredient"><span class="measurement">1/2 tsp</span> cinnamon</li>
3        <li class="ingredient"><span class="measurement">1 cup</span> sugar</li>
4    </ul>
```

Figure 2-2: Using a class selector. CSS rules are on the top-left, HTML is at the bottom, and the resulting output is on the top-right.

Lines 1 and 5 in the CSS show how you can name the class selector by using a dot in front of the name. You can set the name to be anything you want—just like a variable name. In the HTML, you can see how you apply the class by using the `class` keyword. In this case, the `ingredient` class is applied to each of the `li` tags, so the text turns blue, and the `measurement` class is applied to a `span` tag surrounding the amount to use.

Multiple Selectors for the Same Rule

You can specify that the same CSS rule applies to multiple HTML tags by simply adding additional selectors and separating them with commas. Figure 2-3 shows a few examples of this. In line 1, the first four headings are all specified to have the same color, size, and font family. The font family uses a list of fonts, and the first one listed will be used if it is available to the browser. In line 7, you can see an example that mixes an HTML tag and a class tag to turn the text color blue for both the p element and any element with the `ingredient` class applied.

```
1   h1, h2, h3, h4 {
2       color: green;
3       font-size: 12;
4       font-family: Verdana, Arial, sans-serif;
5   }
6
7   p, .ingredient {
8       color: blue;
9   }
```

Figure 2-3: Multiple selectors for a CSS rule.

Selector Types

CSS defines a number of different types of selectors. Some of these you have already seen; others will be described in more detail in the coming sections of this chapter.

- An **element selector** is one that specifies a particular HTML element, as shown in line 1 in Figure 2-3.

- A **class selector** indicates a class that you have specified. You saw two examples of classes in Figure 2-2.

- The **ID selector** is similar to the class selector in that you specify the name of the ID. However, you can only use an ID once in an HTML file, whereas a class selector can be used many times. The ID selector is often used to control the layout of the page or to indicate specific types of content. An ID is written like this in CSS:

  `#sidenav{ ... }`

  And is used like this in HTML:

  `<div id="sidenav">...</div>`

- With a **descendant selector**, you can specify a rule that applies to an HTML tag defined inside another tag. For example, if you wanted to target all `li` tags inside numbered lists (`ol`) but not those inside bulleted lists (`ul`), you could write CSS like this with a space between the two tags:

  `ol li { ... }`

- The **attribute** selector allows you to target HTML elements that have specific attributes or attribute values. For example, this rule targets all links that specify a target attribute:

  `a[target] { ... }`

- **Pseudo-classes** are used to define a special state of an element, such as whether the mouse is hovering over it or whether a link has been visited or not. The pseudo-class is specified after a colon after a regular selector. For example:

  `a:hover { ... }`

- **Pseudo-elements** let you style a specified part of an element, such as the first line or letter of a paragraph of text. A pseudo-element is specified with a double colon after an element to style the first letter inside every paragraph element:

  `p::first-letter { ... }`

2.1.3 Linking CSS Stylesheet to HTML

There are different options for where to write the CSS rules, but the most common is to place them all in a separate file with the extension .css. To link this file to the HTML that they target, add a `<link>` tag in the `<head>` section of the HTML file, as shown in Figure 2-4. As you can see, you can even have multiple CSS files in the same HTML file, so you can bring in rules from different sources. The CSS files can be both local as part of the directory structure of the website you are creating (Line 4) or they can be on a remote server using a URL to point to them, as shown in line 5 (you'll see the Bootstrap CSS used in the next chapter).

```
1   <head>
2       <meta charset="utf-8" />
3       <title>Oatmeal Recipe</title>
4       <link rel="stylesheet" type="text/css" href="stylesheets/styles.css" />
5       <link rel="stylesheet"
        href="https://maxcdn.bootstrapcdn.com/bootstrap/3.3.7/css/bootstrap.min.css">
6   </head>
```

Figure 2-4: Linking CSS stylesheets to an HTML file

Review Questions

1. What are CSS declarations?
2. What's the difference between properties and values in CSS?
3. How is a CSS class declaration applied to an HTML element?
4. How can the same CSS rule be applied to multiple HTML elements?

Tutorial: Getting Started with CSS

In this tutorial, you will see how you can get started with applying CSS and working with styling text elements. While the steps in the tutorial refer to using Visual Studio, this tutorial can also be completed using your favorite code editor. You will be working on a slightly modified version of the Oatmeal recipe you created in Chapter 1.

Figure 2-5 shows the completed page as it will look when you have completed all the tutorials in this chapter.

Step 1: Open Visual Studio Code.

Open the folder Ch02-CSS-Starter in Visual Studio Code.

Step 2: Add the CSS stylesheet.

Add a new folder named **stylesheets** to the project by right-clicking the project name and selecting *New Folder*.

Inside the stylesheets folder, add a Style Sheet named **styles.css** by right-clicking the stylesheets folder and selecting *New File*.

Step 3: Link the CSS to the HTML.

Add the following line inside the `<head>` section of the oatmeal-recipe.html page:

```
<link rel="stylesheet" type="text/css" href="stylesheets/styles.css" />
```

The `href` attribute points to the stylesheets folder and the styles.css file you created previously.

Before making any styling changes, open the page in the browser. It should look similar to Figure 2-6. You'll notice a few changes compared to Chapter 1. The biggest change is that the list of ingredients is an unordered list instead of a definition list.

P&I Recipes

Oatmeal | Pancakes | Dinner | Snacks

Related Recipes

Nutty & Fruity Oatmeal

A healthy breakfast is key to a helthy diet. There are many websites that describe the benefits of oatmeal, including: WebMD Organic facts HHS

Why oatmeal?

Among the many benefits of oatmeal are:

- excellent source of fiber that helps to reduce cholestrol
- loaded with nutrients including vitamins, minerals, and antioxidants
- helps with weight reduction
- reduce constipation

Here is a recipe for a *delicious* oatmeal that increases the high nutritious value of plain oatmeal.

Ingredients

- 1/2 cup quick oats
- 1/2 cup each of almond and vanila soy milk:
- 2 Tbsp chopped walnuts
- 1/2 cup each chopped banana and blueberries
- A pinch each of cardamom and ginger powder
- 1/4 tsp cinnamon powder

Not sure what quick oats are? Quick oats are oat grains that are cut to smaller pieces and then steamed and rolled.

Directions

1. Mix all ingredients in a cooking pot
2. Add one cup of water
3. Bring to a boil and simmer for a couple of minutes. **Important:** Make sure you stir the mixture **frequently** while simmering.

Nutritional value

Nutritional value of 1/2 cup of dry quick oats

Nutrition	Amount
Calories	150 cal (kcal)
Carbohydrates	27 g
Protein	5 g
Fat	3 g
Dietary fiber	4 g

You may find the nutritional value of almost any food from sources like Nutrition Value and What's In Food

Figure 2-5: Completed page after doing all the tutorials. Cookbook photo by Scott Akerman/CC BY 2.0; oatmeal and brown sugar photo by Nate Steiner/CC0 1.0 (public domain); cookies photo by Whitney/CC BY 2.0; pancakes photo by Whitney/CC BY 2.0; stout photo by Amber DeGrace/CC BY 2.0.

The following shows the complete HTML listing for the initial page.

```
1   <!DOCTYPE html>
2   <html>
3   <head>
4       <meta charset="utf-8" />
5       <title>Oatmeal Recipe</title>
6   </head>
7   <body>
8       <header>
9           <h1>P&I Recipes</h1>
10          <img src="img/cookbook-header.png" width="100%" />
11          <nav>
12              <a href="oatmeal-recipe.html">Oatmeal</a>
13              <a href="pancakes.html">Pancakes</a>
14              <a href="dinner.html">Dinner</a>
15              <a href="snacks.html">Snacks</a>
16          </nav>
17      </header>
18      <main>
19        <section>
20          <h1>Nutty & Fruity Oatmeal</h1>
21          <img src="img/Nutty-fruity-oatmeal.jpg"
                    width="200" alt="Nutty & Fruity Oatmeal" />
22          <p>
23              A healthy breakfast is key to a healthy diet. There are many websites that
24              describe the benefits of oatmeal, including:
25              <a href="https://www.webmd.com/diet/oatmeal-benefits#1"> WebMD </a>
26              <a href="https://www.organicfacts.net/oatmeal.html"> Organic facts </a>
27              <a href="https://www.hhs.gov/fitness/eat-healthy/
                        dietary-guidelines-for-americans/index.html"> HHS </a>
28          </p>
29          <h2>Why oatmeal?</h2>
30          <p> Among the many benefits of oatmeal are: </p>
31          <ul>
32              <li>excellent source of fiber that helps to reduce cholestrol</li>
33              <li>loaded with nutrients including vitamins, minerals, and antioxidants</li>
34              <li>helps with weight reduction</li>
35              <li>reduce constipation</li>
36          </ul>
37          <p>
38              Here is a recipe for a <em>delicious</em> oatmeal that increases the high
                nutritious value
39              of plain oatmeal.
40          </p>
41        </section>
42        <section>
43          <h2>Ingredients</h2>
44          <ul>
45              <li> 1/2 cup quick oats </li>
46              <li> 1/2 cup each of almond and vanila soy milk:  </li>
47              <li> 2 Tbsp chopped walnuts </li>
48              <li> 1/2 cup each chopped banana and blueberries </li>
49              <li> A pinch each of cardamom and ginger powder </li>
50              <li> 1/4 tsp cinnamon powder</li>
51          </ul>
52          <p>
53              Not sure what quick oats are?
54              <def>Quick oats</def> are oat grains that are cut to smaller pieces and
                then steamed and rolled.
55          </p>
56        </section>
57        <section>
58          <h2>Directions</h2>
59          <ol>
60              <li>Mix all ingredients in a cooking pot</li>
61              <li>Add one cup of water</li>
62              <li>
63                  Bring to a boil and simmer for a couple of minutes.
64                  <strong>Important:</strong> Make sure you stir the mixture
                    <strong>frequently</strong> while simmering.
```

```
65                    </li>
66                  </ol>
67              </section>
68              <section>
69                  <h2>Nutritional value</h2>
70                  <table border="1" style="color: chocolate; font-size: 14px">
71                      <caption> Nutritional value of 1/2 cup of dry quick oats</caption>
72                      <tr>
73                          <th>Nutrition</th>
74                          <th>Amount</th>
75                      </tr>
76                      <tr>
77                          <td>Calories</td>
78                          <td>150 cal (kcal)</td>
79                      </tr>
80                      <tr>
81                          <td>Carbohydrates</td>
82                          <td>27 g</td>
83                      </tr>
84                      <tr>
85                          <td>Protein</td>
86                          <td>5 g</td>
87                      </tr>
88                      <tr>
89                          <td>Fat</td>
90                          <td>3 g</td>
91                      </tr>
92                      <tr>
93                          <td>Dietary fiber</td>
94                          <td>4 g</td>
95                      </tr>
96                  </table>
97                  <p>
98                      <aside style="margin-left: 30px; font-size: 14px">
99                      You may find the nutritional value of almost any food from sources like
                        <a href="https://www.nutritionvalue.org"> Nutrition Value</a>
                        and <a href="https://www.nutrition.gov/subject/whats-in-food">
                        What's In Food</a>
100                     </aside>
101                 </p>
102             </section>
103         </main>
104         <footer>
105             Copyright @ 2019 P&I Traders. All rights reserved.
106             <address>
107                 Contacts: J. Iversen, iversen@uwosh.edu; G. Philip, philip@uwosh.edu
108             </address>
109         </footer>
110</body>
111</html>
```

Step 4: Create the CSS rules.

For your first CSS rule, you will make the page title stand out using a different font, colors, and border.

Switch to styles.css and add a rule with h1 as the selector, as shown here:

```
1    h1 {
2        font-family: Arial, Helvetica, sans-serif;
3        color: cornflowerblue;
4        background-color: antiquewhite;
5    }
```

This will set the font to one of the listed fonts in the preference shown (i.e., Arial, if installed; otherwise, Helvetica; and as a fallback the default sans serif font on the system). Line 3 sets the text color to a light blue, and line 4 sets the background color to an off white.

Show the page in the browser, and you can see the changes

P&I Recipes

Oatmeal Pancakes Dinner Snacks

Nutty & Fruity Oatmeal

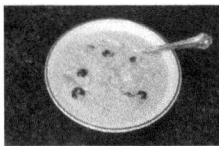

A healthy breakfast is key to a helthy diet. There are many web sites that describe the benefits of oatmeal, including: WebMD Organic facts HHS

Why oatmeal?

Among the many benefits of oatmeal are:

- excellent source of fiber that helps to reduce cholestrol
- loaded with nutrients including vitamins, minerals, and antioxidants
- helps with weight reduction
- reduce constipation

Here is a recipe for a *delicious* oatmeal that increases the high nutritious value of plain oatmeal.

Ingredients

- 1/2 cup quick oats
- 1/2 cup each of almond and vanila soy milk:
- 2 Tbsp chopped walnuts
- 1/2 cup each chopped banana and blueberries
- A pinch each of cardamom and ginger powder
- 1/4 tsp cinnamon powder

Not sure what quick oats are? Quick oats are oat grains that are cut to smaller pieces and then steamed and rolled.

Directions

1. Mix all ingredients in a cooking pot
2. Add one cup of water
3. Bring to a boil and simmer for a couple of minutes. **Important:** Make sure you stir the mixture **frequently** while simmering.

Nutritional value

Nutritional value of 1/2 cup of dry quick oats

Nutrition	Amount
Calories	150 cal (kcal)
Carbohydrates	27 g
Protein	5 g
Fat	3 g
Dietary fiber	4 g

You may find the nutritional value of almost any food from sources like Nutrition Value and What's In Food

Figure 2-6: Initial web page for Tutorial 1. Cookbook photo by Scott Akerman/CC BY 2.0.

Next, you can add a border to the heading by adding this rule:

```
border: 3px dotted blue;
```

Run the page in the browser to show your work so far—it should look similar to this:

P&I Recipes

The text is very close to the borders, so we'll add some space by increasing the padding around the text, as well as center the text and make the corners of the box a little rounded. Add the following rules to the rules for h1:

```
1   text-align: center;
2   padding: 5px 10px;
3   border-radius: 10px;
```

Line 1 centers the text in the browser, and line 2 adds padding around the tag that is being targeted; in this case, it is the h1 tag. You can specify one, two, three, or four values as follows:

- One value: Same padding on all four sides.

- Two values: The first value specifies the top and bottom padding, and the second specifies the right and left.

- Three values: The first value specifies the top padding, the second specifies the right and left, and the third specifies the bottom padding.

- Four values: Specifies all four sides, in this order: top, right, bottom, and left (clockwise).

In this example, we specified two values, so it is 5 pixels on the top and bottom and 10 pixels on the right and left.

You can also specify the padding as a percentage, which will then be a percentage of the overall size of the element.

Line 3 adds rounded corners to the element. Similar to padding, you can specify multiple values to have different rounding on each corner. The number specifies how much is rounded from the corner, and it can be specified in both pixels and percent.

Save the CSS file and refresh the browser to see the result, which should look like this:

P&I Recipes

Notice that both h1 headings on the page have been changed with the same styling.

It's time to practice!
- Change the color of the heading to red and the border to dashed.
- Style the h2 headings according to your own design

2.2 Style Specificity

You may have wonfdered, How did CSS get its name—and why is it called *cascading* style sheets? Well, the name reflects the nature of how a particular HTML element is rendered by the browser. There may be multiple style rules that could affect an HTML element, so the browser has to have a way to determine which rules take effect and how. It uses a system of increasingly important rules in a *cascading* fashion to make this determination. It is important to understand how the browser decides which styles to apply, in order to achieve your desired result. Not properly understanding this is a common cause of frustration, as you may think a certain rule should be applied but then it isn't. For simple sites, it's usually fairly straightforward to determine which rules apply. However, as the complexity of a site grows, it can become confusing. CSS uses a concept called *specificity* to determine which rule will be applied to a given element. With conflicting rules, the more specific rule will be applied. This part of the chapter will discuss how to measure rule specificity.

2.2.1 Cascading Styles

Before we get to the detailed discussion of specificity, it is helpful to understand the different places where CSS rules can be written. Figure 2-7 shows a hierarchy of where rules can be written.

Let's take a quick look at what each of the types of styles means:

1. Inline styles: Written directly into the page (and tags). Inline styles control a single character, a line, or short areas of that page.

2. Embedded styles: Appear in the head portion of the page and provide named styles for that particular HTML document.

3. Linked styles: External style sheets. A link appears in the head portion of the HTML pages that are to be controlled by a particular linked style sheet. This allows for controlling the layout of many HTML pages in a single stylesheet. This is the generally recommended approach to controlling layout.

4. Imported styles: A style sheet referenced from within another style sheet. This is similar to linked styles but is only referred to from within another style (external, embedded, or inline).

5. Browser defaults (user agent): When no style is available, the browser defaults are used.

In general, rules that are further to the outside of the circle will take precedence over rules to the center. That is, if a rule is declared as inline, it will override rules written as embedded or linked styles. However, specifically for imported, linked, and embedded styles, what really matters is the order they appear in—that is, rules written later will take precedence over rules written earlier.

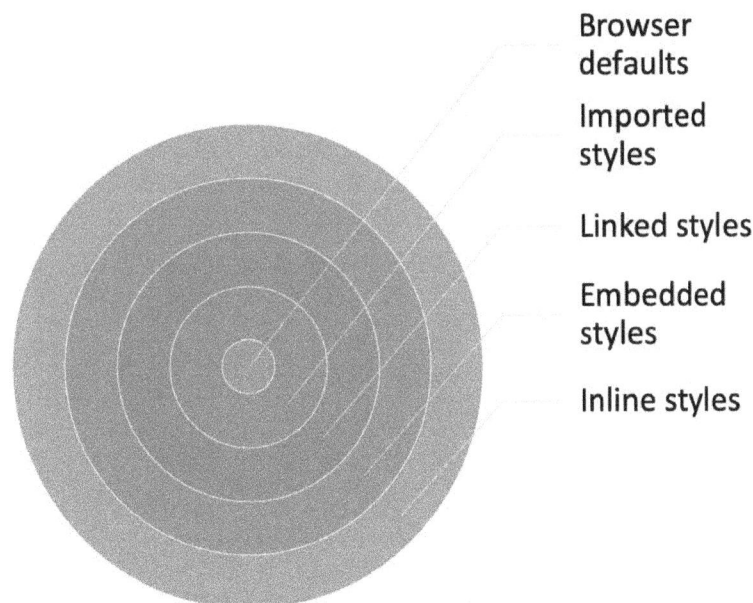

Figure 2-7: Hierarchy of cascading styles

2.2.2 Specificity Rules

Beyond the cascading of the styles, you will also need to consider the *specificity* of the rule. This comes into play when there are multiple rules that target the same HTML element and property. For example, consider the following line of HTML:

```
<h1 class="ingredients">Ingredients</h1>
```

Say the following CSS is applied to it:

```
1   h1 {
2       color: blue;
3   }
4
5   .ingredients {
6       color: green;
7   }
```

Which color will the heading for ingredients be? In line 2, the h1 element is given the color blue, but in line 6, the ingredients class is given the color green. Both rules apply to the heading, so which one "wins"? This is where specificity comes in. In this case, the heading will be green because classes have a higher specificity than regular elements. In order to determine exactly which rules will be applied, you can assign a point value to each type of rule as follows:

- Inline style attributes: 1,000 points
- ID rules: 100 points
- Attribute, class, or pseudo-class: 10 points
- Element name or pseudo-element: 1 point

To apply these rules, consider each rule and assign it a point value. When the browser renders the element, if there are conflicting rules, the one with the highest point value will be applied.

In the previous example, the h1 rule in lines 1–3 gets 1 point, as it is an element. The ingredients class gets 10 points because it is a class. Thus, the class wins and the heading is colored green.

As another example, consider this rule:

```
1   section img:hover .profile #directions {
2       background-color: blue;
3   }
```

This applies a blue background color when the user hovers over an image inside a section element that has the class profile and ID directions applied to it. The specificity of this rule is

- section: 1 pt (element)
- img: 1 pt (element)
- hover: 10 pts (pseudo class)
- profile: 10 pts (class)
- directions: 100 pts (ID)

For a total of 122 points.

If another rule tries to set the background color when the user hovers over a particular image, the specificity of that rule would have to be higher than 122 to take effect.

If two rules have the same specificity, then the last one declared will win. Further, if an HTML element is declared inside another element, then it is considered more specific. For example, a rule that targets the h1 element is more specific than one that targets the body element that the h1 is inside of.

Review Questions

5. What is the difference between inline, embedded, and linked CSS styles with regard to what they can be applied to?

6. How does specificity help determine which CSS rule is applied to an HTML element?

2.3 CSS Box Model

In order to determine how elements are placed and how large they are, it's important to understand the CSS Box Model, which the browser uses to render each HTML element as a set of four concentric boxes. You can specify the size of each of these boxes. Figure 2-8 shows each of the four boxes: **Content**, **Padding**, **Border**, and **Margin**. One way to think about the box model is that you can put a border around each element, and then you can control how much space is between the border and the content using Padding, and between the border and any other elements on the page using Margin. Both the Padding and Margin are transparent.

You can control the size of each of these boxes with the properties shown in Table 2-2.

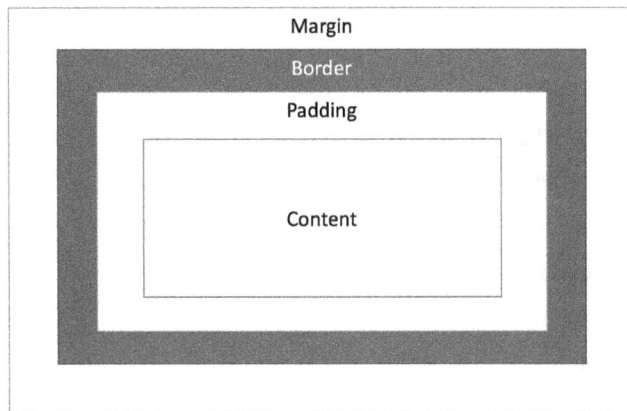

Figure 2-8: CSS Box Model

Table 2-2: Specifying the different sizes of the CSS box

Property	Size attributes	Notes
Content	`width, height, top, bottom.`	Only sets the size for the content, not the entire box. Can also set min and max values for width and height.
Padding	`padding-top, padding-bottom, padding-left, padding-right`	
Border	`border-width`	Can also set border-style, border-color, and specify rounded corners.
Margin	`margin-top, margin-bottom, margin-left, margin-right`	

It's important to note that the total size of an element is determined by the total of all these settings in either the horizontal or vertical direction. For example, consider the following settings for an HTML element:

```
1   .boxmodel{
2       width: 100px;
3       height: 50px;
4       padding: 20px;
5       border-width: 10px;
6       margin-right: 15px;
7   }
```

To calculate the total height of this box, you consider all four boxes, like this:

Content height	50
Padding top	20
Padding bottom	20
Border top	10
Border bottom	10
Margin top	0
Margin bottom	0
Total	**110**

The total width is calculated as follows:

Content width	100
Padding left	20
Padding right	20
Border left	10
Border right	10
Margin left	0
Margin right	15
Total	**175**

There are several ways to specify the size of each box without writing out each of the box properties. You saw in the previous example that padding and border-width were both set with a single value. In that case, the value is set the same on all four sides of the box. You can also use shorthand notation to quickly specify different values on several of the sides. Here are a couple of examples:

- Specify one value to set the same size on all four sides of the box:

  ```
  border-width: 10px;
  ```

- Specify different vertical and horizontal values—first value specifies the vertical (top and bottom) at 10 pixels and the second specifies the horizontal (left and right) at 20 pixels:

  ```
  border-width: 10px 20px;
  ```

- Specify three values to set different top and bottom and same values for left and right in the order of top, horizontal, bottom. In this example, top is 10 pixels, right and left are 20 pixels, and bottom is 30 pixels:

  ```
  border-width: 10px 20px 30px;
  ```

- If you specify all four values, you can have different values on all four sides, in the order of top, right, bottom, left (you can use the first letters—TRBL—as a mnemonic spelling the word *trouble* to help remember the order):

  ```
  border-width: 10px 20px 30px 40px;
  ```

You can specify the lengths of the sides in three different ways:

- **Length:** Use any number of measures, such as pixels (px), cm, mm, em, inches (in), and so forth. The default is pixels.
- **Percent (not valid for the border):** Sets the size as a percentage of the containing element.
- **Auto (not valid for the border):** The browser calculates the size automatically.
- **Inherit (margin and padding only):** The length is inherited from the parent element.

If you specify a background color or a background image for the content of an element, this will extend to the outer edge of the border. So, if the border isn't solid, the background will be visible underneath the margin and the border, but not under the padding. To illustrate, consider this example that has one box (inner) inside another (outer):

```
1  <div class="outer">
2      <div class="inner"> Not sure what quick oats are? Quick oats are oat grains that
   are cut to smaller pieces and then steamed and rolled, whereas old fashioned oats are
   whole oat grains that are steamed and pressed. Steel cut oats are oat grains that are
   cut to two or more pieces.
3      </div>
4  </div>
```

Here is the CSS that defines the two boxes:

```
1  .outer {
2      background-color: yellow;
3      width: 450px;
4      height: 200px;
5      border: 1px solid black;
6  }
7
8  .inner {
9      width:  350px;
10     height: 100px;
11     background-color: lightblue;
12     border: 10px dotted blueviolet;
13     padding: 20px;
14     margin: 20px;
15 }
```

This will generate an output like this:

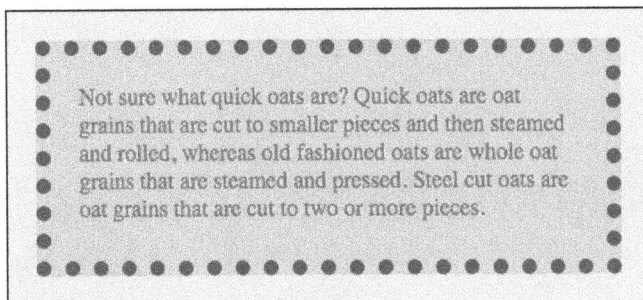

Notice how the light blue background color of the inner box extends underneath the content, padding between content and border, and is visible between the dots in the border. However, the background color does not show in the padding, as you can see from the yellow color. This is the background color of the outer box, which has the exact same size (450×200) as the inner box (width: 350 + 2×10 + 2×20 + 2×20 = 450, height: 100 + 2×10 + 2×20 + 2×20 = 200). The yellow background color of the outer box is thus showing through the 20-pixel margin of the inner box.

Review Questions

7. Which attributes should be added together to determine the total height and width of an HTML element?

8. With this declaration, what will the thickness of each of the four borders be?
```
border-width: 10px 20px 30px;
```

Tutorial: Recipe Description

In this tutorial, you will continue building the recipe page by taking a look at the description of the recipe. The HTML you will be working with is shown in Figure 2-9.

```
1  <main>
2      <section id="recipe-description">
3          <h1>Nutty & Fruity Oatmeal</h1>
4          <img src="img/Nutty-fruity-oatmeal.jpg"
                 width="200" alt="Nutty & Fruity Oatmeal" />
5          <p>
6              A healthy breakfast is key to a helthy diet. There are many websites that
7              describe the benefits of oatmeal, including:
8              <a href="https://www.webmd.com/diet/oatmeal-benefits#1"> WebMD </a>
9              <a href="https://www.organicfacts.net/oatmeal.html"> Organic facts </a>
10             <a href="https://www.hhs.gov/fitness/eat-healthy/dietary-guidelines-for-americans/
    index.html"> HHS </a>
11         </p>
```

Figure 2-9: HTML code for the oatmeal recipe (*continues*)

```
12          <h2>Why oatmeal?</h2>
13          <p>Among the many benefits of oatmeal are:</p>
14          <ul>
15            <li>excellent source of fiber that helps to reduce cholestrol</li>
16            <li>loaded with nutrients including vitamins, minerals, and antioxidants</li>
17            <li>helps with weight reduction</li>
18            <li>reduce constipation</li>
19          </ul>
20          <p>Here is a recipe for a <em>delicious</em> oatmeal that increases the high
            nutritious value of plain oatmeal.</p>
21        </section>
```

Figure 2-9: HTML code for the oatmeal recipe (*continued*)

In the first tutorial, you styled the heading 1 element, which affected all h1 elements on the page. But the page might look better if the h1 with the recipe title is different from the one in the page header. So, you'll start by changing the recipe title so it looks a little different.

Step 5: Add an ID to the HTML.

The first thing to determine when adding a rule is how to target the proper HTML element. In this case, you'll notice in Figure 2-9 that the h1 we want to target is also inside the section element. By adding an id to that element, we can target only elements inside that section of the page. Change line 2 as follows:

```
<section id="recipe-description">
```

Step 6: Add a CSS rule.

Add the following CSS rule to target the h1 element inside the recipe-description:

```
1    #recipe-description h1{
2        color: brown;
3        background-color: burlywood;
4    }
```

Line 1 specifies that the rule applies to any h1 element within the element that has the ID recipe-description applied to it. Then lines 2 and 3 change the text color and background color of the h1 element.

Save the CSS and refresh the page in the browser and you will see that the two heading 1 elements appear differently.

Step 7: Change the body of the page.

Next, we will make a few changes to the body section of the page. Add the following CSS rule:

```
1    body {
2        background-color: honeydew;
3        font-family: 'Segoe UI', Tahoma, Geneva, Verdana, sans-serif;
4        border: 2px dashed green;
5        padding: 10px;
6    }
```

This changes the background color (line 2) and font family (line 3). Line 4 adds a dashed green border around the page. Finally, line 5 adds a little padding to the body element. This has the effect of moving the border in a little so it is more visible. Try loading the page both with and without this last line to observe its effects.

This rule affects the entire `body` element—which means pretty much everything on the page. However, it didn't change any of the `h1` elements, since they were already styled, and those rules are more specific than the `body` rule.

To recap, you have seen how the body rule made changes to the entire page, but a more specific rule (`h1`) was able to override those changes, and a yet more specific rule (`recipe-description`) overrode the `h1` rule.

Step 8: Add a <style> section.

Next, you will see how a `<style>` section in the HTML page affects the rules in the linked CSS stylesheet. Add lines 5-9 of the following to the `<head>` section of the HTML page:

```
1    <head>
2        <link rel="stylesheet" type="text/css" href="stylesheets/styles.css" />
3        <meta charset="utf-8" />
4        <title>Oatmeal Recipe</title>
5        <style>
6            h1 {
7                color: green;
8            }
9        </style>
10   </head>
```

This is what's called **embedded styles**, and it allows you to provide styling to a single page. If you reload the browser with this change, you will see that the `h1` in the header section turns green but not the one in the recipe description, as that `h1` is targeted by an ID rule, which is still more specific than the general `h1` rule in the embedded style section.

You might be wondering why the `h1` in the page header turned green. Why did the `h1` in the embedded section override the linked stylesheet, since they have the same specificity? This is because the style section is written after the link to the stylesheet. If you move line 2 to after line 9, you will see that the stylesheet "wins."

Before moving on, remove the `<style>` section.

Step 9: Add a recipe description box.

We will illustrate the box model in more detail by adding a border around the recipe description section. Add this CSS rule to add a thick blue border around the description:

```
1    #recipe-description {
2        border: 5px solid blue;
3    }
```

When you reload the page, you will see that the border touches the content on the left and right sides. To fix this, you will need to use your knowledge of the box model. You have four parts of the box you can adjust: content, margin, border, and padding. Content and border will not help us add separation, as the border currently extends from the edge of the content and out by the width of the border. Changing the content will just adjust the border with it, and changing the border will just change its width but keep its inside edge stuck to the content. Looking at Figure 2-8, you can see that padding is between the content and the border, while the margin is outside the border. This means if you adjust the margin, the recipe description box will shrink away from the outside border of the body, but to change the spacing between the border and content, you will need to adjust the padding.

As with many of the CSS rules that relate to the box model, you can specify one, two, three, or four different values for padding. In this case, we just need to specify that the right and left sides should be changed. This can be done in two ways. The first is with shorthand by specifying two values:

```
padding: 0px 10px;
```

This sets the padding along the top and bottom to 0 pixels, and the padding along the left and right sides to 10 pixels. You can also use this notation to set the values:

```
padding-left: 10px;
padding-right: 10px;
```

Try both approaches, then change to set the margin instead of the padding and observe the results.

Next, we will add some interesting corners to the box. You saw in the `h1` rule that you can use `border-radius` to add a rounded corner to an element. However, just like with other parts of the box model, you can also pick only some of the corners to round to get an interesting effect. Change the recipe description rule to the following:

```
1    #recipe-description {
2        border: 3px solid blue;
3        padding: 0px 10px;
4        border-radius: 30px 0px;
5        margin-top: 10px;
6        background-color: antiquewhite;
7    }
```

The border is a little thinner (line 2) and the box has a background color (line 5), but the interesting line is line 4, which sets the corners. When specifying two values, you specify the upper-left and bottom-right

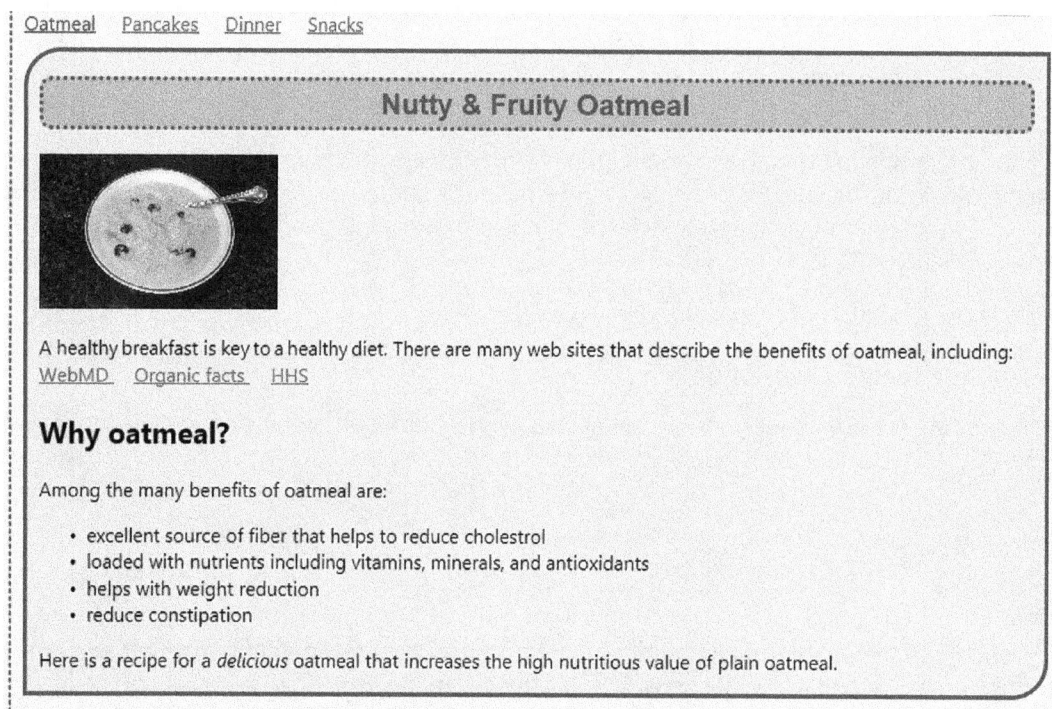

Figure 2-10: Final styling of recipe description

in the first parameter and the upper-right and lower-left in the second parameter. Line 5 sets the margin on the top to move the box a little away from the navigation links above it. The result should look similar to Figure 2-10.

It is worth noting that the background color of the recipe description box stays inside the box even under the rounded corners where the body background color is visible.

It's time to practice! Complete the following exercises:

- Use the specificity rules in Section 2.2 to explain why the recipe description h1 overrode the page-wide h1 CSS rule.
- Change the h2 inside the recipe description box (Why Oatmeal?) to have blue text and the other h2 headings on the page to be red.

2.4 CSS Positioning

Now that you have seen how the box model works to help you understand how elements are sized, we can turn our attention to the placement of various elements on the page.

2.4.1 Normal Flow of HTML

Before diving deeper into controlling the positioning of elements, let's take a look at how HTML positions elements by default. This is called the *Normal Flow of HTML* and is fairly straightforward. Each HTML element is rendered one after the other. Some elements, such as paragraphs and headings, force a line break, and elements are thus placed down the screen in the order they appear in the HTML. For elements that don't cause a line break, such as div and img, the elements are placed horizontally one after the other, and the browser decides how many to fit before moving the next one to the next line. As the browser window is resized, the browser will move things around to keep them from moving outside the screen horizontally.

When you create your HTML document, you should always try to place the various elements in a natural order that takes advantage of normal flow. While you will see various ways in the following sections where the placement of each part of a page can be changed and managed significantly, there are a variety of situations where the content may be rendered in normal flow and not based on your CSS rules. For example, a visually impaired person using a screen reader will get the content presented based on normal flow. This will also happen if your CSS fails to load. So, as you structure the HTML page, think about the logical structure of the document before considering how it may be presented on screen.

2.4.2 Floating

If you have images on your site, you are likely to want to be able to place those images within the text and have the text flow around the image. This can be done by using floats. To do this, you simply specify a CSS rule with the property float and a value of either left or right. Consider the following HTML code:

```
1  <div class="floater"></div>
2  <p>Not sure what quick oats are? Quick oats are oat grains that are cut to smaller
   pieces and then steamed and rolled, whereas old fashioned oats are whole oat grains
   that are steamed and pressed. Steel cut oats are oat grains that are cut to two or more
   pieces.</p>
```

This places an empty div with a CSS class called floater applied to it ahead of a paragraph of text. The following shows the effect of applying some CSS to the .floater div. Lines 2–4 specifies a size of 50×50 pixels and a light blue background. Line 5 is the key line that specifies whether the float is to the left or the right:

``
1 .floater {
2   width: 50;
3   height: 50;
4   background-color: lightblue;
5   float: left;
6 }
``
	``
	1 .floater {
	2 width: 50;
	3 height: 50;
	4 background-color: lightblue;
	5 float: right;
	6 }
	``

Not sure what quick oats are? Quick oats are oat grains that are cut to smaller pieces and then steamed and rolled, whereas old fashioned oats whole oat grains that are steamed and pressed. Steel cut oats are oat grains that cut to two or more pieces.	Not sure what quick oats are? Quick oats are oat grains that are cut to smaller pieces and then steamed and rolled, whereas old fashioned oats are whole oat grains that are steamed and pressed. Steel cut oats are oat grains that are cut to two or more pieces.

As you can see, the float property causes the text to float around the .floater div. If you wanted to create some space between the floated item and the text, you would need to add some margin to the floated item. If you added a margin to the text, it would simply move it in from the side of the container. For example, to keep the left-floated item to the top and left of the container, you would add some margin to the right and bottom sides of the item.

2.4.3 Positioning

In order to control the placement of HTML elements outside of the normal flow, you can use the `position` property. Once you have specified the position property on an element, you can then use `left`, `right`, `top`, and `bottom` to control the placement. The effect of each of those properties depends on which type of position you have chosen. Table 2-3 shows the options available for positioning HTML elements.

Table 2-3: Options for positioning HTML elements

Value of position property	Effect
static	Default value. Elements are placed based on normal flow.
relative	The element is placed relative to its normal position—you can specify how far away in pixels, em, etc. the element should be placed from where it would normally be.
fixed	Places the item in a fixed position on the screen, regardless of how the page is scrolled. You can control where on the screen an item is placed—for instance, you can have an item be fixed at the bottom of the screen.
absolute	Positioned relative to the nearest positioned ancestor (an element placed outside the element). If an absolute element doesn't have any positioned ancestor, it will use the document body.
sticky	This is the newest addition to the positioning family and isn't always well supported. But it allows for an element that can switch between relative and fixed, depending on scroll position. It is often used to have an element scroll up the screen until it reaches a specific point, and then it "sticks." For example, a menu bar may start below a header image, but then as the user scrolls down, the menu bar sticks to the very top of the screen.

Review Questions

9. Why should HTML elements be placed as much as possible according to normal flow?

10. If an element is floated left, where is it placed relative to the remaining content?

Tutorial: Placing HTML Elements

In this tutorial, you will see the effect of changing the positioning of HTML elements relative to each other. This is useful when working with images where you want the text to flow around them.

Step 10: Add a class to the HTML.

The recipe image currently sits above the recipe description text and leaves a lot of blank space to the right of the image. By floating the image, you can reclaim some of that space. Before changing the CSS, we have to determine which element to target. We could target the `img` HTML element, but this would affect all images on the page and might not be what is desired. We could also use the same technique we used with the `h1` element in the recipe description, but there is a third option of adding a class attribute to the element in the HTML. Change the `img` (line 4 in Figure 2-9) to remove the `width` attribute and add a `class` attribute:

```
<img class="recipe-picture" src="img/Nutty-fruity-oatmeal.jpg"
     alt="Nutty & Fruity Oatmeal" />
```

This will allow us to only target images with that class.

Step 11: Add a CSS to float the recipe image.

Add the following rule to the CSS:

```
1    .recipe-picture{
2        width: 200px;
3        float: left;
4    }
```

When you removed the `width` attribute from the picture, it needed to be added to the CSS (line 2) to ensure the width stayed the same. In general, you should control as much of look and feel as possible in the CSS and not in the HTML.

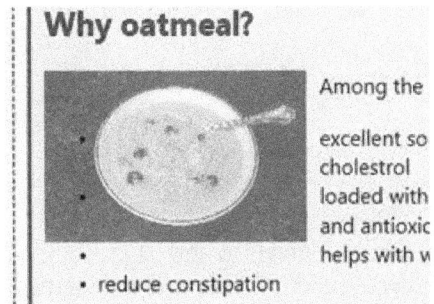

Figure 2-11: The effect of floating the image left

If you reload the page, you will see that the text flows around the image, but there's very little space between the image and the text. You can fix this problem by adding some margin to the right side of the image by adding this line to the `recipe-picture` rule:

```
margin: 0px 10px 0px 0px;
```

Step 12: Adjust floated images and bullets.

When images are floated left of a bulleted list, an unfortunate effect occurs. We can illustrate this by moving the image down to immediately following the `h2` "Why Oatmeal?" heading. When doing this (and setting the opacity of the image to 60%), you will see that the bullets in the bulleted list stayed in their original left-aligned location and didn't move to the right with the text (exactly where they end up depends on the browser; Figure 2-11).

To understand what's going on, add the following rule to add a border around the `ul` element:

```
1    ul {
2        border: 1px solid red;
3    }
```

Reload the page and you will see a red border around the entire `ul` element that crosses over the image:

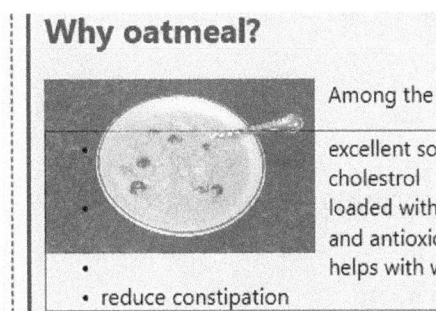

There is a property called `list-style-position` that controls the placement of the bullets. By default, this is set to `outside`, meaning the bullets are placed outside the box that surrounds the actual text of the bulleted list. You can change this to `inside`, which would move the bullets out of the image but would cause subsequent lines in a bullet to start aligned with the bullet and not indented. Here's what this looks like:

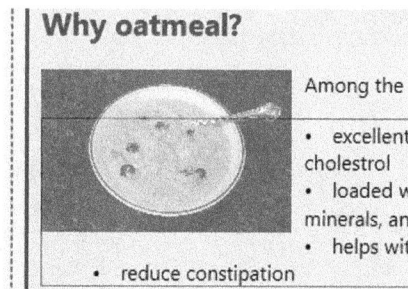

Another approach is to add `ul {overflow: hidden}`. This will fix the bullets but will not let the bulleted list flow around the image. It will instead keep the list going straight down and leave a big blank space below the image.

Yet another possible solution is the following:

```
1    ul {
2        list-style: outside;
3    }
4
5    ul li {
6        position: relative;
7        left: 1em;
8    }
```

This keeps the bullets on the outside, but by giving the `li` element the `relative` position, they are shifted away from their normal position. In this case, it shifts each `li` element to the right so they are outside the image. It may seem confusing that the property is `left` but it moves the element to the right. But that's because "left" refers to the left edge of the element. So, the left side of the element is shifted in a positive direction, which is to the right. The primary problem with this solution is that it doesn't work in Microsoft's browsers (Edge and Internet Explorer).

It is unfortunate to be in a situation of choosing between various imperfect solutions, but that happens from time to time when creating websites. It illustrates the need to test your work in multiple browsers (including mobile, as we will discuss in the next chapter).

One final note on the CSS rules above is that you would want to be more precise in your targeting so you don't change every bulleted list on the site. For the purposes of this tutorial, we will use this rule to target only lists in the recipe description section and stay compatible with all browsers:

```
1    #recipe-description ul {
2        overflow: hidden;
3    }
```

Before moving on, move the image back to its original position and remove the red box around the bulleted list.

Step 13: Create a navigation menu.

The `nav` section of the page has links to other pages on the site. Currently, the links are placed on one line separated by spaces. However, in order to target them for better positioning as a navigation bar, we will first turn them into an unordered list that will be displayed in an alternative way as an actual menu bar that goes across the screen below the header image.

Modify the `nav` section in HTML as follows:

```
1    <nav>
2        <ul>
3            <li><a href="oatmeal-recipe.html">Oatmeal</a></li>
4            <li><a href="pancakes.html">Pancakes</a></li>
5            <li><a href="dinner.html">Dinner</a></li>
6            <li><a href="snacks.html">Snacks</a></li>
7        </ul>
8    </nav>
```

Since the menu is in the `nav` element, we will use that to target the CSS rules. Add the following four rules to the CSS file:

```
1   nav {
2       padding: 0 0 20px 0;
3   }
4
5   nav ul {
6       list-style-type: none;
7       margin: 0;
8       padding: 0;
9       overflow: hidden;
10      background-color: cadetblue;
11  }
12
13  nav li {
14      float: left;
15      border-right: 1px solid black;
16  }
17
18  nav li a {
19      padding: 16px;
20      display: block;
21      text-decoration: none;
22      color: aliceblue;
23  }
```

Starting from the bottom, the last rule (lines 18–23) changes the links inside the `nav` to display as a block, which makes each link behave as a block element (like <p>), and also makes the whole block clickable and not just the text. By default, links are displayed inline. This rule also removes text decorations like underline and changes the color. Adding some padding makes the whole block clickable, not just the link.

The next rule up (lines 13–16) floats each of the list items, which will place them horizontally next to each other with some padding in between. It also adds a border to the right of each element, to show a box for each menu item.

The `ul` element is styled in lines 5–11 to remove the bullets from the list using `list-style-type: none` and setting the margin and padding to 0 to override browser defaults. Setting `overflow: hidden` ensures that if the content doesn't fit in the container, then it won't flow out of it—and it also prevents scrollbars.

The first rule sets some padding on the bottom of the entire nav element to give the menu bar some space before the next element.

Reload the page to see the result, which should look similar to this:

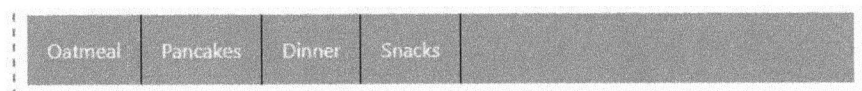

Step 14: Make the navigation bar "sticky."

If you want to have the navigation bar scroll up until it gets to the top of the screen and then stick there until the user scrolls far enough down to make the header visible, you can do that simply by changing the rule for the `nav` element:

```
1   nav {
2       padding: 0 0 20px 0;
3       position: sticky;
4       top: 0;
5   }
```

You can also make the `nav` stick to the bottom of the page by specifying `bottom: 0;` or make elements stick to the side of the browser by specifying `left` or `right`. The value is how many pixels from the edge to stick—in case you don't want to stick to the very edge. The other content will now scroll underneath the `nav` element.

2.5 Pseudo Classes and Elements

Sometimes we need to apply a style when an element is in a certain state or to only a certain part of an element. To accomplish this, we turn to pseudo classes and pseudo elements. In this section, you will see how these are used and some examples of how they are applied.

2.5.1 Pseudo Classes

Traditionally, the most common element for applying pseudo classes has been the anchor element for styling links in different ways. By using pseudo classes, you can control the style of a link, depending on whether it has been visited by the user, whether the user is hovering over it, or whether it has focus. Pseudo classes are specified after a colon from the element they are targeting. Here are examples of styling links using pseudo classes:

```
1  a:link {    /* Unvisited link */
2     color: blue;
3  }
4  a:visited {   /* Visited link */
5     color: green;
6  }
7  a:hover {    /* Hovering over link */
8     color: red;
9  }
10 a:active {  /* Activated link */
11    color: purple;
12 }
```

In this example, the text color is changed for different states of the links—blue for links that haven't been visited, turning to green after they have been visited, while if the user hovers over a link, it will turn red. Tabbing to a link to activate it will turn it purple. You can do any kind of styling of text for any of these states, including background color, bolding, and so forth.

The hover pseudo class can be applied to many different elements, such as <p> or <div>, allowing you to make hovering the mouse affect the look of the page significantly. However, it's worth noting that hovering doesn't work on a mobile platform.

Another useful set of pseudo classes allow for selecting an element based on its position relative to other similar elements, such as the first, last, or all odd- or even-numbered items. Here are a few examples of this:

```
1  li:first-child {  /* First element in list */
2     color: red;
3  }
4  li:last-child {  /* Last element in list */
5     color: red;
6  }
7  li:nth-child(odd) {  /* All odd elements (1, 3, 5 etc) in list */
8     color: red;
9  }
10 li:nth-child(4) {  /* Element number 4 */
11    color: red;
12 }
13 li:nth-child(3n) {  /*Elements 3, 6, 9, etc */
14    color: red;
15 }
16 li:nth-child(3n+2) {  /*Elements 2 (0+2), 5 (3+2), 8 (6+2) etc */
17    color: red;
18 }
```

The first four of these should be fairly self-explanatory based on the comments in the code. However, the notation in the last two requires a little more explanation. The general notation is An+B, where A and B are both integers. To determine which elements the styling applies to, multiply A with n, where n takes on integer values from 0 and up. Then you add the value of B. So, in the last example (3n+2), you get the following values:

$$3 \times 0 + 2 = 2$$
$$3 \times 1 + 2 = 5$$
$$3 \times 2 + 2 = 8$$
$$3 \times 3 + 2 = 11$$

etc.

Of course, these pseudo classes don't just apply to `li` elements but to any kind of element or class. They are commonly used to style tables and make it simple to create a table where every other row should be highlighted.

2.5.2 Pseudo Elements

Pseudo elements allow you to target a specific part of an element to style. Starting with CSS 4, they use a double colon to distinguish them from pseudo classes. As a simple example, you can target the first line of text in a block-level element like this:

```
1   ::first-line {
2      background-color: grey;
3      text-transform: capitalize;
4   }
```

This rule will target all text in a block-level element (including `p`, `div`, `table`). In most cases, you would probably want to be more specific about where on the page this would apply, for instance by limiting to text in paragraphs (`p::first-line`) or to limit to a specific class (`.title::first-line`).

2.5.3 CSS Functions

In recent years, CSS has included sometimes surprisingly sophisticated functionality. Much of this has been added through CSS functions. There is now a large number of functions focused on a range of tasks within CSS. In the next part of the tutorial, we will introduce a few of these functions. Examples include:

- N-th-child
- Blur
- Circle
- Brightness
- Drop-shadow
- Rotate
- Scale
- Translate
- Calc
- Color: rgb, hsl

Review Questions

11. What is the difference between pseudo classes and pseudo elements?
12. What would be a rule to set the color of every third table row as red?

Tutorial: Advanced CSS

In this tutorial, you will get a chance to explore some of the more advanced CSS features, like pseudo classes, pseudo elements, and functions. You will be doing advanced formatting of the menu and the nutrition table.

Step 15: Highlight a menu item when hovering.

We'll start by highlighting the menu item that the user is hovering over with the mouse. This is done by using a pseudo class called `:hover`. Add this rule to the stylesheet after the rule for `nav li a`:

```
1   nav li a:hover {
2      background-color: lightblue;
3   }
```

This is a simple and very common rule for highlighting links by hovering over them to change the background color. Another common approach is to make the link appear in bold on hovering. This can be achieved by changing line 2 to `font-weight: bold`.

Step 16: Move the last menu item to the far right.

Next, we'll use the `last-child` pseudo class to move the last element in the navigation menu to the far right by adding this rule:

```
1   nav li:last-child {
2       float: right;
3       border-right: 0;
4       border-left: 1px solid black;
5   }
```

The first line allows for targeting the last list item inside the `nav` element. The key to moving it right is to float that element to the right (all the other list items are floated left). Lines 3 and 4 fixes the separating lines by removing the border on the right side and adding one on the left side. Figure 2-12 shows the effect of hovering over a menu item, with the last menu item moved to the far right.

Figure 2-12: Hovering over the menu item and the last item moved to the right.

Step 17: Format the nutrition table.

Next, you will work on formatting the nutrition table. The HTML is shown in Figure 2-13. This is a very straightforward table with two columns and six rows. The table as rendered by default is shown in Figure 2-14. This is not particularly attractive but can be easily changed to look like the final result in Figure 2-15.

```
1   <table border="1" style="color: chocolate; font-size: 14px">
2       <caption> Nutritional value of 1/2 cup of dry quick oats</caption>
3       <tr>
4           <th>Nutrition</th>
5           <th>Amount</th>
6       </tr>
7       <tr>
8           <td>Calories</td>
9           <td>150 cal (kcal)</td>
10      </tr>
11      <tr>
12          <td>Carbohydrates</td>
13          <td>27 g</td>
14      </tr>
15      <tr>
16          <td>Protein</td>
17          <td>5 g</td>
18      </tr>
19      <tr>
20          <td>Fat</td>
21          <td>3 g</td>
22      </tr>
23      <tr>
24          <td>Dietary fiber</td>
25          <td>4 g</td>
26      </tr>
27  </table>
```

Figure 2-13: HTML for the nutrition table

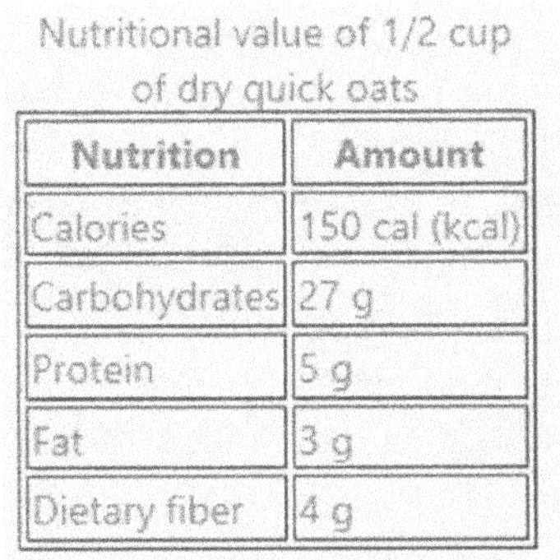

Figure 2-14: Default rendering of the nutrition table

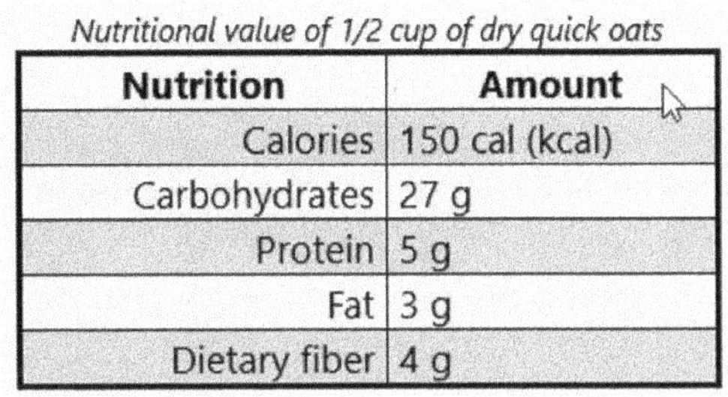

Figure 2-15: A formatted nutrition table

Line 1 of the HTML includes some border and style instructions. By default, the border is added around the content portion of each cell, as well as the table itself. This is why the table has a double border.

Start by replacing line 1 of the HTML with this line:

```
<table id="nutrition-table">
```

Next, add the CSS in Figure 2-16 to the CSS stylesheet one rule at a time, reloading the page after each rule to observe the effect. The first rule in lines 1–5 targets the entire table and defines a 3-pixel black border. It also sets the width of the table to a fixed 300 pixels. The last rule (line 4) collapses the borders between elements inside the table and ensures that you only get a single line between cells.

With only this rule, all the borders both inside and out would have the same width of 3 pixels. So, to ensure the inside borders are narrower, the rule in lines 7–10 targets the th and td elements and changes the border thickness. Since this rule includes both the id and an element, it has higher specificity than the first rule and overrides the border thickness, but only for th and td elements, not for the entire table, so the outside border stays at 3 pixels. The rule also adds a little padding on the left and right sides of each element. This keeps the text away from the center of the table.

```
1  #nutrition-table {
2      border: 3px solid black; /* Outside border */
3      width: 300px;
4      border-collapse: collapse;
5  }
6
7  #nutrition-table th, td {
8      border: 1px solid black; /* Inside borders */
9      padding: 0px 5px;
10 }
11
12 #nutrition-table tr:nth-child(even) { /* Zebra stripes */
13     background-color: lightgray;
14 }
15
16 #nutrition-table td:nth-child(1) { /* First column */
17     text-align: right;
18 }
19
20 #nutrition-table caption {
21     font-style: italic;
22     font-size: small;
23 }
```

Figure 2-16: CSS for the nutrition table

The rule in lines 12–14 shows the use of a CSS function. This one targets every even row of the table by applying `nth-child` to the `tr` element. Line 16 shows how you can target a specific column in a table by targeting `td:nth-child`. The number in parenthesis is the column number. Note the difference between line 12 and 16 is that in line 12, you are targeting the `tr` element, so every even row is given a light gray background color, whereas line 16 targets the `td` element, which appears inside each row. So, the rule is applied once for each row, but just the first `td` element within the row is targeted.

The last rule in lines 20–23 targets the table caption and makes the font small and italicized.

2.6 Page Layout with HTML and CSS

In the next chapter, you will see how to use the Bootstrap framework to control page layout that allows for targeting devices with screens of different sizes. However, CSS is increasingly able to accomplish much of the layout work that developers have trusted Bootstrap to perform, and there is an active debate within the community of whether CSS or Bootstrap are the most appropriate approaches to page layout. In this section, we will briefly look at two of the CSS approaches you can use to lay out the elements of a page: grid layout and flexbox.

2.6.1 Flexbox

The flexbox layout approach allows you to lay out elements in either the horizontal or vertical direction, and then control the placement along the chosen axis. By default, in a left-to-right language, the items in a flexbox are laid out in a row starting on the left and proceeding to the right.

Each item in a flex container can have three properties applied that determine the relative size of each element in the flexbox, as described in Table 2-4.

Table 2-4: Flex item properties

Flex item property	Description
`flex-grow`	If set to a positive value, the item is allowed to grow to take up available space in the flexbox.
`flex-shrink`	If set to a positive value, the item is allowed to shrink if there isn't enough space inside the flexbox to contain all the elements at their basis size.
`flex-basis`	The size of the item along the main axis of the flexbox. If all the elements fit exactly the flexbox with this size, flex-grow and flex-shrink are not applied.

If elements are given different values for flex-grow and flex-shrink, the items with higher values will grow or shrink more than those with lower values.

The flex item properties can be expressed with the flex shorthand property like this:

```
flex: 1 2 100px;
```

Where the first value is flex-grow, the second is flex-shrink, and the last is flex-basis.

As an example, consider the following HTML and CSS:

```
HTML:                                    CSS:
1   <div class="flexbox">                1   .flexbox {
2       <div>1</div>                     2       display: flex;
3       <div>2</div>                     3       border: 4px solid red;
4       <div>3</div>                     4       border-radius: 5px;
5       <div>4</div>                     5       width: 300px;
6   </div>                               6       height: 75px;
                                         7       background-color: lightgray;
Output:                                  8       padding: 5px;
                                         9   }
                                         10
                                         11  .flexbox > div{
                                         12      border: 1px solid black;
                                         13      background-color: lightblue;
                                         14      font-size: 20;
                                         15      padding: 3px;
                                         16      margin: 5px;
                                         17  }
```

Line 2 in the CSS sets the display of the outer div to be a flexbox. The flexbox is given a size of 300×75 pixels (lines 5–6). Line 11 shows how you can target elements inside another element without targeting the element itself. Since the individual flex items haven't been given any flex information, they retain their normal size and can thus fit inside the container, retaining their default placement from left to right.

By adding this rule, you can change the size of the second element by giving it a basis value and allowing it to both grow and shrink:

```
1   .flexbox :nth-child(2){
2       flex: 1 1 200px;
3   }
```

The result is that the second element fills out the entire remaining space inside the container. This is because flex-grow is set for this item and none of the others. If you added the same rule to the third element, items 2 and 3 would be the same size, but items 1 and 4 would remain small, and if you added the rule to all the items, they would be spaced equally and fill the space in the container.

In the tutorial, you will see how to use flexbox to lay out the navigation menu.

2.6.2 Grid Layout

Where flexbox allows for laying out elements along one axis, grid layout is used when you need to place items in both rows and columns.

You set this up by specifying display: grid in a container and then setting up the sizes of rows and columns in the grid—or, as is commonly done, specify the size of columns and then have the grid container automatically create rows. Column sizes are often specified in the new unit called fr, which allows for specifying the relative size of each column. Figure 2-17 shows an example of creating multiple columns using a grid display.

```
1   <div class="gridcontainer">          1   .gridcontainer {
2       <div>1</div>                      2       display: grid;
3       <div>2</div>                      3       grid-template-columns: 1fr 2fr 2fr 1fr;
4       <div>3</div>                      4       border: 4px solid red;
5       <div>4</div>                      5       border-radius: 5px;
6       <div>5</div>                      6       width: 300px;
7       <div>6</div>                      7       background-color: lightgray;
8       <div>7</div>                      8       padding: 5px;
9   </div>                               9   }
                                         10
                                         11  .gridcontainer > div {
                                         12      border: 1px solid black;
                                         13      background-color: lightblue;
                                         14      font-size: 20;
                                         15      padding: 3px;
                                         16      margin: 5px;
                                         17  }
```

Figure 2-17: Multicolumn layout using a grid display

The HTML code on the right sets up seven separate `div`s with a number in each. The CSS on the right formats the `div`s into columns. Line 2 sets up the `display` in a `grid`; then line 3 specifies the relative column widths for the columns. The number of values (four) determines the number of columns. The relative values indicate the width of each column. In this case, the two middle columns will be twice as wide as the outer two columns.

Since just four columns were created, the seven items will be laid out in two rows. Notice that the grid container doesn't specify a fixed height—it will grow to accommodate the items in the grid.

The following tutorial will show you how to lay out the entire page in a three-column layout using a grid.

Review Questions

13. In a flexbox container with four items, if one is given a smaller value for `flex-shrink` than the other three, how will the size of the four elements behave as the container is resized?

14. Similarly, what will happen if one element is given a smaller value for `flex-grow`?

15. In grid layout, how is the number of columns and rows determined?

Tutorial: Page Layout

In this part of the tutorial, you will see how to use the flexbox and grid layout to control the placement of items in different parts of the page. First, we'll take a look at the navigation menu.

Step 18: Modify the CSS for the navigation menu.

You will need to make a few changes to the CSS, as shown in the following listing. Comment out line 11 and add line 13 to make the unordered list into a flexbox. Uncomment line 17, as floats aren't used to control layout in a flexbox. Add line 33, and comment out line 34. By using an auto margin along the main axis, you can split elements into groups that are aligned at opposite ends. If you assigned the auto margin to another element in the middle of the list, all subsequent elements would move to the right.

```
1    nav {
2        padding: 0 0 20px 0;
3        position: sticky;
4        top: 0;
5    }
6
7        nav ul {
8            list-style-type: none;
9            margin: 0;
10           padding: 0;
11           /*overflow: hidden;*/
12           background-color: cadetblue;
13           display: flex;
14       }
15
16       nav li {
17           /*float: left;*/
18           border-right: 1px solid black;
19       }
20
21           nav li a {
22               padding: 16px;
23               display: block;
24               text-decoration: none;
25               color: aliceblue;
26           }
27
28               nav li a:hover {
29                   background-color: lightblue;
30               }
31
32           nav li:last-child {
33               margin-left: auto;
34               /*float: right;*/
35               border-right: 0;
36               border-left: 1px solid black;
37           }
```

Refresh the page and you should see no difference in how the navigation bar looks and works. Try making the Dinner and Snack menu items move to the right.

Step 19: Determine the page layout.

In order to set up the page for grid layout, you have to first determine how you want the page layout to appear. For the recipe page, we have several elements we expect to find for every recipe on the site: a header with an image and navigation bar, a description of each recipe, ingredients, instructions, nutritional information, and a footer. It's also common on websites to have a left sidebar with important links or navigation, so we'll add that into the design. Websites often rely on a design technique called wireframe to sketch out the overall design of a page. Figure 2-18 shows one possible design for the recipe page. You'll use this wireframe as a guide in setting up the layout of the page in the following steps.

The header and navigation bar have already been set up, so you will focus on the rest of the page. This can be envisioned as a grid with three columns and four rows, where the left and right columns span three rows and the footer spans three columns.

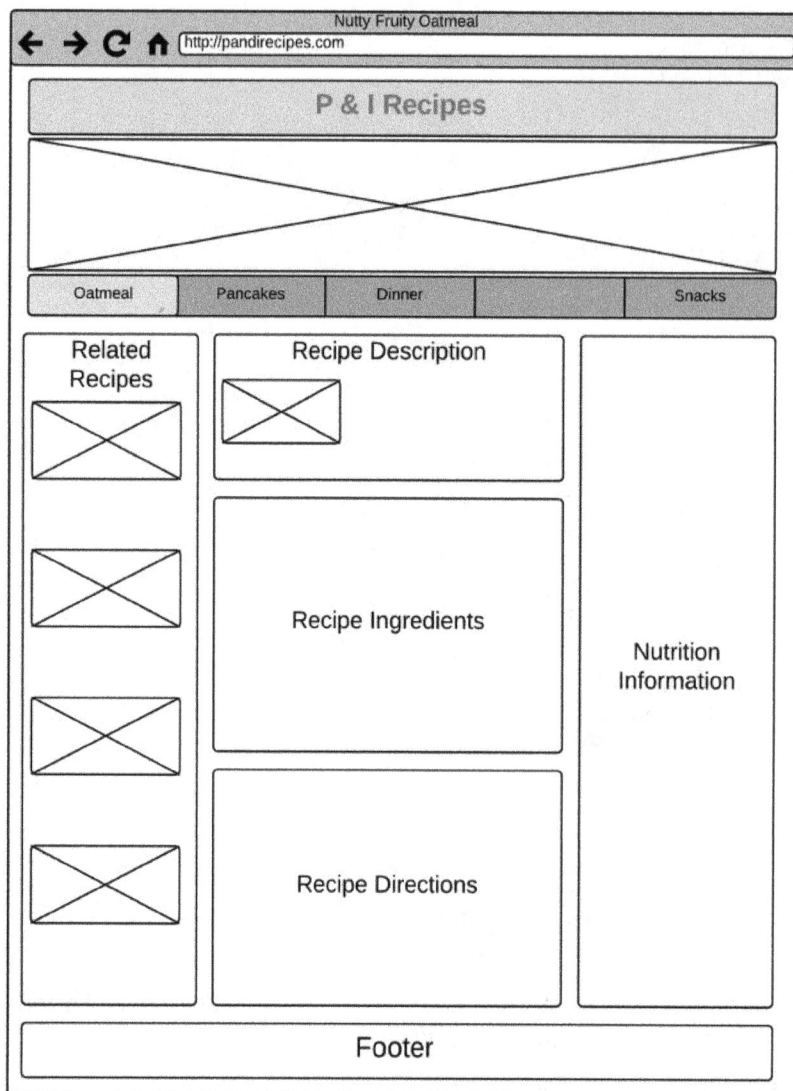

Figure 2-18: Wireframe for the recipe page

Step 20: Determine the structure of the HTML for the page layout.

In order to use the grid, we need to have the HTML elements structured properly so we can place them in the grid. Currently, the HTML is structured as shown in Figure 2-19.

There are a few problems with this structure: First, the sidebar for related recipes is missing. Second, in order to use the grid layout, we need a single container to hold all the elements that will appear in the grid. One way to fix these problems is to add a `div` element to hold all the elements that will be part of the grid. Another approach is to use the `main` element as the grid container and then move the footer inside the main element. We will use the latter approach, as it means fewer changes to the HTML page. The structure of the HTML will then be as shown in Figure 2-20.

Step 21: Restructure the HTML.

Start by moving the closing `</main>` element below the `footer` element.

Next, add the related recipes section at the beginning of the main element by adding these lines just below the line that contains `<main>`:

```
1   <section id="sidebar">Related Recipes.
2   </section>
```

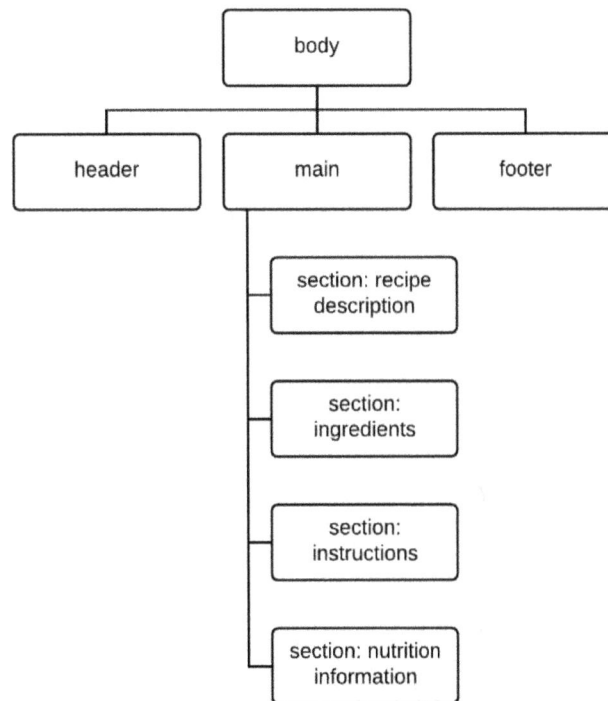

Figure 2-19: Initial structure of the HTML page

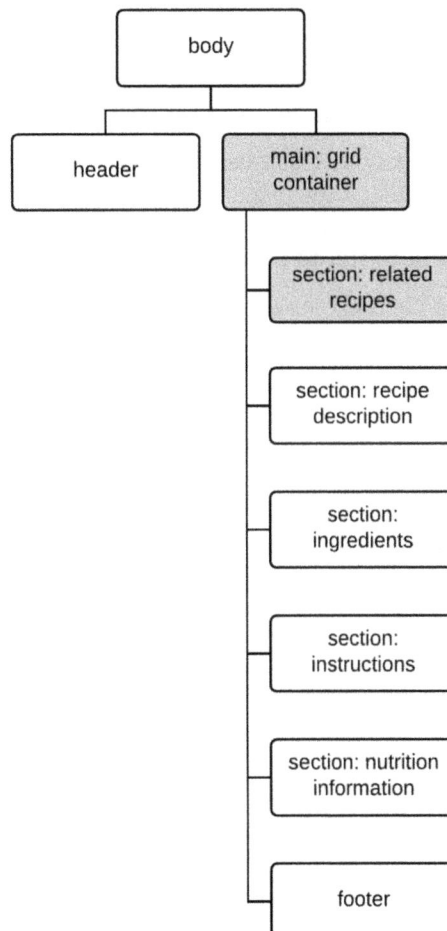

Figure 2-20: HTML structure revised for the grid layout

```
 8   ⊟<body>
 9   ⊞    <header>...</header>
21   ⊟    <main>
22   ⊞        <section id="sidebar">...</section>
25
26   ⊞        <section id="recipe-description">...</section>
48
49   ⊞        <section id="ingredients">...</section>
65
66   ⊞        <section id="directions">...</section>
77
78   ⊞        <section id="nutrition">...</section>
115
116  ⊞        <footer>...</footer>
122          </main>
123      </body>
```

Figure 2-21: Structure of the HTML with id labels

You'll add more content to the sidebar later in the tutorial. Then, add an `id` to each of the `section` elements with the names shown in Figure 2-21.

Step 22: Set up the grid structure.

In the CSS file, add the following rule to create the grid structure in the `main` element:

```
1   main {
2       display: grid;
3       grid-template-columns: 1fr 4fr 1fr;
4   }
```

This sets up the grid, where the middle column is four times wider than either of the side columns.

If you save and reload the browser now, you will see that the content sections are placed in the columns, but not as expected. The next step will fix the placement of the elements on the page.

Step 23: Control the placement of items in the grid.

Add the following CSS rules after the rule for `main`:

```
1   #sidebar {
2       border: 1px solid black;
3       grid-row-start: 1;
4       grid-row-end: 4;
5   }
6
7   #nutrition {
8       border: 1px solid black;
9       grid-column-start: 3;
10      grid-row-start: 1;
11      grid-row-end: 4;
12  }
13
14  #ingredients{
15      grid-column-start: 2;
16  }
17
18  #directions {
19      grid-column-start: 2;
20  }
21
22  footer{
23      border: 1px solid black;
24      grid-column-start: 1;
25      grid-column-end: 4;
26  }
```

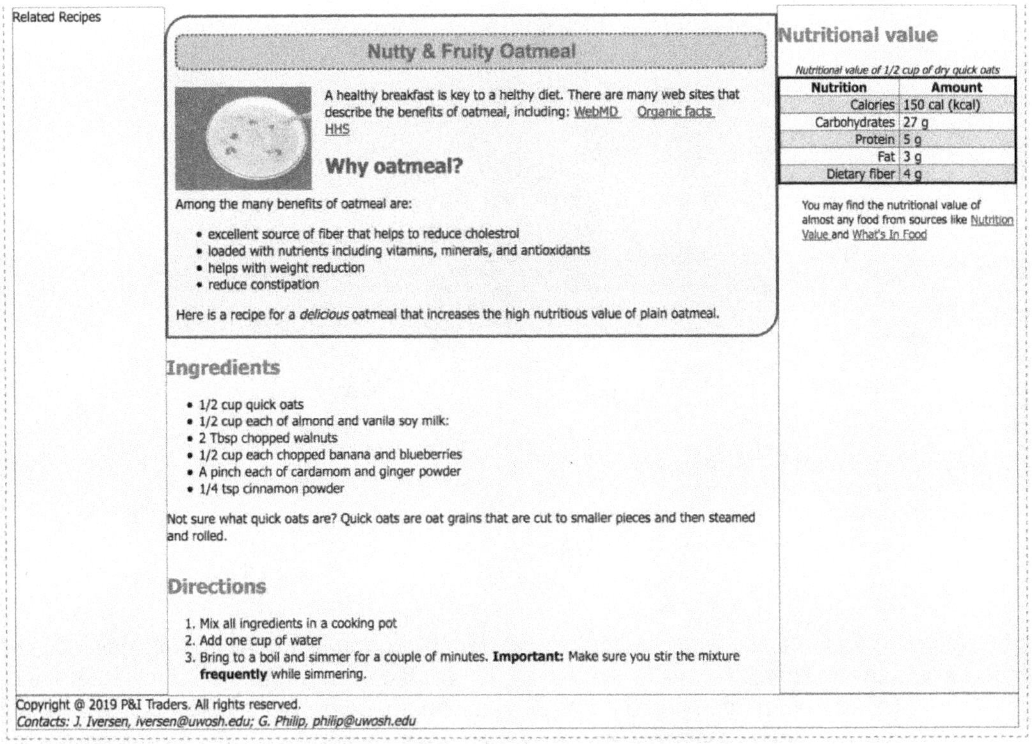

Figure 2-22: The resulting page with the grid layout

The borders around the sidebar, nutrition, and footer are only added so you can see the size of each of those boxes in the layout. Save the CSS and reload the page, and you should see the result as shown in Figure 2-22.

The content in the two side columns spans three rows, and the middle section has three elements in each of the three rows. The footer spans all three columns but makes up a single row.

Looking at the code, you will see that several of the elements have been given instructions on which row and column to start and end at. For example, the ingredients and directions are set to start in column 2. With no row start information, they are placed automatically in a row based on their placement in the HTML, so they end up in the second and third rows.

The related recipes sidebar doesn't have a column start, since it is the first element in the HTML and thus ends up as needed in the first column. It is specified to start at row 1 and end at row 4 (lines 3 and 4). This may be confusing since the footer is the fourth row of the grid. Why doesn't the sidebar extend to the fourth row? This is because the row numbers actually refer to the *grid lines* between the *row tracks,* which are the actual content rows. So, this grid has four row tracks and five horizontal grid lines. Similarly, it has three *column tracks* and four vertical grid lines.

An alternative notation for having a cell span multiple rows, which may be less confusing, is the following:

```
grid-row: 1 / span 3;
```

This indicates that the element starts on row track 1 and spans 3 rows.

The footer code (lines 22–26) shows how to span multiple columns.

Step 24: Add content to the related recipes sidebar.

The left sidebar should have images of related recipes. In the img folder, there are three additional pictures of oatmeal you can add to the sidebar. To control the placement, you will use a flexbox inside the sidebar. Start by changing the HTML for the sidebar as follows to add the images:

```
1    <section id="sidebar">
2        <h3>Related Recipes</h3>
3        <img src="img/pic1.jpg">
4        <img src="img/pic2.jpg">
5        <img src="img/pic3.jpg">
6        <img src="img/pic4.jpg">
7    </section>
```

Then configure the sidebar for flex display in the CSS:

```
1    #sidebar {
2        border: 1px solid black;
3        grid-row: 1 / span 3;
4        grid-row-start: 1;
5        grid-row-end: 4;
6        margin-right: 10px;
7        display: flex;
8        flex-direction: column;
9        justify-content: space-between;
10       align-items: flex-end;
11   }
```

Lines 6–11 are new. Line 6 simply adds a little space between the sidebar and the main content of the page. Line 7 sets the sidebar up for flex display, and line 8 specifies that it's a column and not a row. Line 9 specifies that the flex elements should be displayed with space between them, and line 10 aligns the items to the right side of the container and makes them keep their default size. In this case, the images all have the same size, so they look good like this. If they weren't of equal size, you should specify the same width for each image.

Reload the browser and observe how the flex container behaves when you resize the browser window. The images should stay distributed within the container.

Exercises

Exercise 2-1: Fixing the Recipe Page

Fix the recipe page up with a few finishing touches:

- Remove the borders around the sidebar, nutrition info, and footer.
- Add a background color to the sidebar: background-color: #c0ebe9;
- Add some space on the right side of the images: #sidebar img { margin-right: 10px; }
- Left-align the heading in the sidebar: #sidebar h3 { align-self: flex-start; }
- Add some space on the left side of the nutrition information: margin-left: 5px;
- Add a border and some space above the footer:

```
1    border-top: double;
2    margin-top: 10px;
3    padding-top: 10px;
```

Exercise 2-2: Style the Grand Website

Use a CSS written in an external stylesheet to recreate the Grand Opera House website as shown in Figure 2-23.

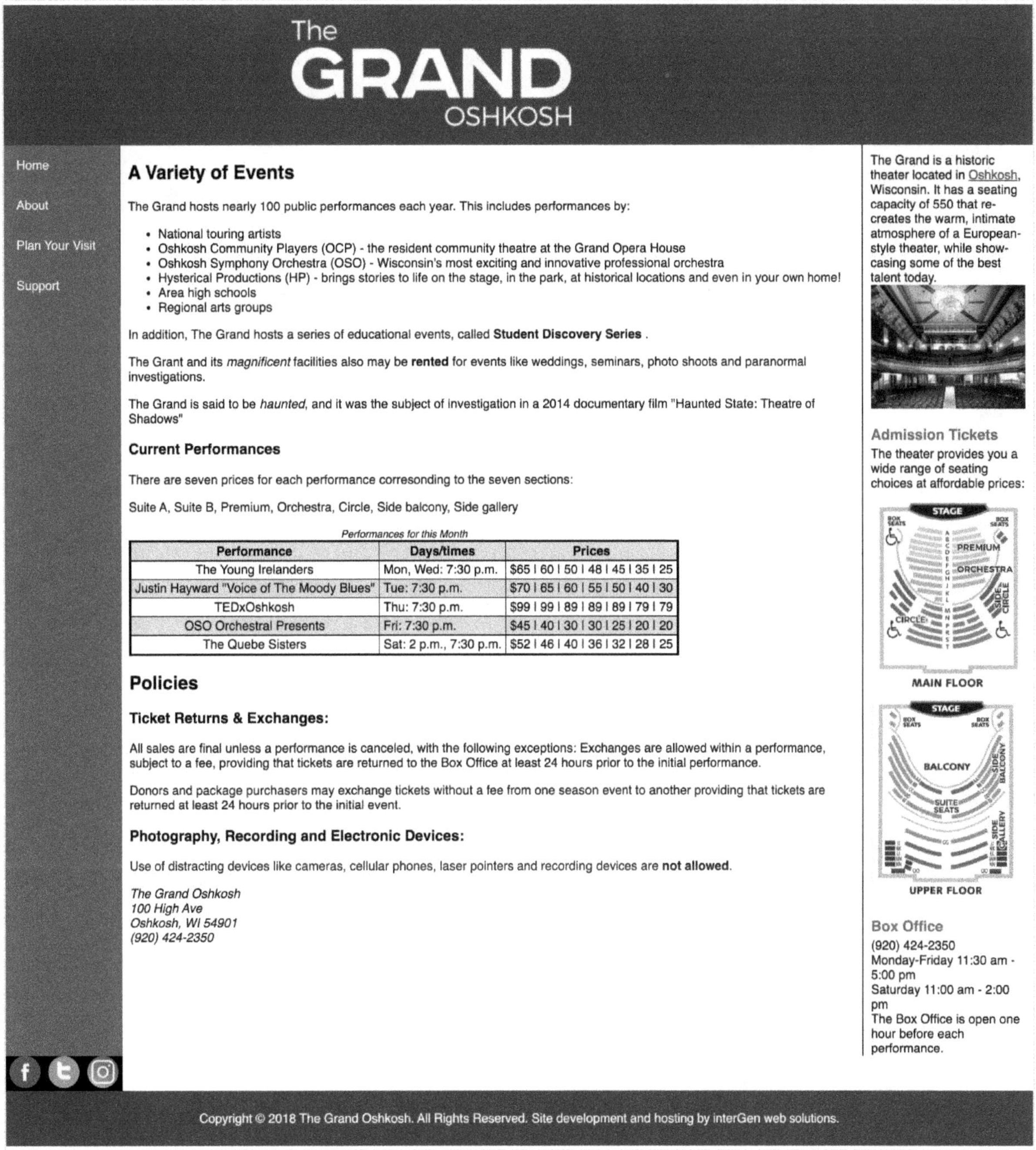

Figure 2-23: Styled Grand Opera House website

The HTML and all graphics for the page are available in the starter project for the exercise.
Use these notes to help you create the styling:

- The navigation bar on the left should show orange lines under each link as the user hovers over each one.
- The social links should always appear on screen, regardless of scrolling.

- Color codes used:
- Header: #72246C
- Left sidebar: #24722b
- Footer: #722b24
- Headers in the right sidebar: #246c72

Chapter 3

Responsive Web Design with Bootstrap

As mobile computing has become increasingly popular, it has become increasingly important for website designers to create websites that work as well on mobile devices as they do on traditional computing devices. What is considered the state-of-the-art approach to mobile web design has shifted several times and is still changing rapidly, with new approaches, frameworks, and tools released regularly and different vendors vying furiously for the attention of web developers.

In this chapter, we will present what has become the current standard approach for many web developers in creating websites that work well for both mobile and traditional devices: responsive web design supported by the Bootstrap framework. By using this approach, developers have shifted to a mobile-first strategy, where the website is designed primarily with a mobile-user experience in mind and then scaled up to a desktop experience.

Learning Objectives

After studying this chapter, you should be able to

- Describe how responsive design uses the actual size of the device screen to determine the layout for a website.
- Effectively use common Bootstrap classes.
- Format tables using Bootstrap.
- Set up common Bootstrap controls like card deck, accordion, modal dialog, and carousel.
- Discuss the advantages and disadvantages of using Bootstrap.

3.1 Introduction to Responsive Design

Web designers considering how to accommodate mobile devices face at least three big issues: (1) Mobile devices have smaller screens than traditional computing device like a laptop or desktop computer. (2) Input options are different due to the lack of a physical keyboard and mouse, and many devices (most notably all iOS devices) don't have access to uploading files from a file system. While it is much easier to access cameras and microphones on a mobile device, mobile web browsers often aren't allowed to access these sensors on the device. (3) The network connection on a mobile device can be less stable than on a traditional device, and data may also be more costly to download.

Responsive web design addresses the first of these concerns by recognizing that a single web design may not be appropriate for all screen sizes. An experience designed to work well on a 24-inch monitor is not likely to be a good experience on a 4-inch mobile phone screen—and vice-versa. With a responsive design approach, the designer creates a different layout for different screen sizes and then, through the use of a media query that detects the screen size of the device, serves up a layout appropriate for that screen size.

This allows for rearranging the content to fit the width of the screen. For example, on a desktop device, you may have a menu that goes across the screen but turn that into a dropdown menu on a mobile device. It could also allow for serving scaled-down and compressed versions of images on smaller screens that aren't able to display as much detail, thus saving on bandwidth.

With responsive design, you set several breakpoints for screen width, and then you can design a layout optimized for screen sizes between the breakpoints. When a device requests the page, the screen size of the device determines which layout its browser will render. For example, Table 3-1 shows the breakpoints used to determine which layouts to use in the Bootstrap framework. Using these breakpoints, a designer could create five different layouts for each of the ranges in the table—or use the same design for several adjacent ranges. For example, you may have one design for phones (extra-small and small), a second for tablets (medium and large), and a third for computers (extra-large).

Table 3-1: Breakpoints for responsive design sizes in Bootstrap 4

Screen size	Bootstrap designation	Typical device	Min breakpoint	Max breakpoint
Extra-small	`*-xs`	Phone—4"	0	575
Small	`*-sm`	Large phone—4.5"+	576	767
Medium	`*-md`	Tablet—portrait	768	991
Large	`*-lg`	Tablet—landscape	992	1199
Extra-large	`*-xl`	Desktop/Laptop	1200	N/A

You can observe the effect of this type of design on most modern websites by using a desktop browser on a screen with a resolution higher than 1200 pixels wide. Load up a modern website, and then gradually drag the edge of the browser to make the window smaller. You should see an effect at various points as menus are collapsed and content is moved around to better accommodate the smaller screen width. Figure 3-1 shows the Canvas website with several course cards displayed at different browser widths. In the widest display, the To Do list is displayed on the right, but as the browser narrows, this element disappears from the side and is moved below the course cards. In the narrowest width, the menu on the right is collapsed behind the hamburger menu in the top bar.

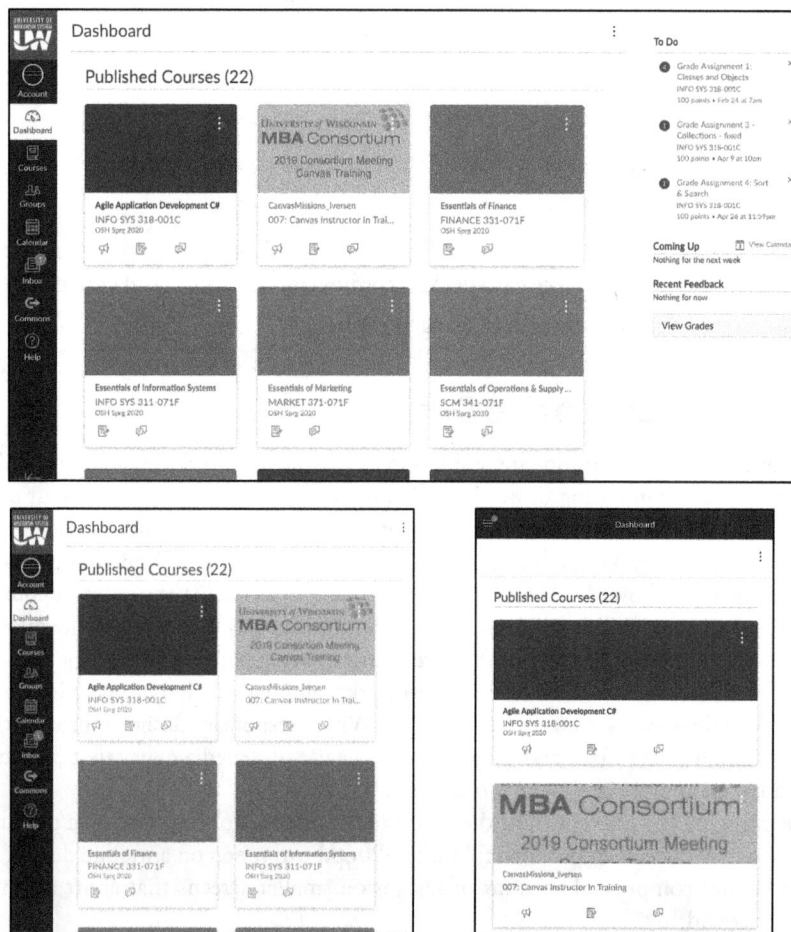

Figure 3-1: Example of a responsive website at three different widths

3.2 Introduction to Bootstrap

The Bootstrap framework was developed by Twitter as a way to standardize internal web development but was released under an open-source license in 2011. The most recent version (Bootstrap 4.5) was released in May 2020. One of the major changes from the previous version is that Bootstrap is now based on CSS Flexbox, which you learned about in the previous chapter. It is estimated that Bootstrap is used on close to 20% of all websites.

Bootstrap supports a mobile-first responsive design approach and includes a large number of user interface elements that can be easily added to an HTML page using CSS classes. Bootstrap allows for fairly easily creating a unified look and feel across all the pages of a website. Themes can be used to control the look and feel. Many themes are available for purchase to further customize your website, so it doesn't look like every other Bootstrap-enabled website out there.

There are two ways to include Bootstrap in your project. You can download Bootstrap (source code and/or compiled) and add it locally, or you can link to an online version of Bootstrap hosted by a content delivery network (CDN). In this book, we will use this method of incorporating access to the Bootstrap controls. If you use Visual Studio and choose an ASP.NET website project type, Bootstrap will automatically be downloaded and installed into your project.

The official website for Bootstrap is www.getbootstrap.com, which contains a wealth of resources as well as descriptions of how to download or link to a CDN-hosted version of the framework.

Review Questions

1. What is meant by "responsive design"?
2. Load several websites and try to determine how many different designs have been implemented and which breakpoints are used to separate them. (Note that if a site doesn't use Bootstrap, the specific breakpoint values may differ from Table 3-1.)
3. What are some benefits and drawbacks to using the CDN version of Bootstrap compared to a downloaded version?

Tutorial: Getting Started with Bootstrap

Step 1: Set up the project.

Open up the Ch03-Bootstrap-Starter directory in Visual Studio Code. You'll find two pages already created: bootstrap-template.html and oatmeal-recipe.html. The former is a blank page that contains all the necessary links to the Bootstrap framework through CDNs. You can copy the content of this page into each new HTML page you create. Figure 3-2 shows the content of this page. The other page is the HTML page you worked on in Chapter 1.

```
1  <!DOCTYPE html>
2  <html lang="en">
3
4  <head>
5      <!-- Required meta tags -->
6      <meta charset="utf-8" />
7      <meta name="viewport" content="width=device-width, initial-scale=1, shrink-to-fit=no" />
8
9      <!-- Bootstrap stylesheet-->
10     <link rel="stylesheet"
11         href="https://stackpath.bootstrapcdn.com/bootstrap/4.5.2/css/bootstrap.min.css"
12         integrity="sha384-JcKb8q3iqJ61gNV9KGb8thSsNjpSL0n8PARn9HuZOnIxN0hoP+VmmDGMN5t9UJ0Z"
13         crossorigin="anonymous" />
14
15     <title></title>
16 </head>
17
18 <body>
19
20     <!-- JavaScripts needed for some Bootstrap functionality -->
21     <script src="https://code.jquery.com/jquery-3.5.1.slim.min.js"
22         integrity="sha384-DfXdz2htPH0lsSSs5nCTpuj/zy4C+OGpamoFVy38MVBnE+IbbVYUew+OrCXaRkfj"
23         crossorigin="anonymous"></script>
24     <script src="https://cdn.jsdelivr.net/npm/popper.js@1.16.1/dist/umd/popper.min.js"
25         integrity="sha384-9/reFTGAW83EW2RDu2S0VKaIzap3H66lZH81PoYlFhbGU+6Bzp6G7niu735Sk71N"
26         crossorigin="anonymous"></script>
27     <script src="https://stackpath.bootstrapcdn.com/bootstrap/4.5.2/js/bootstrap.min.js"
28         integrity="sha384-B4gt1jrGC7Jh4AgTPSdUtOBvfO8shuf57BaghqFfPlYxofvL8/KUEfYiJOMMV+rV"
29         crossorigin="anonymous"></script>
30
31     <!-- Font Awesome icons -->
32     <script src="https://use.fontawesome.com/13047ec227.js"></script>
33 </body>
34
35 </html>
36
```

Figure 3-2: Bootstrap template page

Line 10 contains the link to the main Bootstrap stylesheet in version 4.5.2 in this file. Just before the end of the `body` element, you will see references to four JavaScripts (lines 20–32), jQuery, Popper, Bootstrap, and Font Awesome. The first three are required for some of the Bootstrap components. The last one references the free version of a set of icons from Font Awesome that you will be using to make the website look modern with standard icons. The JavaScript references should always stay at the end of the code file, so place your content before these three lines. To ensure you have the latest version of Bootstrap, you can go to GetBootstrap.com to get the latest CDN URLs and copy them into your files. For Font Awesome, you can go to FontAwesome.com.

Step 2: Add Bootstrap to the recipe page.

Open the oatmeal-recipe.html page in VS Code and then display it in a browser. You should see a plain version of the oatmeal recipe rendered using browser defaults.

Copy lines 5–13 in Figure 3-2 from bootstrap-template.html to the `head` element in the recipe page. Copy lines 20–32 to the bottom of the recipe page just before the closing `body` tag.

Save the recipe file and reload the page in the browser, and you should see some subtle differences like the specific font used. This indicates some default settings used in the Bootstrap framework (such as which fonts are used) were applied.

Step 3: Set up the navigation bar.

The navigation section of the page needs some work, as it is just a set of links in the default page. There are many ways to set up a navigation bar using Bootstrap, so this is just an introduction that illustrates some of the options. You can learn about other options on the GetBootstrap website or on W3Schools.

For a basic navigation bar, add the .nav class to the nav element, then add .nav-item and .nav-link to each navigation link, as shown in Figure 3-3.

```
1    <nav class="nav">
2        <a class="nav-item nav-link" href="oatmeal-recipe.html">Oatmeal</a>
3        <a class="nav-item nav-link" href="pancakes.html">Pancakes</a>
4        <a class="nav-item nav-link" href="dinner.html">Dinner</a>
5        <a class="nav-item nav-link" href="snacks.html">Snacks</a>
6    </nav>
```

Figure 3-3: Code for the basic navigation bar

This results in a navigation bar that looks like Figure 3-4.

Oatmeal   Pancakes   Dinner   Snacks

Figure 3-4: Basic navigation bar

When using Bootstrap, you apply CSS classes to HTML elements, similar to what you did in Chapter 2. However, the Bootstrap framework defines many classes for you, so in order to use Bootstrap, you need to apply the standard Bootstrap classes rather than make your own. Changing the look, feel, or behavior of the page can then be done by overriding the Bootstrap classes in your own CSS stylesheet. To keep things simple, in this chapter, you will only do very minimal overriding of the Bootstrap classes.

The navigation bar can be further enhanced by turning the navigation items into tabs and marking the current page as being active. Add nav-tabs after nav in line 1, and add active after nav-link in line 2, as shown in Figure 3-5.

```
1    <nav class="nav nav-tabs"  >
2    <a class="nav-item nav-link active" href="oatmeal recipe.html">Oatmeal</a>
```

Figure 3-5: Code for adding tabs to a navigation bar

You should get the result shown in Figure 3-6, where the active page (Oatmeal) is shown as a tab with a different font color and a line surrounding it.

Oatmeal   Pancakes   Dinner   Snacks

Figure 3-6: Navigation bar with tabs

If you want the navigation bar to stretch across the entire page and equally divide each link into the available space, you can add .nav-justified to the nav element.

The navbar currently is very close to the left of the screen and the image above it. To add some margin around the navbar, you can add the class m-3 to the nav element. Later in the chapter, we will discuss the details of how you can use Bootstrap to manage the sizes of the various parts of the CSS box model.

Step 4: Make the navigation bar responsive.

The current navigation bar will stay horizontal, with navigation elements wrapping to the next line if the window becomes too narrow. This is not a good experience on smaller screens, so we will make the navigation more responsive. To do this, turn the navigation into an actual navigation bar using the .navbar class. This works similar to the .nav class but allows for some more sophisticated functionality, like turning the navigation vertical or collapsing it into a button at smaller screen sizes.

The navigation bar requires a div surrounding the navigation elements inside the nav element. By default, the navigation bar is vertical, so to create a horizontal navigation bar, add the class .navbar-expand-*, replacing the asterisk with one of the breakpoint designations, above which the navigation is horizontal and below which the navigation is vertical. Figure 3-7 shows a navigation bar that is horizontal for screens above 768 pixels (md). Table 3-1 shows all the breakpoints you can use.

```
1   <nav class="navbar navbar-expand-md bg-light m-3">
2       <div class="navbar-nav nav-tabs">
3           <a class="nav-item nav-link active" href="oatmeal-recipe.html">Oatmeal</a>
4           <a class="nav-item nav-link" href="pancakes.html">Pancakes</a>
5           <a class="nav-item nav-link" href="dinner.html">Dinner</a>
6           <a class="nav-item nav-link" href="snacks.html">Snacks</a>
7       </div>
8   </nav>
```

Figure 3-7: Horizontal navbar on screen widths above medium (md)

In line 1, the bg-light class gives the navbar a light-gray background. You can also use bg-dark for a darker look. Line 2 has the div that surrounds the navigation items. Note the use of navbar-nav instead of nav, as you did above where you didn't use navbar. This also has the active class, to highlight which page is currently displayed.

Now, when the screen is resized, you will observe that at a certain point, the navigation bar switches from horizontal to vertical and vice versa.

Step 5: Collapse the navigation bar.

In many instances, it is preferable to hide the navigation bar on smaller screens to avoid it taking up too much space. To do this, you will need to add a button that the user can click or tap to expand the collapsed menu. The button is then set up to reference the navigation items to collapse. Figure 3-8 shows how this is typically structured.

```
1   <nav class="navbar navbar-expand-md bg-light">
2       <!-- Toggler/collapsibe Button -->
3       <button class="navbar-toggler" type="button"
4               data-toggle="collapse" data-target="#collapsibleNavbar">
5           <span class="fa fa-bars"></span>
6       </button>
7       <div class="navbar-nav nav-tabs collapse navbar-collapse"
8            id="collapsibleNavbar">
9           <a class="nav-item nav-link active" href="oatmeal-recipe.html">Oatmeal</a>
10          <a class="nav-item nav-link" href="pancakes.html">Pancakes</a>
11          <a class="nav-item nav-link" href="dinner.html">Dinner</a>
12          <a class="nav-item nav-link" href="snacks.html">Snacks</a>
13      </div>
14  </nav>
```

Figure 3-8: Button for collapsing the navbar

The button is set up in lines 3–6. The `data-target` points to the ID that is added to the `div` surrounding the navigation items in line 8. Line 5 contains the content of the button. In this case, we are using an icon from Font Awesome called fa-bars, which is a typical hamburger menu with three horizontal bars. Figure 3-9 shows the result.

Figure 3-9: Hamburger menu icon hiding a collapsible menu

If you wanted to exclude some navigation items from being collapsed, you can end the `div` before those items and then surround the items you don't want to have collapse in their own `div` that would have to have the `navbar-nav` class applied.

Step 6: Control the placement of the navigation bar.

You can easily place a navigation bar in a fixed location at the top or bottom of the window by applying `fixed-top` or `fixed-bottom` to the `nav` element. You can also make the navigation bar sticky at either the top or bottom, where the navigation bar will scroll with the content and then stick at the top or bottom when the navigation bar reaches the edge by applying `sticky-top` or `sticky-bottom`. For the recipe page, you can add `sticky-top`. One issue with doing this in the current page is that the `nav` element is inside the `header` element, so it will only behave stickily until the end of the `header` element. To solve this problem, you can move the `nav` element between the `header` and `main` elements.

Place the navigation bar both at the top and bottom, and observe the results of each option. Then experiment with making it sticky.

It's time to practice! Complete Exercise 2 at the end of the chapter.

3.3 Contextual Classes and Colors

Throughout the Bootstrap framework, color is used to convey specific meaning—success, danger, warning, information, and so forth. This use of color is embedded in a set of contextual classes that can be applied to text, background, buttons, and other elements. Table 3-2 shows examples of some of these contextual classes, how they are used on various elements, and how they end up looking. It's difficult to read the light text in the last row on a white background, so this works better on web pages with a darker background.

You could achieve a similar result by creating custom CSS classes, but you would have to create multiple classes to completely match what is in Bootstrap. For example, if you created a CSS class called `success`, and specified that text color would be green, that same class couldn't also be used to style a button with green background and white text, as you would either end up with green text on a green background or have conflicting instructions for text color (green and white).

Table 3-2: Examples of contextual classes in Bootstrap

Contextual description	Class examples	Color examples (button and text)

Primary	`.btn-primary`
	`.bg-primary`
	`.text-primary`
	`.table-primary`
		Primary
		Read carefully

Secondary	`.btn-secondary`
	`.bg-secondary`
	`.text-secondary`
	`.table-secondary`
		Secondary
		In second place

Success	`.btn-success`
	`.bg-success`
	`.text-success`
	`.table-success`
		Success
		Congratulations!

Info	`.btn-info`
	`.bg-info`
	`.text-info`
	`.table-info`
		Info
		Just to let you know

Warning	`.btn-warning`
	`.bg-warning`
	`.text-warning`
	`.table-warning`
		Warning
		Watch out!

Danger	`.btn-danger`
	`.bg-danger`
	`.text-danger`
	`.table-danger`
		Danger
		Danger ahead!

Dark	`.btn-dark`
	`.bg-dark`
	`.text-dark`
	`.table-dark`
		Dark
		Dark and stormy

Light	`.btn-light`
	`.bg-light`
	`.text-light`
	`.table-light`
		Light
		See the light.

3.4 Formatting Tables with Bootstrap

Bootstrap contains a number of classes that make it easy to apply styling to tables. Table 3-3 shows several of the classes that are used to format tables in Bootstrap.

Table 3-3: Classes for formatting tables in Bootstrap

Table formatting class	Effect
`.table`	Basic class to apply to tables and adds basic styling with bold headers and lines between rows. The following classes are applied in addition to this one.
`.table-striped`	Adds gray zebra stripes to a table
`.table-bordered` / `.table-borderless`	Adds or removes borders for the table
`.table-hover`	Adds a hover effect that highlights the row the user is hovering over
`.table-sm`	Reduces the cell padding by half to produce a more compact table
`Contextual classes (.table-primary, .table-success, .table-danger etc)`	Adds background color to a table row or cell based on the contextual colors as shown in Table 3-2
`.thead-dark` / `.thead-light`	Adds dark or light background color to the `thead` element of a table
`.table-responsive-*`	Makes a table responsive by adding a horizontal scrollbar to the table. You can control the width at which the scrollbar is added by specifying a breakpoint. This class is added to a `div` surrounding the `table` element.

In the next tutorial, you will see how these classes can be applied to the nutrition value table in the recipe page.

Tutorial: Formatting Tables and Images

In this part of the tutorial, you will work with the table and contextual classes introduced above to improve the design of the Nutritional Values table in the Oatmeal Recipe page, while also adding some formatting to the image in the recipe description.

Step 7: Format the table.

Modify the Nutritional Value table as shown in Figure 3-10.

```
1   <h2>Nutritional value</h2>
2   <table id="nutrition-table"
3           class="table table-striped table-bordered table-hover">
4       <caption> Nutritional value of 1/2 cup of dry quick oats</caption>
5       <thead class="thead-dark">
6           <tr>
7               <th>Nutrition</th>
8               <th>Amount</th>
9           </tr>
10      </thead>
11      <tr>
12          <td>Calories</td>
13          <td class="table-primary">150 cal (kcal)</td>
14      </tr>
15      <tr>
16          <td>Carbohydrates</td>
17          <td>27 g</td>
18      </tr>
19      <tr>
20          <td>Protein</td>
21          <td>5 g</td>
```

Figure 3-10: Code for the nutrition value table (*continues*)

```
22      </tr>
23      <tr class="table-success">
24          <td>Fat</td>
25          <td>3 g</td>
26      </tr>
27      <tr>
28          <td>Dietary fiber</td>
29          <td>4 g</td>
30      </tr>
31 </table>
```

Figure 3-10: Code for the nutrition value table (*continued*)

Add the classes in line 3 one at a time, reloading the page for each one to observe the result. Add lines 5 and 10 to create a header for the table. It is worth noting that you can turn any row in a table into a header—perhaps you would use `thead-dark` for the overall page header and `thead-light` for each new grouping of content. Line 13 shows how you can apply one of the contextual classes to a single cell, and line 23 shows how you can apply it to an entire row.

The code provided in Figure 3-10 will produce the output shown in Figure 3-11.

Nutrition	Amount
Calories	150 cal (kcal)
Carbohydrates	27 g
Protein	5 g
Fat	3 g
Dietary fiber	4 g

Nutritional value of 1/2 cup of dry quick oats

Figure 3-11: Nutritional Value table formatted using Bootstrap

Step 8: Float a recipe image.

To format and float an image, apply special Bootstrap classes to the `img` element.

Add `float-right` to the `img` class property. This will float the image right at all resolutions. By adding a breakpoint instruction, you can control how the image is floated at different resolutions. For example, you can float the image right at large resolutions, left at medium resolutions, and have no floating at small (and extra-small) resolutions by adding these three classes: `float-lg-right float-md-left float-sm-none`

Add those classes to the oatmeal image as shown in Figure 3-12 and observe the results of changing the browser size. You should see the image move to the right at the widest width, move to the left at medium widths, and have no text on the side of the image at the narrow width.

```
1  <img src="img/Nutty-fruity-oatmeal.jpg"
2      width="200" alt="Nutty & Fruity Oatmeal"
3      class="m-3 float-lg-right float-md-left float-sm-none" />
```

Figure 3-12: Floating image options

Step 9: Format the image's appearance.

You can add a rounded corner to an image by using the `rounded` class. This has variants for each of the sides, so you can add rounded corners on the top corners by `rounded-top` or on the right by `rounded-right`.

You can turn the image into a thumbnail appearance by using `img-thumbnail`. This adds a thin border with rounded corners. Adding this class negates rounded corners.

Leave the image as a thumbnail. Figure 3-13 shows the result of applying `img-thumbnail` to the image. At this width, it is also floated left.

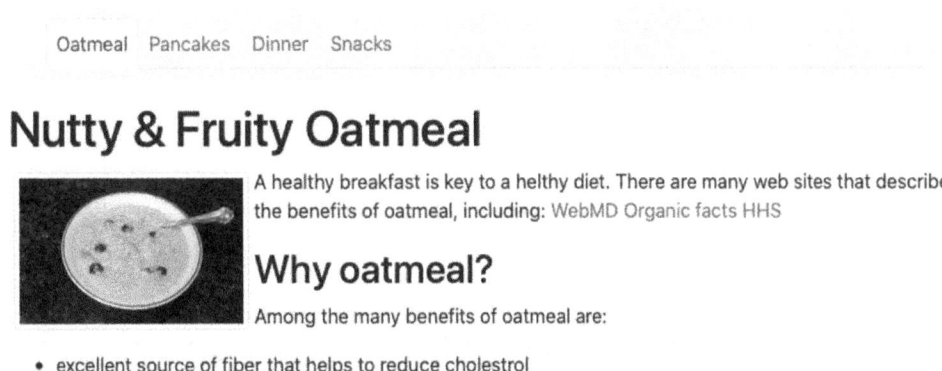

Oatmeal   Pancakes   Dinner   Snacks

Nutty & Fruity Oatmeal

A healthy breakfast is key to a helthy diet. There are many web sites that describe the benefits of oatmeal, including: WebMD Organic facts HHS

Why oatmeal?

Among the many benefits of oatmeal are:

- excellent source of fiber that helps to reduce cholestrol

Figure 3-13: Image with a thumbnail and floated left

3.5 Cards

Bootstrap cards allow for a simple way to present information that includes images and text in boxes that can be ordered in different ways. One example of a popular website using the card approach is Pinterest. Figure 3-14 shows an example of a card with an image, text, and a button with a link to more information. Cards are created as a `div` with the `card` class applied to it. Inside the card, you then can place whatever content you like. There are classes for a header, body, and footer, though you can also place content outside any of these, as you will see in the next tutorial.

To create a card, add the `card` class to a `div`. Inside that `div`, you can add additional `div`s where you apply `card-header`, `card-body`, and `card-footer`. Figure 3-15 shows a basic outline of a `card` with simple text content.

Oatmeal Cookies

Crunchy on the outside - soft on the inside.

Recipe

Figure 3-14: Example of a card with an image. Photo by Whitney/CC BY 2.0.

```
1  <div class="card">
2      <div class="card-header">
3          <h3 class="card-title">Card Header</h3>
4      </div>
5      <div class="card-body">
6          <div class="card-text">Text for the card body.</div>
7      </div>
8      <div class="card-footer">Card Footer</div>
9  </div>
```

Figure 3-15: Code for a card

Tutorial: Cards and Card Decks

In this tutorial, you will see how to create several cards with related recipes and organize them in a card deck.

Step 10: Add a single card.

In the oatmeal recipe page, add a new `div` immediately below the end of the recipe directions `section` element and add the code as shown in Figure 3-16.

```
1   <div>
2       <h3>Related Recipes</h3>
3       <div class="card">
4           <img class="card-img-top" src="img/pic2.jpg" />
5           <div class="card-body">
6               <h4 class="card-title">Oatmeal Cookies</h4>
7               <div class="card-text">
                        Crunchy on the outside - soft on the inside.</div>
8           </div>
9           <div class="card-footer">
10              <a class="btn btn-outline-primary" href="#">Recipe</a>
11          </div>
12      </div>
13  </div>
```

Figure 3-16: Single card for a related recipe

This adds the code for a single card (the one shown in Figure 3-14). Line 3 sets up the card. Line 4 adds an image placed at the top of the card, as indicated with the `card-img-top` class. Perhaps not surprisingly, you can also place the image at the bottom of the card with `card-img-bottom`. By placing the image outside the card body and header, it fills out to the edges of the card. If you place the image inside one of the card sections, it will add a border around the image.

Lines 5–8 show how the card body is set up using the `card-title` and `card-text` classes to control the formatting of the text. Lines 9–11 show how to add a link button in the footer with an outline using the primary color. By pointing the link to `"#"`, the page will simply reload when the link is clicked. This is useful when initially developing the page, until you know what the actual link destination is.

If you reload the page at this point, you will find that the card takes up the entire width of the page. We will fix that later, but if you want to see the card in a smaller size, you can add a `style="width: 200px;"` statement on the `card div`.

Step 11: Add a card deck.

If you add multiple cards next to each other, by default they will line up flush against their neighbors and may not have the same width. That can be fixed by placing cards inside a card deck. Cards in a deck will have equal height and width, and their footers will line up with each other.

Modify the code you added for the card with the code in Figure 3-17. This adds three more cards and lines them up in a row, with some space in between. Line 1 sets up the card deck in a `div` around all the cards. Each card follows the same pattern as the first one, just with different images and text.

Once you have entered the code, the output in the browser should look like what you see in Figure 3-18. If you resize the browser window, the cards will shrink and grow to fill the available space, until you get to the small width, at which point they will stack vertically.

```
1  <div class="card-deck">
2      <div class="card">
3          <img class="card-img-top" src="img/pic2.jpg" />
4          <div class="card-body">
5              <h4 class="card-title">Oatmeal Cookies</h4>
6              <div class="card-text">
7                  Crunchy on the outside - soft on the inside.
8              </div>
9          </div>
10         <div class="card-footer">
11             <a class="btn btn-outline-primary" href="#">Recipe</a>
12         </div>
13     </div>
14
15     <div class="card">
16         <img class="card-img-top" src="img/pic1.jpg" />
17         <div class="card-body">
18             <h4 class="card-title">Fruity Oatmeal</h4>
19             <div class="card-text">
20                 Oatmeal with tropical fruit - comforting and refreshing.
21             </div>
22         </div>
23         <div class="card-footer">
24             <a class="btn btn-outline-primary" href="#">Recipe</a>
25         </div>
26     </div>
27
28     <div class="card">
29         <img class="card-img-top" src="img/pic3.jpg" />
30         <div class="card-body">
31             <h4 class="card-title">Oatmeal Pancakes</h4>
32             <div class="card-text">
33                 Serve with smoked maple syrup for a great camp experience.
34             </div>
35         </div>
36         <div class="card-footer">
37             <a class="btn btn-outline-primary" href="#">Recipe</a>
38         </div>
39
40     </div>
41     <div class="card">
42         <img class="card-img-top" src="img/pic4.jpg" />
43         <div class="card-body">
44             <h4 class="card-title">Oatmeal Stout</h4>
45             <div class="card-text">
46                 This is a great drink on cold nights.
47             </div>
48         </div>
49         <div class="card-footer">
50             <a class="btn btn-outline-primary" href="#">Recipe</a>
51         </div>
52     </div>
53 </div>
```

Figure 3-17: Card deck with four cards

Related Recipes

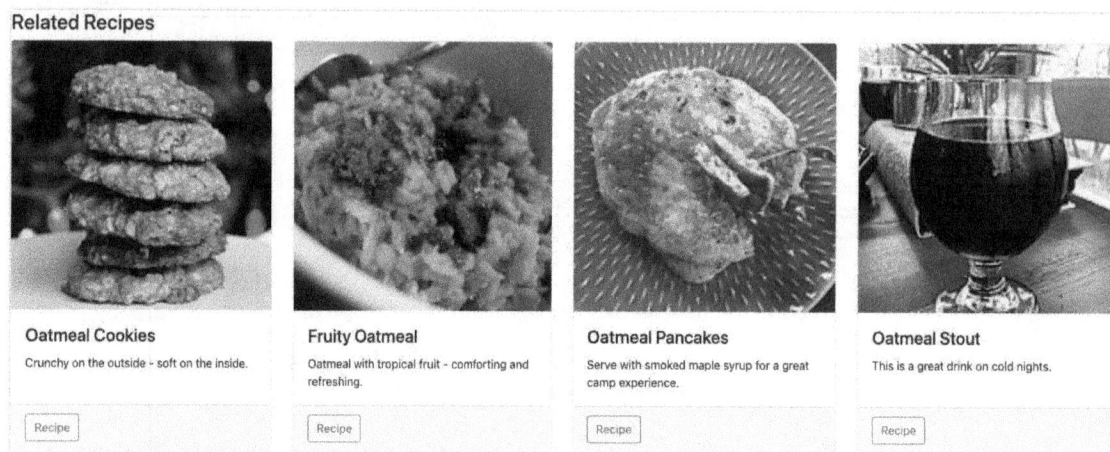

Oatmeal Cookies

Crunchy on the outside - soft on the inside.

Recipe

Fruity Oatmeal

Oatmeal with tropical fruit - comforting and refreshing.

Recipe

Oatmeal Pancakes

Serve with smoked maple syrup for a great camp experience.

Recipe

Oatmeal Stout

This is a great drink on cold nights.

Recipe

Figure 3-18: Card deck with four cards rendered in the browser. Cookies photo by Whitney/CC BY 2.0; fruity oatmeal photo by Nate Steiner/CC0 1.0 (public domain); oatmeal pancakes photo by Whitney/CC BY 2.0; stout photo by Amber DeGrace/CC BY 2.0.

3.6 The Bootstrap Grid

Now that you have seen several different ways of working with Bootstrap and some of the considerations with regard to responsiveness that you have to be concerned with, it's time to take a look at one of the most fundamental concepts in Bootstrap when it comes to controlling the placement of elements: the Bootstrap grid. In Bootstrap, you can use a 12-column grid to decide how to place elements at different screen widths. You simply have to decide how many columns each element is allowed to span at each screen width. For example, you might decide that if you have three pieces of content at the largest screen resolution, they are organized with 25%, 50%, and 25% of the width of the screen, respectively, but at the medium resolution, it looks better to give them an equal size, and at the smallest screen size, they should stack vertically. You would then specify that at the largest resolution, they would span 3, 6, and 3 columns, respectively, and at the medium width, they would each span 4 columns. At the smallest width they would span all 12 columns.

When specifying the column widths, you use a set of classes that follow this pattern:

```
col-[size]-[columns]
```

where `size` is one of the Bootstrap breakpoints (Table 3-1) and `columns` is the number of columns to span. The number of columns you specify is applied at the specified size and above. So, for example, `col-md-3` will span a piece of content over 3 columns at the medium resolution or higher. If you don't specify a size lower, the content will take up all 12 columns and will thus stack vertically. You can specify more than one column value on the same element at different sizes to specify different layouts at different screen resolutions. For example:

```
<div class="col-lg-3 col-md-4">
```

This specifies that at the large resolution and above, the element will span 3 columns, but at the medium resolution and higher, the element will span 4 columns. At resolutions below medium, the element will span the full 12 columns.

The elements you wish to organize must be inside a `div` element with the `row` class applied, and that in turn must be inside an element with `container` or `container-fluid` applied. Figure 3-19 shows an example of the basic structure.

```
1   <body class="container-fluid">
2       <div class="row">
3           <div class="col-lg-3 col-md-4">First box</div>
4           <div class="col-lg-6 col-md-4">Second box</div>
5           <div class="col-lg-3 col-md-4">Third box</div>
6       </div>
7   </body>
```

Figure 3-19: Fluid container

In this case, the entire body of the page is a `container-fluid`, which means that it fills out the entire width of the browser. A `container`, on the other hand, is given a specific width. This shows three boxes organized in a row (line 2). At the large resolution and above, the boxes span 3, 6, and 3 columns, respectively, whereas if the browser is narrower at the medium resolution, the boxes will have an equal width of 4 columns each. Since nothing is specified below medium, if the browser is narrower, the boxes will all span 12 columns, and will thus stack vertically.

Review Questions

4. How many sections of content can be added to a single Bootstrap card?

5. How would you specify that a layout of four columns with equal width up to and including the medium resolution and above, where the two outer columns would fill up one-third of the screen and the two inner columns would be of equal size, filling up the remaining space?

3.7 Bootstrap Utilities

Bootstrap includes a number of utility classes that allow for quickly applying a variety of different formatting options. This section describes some of the common utility classes you will encounter. Many of these simply allow for adding CSS styling to an element, allowing you to avoid creating a separate style rule.

3.7.1 Border

By applying the `.border` class to an element, you can quickly add a simple border to that element. You can also control which border to add by specifying `border-top`, `border-left`, and so forth. You can also specify a color by specifying `border-primary`, `border-danger`, and so on, using the standard contextual classes. Here's an example of an element with borders on the top and right sides in the primary color:

```
<div class="border-top border-right border-primary">This is a bordered box.</div>
```

You can also specify rounded corners by using the `.rounded` class with variations for which corner(s) to round, such as `rounded-top` or `rounded-left`, which will round the two corners on the specified side. The utility classes only allow for specifying two or four corners to round:

```
<div class="border border-primary rounded-top">This is a bordered box.</div>
```

3.7.2 Display

The display utility allows you to quickly specify which display to use for an element. This class is responsive, so you can apply different displays at different screen widths. The display utility uses this syntax:

```
d-{breakpoint}-{value}
```

The *breakpoint* value is one of the standard Bootstrap values above `xs` (`sm`, `md`, `lg`, and `xl`). For screen sizes at `xs` and smaller, you simply omit the breakpoint.

The *value* is one of the standard CSS displays: `none`, `inline`, `inline-block`, `block`, `table`, `table-cell`, `table-row`, `flex`, or `inline-flex`.

The display utility can also specify whether an element should be printed or not by specifying `d-print-{value}` using the same values as the regular display.

3.7.3 Screen Reader

For users with screen readers, you can specify whether an element should be read by that device. Some elements may only be relevant to screen readers, so you can apply the class `.sr-only` to have the element be read aloud but not show up on the screen. This might be useful if you have a graphical representation of a control with a textual representation for screen readers.

3.7.4 Height and Width

You can control height and width relative to the parent element by using the height and width utilities. This allows you to specify in increments of 25% how wide or tall an element should be. For example, w-50 specifies that the element should be 50% of the parent element, and h-75 specifies 75% of the height of the parent. In addition to the percent values (25, 50, 75, and 100), you can also specify h-auto and w-auto to let the content and display setting for the element dictate the height and width. Finally, you can cap height and width at a certain percentage by using mh-* and mw-* (for example, mh-25 and mw-50).

3.7.5 Spacing: Margin and Padding

As you saw in the previous chapter, controlling margin and padding are important to controlling the placement of elements on the page. Bootstrap includes utility classes to control many of the elements of both margin and padding.

To set margins, use the form

```
m{sides}-{breakpoint}-{size}
```

And to set padding, use the form

```
p{sides}-{breakpoint}-{size}
```

As with the display utility, breakpoint is omitted for the xs screen size. For both margins and padding, sides can be one of the following values:

- t: top
- b: bottom
- l: left
- r: right
- x: both left and right
- y: both top and bottom
- blank: all four sides

The size is a number from 0 to 5 or auto. If size is set to 0, the margins are eliminated, and for 1 through 5, the size is relative to the font used and is a multiple of the *rem* unit, which is the size of the basic html font-size (root element). For most browsers, this is 16 px (1rem = 16 pixels). Size for 1 through 5 is:

- 1: 0.25rem = 4 pixels
- 2: 0.5 rem = 8 pixels
- 3: 1 rem = 16 pixels
- 4: 1.5 rem = 24 pixels
- 5: 3 rem = 48 pixels

Here are a few examples of using the spacing utilities:

```
1  <div class="w-50 border">
2      <div class="p-4 bg-info">
3          No margin.<br />
4          Padding all around (1.5 rem = 24 pixels)
5      </div>
6      <div class="mt-3 mr-2 pb-4 bg-success">
7          Margin on top (1 rem =  16 pixels) and right (0.5 rem = 8 pixels).<br />
8          Padding on bottom (1.5 rem = 24 pixels).
9      </div>
10     <div class="ml-5 py-2 bg-warning">
11         Margin on left (3 rem = 48 pixels).<br />
12         Padding on top and bottom (0.5 rem = 8 pixels).
13     </div>
14 </div>
```

This gets rendered as shown in Figure 3-20.

3.8 Accordion/Collapse

The collapse control allows for hiding and showing content on the page. This could be just a single section of content that can be displayed when the user clicks a button or link, or the collapsible content might be organized in an "accordion" where only one section is visible at a time.

To set this up, you need two elements: a button or link that the user will click to show/hide the content, and the actual content to be shown/hidden. The button or link will need a

No margin.
Padding all around (1.5 rem = 24 pixels)

Margin on top (1 rem = 16 pixels) and right (0.5 rem = 8 pixels).
Padding on bottom (1.5 rem = 24 pixels).

Margin on left (3 rem = 48 pixels).
Padding on top and bottom (0.5 rem = 8 pixels).

Figure 3-20. Using the margin and padding utilities

`data-toggle` property with a value of `"collapse"` and, in the case of a button, a `data-target` property with a value with the id of the content to show/hide. When using a link, the `data-target` property is replaced with the `href` property. Figure 3-21 shows an example of the controls used to manage collapsing content.

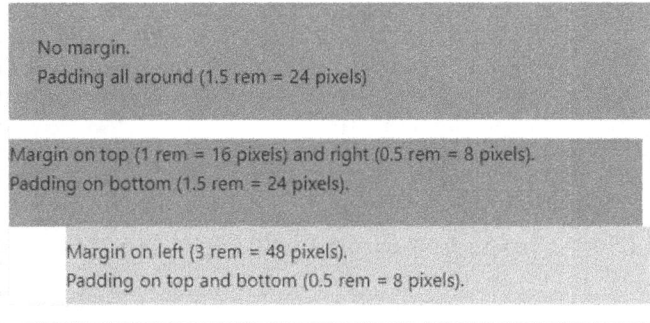

```
1  <button data-toggle="collapse" data-target="#description"
       class="btn btn-primary">Description (button)</button>
2  <a data-toggle="collapse" href="#description"
       class="btn btn-primary">Description (link)</a>
```

Figure 3-21: Code for a button to control collapsing an accordion

In this case, both the link and the button are formatted as Bootstrap buttons styled using the primary style. Figure 3-22 shows content being controlled by the controls in Figure 3-21.

```
1  <!-- Content Being Controlled-->
2  <div id="description" class="collapse">
3      <img class="float-left"
         src="img/Nutty-fruity-oatmeal.jpg" alt="Nutty & Fruity Oatmeal" width="200"
/>
4      A healthy breakfast is key to a healthy diet. Oatmeal has many benefits including
       its ability to lower cholesterol and high content of dietary fiber.
5      </ul>
6  </div>
```

Figure 3-22: Content controlled for an accordion collapse

The key here is the id of `description`, which matches the `data-target` in the button or link, and the `.collapse` class applied in line 2.

Tutorial: Accordion

In this tutorial, you will see how to set up the accordion with the oatmeal recipe. This will allow you to collapse different parts of the content. Figure 3-24 shows the code for the entire accordion, and Figure 3-25 shows the output of the complete accordion control. The steps in this section will walk you through an approach to setting this up. You may find it easiest to create this as a new section in the code and then copy the existing code into the new section. The code goes immediately after this line of code:

```
<h1>Nutty & Fruity Oatmeal</h1>
```

Step 12: Set up the overall accordion.

Add a div with class=accordion and id=recipeAccordion surrounding all the content to include in the accordion. This is used in each collapsible section to point back to it as their data-parent (lines 13, 39, and 63 in Figure 3-24).

Step 13: Create the cards.

Add three cards inside the accordion, as shown in Figure 3-23.

```
1  <div class="accordion" id="recipeAccordion">
2      <div class="card">
3          <div class="card-header" id="headingOne">
4
5          </div>
6          <div class="card-body">
7
8          </div>
9      </div>
10     <div class="card">
11         <div class="card-header" id="headingTwo">
12
13         </div>
14         <div class="card-body">
15
16         </div>
17     </div>
18     <div class="card">
19         <div class="card-header" id="headingThree">
20
21         </div>
22         <div class="card-body">
23
24         </div>
25     </div>
26 </div>
```

Figure 3-23: Code for cards for the accordion

Step 14: Create the card headings.

Add the code for the headings. The content for the first heading is in lines 4–10 in Figure 3-24. The header is set up with an h5 element (line 4), so the text of the heading will get whatever styling this heading style is set to. Lines 5–7 contain the button that will control the collapsing. In this case, we are styling it as a link button (btn-link). The button controls the collapsing of the content for the section by setting the data-target to the id of the corresponding card body.

Here we also show two ARIA accessibility controls that will tell screen readers how to manage the content. The aria-expanded property indicates whether the content is actually expanded or not. Line 8 contains the text that will show up in the header area. Each header follows the same pattern, except it changes which part of the content is being controlled.

Step 15: Add a div element between the header and body.

The div elements between the card header and body point back to the overall accordion as the data-parent. They also control which content section(s) are currently displayed.

Step 16: Add the content to be collapsed.

Finally, you can add the content of each card. This is the content of the existing page, which you can just copy into each card body.

You can now save the code and display it in the browser.

```
1   <div class="accordion" id="recipeAccordion">
2       <div class="card">
3           <div class="card-header" id="headingOne">
4               <h5>
5                   <button class="btn btn-link" data-toggle="collapse"
6                           data-target="#collapseOne" aria-expanded="true"
7                           aria-controls="collapseOne">
8                       Description
9                   </button>
10              </h5>
11          </div>
12          <div id="collapseOne" class="collapse show" aria-labelledby="headingOne"
13                  data-parent="#recipeAccordion">
14              <div class="card-body">
15              <img class="img-thumbnail float-left"
16                  src="img/Nutty-fruity-oatmeal.jpg"
17                  alt="Nutty & Fruity Oatmeal" width="200" />
18              A healthy breakfast is key to a healthy diet. Some benefits of oatmeal:
19                  <ul>
20                      <li>excellent source of fiber that helps to reduce cholestrol</li>
21                      <li>loaded with nutrients</li>
22                      <li>helps with weight reduction</li>
23                      <li>reduce constipation</li>
24                  </ul>
25              </div>
26          </div>
27      </div>
28      <div class="card">
29          <div class="card-header" id="headingTwo">
30              <h5>
31                  <button class="btn btn-link collapsed" data-toggle="collapse"
32                          data-target="#collapseTwo" aria-expanded="false"
33                          aria-controls="collapseTwo">
34                      Ingredients
35                  </button>
36              </h5>
37          </div>
38          <div id="collapseTwo" class="collapse" aria-labelledby="headingTwo"
39                  data-parent="#recipeAccordion">
40              <div class="card-body">
41                  <ul>
42                      <li> 1/2 cup quick oats </li>
43                      <li> 1/2 cup each of almond and vanila soy milk </li>
44                      <li> 2 Tbsp chopped walnuts </li>
45                      <li> 1/2 cup each chopped banana and blueberries </li>
46                      <li> A pinch each of cardamom and ginger powder </li>
47                      <li> 1/4 tsp cinnamon powder</li>
48                  </ul>
49              </div>
50          </div>
51      </div>
52      <div class="card">
53          <div class="card-header" id="headingThree">
54              <h5>
55                  <button class="btn btn-link collapsed" data-toggle="collapse"
56                          data-target="#collapseThree" aria-expanded="false"
57                          aria-controls="collapseThree">
58                      Directions
59                  </button>
60              </h5>
61          </div>
62          <div id="collapseThree" class="collapse" aria-labelledby="headingThree"
63                  data-parent="#recipeAccordion">
64              <div class="card-body">
65                  <ol>
66                      <li>Mix all ingredients in a cooking pot</li>
```

Figure 3-24: Code for the complete accordion control (*continues*)

```
67                          <li>Add one cup of water</li>
68                          <li>Bring to a boil and simmer for a couple of minutes.</li>
69                      </ol>
70                  </div>
71              </div>
72          </div>
73  </div>
```

Figure 3-24: Code for the complete accordion control (*continued*)

In the browser, the accordion will look like what's shown in Figure 3-25. Clicking the links will show the content for a section or hide it if the content is already displayed.

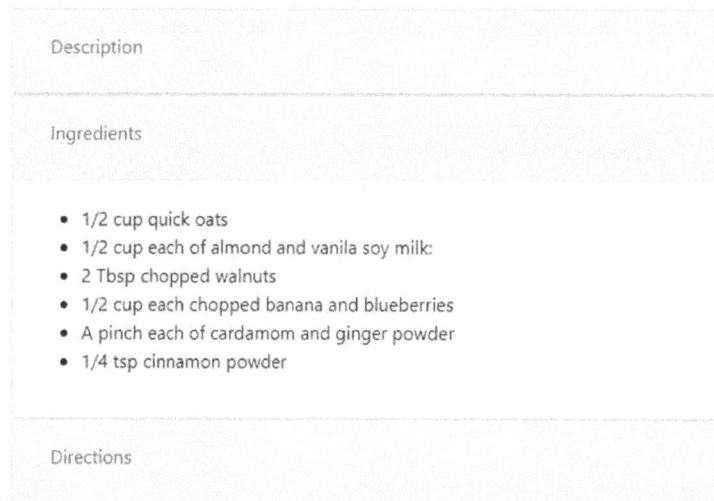

> Description
>
> Ingredients
>
> • 1/2 cup quick oats
> • 1/2 cup each of almond and vanilla soy milk:
> • 2 Tbsp chopped walnuts
> • 1/2 cup each chopped banana and blueberries
> • A pinch each of cardamom and ginger powder
> • 1/4 tsp cinnamon powder
>
> Directions

Figure 3-25: Bootstrap accordion combining card components and collapsible content

Review Questions

6. Describe how you can control the collapsing action in the accordion.

7. How would you set up the accordion to have one of the sections opened by default?

3.9 Modal

A modal dialog box pops up in front of the regular content and must be dismissed before the user can interact with the regular content again. This is often used for warning messages or important content that you want to make sure the user has seen and perhaps acknowledged seeing—for example, a disclaimer of some sort. A Bootstrap modal has three sections: header, body, and footer.

Similar to collapsible content, the modal is activated from a button or link that the user clicks. Figure 3-26 shows the code you would use to launch a modal—showing both a button (lines 2–4) and a link (lines 5–7). In both cases, the modal is identified by an id with the value myModal. In both cases, the key is to add the data-toggle property with the value "modal".

```
1   <!-- Button or link launching the Modal -->
2   <button type="button" class="btn btn-primary" data-toggle="modal"
    data-target="#myModal">
3       Launch Modal (button)
4   </button>
5   <a data-toggle="modal" href="#myModal" class="btn btn-primary">
6       Launch Modal (link)
7   </a>
```

Figure 3-26: Code used to launch a modal dialog box

Figure 3-27 shows an example of a simple modal that would be launched by the link or button in Figure 3-26.

```
1   <!-- Bootstrap Modal -->
2   <div class="modal" id="myModal">
3       <div class="modal-dialog">
4           <div class="modal-content">
5
6               <!-- Header -->
7               <div class="modal-header">
8                   <h5 class="modal-title">Modal Header</h5>
9                   <button type="button" class="close" data-dismiss="modal">
10                      &times;
11                  </button>
12              </div>
13
14              <!-- Body -->
15              <div class="modal-body">
16                  This is the body of the modal where you would put your content.
17              </div>
18
19              <!-- Footer -->
20              <div class="modal-footer">
21                  <button type="button" class="btn btn-primary" data-dismiss="modal">
22                      Close
23                  </button>
24              </div>
25          </div>
26      </div>
27  </div>
```

Figure 3-27: Code for modal dialog

Line 2 specifies that this is a modal by using the class modal, and it connects it to the launching link or button with the id myModal. Line 3 sets up another div inside with the actual dialog, and then the content for the modal is nested inside of it (lines 4–25).

The header is shown in lines 6–12. It is recommended that you add a button to dismiss the modal in the header. The modal can also be dismissed by the escape key or by clicking outside the modal. Line 10 is an example of a Unicode character that produces the times, or multiplication, symbol, which is similar to an X but is symmetrical. There are thousands of Unicode characters and symbols that you can include in a similar fashion. Finding which ones to use can be difficult, but various websites provide lists and search capabilities to find different symbols. One particularly useful website is Amp What (http://www.amp-what.com), which contains a search engine for characters and shows how to write the resulting character using five different notations. For example, ✗ gives a more slanted and handwritten look, as shown in Figure 3-28.

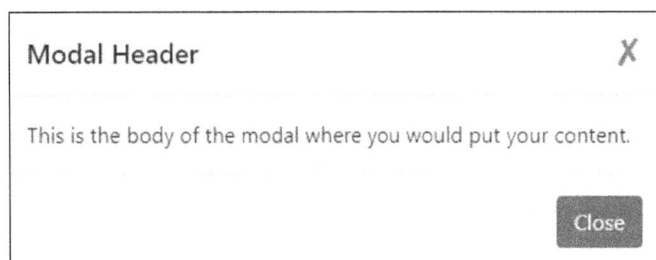

Figure 3-28: Modal dialog

To add a fade effect as the modal appears and is dismissed, you can add the class .fade to .modal in line 2:

```
2   <div class="modal fade" id="myModal">
```

By default, the modal dialog shows up towards the top of the page, if you want to center the dialog vertically, you can add modal-dialog-centered in line 3:

```
3       <div class="modal-dialog modal-dialog-centered">
```

3.10 Carousel

The carousel is a commonly used control that allows for cycling through images or other content as a slideshow. The Bootstrap Carousel can be set up to cycle through the images automatically on a timer or allow the user to click to advance the images. You can also add text and links as an overlay to each image, as well as an indicator of which slide is being displayed.

Tutorial: Setting Up a Carousel

In this tutorial, you will create a carousel with three images in the header of the recipe page.

Step 17: Add the carousel.

Add the code shown in Figure 3-29 in place of the image in the header. This will create a carousel with three images. The three images used here are already available in the img folder.

```
1   <div id="RecipeCarousel" class="carousel slide" data-ride="carousel"
        data-interval="3000">
2       <div class="carousel-inner">
3           <div class="carousel-item">
4               <img class="d-block w-100"
                    style="object-fit: cover"
                    height="300" src="img/carousel01.jpg" alt="First slide">
5           </div>
6           <div class="carousel-item active">
7               <img class="d-block w-100"
                    style="object-fit: cover"
                    height="300" src="img/carousel02.jpg" alt="Second slide">
8           </div>
9           <div class="carousel-item">
10              <img class="d-block w-100"
                    style="object-fit: cover"
                    height="300" src="img/carousel03.jpg" alt="Third slide">
11          </div>
12      </div>
13      <a class="carousel-control-prev" href="#RecipeCarousel"
            role="button" data-slide="prev">
14          <i class="fa fa-3x fa-chevron-circle-left"></i>
15          <span class="sr-only">Previous</span>
16      </a>
17      <a class="carousel-control-next" href="#RecipeCarousel"
            role="button" data-slide="next">
18          <i class="fa fa-3x fa-chevron-circle-right"></i>
19          <span class="sr-only">Next</span>
20      </a>
21  </div>
```

Figure 3-29: Carousel with three images

To set up the carousel, you need three nested div elements. The outermost (line 1) controls the overall functioning of the carousel. The id allows for later controls to reference the carousel. The two classes, carousel and slide, set up the carousel, whereas the data-ride attribute set to carousel starts the animation immediately when the page is loaded. The data-interval attribute specifies how many milliseconds each slide is displayed. In this case, each image will be displayed for 3 seconds.

The next div (line 2) is given the class carousel-inner and encompasses all the carousel slides. Each slide is in a div nested inside carousel-inner. In this example, each slide simply is made up of an image. However, a caption may also be added, as you'll see later.

The images in this case are set to block display (d-block) and a width of 100% (w-100). By applying the style object-fit: cover, the image will be centered and cropped to fit into the height and width specified. The active class applied in line 6 must be applied to one of the slides in order for the carousel to show up. This is also the slide that shows up first when the page is loaded.

The last part of the code sets up previous (lines 13–16) and next (lines 17–20) controls on the left and right side of the image. Notice that the href attribute points to the id of the carousel. These controls use Font Awesome icons at 3× size to indicate the user can click previous (line 14) and next (line 18).

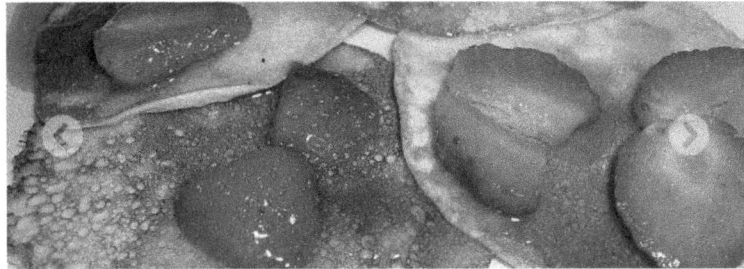

Figure 3-30: Carousel with an image and previous/next controls

Lines 15 and 19 set up text that can be read aloud by a screen reader but won't show up on the page as an alternative to the Font Awesome icons. Figure 3-30 shows an example of what the carousel looks like with an image and controls.

Step 18: Add a caption.

To add a caption to the carousel, you need to add a div with class carousel-caption inside the carousel-item. You can then add whatever content you want. Figure 3-31 shows a heading with an h3 element and a regular paragraph of text, with the result shown in Figure 3-32.

```
1   <div class="carousel-item">
2       <img class="d-block w-100" style="object-fit: cover"
          height="300" src="img/carousel03.jpg" alt="Third slide">
3       <div class="carousel-caption">
4           <h3>Strawberry Pancakes</h3>
5           <p>Sweet and Savory!</p>
6       </div>
7   </div>
```

Figure 3-31: Code for the carousel image caption

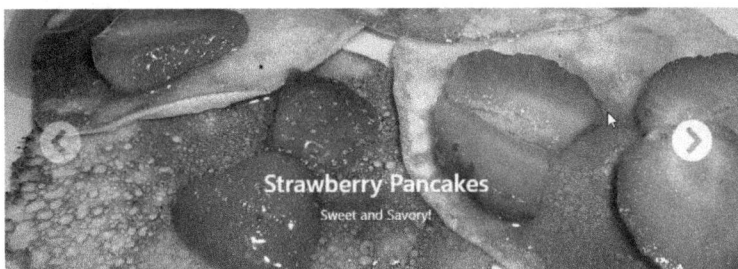

Figure 3-32: A carousel item with a caption

Step 19: Add slide indicators.

To include indicators of which slide is being shown, add the code shown in Figure 3-33 just after the first line of the carousel.

```
1  <ol class="carousel-indicators">
2      <li data-target="#RecipeCarousel" data-slide-to="0"></li>
3      <li data-target="#RecipeCarousel" data-slide-to="1" class="active"></li>
4      <li data-target="#RecipeCarousel" data-slide-to="2"></li>
5  </ol>
```

Figure 3-33: Code for slide indicators

Figure 3-34 shows the slide with indicators. This is the third of three slides in the carousel.

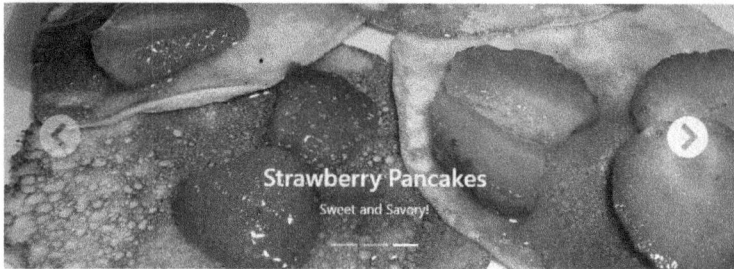

Figure 3-34: Carousel slide with indicators

Step 20: Change the appearance of the slide indicators.

To change the appearance of the indicators, you can use a CSS rule similar to the one shown in Figure 3-35. Since this project doesn't have a separate CSS stylesheet, you can add this rule in a `style` section in the `head` element of the HTML file.

```
1  .carousel-indicators li {
2      border-radius: 50%;
3      width: 1.5rem;
4      height: 1.5rem;
5      background-color: darkgray;
6  }
```

Figure 3-35: CSS for carousel indicators

This targets the `li` elements inside `carousel-indicator`. Setting `border-radius` to 50% produces a circle. The width and height are increased to make the indicators more prominent. The `background-color` applies to the non-active indicators. In order to change the color of the active indicator, you would have to use something like this rule:

```
1  .carousel-indicators .active {
2      background-color: blue;
3  }
```

Taken together, the two rules provide the output seen in Figure 3-36.

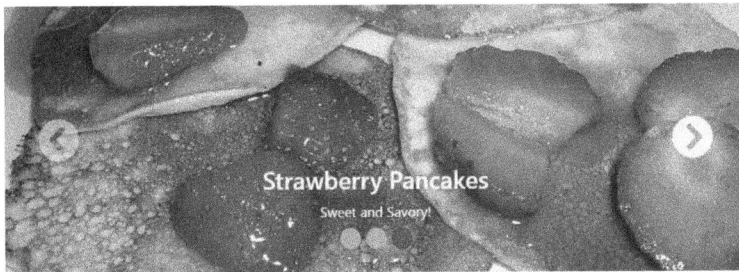

Figure 3-36: Formatted indicators

3.11 Issues with Bootstrap

While Bootstrap provides a number of great features and abilities that allow you to create a responsive experience for users, it also has some drawbacks you should be aware of.

As you saw in Chapter 2, one of the design principles of websites is to separate content from presentation. During the history of the development of HTML, we have increasingly been able to only let HTML be concerned with the content, and instead move all the presentation concerns to CSS, thus simplifying the resulting HTML. However, Bootstrap represents a step back in that you have to add a lot of classes directly into the HTML. This also leads to poor reusability. The way you set up a control on one page cannot easily be replicated on other pages in the site.

A related concern is if you inherit a project that didn't use Bootstrap from the beginning. In that situation, it can be a quite a task to convert it to Bootstrap, since you will need to touch every single page to apply the proper classes to all the relevant elements.

Another concern is that the Bootstrap framework itself represents a significant download size, which can cause problems for those with slow or unreliable Internet connections. Some developers find that they only use the grid system of Bootstrap, which can be relatively easily replicated by the newer CSS controls you saw in the previous chapter and thus feel like including the entire Bootstrap framework just isn't worth the hassle.

Finally, since Bootstrap is very popular with developers, there's a definite risk that your website will end up looking just like every other website using Bootstrap. Some discerning users are able to detect those similarities and might find that the developers haven't put in a lot of effort into making the site look unique.

You will need to determine if the benefits of the various controls, the grid system, and the responsive design in the Bootstrap outweigh the drawbacks, and then decide how to move forward.

This chapter has explored how Bootstrap classes can be used to make a page look nice, be responsive, and have some advanced functionality, like navigation bars, accordions, and modal dialogs. In Chapter 12, you will see how Bootstrap can also be applied to forms to give them some of the same features.

Review Questions

8. What are some examples of when a modal dialog could be used?

9. What are some examples of how Bootstrap breaks the principle of separating content and presentation?

10. How are card decks and carousels similar and different?

Exercises

Exercise 3-1

Redesign the Grand Oshkosh website you created in Chapter 2 using Bootstrap to control the layout and increase the functionality of the site.

Exercise 3-2

Observe the results of modifying the navigation bar using each of the following four classes: fixed-top, fixed-bottom, sticky-top, and sticky-bottom. Describe the results you obtained and discuss when it would be appropriate to use each placement.

Exercise 3-3

Adjust the nutrition value table to take up only 50% of the space on the screen. Then adjust it to take up a fixed amount of space, such as 200 pixels.

Exercise 3-4

Add a section to the accordion with a description of the benefits of oatmeal.

Exercise 3-5

Use the Bootstrap utilities to set the width of the accordion to 50% of the page width and increase the margin around the content of each card.

Exercise 3-6

Adjust the carousel as follows:

- Add another image to the carousel.
- Find a different set of arrows to use as the next and previous indicators.
- Change the look and feel of the slide indicators.

## Chapter 4

Interactive Web Pages: HTML Web Forms

In Chapter 1, you learned how to use HTML to develop **static** web pages. Almost all web applications today (e-commerce, social media, blogs, weather maps, and games) let users interact with web pages and dynamically change the contents of a page. A user may look up the price of a product on an e-commerce site, communicate with friends on Facebook, click on a weather map to get the forecast for a region, or interact with images to play games.

Interactive web pages typically accept user input and validate, share, look up, and process data. Two different approaches to developing such pages are **HTML web forms** and **ASP.NET web forms**. HTML web forms are developed with HTML code, giving you more control over the code. HTML web forms are an integral part of newer web development frameworks like Razor pages and MVC (model view controller). You can develop HTML forms in any text editor, like Visual Studio Code, thus reducing the overhead and avoiding dependence on a specific platform. ASP.NET web forms, which is a more traditional approach, uses drag-and-drop server-side controls, typically using Visual Studio.

To make web pages interactive, you use client-side code (script) that runs on the client and server-side code that runs on the server. Client-side script is typically developed in JavaScript, and server-side code is developed in languages like C#, PHP, Node.js, and Python, as explained in later chapters.

Understanding HTML forms will help you learn the ASP.NET web forms presented in Chapters 6–10 and Razor pages and MVC presented in Chapters 16, 17, and 18. Chapters 4 and 5 present HTML web forms. This chapter uses HTML to develop the user interface, and Chapter 5 uses JavaScript to do client-side processing.

Learning Objectives

After studying this chapter, you should be able to

- Develop HTML forms.
- Create and use different types of elements to input data and display output.
- Work with different types of buttons.
- Validate data.
- Display one page from another.

4.1 Introduction to HTML Forms

Web forms have become an integral part of websites that let users submit data to the site. They are typically used for applications such as placing an order, making a reservation, registering personal information, and providing feedback. In this chapter, you will learn to develop **HTML web forms**, which use HTML elements to create controls like text fields, drop-down lists, and checkboxes to accept user input.

To help you understand the use of HTML forms, you will develop an HTML page that lets users make reservations for performances at The Grand Oshkosh theater, described in Exercise 1-1 in Chapter 1. You may review Exercise 1-1 for a detailed description of the theater. The reservation page lets the user select a performance (event), the performance date, the seating zone, and membership status, and enter the number of tickets. The page also displays the total cost for the reservation. Figure 4-1 shows this page.

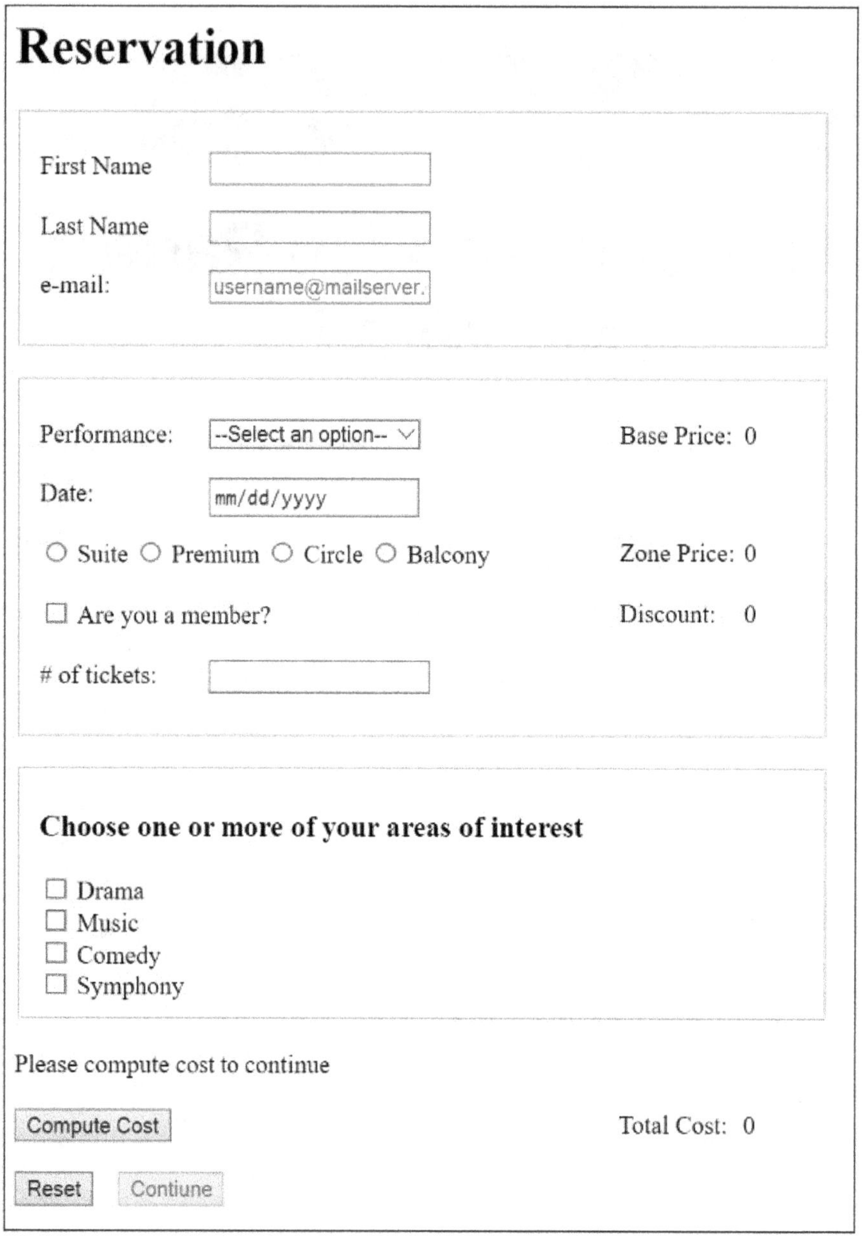

Reservation

First Name

Last Name

e-mail: username@mailserver.

Performance: --Select an option-- ⌄   Base Price: 0

Date: mm/dd/yyyy

○ Suite ○ Premium ○ Circle ○ Balcony   Zone Price: 0

☐ Are you a member?   Discount: 0

# of tickets:

Choose one or more of your areas of interest

☐ Drama
☐ Music
☐ Comedy
☐ Symphony

Please compute cost to continue

[Compute Cost]   Total Cost: 0

[Reset] [Contiune]

Figure 4-1: The reservation page

This is how the total cost is computed:

Total cost = Number of tickets × Cost per ticket
Cost per ticket = Base price + Zone price − Member discount

Each performance may have a different base price. In addition, each zone may have an additional price, over and above the base price. Members get a discount.

In addition to the reservation page, you will also develop a confirmation page, as shown in Figure 4-2, which displays the reservation data and asks the user to confirm the reservation. This page will help you understand how data is shared between pages using query strings.

Confirm reservation

First Name: Sam
Last Name: Adams
E-mail: null
Performance: Justin Hayward
Date: 2022-02-22
Zone: premium
Tickets: 2
Total Cost: 130.00

[Edit Reservation] [Confirm]

Figure 4-2: The confirmation page

4.1.1 HTML Form Elements

To create an HTML form, first create a **<form>** element that serves as a container of one or more elements that make up the form. Then, add HTML elements that create controls to accept user input and display labels, captions, and output. Such elements that are placed within a <form> element are referred to as **HTML form elements**. Table 4-1 provides a list of form elements and their descriptions.

Table 4-1: HTML form elements

Element	Description
<form>	Defines an HTML form
<input>	Defines a variety of input controls, including text field, checkbox, radio button, and submit buttons
<textarea>	Defines a multiline input control
<label>	Defines a label for an input control
<fieldset>	Groups related elements in a form
<legend>	Defines a caption for a <fieldset> element
<select>	Defines a drop-down list
<optgroup>	Defines a group of related options in a drop-down list
<option>	Defines an option in a drop-down list
<button>	Defines a clickable button
<datalist>	Specifies a list of options for autocomplete
<output>	Defines the result of a calculation

Let's take a look at the more commonly used elements, in more detail.

4.1.2 The <form> Element

The <form> element is a container for one or more elements that create controls (fields) to accept user input.

Figure 4-3 shows an example of a <form> element that contains two input elements to let the user enter a first name and a last name, along with a button to submit the form.

```
1   <form action="confirmation.html" id="reservation" method="post">
2       First Name: <input type="text" name="fName" id="fName"><br><br>
3       Last Name: <input type="text" name="lName" id="lName"><br><br>
4       <button type="submit"> Continue </button>
5   </form>
```

Figure 4-3: HTML code for a simple form

Figure 4-4 shows how the browser displays the form. For now, don't worry about the appearance; later, we will use CSS to format it.

The <form> element presented herein has three attributes: **action, id,** and **method.** The id attribute is a global attribute, which is common to all HTML elements, as described shortly.

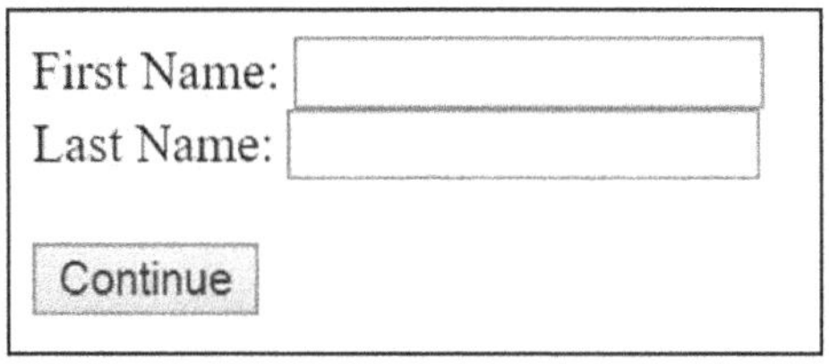

Figure 4-4: Form display in the browser

The Action Attribute

The `action` attribute specifies where the data should be sent when you submit the form. Typically, the value of the `action` attribute is the URL of another web page that processes or uses the data. The URL could be an absolute URL or a relative URL. An absolute URL specifies the entire address using the pattern *protocol://host.domain/path/page*. You typically use an absolute URL to refer to a page in another website, like https://www.thegrandoshkosh .org/grand-opera-house/rent.html.

Specifying the absolute URL for pages within the same domain makes it necessary to change the URL if the domain changes. A relative URL, like *confirmation.html*, points to a page within the same domain. Relative URLs make it easier to code the URL and avoid the need to change the URLs when the domain changes.

You may submit the data to the same page by specifying the page name as the value of the action attribute. Leaving out the action attribute has the same effect; however, it is not recommended because of security risks and unpredictable results with certain browsers.

The Method Attribute

The `method` attribute specifies how the data is to be sent to the location on the server, which is specified in the action attribute. The method attribute can have one of several values, including **get** and **post**, which are the two most commonly used values. The value *get* is typically used to fetch information, whereas *post* is used to update data. You set a value for the method as shown here:

```
method="get"
```

A value of **get** means the browser will send the data to the server by appending the data to the URL as a query string. For example, you may use *get* to send a product id to the server to look up product information when you click a link.

In a query string, each data item is sent as a name-value pair. If you enter the first name *Luke* and last name *Skywalker* and submit the form by clicking the Continue button, the URL of the displayed page will appear as follows.

file:///C:/WebProjects/html-forms/confirmation.html?fName=Luke&lName=Skywalker

The URL is followed by a question mark ("?") and the two name-value pairs for last and first names. In the next chapter, you will learn how to retrieve data from a query string using JavaScript.

Because the data sent through the query string is part of the URL, it is visible on the page, making it unsuitable for sending sensitive data or data that should not be tampered with. The user can easily change the data displayed with the URL.

The amount of data that can be sent is limited. However, the get method has the advantage that it doesn't take up room on the server to store the data, making it suited for relatively simpler forms that don't have sensitive data.

A value of **post** for the `method` attribute means the browser sends the data by adding it to the body of the HTTP request, making it more suitable for sensitive and larger sets of data. The post method, for example, would be appropriate for sending the data to the server to create a new reservation record in a database. We will later use the *post* method to send the form data to the server.

Tutorial: Reservation Form

This tutorial creates the reservation and confirmation pages presented in Figures 4-1 and 4-2, respectively. You will use a query string to share data between pages and JavaScript to process data.

It is important to note that there may be differences in the way HTML elements are rendered by different browsers and by different versions of the same browser. You should try to minimize problems due to such differences. For example, differences may exist between different browsers in the default value of an attribute of an element. One way to address this problem is to explicitly specify the value of the attribute, rather than relying on its default value.

You will continue to use **Visual Studio Code** to develop *.html* files. (See Chapter 1, Section 1.2, for instructions on setting up and using Visual Studio Code.) Instead of creating a workspace, you will work with a folder named *html-forms*.

Step 1: Create a new folder named *html-forms* within the folder *WebProjects*.

Step 2: Open the html-forms folder within VS Code:

Open Visual Studio Code

Select *File > Open Folder*

Select *html-forms*, and click *Select Folder*.

Step 3: Create two new HTML files: Reservation-1.html and Confirmation.html.

Click the *HTML-FORMS* folder in the Explorer window.

Click the *New File* button next to the folder name and enter *Reservation-1.html* for the filename. Press the *Enter* key.

Type in a single exclamation symbol on line 1 and press the *Enter* key.

Change the title of the page to *Reservation-1*.

Repeat the process to create the *Confirmation.html* file.

Step 4: Add a form to the reservation page.

Type in the heading *Reservation* and the HTML code in the <body> section, as shown in Figure 4-5.

```
1   <h1>Reservation</h1>
2   <form action="Confirmation.html" id="reservation" method="post">
3       First Name: <input type="text" name="fName" id="fName"> <br><br>
4       Last Name: <input type="text" name="lName" id="lName" > <br><br>
5       <button type="submit" id="continue"> Continue </button>
6   </form>
```

Figure 4-5: Initial code for the reservation page

For now, just type in the code. The `<input>` and `<button>` elements will be explained later.

Step 5: Display the reservation page in the browser.

Display the VS Code Explorer window (View > Explorer). Select *Reservation1.html*.

Press **Alt-b** to display the page in the default browser.

Enter the values for first and last names, and click the Continue button to submit the form.

You should see a Confirmation page with no contents.

Step 6: Examine the effect of setting the method attribute of the <form> element to "get."

Change the value of the method attribute to `"get"`, as shown here.

```
<form action="Confirmation.html" id="reservation" method="get">
```

Display the page in the browser and enter values for first and last names.

Click the Continue button to submit the form.

You should see the Confirmation page.

As discussed earlier, the *get* method sends the data to the server by appending the name and value of each data item to the URL, whereas the post method sends the data within the HTTP request.

Notice that the URL of the Confirmation page includes the name/value of input fields as follows:

WebProjects/html-forms/confirmation.html**?fName=Darth&lName=Vader**

Global Attributes of HTML Elements

Before we study the form elements and their attributes, let's look at attributes that are common to all HTML elements. Table 4-2 shows examples of such attributes, called *global attributes*.

Table 4-2: Global attributes of HTML elements

Global attribute	Description
accesskey	Specifies a shortcut key to activate/focus an element
class	Specifies a class name(s) for one or more element(s). The class name is commonly used to refer to a group of elements in CSS.
hidden	Specifies that an element is not yet or is no longer relevant, so that it doesn't show up
id	Specifies a unique id for an element. The id is commonly used to refer to an element in JavaScript and CSS.
style	Specifies an inline CSS style for an element
tabindex	Indicates whether the element is focusable and its sequential position, if any, in keyboard navigation

Next, we will study the elements that are commonly used in a form.

4.2 The <input> Element

The <input> element is the most commonly used element in HTML forms. It can be used to accept different types of data by specifying appropriate values for the type attribute. We will look at the input elements from Figure 4-3, which lets the user enter a first name.

```
<input type="text" name="fName" id="fName">
```

Note that Figure 4-3 has the text "First Name:" to the left of the input element to provide a caption, as shown here:

```
First Name: <input type="text" name="fName" id="fName">
```

This is a simple way to provide a caption. However, a preferred method to display a caption is to use the **label element** and link it to the input element, as shown here:

```
1   <label for="fName">First Name:</label>
2       <input type="text" name="fName" id="fName">
```

4.2.1 The <label> Element

The <label> element specifies a caption for an input field and associates the caption with the field. It is linked to the <input> by setting the **for** attribute of the <label> to the id of the <input>. In the browser, the caption looks the same as plain text typed next to the field. However, linking the caption to the input field provides certain benefits. Assistive technologies can associate the caption with the field. Screen readers, for example, can read the caption to help the user enter the right data. Further, clicking the caption will put the focus on the input field, making it easier to select the field. So, to provide a caption, it is preferable to use a <label> than typing in the text. It also makes it simpler to provide formatting in Bootstrap.

Step 7: Add a prompt to enter a first name.

Modify the <input> element for first name in the reservation form to use the <label> element for the prompt as follows:

```
1   <label for="fName">First Name:</label>
2      <input type="text" name="fName" id="fName">
```

Repeat this for the last name field.

4.2.2 Attributes of the <input> Element

The <input> element shown in the current example has three commonly used attributes. The **id** attribute, as described earlier, is a global attribute commonly used to refer to an element in style sheets.

The **name attribute**, which also is common to all form elements, is used by the browser to identify each data item sent to the server when the form is submitted. The value of each <input> element is sent as a key-value pair, where the name becomes the key and the value of the <input> element becomes the value in the key-value pair. So, the name attribute is important for input fields on a form because typically the input data is submitted to the server. However, the name attribute is not needed for an output field where a result is displayed.

The **type attribute** specifies the type of input that is accepted by the <input> element. A value of "text" specifies that any text is accepted. Other types include number, date, email, password, checkbox, and radio. A list of different types with a brief description of each is presented in Table 4-3.

Table 4-3: Values of the type attribute

Value of type attribute	Description
button	Defines a clickable button (<button> element is preferable)
checkbox	Defines a checkbox
color	Defines a color picker
date	Defines a date control (year, month, day [no time])
datetime-local	Defines a date and time control (year, month, day, time [no timezone])
email	Defines a field for an e-mail address
file	Defines a file-select field and a "Browse" button (for file uploads)
image	Defines an image as the submit button
month	Defines a month and year control (no timezone)
number	Defines a field for entering a number
password	Defines a password field
radio	Defines a radio button
range	Defines a range control (like a slider control)
reset	Defines a reset button (<button> element is preferable)
search	Defines a text field for entering a search string
submit	Defines a submit button (<button> element is preferable)
tel	Defines a field for entering a telephone number
text	Default; defines a single-line text field
time	Defines a control for entering a time (no timezone)
url	Defines a field for entering a URL

In addition to the id, name, and type, the <input> element has a large number of attributes. However, the set of attributes applicable to an <input> element vary with the value of the type attribute.

Table 4-4 provides a list of attributes of the <input> element. Attributes that are applicable to all types of <input> elements are highlighted in gray. Other attributes work with only certain types, as specified in the description.

Table 4-4: Attributes of the input element

Attribute of \<input\>	Description
`autocomplete`	Enable/disable autocomplete
`autofocus`	Specifies what \<input\> element should get focus when the page loads
`disabled`	Disables the control
`form`	Specifies the form element the \<input\> element belongs to
`max`	Specifies the maximum value. The max and min attributes work with the following input types: number, range, date, datetime, datetime-local, month, time, and week.
`maxlength`	Specified the maximum number of characters allowed in a text field
`min`	Specifies the minimum value
`list`	Specified a \<datalist\> element that contains options for the user to choose from
`name`	Assigns a name to the input control
`pattern`	Specifies a pattern for the input data. It works with text, date, search, url, tel, email, and password types.
`placeholder`	Specifies a short hint that describes the expected value. It works with the input types: text, search, url, tel, email, and password.
`readonly`	Sets the input control to read-only
`required`	Specifies that a value must be entered before submitting the form. It works with text, search, url, tel, email, password, date pickers, number, checkbox, radio, and file types, and the elements \<select\> and \<textarea\>.
`size`	Specifies the width of the control in characters. It works with text, search, tel, url, email, and password.
`step`	Specifies the interval that determines the valid numbers for the input field. The default value is zero. The step attribute works with number, range, date, datetime, datetime-local, month, time, and week types.
`type`	Specifies the type of control
`value`	Specifies the value that is sent to the server when the form is submitted

4.2.3 Types of Input Element

Next, you will learn how to use several commonly used types of \<input\> elements and apply them to the reservation page.

\<input type="email"\>

\<input\> elements with `type="email"` lets the user enter an email address and validates that it has the format *username@mailserver.top-leveldomain*. If the `multiple` attribute is used, you may enter multiple e-mail addresses separated by commas. Here is an example:

```
1   <label for="emailaddress">e-mail:</label>
2      <input type="email" name="emailAddress" id="emailAddress"
3          required placeholder="e.g. username@mailserver.com">
```

This is how the field appears in the browser:

> e-mail: e.g. username@mailserv

This example introduces two attributes, `required` and `placeholder`.

The Required Attribute

The *required* attribute specifies that the field must be filled out before submitting the form. This attribute is applicable to a variety of <input> elements: text, search, url, tel, email, password, date pickers, number, checkbox, radio, and file. It also can be used with <select> and <textarea>.

The Placeholder Attribute

The placeholder attribute serves as a hint describing the expected value.

> **Step 8: Add an input field to the reservation form to enter the user's e-mail.**
>
> Use the code presented above to create the e-mail field. This field and others to be added are to be placed above the *Continue* button.
>
> In addition, do your own research on the use of the *pattern* attribute, and use it in the <input> element to specify that only e-mail addresses with "prospectpressvt.com" as the domain are valid (that is, the e-mail address has the pattern username@prospectpress.com). Use the title attribute to display the message "The domain must be prospectpressvt.com."
>
> Test the code by displaying the form in the browser.

<fieldset> Element

The fieldset element groups related elements and draws a box around it.

> **Step 9: Insert a comment and a <fieldset> element as shown in Figure 4-6.**

```
1   <!-- Input fields to enter personal information-->
2   <fieldset width="400px">
3       <label for="fName">First Name</label>
4       <input type="text" name="fName" id="fName"> <br><br>
5       <label for="lName">Last Name</label>
6       <input type="text" name="lName" id="lName"><br><br>
7       <label for="emailAddress">e-mail:</label>
8       <input type="email" name="emailAddress" id="emailAddress"
9           placeholder="username@mailserver.com"><br><br>
10  </fieldset><br>
```

Figure 4-6: The fieldset element for personal information

Display the page to test the code.

<input type="date">

The date type of the input field lets the user enter a valid date into the input field, typically using a date picker, which varies with the browser. Here is an example of the html code and the corresponding display in the browser:

```
1   <label for="performDate">Date:</label>
2       <input type="date" name="performDate" id="performDate" required><br><br> 2
```

Date: mm/dd/yyyy

To display date and time, set type="datetime-local".
Next, you will add fields to enter information related to the performance.

Step 10: Add a second <fieldset> element to the reservation form to group elements related to the performance.

Use the code in Figure 4-7 to add the element immediately following the closing tag of the first `fieldset`.

```
1    <!-- Input fields to enter performance information -->
2    <fieldset> <br>
3
4    </fieldset> <br>
```

Figure 4-7: The `fieldset` element for performance

Step 11: Add a date field within the `<fieldset>` element to let the user select the date of the performance.

Use the code presented earlier to add a date field. Display the form in the browser, and click the label or the field. You should see the date picker used by the browser.

You may choose a different browser (Shift-Alt-b) and display the form again. The browser may use a different date picker.

<input type="radio">

The `radio` type creates a radio button. Radio buttons are used as a group to let the user make a single selection from a group of limited options (typically under 8 to 10)—for example, selecting the method of shipping or the title of a person. Here is an example of a radio button that lets you select the zone "suite" for a theater reservation:

```
1    <input type="radio" name="zone" id="suite" value="suite">
2        <label for="suite">Suite</label>
```

This is how it appears in the browser:

◯ Suite

The `value` attribute specifies the value that is sent to the server if the user selects the button from a group and submits the form. If the selected button has no `value` attribute, then the value `on` will be sent to the server, as in zone=on. Later, you will learn how to use JavaScript to process such values when the form is submitted.

To specify that multiple radio buttons belong to the same group, you give the same `name` for all buttons. Note that it is the name, not the proximity of the buttons, that determines the group. In Figure 4-8, the names of all three radio buttons are set to `zone`, making them members of the same group so that a user can select only one.

```
1    <input type="radio" name="zone" id="suite" value="suite" >
2        <label for="suite">Suite</label>
3    <input type="radio" name="zone" id="premium" value="premium">
4        <label for="premium" >Premium</label>
5    <input type="radio" name="zone" id="circle" value="circle">
6        <label for="circle" >Circle</label>
```

Figure 4-8: A group of radio buttons

The browser will display the buttons as follows:

◯ Suite ◯ Premium ◯ Circle

You may create multiple groups of radio buttons in the same form by giving a different name to the buttons in each group.

If the user selects a button and submits the form, the name of the selected radio button and its value will be sent to the server. For example, if *Premium* is selected, the name *zone* and the value *premium* will be sent.

To have a radio button selected, by default, specify the checked attribute as follows:

```
<input type="radio" name="zone" id="circle" value="circle" checked>
```

It may sound strange, but whether you assign a value to the checked property or what value you assign doesn't matter. Thus *checked*, *checked="checked"* and *checked=false* all have the same effect of making the button checked.

Step 12: Create a set of three radio buttons below the date field to let the user select a zone from the group Suite, Premium, and Circle.

Use the code presented previously to create the radio buttons.

It's time to practice: Do the following step:

Step 13: Add a fourth radio button.

Add a fourth radio button to the group to include the zone "Balcony". The browser should display the options as follows:

> ⊙ Suite ⊙ Premium ⊙ Circle ⊙ Balcony

<input type="checkbox">

The checkbox type creates a checkbox, which can be checked or unchecked by clicking the control or its label. The following code creates a checkbox that lets the user specify whether a person is a member:

```
1  <input type="checkbox" name="memeberStatus" id="member" value="member">
2  <label for="member">Are you a member?</label> <br><br>
```

This is how the browser displays the checkbox:

> ☐ Are you a member?

If the checkbox is checked, submission of the form sends the name of the element (*memberStatus*) and the value (*member*) to the server, as in memberstatus=member. If the value attribute is omitted, the value *on* is sent, as in memberstatus=on.

If the checkbox is unchecked, no value is sent.

Step 14: Use the code presented here to create a checkbox below the radio buttons.

Display the form in the browser to test the code.

Using a Group of Checkboxes

You may use a group of checkboxes with the same name but different values to let the user select multiple options from a group. Figure 4-9 shows the code and the corresponding display for a set of four checkboxes to let the user select one or more areas of interest.

```
1   <!-- Checkboxes to enter preferences -->
2   <fieldset>
3       <h3>Choose one or more of your areas of interest</h3>
4       <input type="checkbox" name="area" id="drama" value="drama">
5           <label for="drama">Drama</label><br>
6       <input type="checkbox" name="area" id="music" value="music">
7           <label for="music">Music</label><br>
8       <input type="checkbox" name="area" id="comedy" value="comedy">
9           <label for="comedy">Comedy</label><br>
10      <input type="checkbox" name="area" id="symphony" value="symphony">
11          <label for="symphony">Symphony</label>
12  </fieldset><br>
```

Figure 4-9: A group of checkboxes

Figure 4-10 shows how the checkboxes will be displayed in the browser.

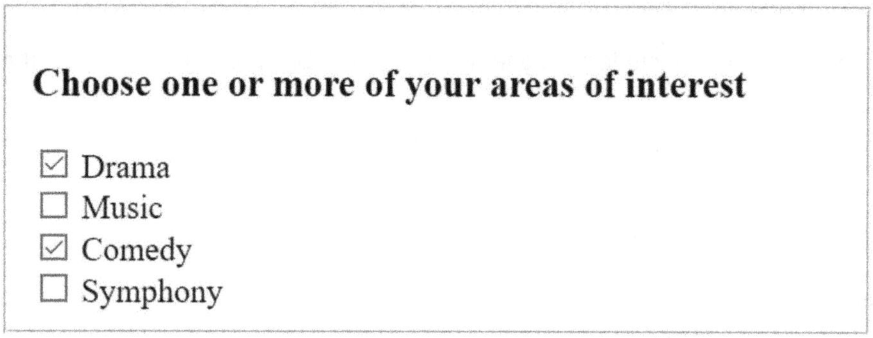

Figure 4-10: Browser display of checkboxes

If you select Drama and Comedy, submitting the form will send both values to the server, as shown here:

```
&area=drama&area=comedy
```

Step 15: Add a set of checkboxes to let the user select one or more areas of interest.

Create a third `fieldset` above the Continue button.

Use the code presented herein to create a set of four checkboxes within the `fieldset`.

Display the form to view the checkboxes

<input type="number">

The `number` type specifies an input field for numbers. This type of input field has a built-in validator that accepts only numeric digits and a decimal point. Validation takes place when the form is submitted. You also may specify a range for the input data using the `min` and `max` attributes, and the `step`, as shown here.

```
1   <label for="tickets"># of tickets:</label>
2       <input type="number" name="tickets" id="tickets" min=1 max=10 step=1
            required> <br>
```

It is important to keep in mind that client-side validation is not fully reliable, as discussed in the next section.

The Step Attribute

The step attribute that specifies an interval provides a powerful way to specify allowable values. For example, if step=2, and there is no min attribute specified, the valid entries are 0, 2, 4, 6, ..., and -2, -4, -6, ...

The default value for step is 1. If step and min attributes are not specified, then any whole number is valid.

To accept decimals, you have to specify a step size that is less than one. For example, step=0.01 lets you enter 0.01, 0.02, 0.03, ..., 1.01, 1.02, ...

If the min attribute is specified, then the value of step is added to the value of min to determine the valid entries. For example, min=0, max=4, and step=0.1 could be used to specify that the valid entries of GPA, for example, are 0, 0.1, 0.2, ..., 3.8, 3.9, 4.0. Thus the step along with min/max specifies a range of acceptable entries.

The Value Attribute

You may use the value attribute to specify an initial value for a field. However, you should exercise caution in providing initial/default values because it makes it difficult to check whether the user forgot to enter a value or meant to keep the default value.

> **Step 16: Add an input field to the reservation form to let the user enter the number of tickets.**
>
> Use the code presented herein to create the input field within the second fieldset, below the checkbox.
>
> Display the form in the browser.
>
> Enter an integer within the range 1–10 and click Continue to submit the form.
>
> The browser should display the Confirmation page, indicating that the form was submitted without any error.
>
> Next, enter each of the following types of invalid data and click Continue to submit the form:
>
> non-numeric data, non-integer, integer outside the range 1–10
>
> You should get an error message for each type.

4.2.4 Data Validation

Validation of input data is a crucial aspect of ensuring the quality of information obtained from a website. Validation includes making sure that

- All required data is entered.
- The data is of the right type (e.g., an order quantity is a number).
- The data is entered in the right format (e.g., an order date is entered in a format that the software understands).
- The data is within the expected range or a subset of a list (e.g., the percent discount is within the range 0 to 20; the month name is one of January through December).

There are two different types of data validation: client-side and server-side.

Client-Side Validation

Client-side validation takes place on the client before the data is submitted to the server. As you have seen, HTML input elements have easy-to-use built-in features to validate the data entered into elements. If you want more flexibility, you also may use JavaScript for client-side validation. For example, in the current application, you will use client-side JavaScript to make sure that the date entered is not a past date.

An advantage of client-side validation is that immediate feedback can be given to the user when each data item is entered, without having to wait until all data is entered and submitted to the server.

Server-Side Validation

Server-side validation is done on the server using server-side script after an entire form is submitted. Thus the user doesn't get any feedback on individual data items as the data is entered. However, server-side validation is necessary in many applications because client-side validation may not be completely reliable. Server-side validation is also needed to guard against security issues that may arise from users attempting to submit malicious data. If data is sent to the server using a query string, it would be fairly simple to circumvent the client data validation by changing the data that is displayed with the URL. You will use server-side validation in Chapters 6 and 17.

Review Questions

1. What are two differences between HTML web forms and ASP.NET web forms?
2. What is the difference between a form with a *get* method and a form with a *post* method?
3. What are the global attributes of HTML elements? List three of them.
4. How do you specify that multiple radio buttons belong to the same group?
5. What is the significance of the value attribute of radio buttons in a group?
6. How do you use attributes of an <input type="number" ...> element to specify that the user should be allowed to enter only whole even numbers between 0 and 10?
7. What is a drawback of providing default values for input elements on a form?
8. How does client-side validation differ from server-side validation?

4.3 The <output> Element

The <output> element is designed to display the result of a calculation. You can use the optional *for* attribute of the <output> element to show the elements that went into the calculation. Here is an example of how you would create an <output> element, which initially displays the value zero, to be used to display the total cost.

```
<output id="zonePrice"> 0 </output>
```

You can use the *for* attribute to show the elements used in the calculation to enhance machine reading.

```
<output id="zonePrice" for="zone">0</output>
```

The *for* attribute lists the id's of the elements used in the calculation, with spaces between the id's.
You may provide a caption for the output using a <label> element, as shown here:

```
1  <label for="zonePrice"> Zone Price: </label>
2      <output id="zonePrice" for="zone">0</output>
```

In the next chapter, you will learn how to use JavaScript to display the total cost in this element.

Step 17: Add an <output> element to display the zone price.

Use the code from lines 3–5 in Figure 4-11 to create the output element between the radio button group and the checkbox.

```
1  <input type="radio" name="zone" id="balcony" value="balcony" checked>
2      <label for="balcony">Balcony</label>
3  <!--Output element to dislay the zone price-->
4  <label for="zonePrice" class="label"> Zone Price: </label>
5      <output id="zonePrice " for="zone" class="output" >0</output><br> <br>
6  <!-- Checkbox to specify member status -->
```

Figure 4-11: The output element

Note that the `class` attribute is specified for the `label` and `output` elements to help display them to the right of the radio buttons using CSS.

Step 18: Specify the CSS style for the label and the output elements.

Enter the styles shown in Figure 4-12 in the Styles.css file.

```
1   fieldset{
2   width: 31em;
3   }
4   .label{
5       position: absolute;
6       left: 25em;
7   }
8   .output{
9       position: absolute;
10      left: 30em;
11  }
12  #fName, #lName, #emailAddress, #performance, #performDate, #tickets{
13      position: absolute;
14      left: 10em;
15  }
```

Figure 4-12: The CSS style

Provide a link to the Styles.css file by inserting the following statement at the end of the head section:

```
<link rel="stylesheet" href="styles.css">
```

Display the reservation page. The `output` field should be displayed to the right of the radio buttons, as shown here:

> ○ Suite ○ Premium ○ Circle ⦿ Balcony   Zone Price: 0

If the link to the .css file doesn't work, you may add a `<style>` element in the head section and enter the above css styles within the `<style>` element.

Step 19: Add an <output> element to display the discount.

The discount should be displayed to the right of the checkbox, with a caption, as follows:

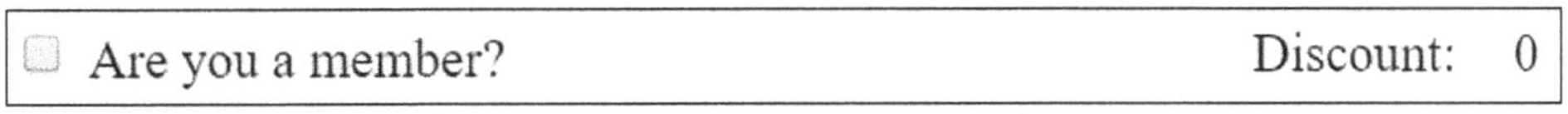

> ☐ Are you a member?   Discount: 0

4.4 The <button> Element

The `<button>` element can be used to create three different types of buttons. The value of the type attribute determines the type of button, as shown in Table 4-5.

Table 4-5: Values of type attribute of the `<button>` element

Type of <button> element	Description
type="submit"	Submits the form when you click the button. This is the default type.
type="reset"	Resets the fields in the form when you click the button
type="button"	Creates a clickable button that is commonly used to execute code when you click it. This type of button doesn't submit the form.

4.4.1 <button type="submit">

The submit type of <button> element creates a "submit" button, which submits all the input data on a form when the button is clicked. The submit button that you already created is shown here.

```
<button type="submit" id="continue"> Continue <button>
```

The code renders a button that looks as follows:

```
Continue
```

The content, which is displayed on the button, could be text, like "Compute Cost," or other HTML elements like image. It also is easier to style buttons rendered by the <button> element, compared to buttons rendered by the older type of buttons created using <input type="button">.

If the name and value attributes are provided, the name-value pair (key-value pair) is sent to the server when the button is clicked.

In addition to the global attributes, submit buttons have certain additional attributes, as shown in Table 4-6.

Table 4-6: Attributes of the <button type="submit"> element

Attribute	Description
formaction	Specifies a URL that overrides the URL specified in the form's action attribute; useful when you have multiple buttons in a form
formenctype	Specifies the encoding type to use for the form data
formmethod	Specifies the HTTP method (get or post) to override the form's method
formnovalidate	Specifies that the form's fields will not be subjected to constraint violations before submitting the data
formtarget	The browsing context into which to load the response returned by the server

The formaction and formmethod attributes are particularly useful when you need to use multiple submit buttons in a form as described next.

Multiple Submit Buttons in a Form

A form may use multiple submit buttons to take different actions, depending on the button clicked by the user. For example, you may want to use four different buttons to create, read, update, and delete (CRUD) data on a form. The code to create these buttons and how the buttons are displayed in the browser is shown in Figure 4-13.

```
1  <form action="createreservation.html" id="reservation" method="post">
2  <button type="submit" name="crud"> Create </button>
3  <button type="submit" name="crud" formaction="readreservation.html"
4      formmethod="get"> Read </button>
5  <button type="submit" name="crud" formaction="updatereservation.html">Update</button>
6  <button type="submit" name="crud" formaction="deletereservation.html">Delete</button>
7  </form>
```

Figure 4-13: Multiple submit buttons in a form

This is how the browser displays the buttons:

```
Create   Read   Update   Delete
```

The Create button doesn't use the *formaction* attribute because it submits data to the same page specified in the *action* attribute of the form (createreservation). Thus there is no need to override the action specified in the form.

The Read, Update, and Delete buttons override the pages specified in the form's *action* attribute, using the *formaction* attribute.

In addition, the Read button overrides the value of the method attribute of the form (post) and set it to *get*, using the formmethod attribute of the button.

4.4.2 The <button type="reset"> Element

The reset button resets every input data item on a form to its initial value. Here is an example:

```
<button type="reset" id="reset"> Reset </button>
```

Step 20: Add a reset button to the reservation form.

Add a reset button on the left of the Continue button, inside a fourth fieldset, as shown in Figure 4-14.

```
1    <fieldset>
2        <button type="reset" id="reset"> Reset </button>
3        <button type="submit" id="continue"> Continue </button><br><br>
4    </fieldset>
```

Figure 4-14: The reset button

Display the form, enter some data and click the reset button to test it.

4.4.3 The <button type="button"> Element

The <button type="button"> element creates a clickable button that is commonly used to execute code when you click it. It doesn't submit the form data or reset it when the button is clicked. You can use it to run JavaScript. This is how you create a button to compute cost:

```
<button type="button" id="cost" onclick="ComputeCost()">Compute Cost</button>
```

The onclick attribute specifies the name of a function (to be created) to compute the cost, as explained in the next chapter.

Step 21: Add a button to compute the total cost of the reservation and an <output> element to display the total cost.

Add the button and output element as shown in Figure 4-15.

The fieldset containing the buttons should resemble Figure 4-16.

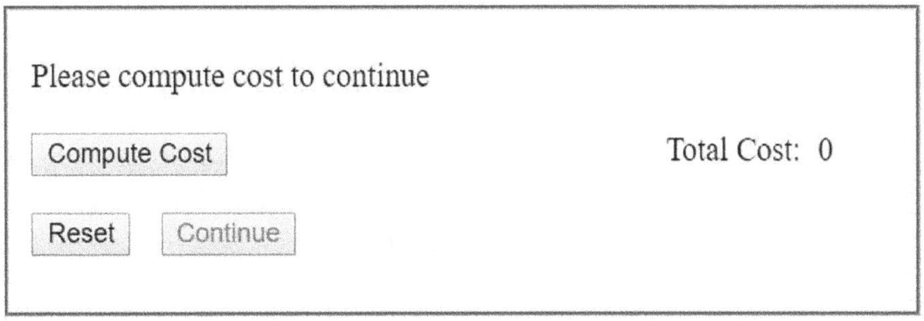

Figure 4-15: Displaying the total cost

```
1  <fieldset><br>
2      <!-- Button to compute total cost -->
3      <div id="msg1">Please compute cost to continue</div> <br>
4      <button type="button" id="cost" onclick="ComputeCost()">Compute Cost
         </button>
5      <!-- Field to display total cost -->
6      <label for="totalCost" class="label"> Total Cost: </label>
7      <output id="totalCost" for="performance zone status tickets"
8              class="output">0</output> <br><br>
9
10     <!-- Buttons to reset and continue -->
11     <div id="msg2"></div>
12     <button type="reset" id="reset"> Reset </button>
13     <button type="submit" id="continue" disabled=true> Continue </button><br>
    <br>
14 </fieldset>
```

Figure 4-16: The `fieldset` for buttons

Line 4 shows the code to create the button, and lines 6 and 7 show the `output` element.

The total cost is computed using the values of elements that have these id's: *performance*, *zone*, *status, and* *totalCost*. Hence, these id's are specified in the *for* attribute of the `<output>` element in line 7.

The Continue button is initially disabled. You will use JavaScript to enable it when the user clicks the Compute Cost button. In addition, the instruction to click the Continue button will be displayed in the empty `<div>` element.

Display the form to test the code.

It should be noted that the `<input>` element also can be used to create the three types of buttons. However, the `<button>` element is recommended because of its advantages, including ease of styling.

Next, you will learn how to use the `<select>` element to create a drop-down list.

Review Questions

9. What is the difference between the elements `<input type="submit">` and `<button type="submit">`?

10. What is the difference between the elements `<button type="button">` and `<button type="submit">`?

11. Consider a form that has multiple submit buttons that submit the form to different pages. What attribute of the form element would you use to specify the page corresponding to each button?

4.5 The <select> Element

The `<select>` element creates a drop-down list with multiple options that the user can choose from. Each option within the drop-down list is specified by the `<option>` element. The code for a drop-down list to let the user select a performance is shown in Figure 4-17.

```
1  <label for="performance">Performance:</label>
2  <select name="performance" id="performance">
3      <option hidden>--Select an option--</option>
4      <option value="Young Irelanders">Young Irelanders</option>
5      <option value="Justin Hayward">Justin Hayward</option>
6      <option value="TedxOshkosh">TedxOshkosh</option>
7      <option value="OSO Orchestral">OSO Orchestral</option>
8  </select>
```

Figure 4-17: The select element

This is how the browser displays the drop-down list:

Performance: --Select an option-- ▼

The value attribute of the option element specifies the data submitted to the server; it is not displayed on the page. It is the content (between the start and end tags) that is displayed. The hidden attribute of the first option makes it invisible from the list after the user makes a selection.

Table 4-7 presents the attributes of the <select> element.

Table 4-7: Attributes of the select element

Attributes of <select>	Description
autofocus	Specifies that the drop-down list should automatically get focus when the page loads
disabled	Specifies that the drop-down list should be disabled
form	Defines one or more forms the select field belongs to
multiple	Specifies that multiple options can be selected at once
name	Defines a name for the drop-down list
required	Specifies that the user is required to select a value before submitting the form
size	Defines the number of visible options in a drop-down list

Except for the **multiple** attribute, the attributes of the <select> element are a subset of the attributes of the <input> element. Further, the *size* attribute in <select> specifies the number of options displayed in the list.

For example, if you specify size=3 (**<select** name="performance" id="performance" size=3>), the list will display three options, as shown here:

Performance: Young Irelanders ∧
            Justin Hayward
            TedxOshkosh   ∨

Adding the multiple attribute (**<select** name="performance" id="performance" size=3 multiple>) lets you select multiple options using the alt key to select non-contiguous items and the shift key to select a group of contiguous items.

> **Step 22: Create a drop-down list to let the user select a performance.**

Use the code presented in Figure 4-18 to add a <select> element above the <date> element.

```
1   <fieldset> <br>
2       <!-- Drop-down list to select a performance-->
3       <label for="performance">Performance:</label>
4       <select name="performance" id="performance">
5           <option hidden>--Select an option--</option>
6           <option value="Young Irelanders">Young Irelanders</option>
7           <option value="Justin Hayward">Justin Hayward</option>
8           <option value="TedxOshkosh">TedxOshkosh</option>
9           <option value="OSO Orchestral">OSO Orchestral</option>
10      </select>
11      <!-- Input field for performance date -->
```

Figure 4-18: The fieldset and the select element

It should be noted that there is a related HTML element, **<datalist>,** that lets you specify a list of items which are only used to autocomplete when you start typing the first letter(s), if there is an exact match. If there is no match, the user can type any value.

Step 23: Add an <output> element to display the base price.

The base price should be displayed to the right of the drop-down list, as shown here (see the <output> element that was used to display zone price):

| Performance: | --Select an option-- ▼ | Base Price: 0 |

Set the id of the element to "basePrice".

The reservation form should appear as shown in Figure 4-19.

Figure 4-19: The reservation form

Display the reservation page and make sure the drop-down list includes the four performance names.

The complete HTML code is shown in Figure 4-20.

```
1   <body>
2   <h1>Reservation</h1>
3   <form action="confirmation.html" id="reservation" method="get"
    onsubmit="return ValidData()" onreset="ResetOutput()" onshow="DisplayOutput()">
4       <!-- Input fields to enter personal information-->
5       <fieldset width="400px"><br>
6       <label for="fName">First Name</label>
7           <input type="text" name="fName" id="fName"> <br><br>
8       <label for="lName">Last Name</label>
9           <input type="text" name="lName" id="lName"><br><br>
10      <label for="emailAddress">e-mail:</label>
11          <input type="email" name="emailAddress" id="emailAddress" required
        placeholder=username@mailserver.com pattern=".+@prospectpressvt.com"
        title="The domain must be pospectpressvt.com"> <br><br>
12      </fieldset><br>
13
14      <fieldset><br>
15      <!-- Drop-down list to select a performance-->
16      <label for="performance">Performance:</label>
17      <select name="performance" id="performance" >
18          <option >--Select an option--</option>
19          <option value="Young Irelanders">Young Irelanders</option>
20          <option value="Justin Hayward">Justin Hayward</option>
```

Figure 4-20: The complete HTML code for the reservation form (*continues*)

```
21          <option value="TedxOshkosh">TedxOshkosh</option>
22          <option value="OSO Orchestral">OSO Orchestral</option>
23      </select>
24      <!--Output element to dislay base price-->
25      <label for="basePrice" class="label"> Base Price: </label>
26      <output id="basePrice" for="performance" class="output" >0</output><br> <br>
27
28      <!--input field to enter date-->
29      <label for="performDate">Date:</label>
30      <input type="date" name="performDate" id="performDate" required> <br><br>
31
32      <!-- Radio buttons to select a zone-->
33      <input type="radio" name="zone" id="suite" value="suite" onchange="ZonePrice()">
34          <label for="suite">Suite</label>
35      <input type="radio" name="zone"id="premium"value="premium"onclick="ZonePrice()">
36          <label for="premium">Premium</label>
37      <input type="radio" name="zone" id="circle" value="circle" onclick="ZonePrice()">
38          <label for="circle">Circle</label>
39      <input type="radio" name="zone" id="balcony"value="balcony"onclick="ZonePrice()">
40          <label for="balcony">Balcony</label>
41      <!--Output element to dislay the zone price-->
42      <label for="zonePrice" class="label"> Zone Price: </label>
43      <output id="zonePrice" for="suite premium circle balcony"
            class="output" >0</output><br> <br>
44
45      <!-- Checkbox to specify member status -->
46      <input type="checkbox" name="memeberStatus" id="status" onclick="Discount()">
47      <label for="status">Are you a member?</label>
48      <!--Output element to dislay the discount-->
49      <label for="discount" class="label"> Discount: </label>
50      <output id="discount" for="status" class="output" >0</output><br> <br>
51
52      <!-- Input field to enter number of tickets -->
53      <label for="tickets"># of tickets:</label>
54      <input type="number" name="tickets" id="tickets" min="1" max="10" step="1"
55          onfocus="this.style.backgroundColor='lightgrey'"> <br><br>
56      </fieldset><br>
57      <!-- Checkboxes to enter preferences -->
58      <fieldset>
59      <h3>Choose one or more of your areas of interest</h3>
60      <input type="checkbox" name="area" id="drama" value="drama">
61      <label for="drama">Drama</label><br>
62      <input type="checkbox" name="area" id="music" value="music">
63      <label for="music">Music</label><br>
64      <input type="checkbox" name="area" id="comedy" value="comedy">
65      <label for="comedy">Comedy</label><br>
66      <input type="checkbox" name="area" id="symphony" value="symphony">
67      <label for="symphony">Symphony</label>
68      </fieldset><br>
69
70      <!-- Button to compute total cost -->
71      <div id="msg1">Please compute cost to continue</div> <br>
72      <button type="button" id="cost" onclick="ComputeCost()">Compute Cost</button>
73      <!-- Field to display total cost -->
74      <label for="totalCost" class="label"> Total Cost: </label>
75      <output id="totalCost" class="output" for="performance zone status tickets">0
            </output> <br><br>
76      <input type="hidden" name="hiddenTotalCost" id="hiddenTotalCost" value="0">
77
78      <!-- Buttons to reset and continue -->
79      <div id="msg2"></div>
80      <button type="reset" id="reset"> Reset </button>
81      <button type="submit" id="continue" disabled=true> Continue </button>
82  </form>
83  </body>
```

Assignments

E-Commerce Assignment: Part A

P&I Laptops is a small business that assembles and sells online a limited set of laptop models. In this assignment, you will develop the user interface for a web page that is part of an e-commerce system.

Develop a web page named **laptop-selection** within the WebProjects\Assignments folder. The web page allows the user to specify the model, software, amount of extra RAM, whether a warranty is needed, and the quantity, as shown below. Choose an appropriate layout. What is shown here is just an example. Part B of the assignment in Chapter 5 will compute and display the costs based on the selections made by the user.

Laptop selection

Model:
| S330 |
| S380 |
| S580 |

Hardware: 0

Software:
  ◯ MS 365 Personal ◯ MS 365 Family ◯ Office Home & Student

Software: 0

Extra RAM: ◯ 4 GB ◯ 8GB

RAM: 0

  ☐ Ext. Warranty?

Warranty: 0

Quantity: [   ]

[Compute Cost]   Total Cost: 0

[Reset] [Select]

The form should have the following features:

- A drop-down list that lets the user select a model. The current set of models, which is likely to change in the future, are:
- S330, S380, S580, G780, B780
- Let the user select software from the following list:
- Microsoft 365 Personal, Microsoft 365 Family, Microsoft Office Home & Student
- Let the user select extra RAM. The options are 4 GB and 8 GB.
- Let the user select an extended warranty, if needed.
- Let the user enter the number of units ordered. When the user enters the number of tickets, change the background color of the input field.
- For now, display zeros for costs.
- A Reset button that resets all input data.
- A button to compute the total cost. The button should be disabled until the user enters the number of tickets.
- A button with the caption *Select*, to submit the input data and display a page named Cart. For now, create an empty web page named Cart for testing. The Select button should be disabled until the total cost is computed.

Auto Rental Assignment: Part A

This assignment develops a web form for "Ace Auto Rentals," which is a small business that rents automobiles. Ace would like you to develop a web form that lets the user select the model of the vehicle, specify related information and compute and display the costs, and create a second page to let the user confirm the selection.

Develop a web form named **Auto-rental** within WebProjects\Assignments folder. The form provides the user interface, as shown in the following figure. You will use JavaScript to compute costs and to confirm the selection in Part B of the assignment in the next chapter.

Rental Reservation

Model:

SUV: Honda CR-V
Minivan:Chrysler Pacifica
Compact car: Ford Escape

Daily Rate: 0

Start date: mm/dd/yyyy

Special rate:
  ○ Senior ○ Government ○ Corporate ○ AAA

Discount: 0

☐ Do you want insurance?

Insurance: 0

Additonal drivers: 0

Drivers: 0

Cost/day: 0

# of days: 0

Total Cost: 0

[Reset] [Continue]

The form should have the following features:

- Let the user select a model, from the following list of models, which is likely to change over time.

SUV: Honda CR-V
Minivan: Chrysler Pacifica
Compact car: Ford Escape
Full-size car: Chevy Impala
Luxury Car: Tesla 3

- Let the user select a special rate category, if any, from the following list, which is not likely to change: Senior, Government, Corporate, AAA.
- Let the user select insurance, if needed.
- Let the user enter the start date, number of extra drivers, and number of days.
- Display the costs. For now, display zeros for costs.

- A Reset button that resets all input data
- A Continue button to submit the input data and display the Confirmation page. Initially, the Continue button should be disabled. For now, create an empty web page named Confirmation for testing.

Choose an appropriate layout. The layout shown here is an example.

## Chapter 5

Interactive Web Pages: JavaScript and Client-Side Scripting

In the previous chapter, you learned how to use HTML to create the user interface for web forms and to send data to another page. This chapter describes the fundamentals of the JavaScript language, which is currently the most popular language to develop client-side code. You will create and run JavaScript in Visual Studio Code. You will use JavaScript to access input data from different types of fields, validate and process data, dynamically change the contents of a page, move to another page, and share data between pages. You also will learn how to use jQuery to simplify JavaScript code. In later chapters, you will use JavaScript in ASP.NET web forms, Razor pages, and MVC to do client-side processing.

To practice what you learn, you will continue to develop the reservation page, starting with a copy of the Reservation-1.html form from Chapter 4, named Reservation-2.html.

Learning Objectives

After studying this chapter, you should be able to

- Develop simple JavaScript programs involving decision and iterative structures and functions.
- Work with different types of data, including dates, arrays, and Maps.
- Debug JavaScript programs.
- Use client-side scripts in web forms.
- Access HTML elements from JavaScript.
- Display one page from another and share data using query string.
- Use some common jQuery functions.

5.1 Introduction

To make web pages interactive and dynamic, web applications run both client-side scripts, which run on the client, and server-side scripts, which run on the server. Strictly speaking, a script is a type of code that is written in interpretive scripting languages like JavaScript, PHP, and CGI, to control the behavior of applications. For instance, you may validate input data on a web form using a script written in JavaScript and executed by the browser. When an interpretive language is run, each statement in the program is translated and run without creating an executable unit of the whole program.

A program, on the other hand, is code written in programming languages like Java and C# that are compiled. When a program is compiled, an executable unit of the entire program is created before running the program.

However, as the difference between scripting and programming languages is becoming blurred, the term *script* is generally used to represent any code that works with a web page, including code written in a programming language like C#.

5.1.1 Client-Side and Server-Side Scripting

Client-side scripts are typically used to make web pages more interesting, user-friendly, and interactive. Applications of client-side scripts include immediate validation of data, running multimedia, and creating gadgets like calculators. Running client-side script is simpler and quicker because it doesn't involve the server.

Client-side script could be embedded in the HTML code for a page or in a separate file. It is run on the client (in the browser) when the browser receives the web page from the server. That would mean the user can view the script by displaying the source code of a page in the browser and the code is less secure.

The process of displaying both dynamic and static web pages involves an HTTP request by the client and a response from the server. However, displaying dynamic pages involves additional processing on the server because they are created on-demand. That is, segments of the page are dynamically generated using additional code (script) embedded in HTML or stored in a separate file(s). The scripts may do tasks like look up inventory levels from a database, compute order totals, validate data, and customize the web page. When a web server receives a request for a web page that includes a server-side script, the server executes the script, generates the HTML code for the web page, and sends it to the browser.

Because the server-side script is run on the server and only the result is sent to the browser, server-side scripting has the advantage that the users won't be able to view the script in the browser and it is more secure. However, it puts an additional burden on the server.

Server-side scripts typically include code written in scripting languages like PHP, Python, CGI, and JSP, which are not compiled, and programs written in programming languages like C#. PHP is by far the most popularly used server-side scripting language. You will use server-side code in ASP.NET web forms, Razor pages, and MVC.

5.1.2 Components of a Web Project

In addition to the web server that handles HTTP requests, a website may include an **application server** and a **database server**. The application server provides an environment to execute scripts and create dynamic web pages, while the database server executes the commands that access the database. The use of these three servers is illustrated in Figure 5-1. It should be noted that a single physical server may perform the role of all three servers.

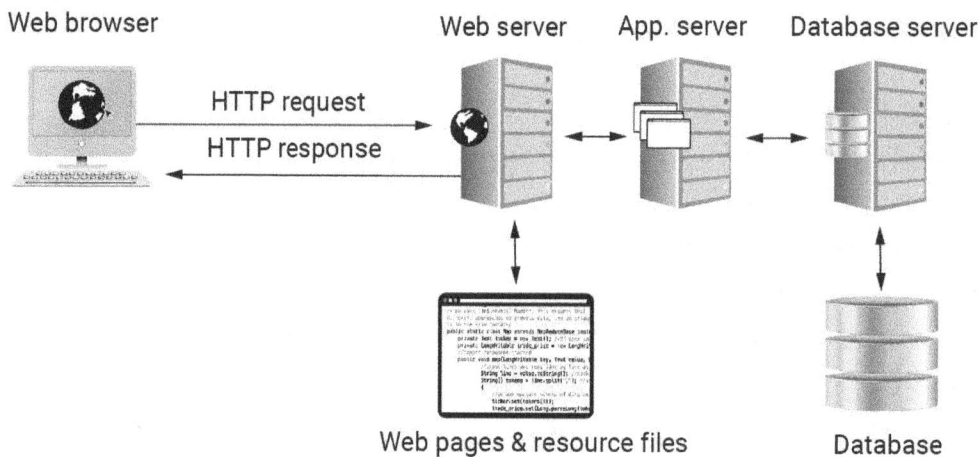

Figure 5-1: Components of a web project

5.1.3 JavaScript

JavaScript (JS) is currently the most popular language used to develop client-side code. Typically, JavaScript is run in the browser on the client to do tasks like validate data, process data, and run multimedia and animation. However, there are versions of JavaScript like Node.js that runs on the server to do tasks like communicating with databases and processing data on the server. In this chapter, we will focus on client-side use of JavaScript.

Why is JavaScript popular? It is a lightweight and free-form language that is easier to learn compared to programming languages like Java and C#. It has less stringent rules and is a lot more forgiving. For example, you don't have to declare the type of variables. Further, JavaScript has fewer data types.

Those who have a basic knowledge of a language like Java, PHP, or C# would find it fairly easy to learn JavaScript because of the similarities in many areas, including syntax of expressions, naming conventions, decision structures, Loops, arrays, and functions.

JavaScript is an object-based language, which uses objects, but it is not a fully object-oriented language. Further, compared to major programming languages, it has limited features.

In this chapter, you will find a basic introduction to programming in JavaScript and how it is used in web pages. If you don't have prior experience with programming, you should be able to follow along. To be able to use JavaScript in your own projects, you may need to study it in some more detail. You can find several good sources of tutorials and descriptions of JavaScript coding, including those from W3Schools, Mozilla Developer Network, and Codecademy.

Review Questions

1. How does the server process the request for a dynamic web page?
2. What are the different types of servers used in a web application?
3. What are the two differences between server-side and client-side scripts?

5.1.4 Running JavaScript in VS Code

Almost all browsers can run JavaScipt programs and help debug JavaScript. In this book, we will use client-side JavaScript to work with web pages. To do this, JavaScipt can be embedded within an HTML file using the `<script>` element or kept in a separate file and the file name specified in the `<script>` tag. You can also run JavaScipt in VS Code independent of web pages, using VS Code extensions like Code Runner, or other tools.

Typically, JavaScript programs are run in response to events like the user clicking a button or loading a web page. For example, you will create a JS program to compute cost when the user clicks a button.

First, to help learn JavaScript, you will create a program that is not associated with any specific event and later apply JS to the HTML form.

Figure 5-2 shows a single-line JavaScript program embedded in an HTML document.

```
1   <!DOCTYPE html>
2   <html>
3       <head>
4          <script>
5              document.write("Welcome to JavaSript programming!");
6          </script>
7       </head>
8   </html>
```

Figure 5-2: Welcome program

Because the purpose of this HTML document is to run the JS program, it doesn't have the `<body>` element. Line 5 is the entire JS program, which is specified within the `<script>` element.

5.1.5 document.write() Function

In line 5, `document.write()` is a JavaScript **function** that writes the specified string into the HTML document at the location where the `<script>` element is placed. A function (also called a **method**) is a block of code that performs a specific task. The string you want to insert into the HTML document is enclosed in quotes. You may use a pair of double quotes or single quotes (apostrophes). In JavaScript, there is no difference between double quotes and single quotes, except that when you want to nest a pair of quotes inside another, the inner pair should be single quotes.

Keep in mind that JavaScript, like C# and Java, is case sensitive. So, `Document.write()` is not the same as `document.write()`.

The program will run any time the HTML page is displayed in the browser. If you display the page, you will see the content:

> Welcome to JavaSript programming!

You will continue to use **Visual Studio Code** to develop .html files. Please see Chapter 1 (Section 1.2) for instructions on installing and using Visual Studio Code.

You will create an HTML file named Sandbox.html to run the above program and then reuse the same file to test a few more blocks of code.

Tutorial: Sandbox file to test JavaScript code

Step 1: Open Visual Studio Code and create a new HTML file named Sandbox.html.

Step 2: Create and run the JavaScript program.

Type in the code from Figure 5-2 into the Sandbox.html file.

Display the page to view the output: Press Alt-b, or double-click Sandbox.html in Windows Explorer (or in Finder on Mac computers).

There is your first JavaScript program. Close the browser window.

5.2 Introduction to Programming

To help you learn some basics of JavaScript programming, let's look at a simple program to compute the total cost for a reservation. The program can be specified as shown in Table 5-1.

Table 5-1: Program specs

Purpose	Compute and display the total cost of a reservation
Output	Total cost
Input data	Cost per ticket, number of tickets
Process (how?)	Get input data and compute the total cost. Total cost = Cost per ticket × Number of tickets

The program to compute the cost is shown in Figure 5-3.

```
1   <!DOCTYPE html>
2   <html>
3       <head>
4           <script>
5
6               // Declare variables to hold data
7               var costPerTicket, tickets, totalCost;
8               // Get input data
9               costPerTicket = 17.5;
10              tickets = prompt("Number of tickets:");
11              // Compute total cost
12              totalCost = costPerTicket*tickets;
13              // Display result
14              document.write("Total cost: " + totalCost);
15          </script>
16      </head>
17  </html>
```

Figure 5-3: Code to compute total cost

The program in lines 6-14 consists of a set of statements with the end of each statement denoted by a semicolon. As indicated by the comments (statements starting with //), the program consists of four segments:

Line 7: Declare the variables.

Line 9: Set the cost per ticket.

Lines 10: Get input data.

Line 12: Compute the total cost.

Line 14: Display the result.

In the next couple of steps, you will create the program and run it so that you can get a feel for what the program does.

Step 3: Type the code from Figure 5-3 in the Sandbox.html file.

Delete the `document.write(…)` statement from the Sandbox.html file.

Type in the JS code from lines 5–14 in Figure 5-3.

You may use the IntelliSense feature, which helps you complete variable names and keywords by providing a list of potential names as you type. You can select a name/word by clicking it from the list or pressing the tab key when it is selected.

Step 4: Display the page in the browser.

Press Alt-b to display the page.

You will be asked to enter the number of tickets.

After entering the data, the program should process the data and display the results as follows:

Total cost: 35

If the browser doesn't show any result, you may have one or more errors in the program. You may choose to jump to the section on debugging to figure out what the error is or wait until you learn more about the statements.

Before we get into the details, let's look at the types of data used in JavaScript.

5.2.1 Data Types

It is important to differentiate between the different types of data used in a program. For example, what you can do with a date is different from what you can do with a number or text. Common data types in JavaScript include number, string, boolean, object, undefined, function, BigInt, and symbol, as described in Table 5-2. The object data types include array, Date, and null. The current program uses only the number and string types.

Table 5-2: Data types in JavaScript

Data type	Description
number	*number* includes whole numbers and decimal numbers (e.g., 17.5).
string	A *string* is a text consisting of a set of characters like the performance name "Rock of Ages" or the zip code "54901." A string is enclosed in pair of double or single quotes.
boolean	The *boolean* type represents Boolean data that have the values **true** or **false**.
object	An *object*, which is discussed in more detail later, has a set of properties representing one or more related data items, and methods (functions) that perform actions on the object. *array* is a special type of object that can store a list of data items. *Date* and *function* are also objects. *null* is an object that has the value null, which means "nothing."
undefined	A variable that is not assigned a value has the value **undefined**.
function	A *function* is a block of statements that does a particular task.
BigInt	*BigInt* can store very large integers.
symbol	*symbol* type is used to create a unique identifier for an object.

Let's look at the code in Figure 5-3 in more detail.

Comments

Lines starting with // are comments for the programmers and are ignored when the program is run. You may comment out multiple JS statements by placing the statements between /* and */.

The prompt() Function

The prompt function, used in line 10, is a function that displays a pop-up window showing a specified prompt and lets the user enter a value, as shown in Figure 5-4.

When the user enters a value and clicks the OK button, the window closes and the program continues running. The value entered by the user is stored in the variable, *performance*, as explained shortly.

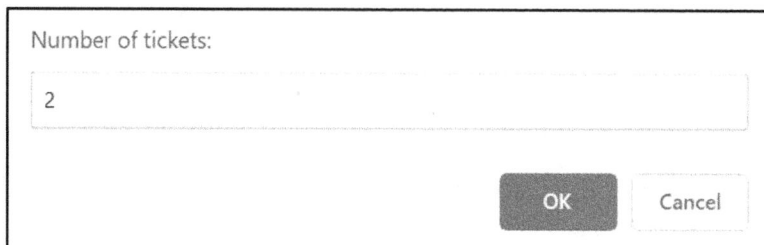

Figure 5-4: The prompt window

5.2.2 Variables

Variables store data items temporarily while the program is running. Variables are created (declared) by specifying the keyword *var* followed by the name of the variable, as in

```
var costPerTicket;
```

Line 7 declares (creates) three variables named costPerTicket, tickets, and totalCost to hold the performance name, cost per ticket, number of tickets, and total cost, respectively.

```
var costPerTicket, tickets, totalCost;
```

A variable created using the keyword *var* can hold different types of data like a number, string, or object. Unlike in many other languages, when you declare a variable, you don't have to specify the type of data that you want to store in the variable.

If a variable is to be used only within a block like a *for loop* or *if-else* statement (to be explained later), you may declare it with the *let* keyword instead of *var*, as in this example.

```
let index;
```

Variables declared with *let* can be used only within the block (within the curly brace, {}) where it is declared, whereas variables declared with *var* within a block can be used within the entire function that contains the block. Variables declared with *var* outside the functions can be used in all functions.

Variable names in JavaScript must start with a letter, dollar sign ($), or underscore (_), and can contain only letters, digits, dollar signs, and underscores. We will use the **Camel Case** for variable names: The first letter of a variable is lowercase, and the first letter of subsequent words, if any, are capitalized, as in totalCost and averageRetailPrice.

Until you store a value in a variable, it has the value, *undefined*. You can assign a value to a variable using the equal (=) operator in a statement of the form

Variable name = value

Line 9 is an example of a simple assignment statement.

```
costPerTicket = 17.5;
```

Executing this statement has the effect of assigning the value on the right-hand side of the equal sign (17.5) to the variable name on the left-hand side. In an assignment statement like this, the left-hand side can only consist of a variable name, but the right-hand side could be an expression (formula).

You may declare a variable and assign a value to it in a single statement, as in

```
var costPerTicket = 17.5, tickets = 2;
```

This declares two variables, `costPerTicket` and `tickets`, and assigns a value to each (`17.5` and 2, respectively).

In the current program, for the sake of simplicity, we hard-code the cost per ticket (17.5), assuming that it is a flat rate that is independent of the performance. In real-world applications, input data is often read from databases/files or provided by the user when the program is run. You will use databases in later chapters.

Undeclared Variables

JavaScript lets you assign a value to a variable without declaring it. However, using undeclared variables increases the chances of producing wrong results by using the wrong variables. For example, by mistake, you may use the variable name `ticket` (instead of `tickets`) in line 10 as shown here:

```
ticket = prompt("Number of tickets:");
```

In this case, the variable `tickets` used in calculating total cost in line 12 (`totalCost = costPerTicket*tickets;`) won't have a value assigned to it. As a result, the expression `costPerTicket*tickets` will yield the result NaN (Not a Number). The output of the program will appear as follows.

```
Total cost: NaN
```

The result NaN indicates an incorrect operation.

TRY IT

In line 10, change the variable name `tickets` to `ticket`, and display the page. You should see the value NaN for the total cost.

Assigning values to undeclared variables is highly discouraged. You can prevent programs from executing statements that assign values to undeclared variables, by adding the following directive to the program:

```
"use strict";
```

The strict mode will cause the program to break at the statement that assigns a value to an undeclared variable and give a run-time error, described later.

TRY IT

Insert the statement `"use strict"` (with the quotes) in line 5 and display the page.

Enter values for performance and number of tickets when prompted.

No result will be displayed, because the program breaks at line 10 due to the use of the undeclared variable, `ticket`.

Make sure you change the variable name in line 10 back to `tickets`.

Line 10 gets the number of tickets and stores the value in the variable, `tickets`.

```
tickets = prompt("Number of tickets:")
```

It is important to note that the number you entered in response to a prompt is returned by the `prompt()` function as a string ("17.5"), not as a number (17.5).

Line 12 multiplies the cost per ticket (stored in `costPerTicket`) by the number of tickets (from tickets) and stores the value in the variable, `totalCost`.

```
totalCost = costPerTicket*tickets;
```

Note that the value of `costPerTicket` is a number because the statement in line 9 (`costPerTicket = 17.5;`) assigns a number to the variable. However, the value of tickets is a string because the `prompt()` function always returns a string. But it doesn't create a problem in JavaScript because if you mix a number and string in an expression, JavaScript automatically converts the string to a number, except when the operation is addition. This is further explained in the next section on doing calculations.

5.2.3 Doing Calculations

Table 5-3 shows ommon mathematical operators used in an expression to do calculations.

Table 5-3: Math operators to do calculations

Operator	Operation	Example
*	multiply	Multiply 17.5 by 3: 17.5*3
/	divide	Divide 17.5 by 3: 17.5/3
+	add	Add 17.5 and 0.55: 17.5+0.55
-	subtract	Subtract 1.25 from 12.5: 12.5-1.25
-	unary minus	Negative 2.5: -2.5
%	Modulus (remainder)	32%10 gives the result 2

Precedence of Operators

JavaScript follows common mathematical practices that are also used in many other programming languages when deciding the order to execute operations in an expression. Operations within parentheses are done first. Further, multiplications and divisions are done before additions and subtractions, and then calculations take place from left to right. Consider the following expression:

5+4*3-2/2

The above expression would evaluate to 5+(4*3)-(2/2), yielding 16.

Here is another example:

12/4*2

This expression would be evaluated from left to right as (12/4)*2, yielding 6. It is always a good practice to use parentheses to make the order of calculations clear and to avoid unintended errors.

Mixing Strings and Numbers

As stated earlier, if you enter a number into a prompt box, the `prompt()` function returns it as a string. Similarly, if you enter a number into an input field on a form, it is stored as a string. So, when using data obtained from a `prompt()` or from form fields to do calculations, it is easy to inadvertently mix string data with numbers.

Unlike many other languages, JavaScript won't raise an error if you mix strings and numbers in an expression. This could be problematic if the operation is an addition. If you add a number and string using the "+" operator, JavaScript will convert the number to a string and concatenate the two strings together. For example,

```
6 + "2" -> 62
"Hello" + 23 -> Hello23
```

Note that, like in many other languages, the "+" operator is used in JavaScript also to concatenate two strings together, as in the following example:

```
"John" + " " + "Adams" -> "John Adams"
```

To do the calculation right, you will have to convert strings to numbers using a function like `Number()`, as described shortly.

However, if you do other types of operations like multiplication and division with a number and a string, the string is converted to a number, as in

```
6 * "2" -> 12
```

Note that calculations take place from left to right within operations of the same precedence group. So, the position of the string in an expression will impact the result, as shown in the following examples.

```
1)  6 + 4 + "2" -> 102
```

The first operation (6+4) between two numbers yields 10. The next operation (10+"2") that involves a string yields 102.

```
2)  6 + "2" + 4 -> 624
```

The first operation (6+"2") involving a string yields the string "62." The next operation ("62"+4) yields 624.

5.2.4 Converting String to Number: Number() Function

The `Number()` function converts a string to a number, as in Figure 5-5.

```
1  <script>
2      var orderTotal = prompt("Please enter order total");
3      var salesTax = prompt("Pleae enter sales tax");
4      var amtDue = Number(orderTotal) + Number(salesTax);
5      document.write("Amount Due: " + amtDue);
6  </script>
```

Figure 5-5: The Number() function

Because the values of `orderTotal` and `salesTax` are strings, adding the two variables will result in concatenating the two strings together unless they are converted to numbers.

TRY IT

Comment out the `document.write()` statement in line 14 in Sandbox.html and then type in the code in Figure 5-5 (lines 2–5) immediately below it.

Display the page and enter the number of tickets, order total, and sales tax when prompted. You should see the correct total of the two numbers you entered.

Now, remove the `Number()` function by deleting the word *Number* in the two places in line 4. Display the page and enter the values.

The number displayed will be a concatenation of the two strings.

Remove the newly added statements and uncomment the `document.write()` statement to reverse the changes.

Returning to our program that computes total cost, the statement in Figure 5-3, line 14, shown below, writes the string formed by concatenating (combining) two different strings using the "+" operator. The "+" operator is used to concatenate strings and also add numbers.

```
document.write("Total cost: " + totalCost);
```

The first string `"Total Cost:   "` is combined with the value stored in the variable, `totalCost`. The combined string written into the HTML code would appear as follows:

Total cost: 35

5.2.5 Finding and Fixing Errors in JavaScript Code

Most browsers have a debugger that you can install to debug errors in code.

Installing Debuggers

You can install the debugger for a browser as follows:

Step 5: Install the Debugger for your browser.

In VS Code, click *View > Extensions*, or click the *Extensions Manager* icon in the left-most column.

Type the name of your browser in the search box (e.g., Chrome, Edge, Firefox, etc.).

A list of options will be displayed, as shown in Figure 5-6.

Click the *Install* button for your browser. (If the debugger is already installed, you will see an asterisk instead of the Install button.)

Change Debug Configuration

In VS Code, the debug configuration information is kept in a file named **launch.json.** You need to make a few changes to this file, as specified in the next step.

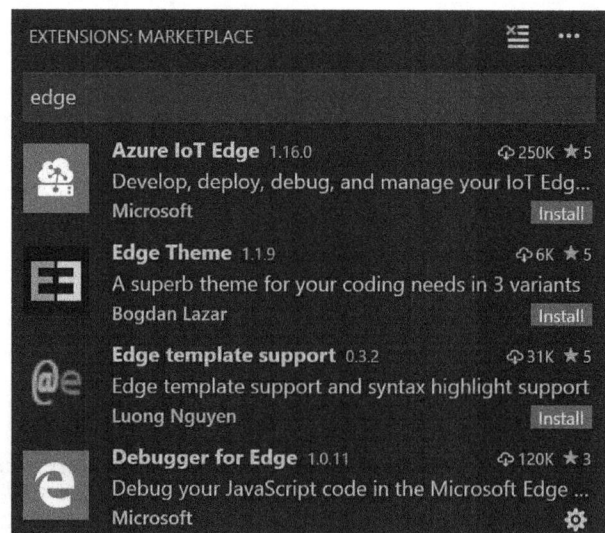

Figure 5-6: Options to install the debugger

Step 6: Change the configuration to specify the Workspace folder.

To view the configuration file, select *Run > Add Configurations*.

Select the launch option for your browser – for example, select *Chrome: Launch* for Chrome and select *Firefox: Launch (file)* for Firefox.

You will see the current configuration that looks like the one shown in Figure 5-7, except it will have an additional line that specifies the URL of the server ("url": http://localhost:8080), and "webRoot" in place of "file" in the last line. Firefox will have "/index.html" following "${WorkspaceFolder}".

Because debugging client-side JavaScript doesn't involve a server, delete the line starting with "`url`."

Change "`webroot`" to "`file`."

The configuration should match Figure 5-7.

```
.vscode ▸ {} launch.json ▸ Launch Targets ▸ {} Launch Chrome
 1   {
 2       // Use IntelliSense to learn about possible attributes.
 3       // Hover to view descriptions of existing attributes.
 4       // For more information, visit: https://go.microsoft.com/
 5       "version": "0.2.0",
 6       "configurations": [
 7           {
 8               "type": "chrome",
 9               "request": "launch",
10               "name": "Launch Chrome",
11               "file": "${workspaceFolder}"
12           }
13       ]
14   }
```

Figure 5-7: The Configuration file

Select File > Save.

Note that the "`file`" is set to the `workspaceFolder` (project folder). This lets you select any file from the project folder for debugging.

Types of Errors

Syntax errors, which are caused by not following the rules of the language, are indicated in VS Code by a red squiggly line under the part of the code causing the error. Examples of syntax errors include a missing closing parenthesis and a missing comma between variables in a declaration statement. A missing comma between variable names is indicated by the line under the name `tickets`, as shown here.

```
var costPerTicket tickets, totalCost;
```

You can hover the cursor over the red line to display the specific error.

If you run a program that has syntax errors, the program won't run, and you won't see any output in the browser. To identify the first error in the browser, you use the shortcut keys or the browser's menu options, as shown in Table 5-4. For example, in Chrome, on Windows computers, press Shift-Ctrl-i.

Table 5-4: Identifying errors in browsers

Browsers	Keyboard shortcut	How to display errors
Chrome	Windows: Ctrl-Shift-i Select the tab, *Console*. Mac: Option-Cmd-i	Open the browser menu and select *More Tools > Developer tools*. Select the tab *Console*.
Firefox	Windows: Ctrl-Shift-k Mac: Option-Cmd-k	Open the browser menu and select *Web Developer > Web Console*.
Safari	Mac: Option-Cmd-c	Select *Safari, Preferences, Advanced*. Check *Show Develop menu in menu bar*. From the *Develop* option, choose *Show JavaScript Console*.
Edge	Windows: F12 > Ctrl-2	Open the menu and select *Developer Tools*. Select the tab *Console*.

The error message is displayed as follows:

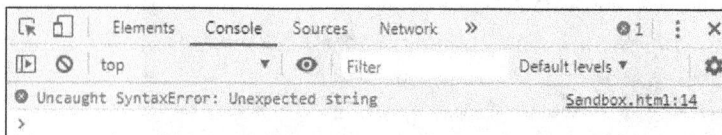

Run-time errors, like an invalid keyword, cause the program to break at the location of the first such error. Note that the statements before the first run-time error will be executed. An example of a run-time error would be to try to call a nonexistent function like `number()` (with a lowercase n). Another example would be to access the value of an undeclared variable without assigning any value to it.

Note that assigning a value to an undeclared variable in regular mode won't raise any error. But it will raise a run-time error in `strict` mode.

Step 7: Test the effect of accessing undeclared variables.

Change the statement in line 12 to use the wrong variable name, `CostPerTicket` *(with a capital "C")*.

```
totalCost = CostPerTicket * tickets;
```

Run the program. You will not be prompted to enter any values because the program breaks at line 12, which accesses an undeclared variable.

Open the Console to see an error message like this.

```
Uncaught ReferenceError: CostPerTicket is not defined
    at Sandbox.html:72
```

Make sure you change the variable name back to `costPerTicket`.

Step 8: Test the effect of an invalid function name.

Change the word `prompt` in line 10 to `Prompt` (with a capital "P") and run the program.

The program will break in line 10 because `Prompt` is an invalid keyword.

Open the Console.

You should see an error message like this:

```
Uncaught ReferenceError: Prompt is not defined
    at Sandbox.html:69
```

Correct the error and run the program again.

If there are more errors, repeat the process until you get the expected results.

Logical errors cause the program to produce incorrect or unintended results. A reservation program that gets the wrong number for cost per ticket from a database or uses the wrong expression to compute the cost are examples of logical errors.

Setting Breakpoints

An effective way to debug logical errors is to set breakpoint(s) in the program, which lets you check the values of variables and continue executing one statement at a time.

By default, you cannot set breakpoints in a script written in an HTML file; you can set breakpoints in a JavaScript (.js) file. To allow setting breakpoints in an HTML file, you change the settings in VS Code as follows.

Step 9: Change the settings to allow breakpoints in an HTML file.

In Visual Studio Code, select File > Preferences > Settings.

Type breakpoint in the Search box to find the Checkbox "Allow setting breakpoint in any file."

Check the Checkbox, as shown in Figure 5-8.

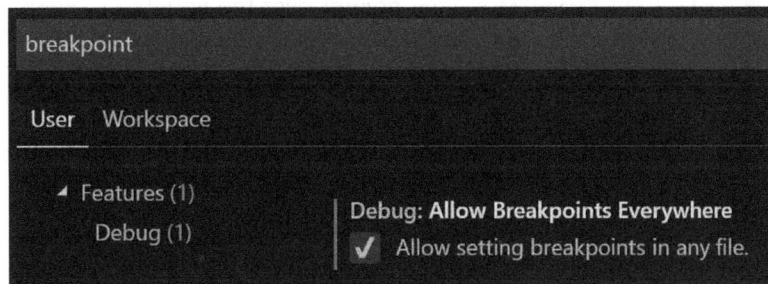

Figure 5-8: Allowing Setting breakpoints

Let's assume that you inadvertently made an error by entering a "+" operator instead of the "*" operator in the expression in line 12, as shown here.

```
totalCost = costPerTicket + tickets;
```

This is a logic error that obviously gives the wrong total cost. To help identify the error using breakpoints, you need to install the debugger for your browser and edit the debug configuration.

Next, you will use breakpoints to examine the values of variables at run time.

Step 10: Create a logic error in the program.

Change the expression in line 12 to add the two values instead of multiplying them, as follows:

```
totalCost = costPerTicket + tickets;
```

Run the program and enter 2 for tickets. The output will show a total cost of 19.5 (17.5 + 2), whereas the expected cost would be 35 (17.5*2).

Step 11: Set breakpoints on lines 12 and 14 by clicking the margin to the left of the line numbers.

You will see a red circle on each of the two lines as shown in Figure 5-9, indicating that the program will stop at each of those lines.

```
11          // Compute total cost
12          totalCost = costPerTicket * tickets;
13          // Display result
14          document.write("Total cost: " + totalCost);
```

Figure 5-9: Breakpoints

Step 12: Run the program in debug mode.

Select Debug > Start debugging (or press F5).

VS Code will display the project folder showing the files within the folder (Figure 5-10).

Select Sandbox.html and enter the number of tickets as prompted.

When the first breakpoint is hit, Visual Studio Code should be brought back to the foreground with the first breakpoint highlighted with a yellow arrow, as follows:

Index of C:\WebPr

🖼 [parent directory]

Name	Size
📁 .vscode/	
Confirmation.html	2.3 kB
Reservation-1.html	3.8 kB
Reservation-2.html	5.1 kB
Reservation-3.html	7.5 kB
Sandbox.html	5.7 kB
Scripts.js	518 B
Styles.css	152 B

Figure 5-10: The WebPojects folder

```
12          totalCost = costPerTicket + tickets;
```

Move the pointer over the variable `tickets` in line 10 to view its value. Move the pointer over `totalCost`. Note that its value is undefined because the program broke before executing this statement.

Click the Continue button (the first one in the figure shown here) to continue the execution of the program.

The program breaks at line 14. Again, move the pointer over `totalCost` in line 12 to view its value.

You may press the Step Over button (the second button) on the Toolbar (or press F10) to execute the current line and move to the next statement.

Press the Stop button to stop debugging.

This process is particularly helpful in larger programs to identify the cause of the error.

Make sure you change the "+" sign to "*" in line 12.

It's time to practice! Do Exercise 5-1 (see the end of the chapter).

Review Questions

4. What is the Camel Case convention to name variables?

5. Which one of these variable names is invalid?

    A. Total-cost

    B. _qty

    C. $expense

    D. Group_1

    E. class2

6. What is the value of a variable that is declared but is not assigned a value?

7. What is the effect of adding the directive `"use strict"` to a program?

8. What would be the output of the following code?

```
var unitPrice = 15, quantity = 2;
totalCost = unitCost*qty;
document.write("Total cost: " + totalCost);
```

Evaluate the following expressions:

9. 10/2+3*4

10. 12-6/2*3+1

11. "2"+6+4

12. A missing closing parenthesis in a program statement is an example of what type of error?

13. What is the effect of a logical error in a program?

5.3 Client-Side Scripting in an HTML Web Form

In this section, you will learn how to use JavaScript in a web form to take actions in response to events like clicking a button or moving focus away from a field. You will also learn how to use JavaScript to get input data from HTML elements, validate data, process data, and display output in HTML elements.

Specifically, you will use JavaScript code in the reservation form to get input data like the number of tickets and cost per ticket from the form elements (instead of using the `prompt` function). You will validate the date and display prices when the user makes selections. In addition, the reservation cost will be computed when the user clicks the *Compute Cost* button. Figure 5-11 shows what the Reservation-2 page would look like.

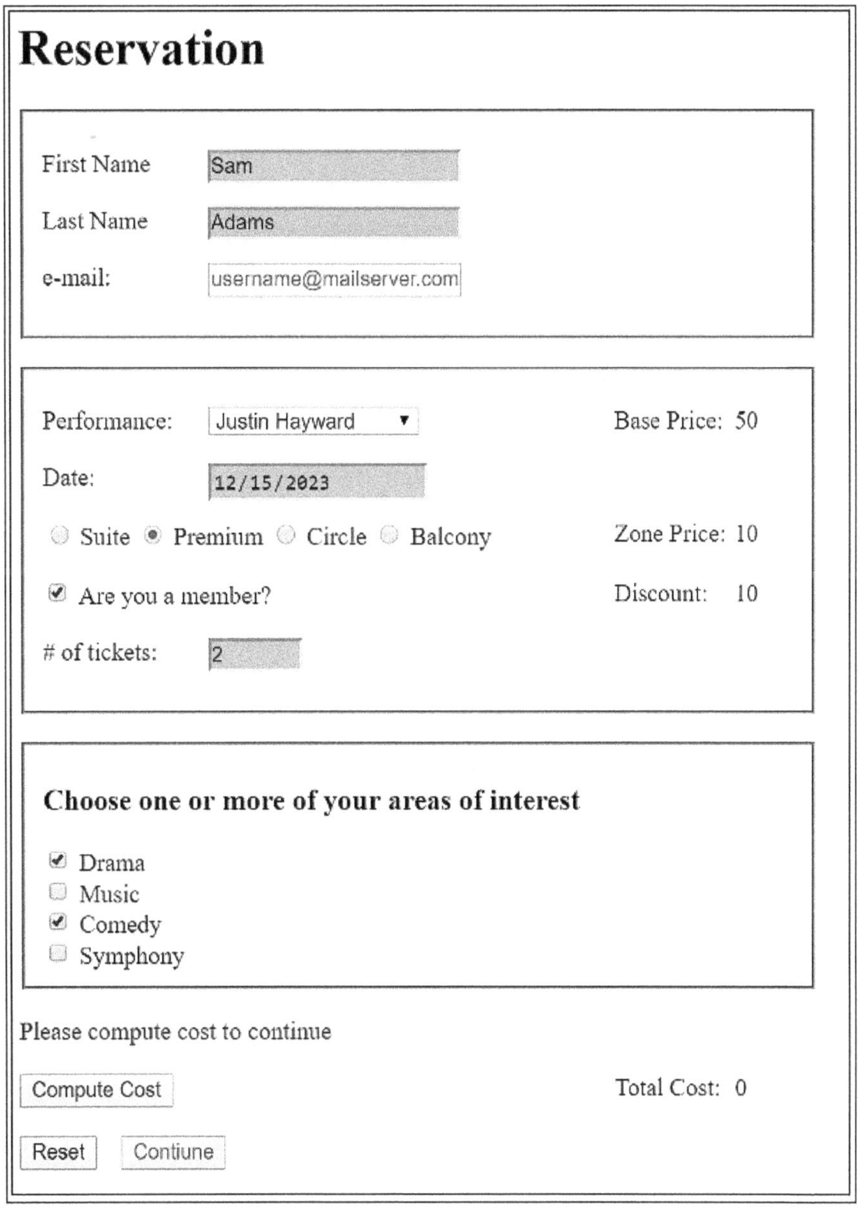

Figure 5-11: Reservation-2 page

Confirm reservation

First Name: Sam
Last Name: Adams
E-mail: null
Performance: Justin Hayward
Date: 2022-11-03
Zone: premium
Tickets: 2
Total Cost: 130.00

| Edit Reservation | Confirm |

Figure 5-12: The confirmation page

On the confirmation page, you will use JavaScript to retrieve reservation data and to let the user go back to the reservation page to make changes, if any. Figure 5-12 shows the confirmation page.

To learn the use of JavaScript in web forms, you will work with a copy of the Reservation-1.html form named **Reservation-2.html**. You will start by creating a JavaScript function to compute the reservation cost. The function will get the input data from the <input> elements on the form and display the output in an HTML element. The function will be called (invoked or executed) when the user clicks the *Continue* button.

5.3.1 JavaScript Functions

A function is just a block of code that does a particular task. It should be noted that a function that is a part of an object is called a **method**. The code within a function can be executed by "calling" or "invoking" the function using its name. The function has the following structure:

```
1  function name(optional parameters separated by commas) {
2     (code to be executed)
3  }
```

A program that calls a function can pass values to it using parameters.

In the following step, you will create a function named ComputeCost using the code you already developed in the Sandbox.html file. The function would resemble Figure 5-13.

```
1  function ComputeCost()
2  {
3     "use strict";
4     // Declare variables to hold data
5     var costPerTicket, tickets, totalCost;
6     // Get input data
7     costPerTicket = 17.5;
8     tickets = prompt("Number of tickets:");
9     // Compute total cost
10    totalCost = costPerTicket * tickets;
11    // Display result
12    document.write("Total cost: " + totalCost.toFixed(2));
13 }
```

Figure 5-13: ComputeCost() function

Note that the statement in line 12 uses the toFixed() function, which formats a number to display a specified number of decimal places.

A function, like ComputeCost(), may be placed in a <script> element within the <head> element, or placed in a separate file and the name of the file specified in the src attribute of the <script> element. For example, if the function is placed within a file named Scripts.js, the <script> element should have the src attribute set to the name of the file as follows:

```
<script src="Scripts.js"></script>
```

When you call a function, JavaScript will look for the function both within the <head> element and the file specified by the src attribute. Initially, you will keep the function in the head section and later move it into a file.

Tutorial: Reservation and confirmation forms

Step 13: Create a copy of Reservation-1.html and name it Reservation-2.html.

Open the Explorer window in VS Code (*View > Explorer*).

Right-click *Reservation-1.html* and select *Copy*.

Right-click within the *HTML-FORMS* window, in the area below the list of files.

Select *Paste*. Rename the new file as *Reservation-2.html*, as shown here.

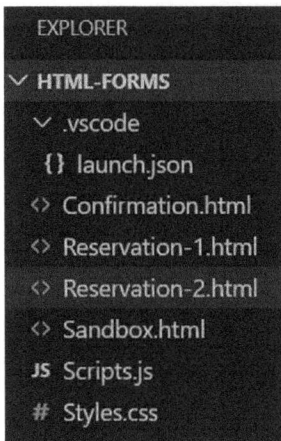

```
EXPLORER

∨ HTML-FORMS
  ∨ .vscode
    {} launch.json
  <> Confirmation.html
  <> Reservation-1.html
  <> Reservation-2.html
  <> Sandbox.html
  JS Scripts.js
  # Styles.css
```

Step 14: Create the function ComputeCost().

Add a `<script>` element within the `<head>` element of Reservation-2.

Insert the function `ComputeCost()` from Figure 5-13 into the <script> element, as shown in Figure 5-14.

```
1   <head>
2       ...
3       <script>
4           function ComputeCost()
5           {
6               "use strict";
7               // Declare variables to hold data
8               var costPerTicket, tickets, totalCost;
9               ...
10              document.write("Total cost: " + totalCost.toFixed(2));
11          }
12      </script>
13  </head>
```

Figure 5-14: ComputeCost function within the HTML document

5.3.2 Accessing HTML Elements Using GetElementById()

Currently, the function `ComputeCost()` gets user input using the `prompt()` function. You will modify the `ComputeCost()` function by deleting the statement involving the `prompt()` function and getting the number of tickets from the `<input>` element on the reservation form. Let's look at how to access HTML elements from JavaScript.

The parts of the reservation form from Chapter 4 that are relevant to the discussion in this chapter are shown in Figure 5-15.

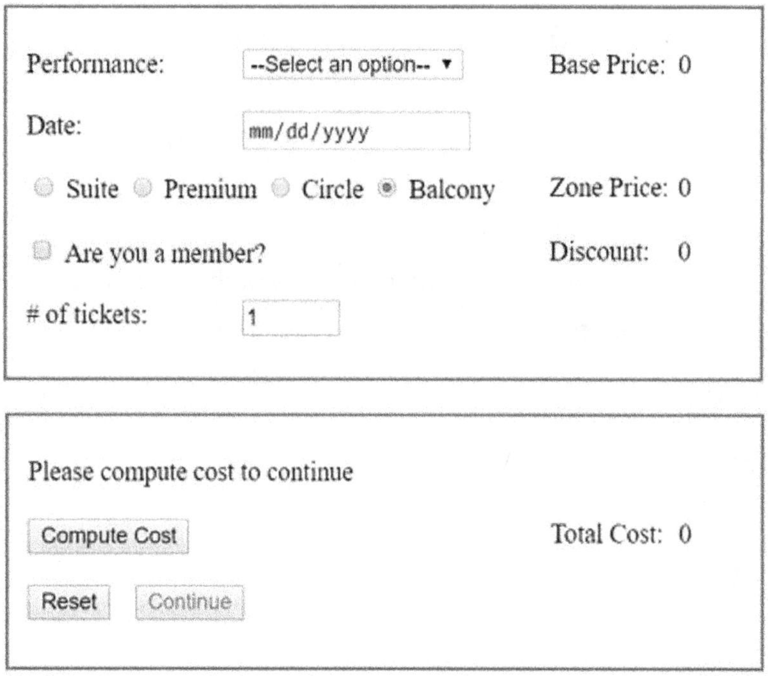

Figure 5-15: Parts of the reservation form

For ease of reference, the HTML code for the input field for the number of tickets, the output field to display total cost, and the <div> elements to display instructions for the user is presented in Figure 5-16.

```
1    <!-- Input field to enter number of tickets -->
2    <label for="tickets"># of tickets:</label>
3    <input type="number" name="tickets" id="tickets" min="1" max="10"
4         step="1" value="1"> <br><br>
5    <!-- Button to compute total cost -->
6    <div id="msg1">Please compute cost to continue</div> <br>
7    <button type="button" id="cost" onclick="ComputeCost()">Compute Cost
8    </button>
9    <!-- Field to display total cost -->
10 <label for="totalCost" class="label"> Total Cost: </label>
11 <output id="totalCost" for="performance zone status tickets"
12        class="output" >0</output> <br><br>
13 <!-- Buttons to reset and continue -->
14 <div id="msg2"></div>
```

Figure 5-16: A segment of the HTML code

You access an element on a web form from JavaScript using the getElementById() method by specifying the id of the element as a string parameter. The function returns the specified element as an *Element* object, if it exists; if not, it returns null. This is how you would get the number of tickets from the <input> element whose id attribute is set to "tickets".

```
var tickets = document.getElementById("tickets").value;
```

This statement gets the *value* of the element whose id is tickets. Note that getElementById() is a method of the global object, document.

In general, you can use the getElementById() method to get the value of any attribute of an element, like its name, type, and class, using the syntax

```
document.getElementById("id of the element").attribute;
```

Changing Attribute Values of an Element

You can also set a value to the attribute of an element using getElementById() method. For example, you can display data in an <output> element by setting its value attribute. The following statement gets the element whose id is "totalCost" and sets its value attribute to the value of the variable totalCost.

```
document.getElementById("totalCost").value = totalCost;
```

In effect, the statement displays the value of totalCost in the <output> element.

Changing the Content of an Element

In addition to getting or changing the attributes of an element, you can use getElementById() to change the content of an element, which is specified between the start and end tags. You use the innerHTML property of the element to get or change the content.

For example, there are two <div> elements on the reservation form to display instructions for the user:

```
<div id="msg1">Please compute cost to continue</div> <br>
...
<div id="msg2"></div>
```

Initially, the first div (id="msg1") instructs the user to compute the cost, and the second div (id="msg2") has no content. After computing the cost, you can remove the first instruction and display an instruction in the second div as follows:

```
1   document.getElementById("msg1").innerHTML = "";
2   document.getElementById("msg2").innerHTML = "Please click Coninue button to proceed";
```

You can get the content of any element using the syntax

```
document.getElementById("id of the element").innerHTML;
```

The statements discussed previously, which use the getElementById() function to get the number of tickets, display the total cost and change the instructions that can be incorporated into the ComputeCost() function, as shown in Figure 5-17.

```
1   <script>
2       function ComputeCost() {
3           "use strict";
4           // Declare variables to hold data
5           var costPerTicket, tickets, totalCost;
6           // Get input data
7           costPerTicket = 17.5;
8           tickets = document.getElementById("tickets").value;
9           // Compute total cost
10          totalCost = costPerTicket * tickets;
11          // Display result
12          document.getElementById("totalCost").value = totalCost.toFixed(2);
13          // Modify the instructions for the user
14          document.getElementById("msg1").innerHTML = "";
15          document.getElementById("msg2").innerHTML =
16                                  "Please click Continue button to proceed";
17      }
18  </script>
```

Figure 5-17: ComputeCost() function using getElementById()

Line 8 gets the number of tickets from the form, line 12 displays the total cost on the form, and lines 14 and 15 change the messages in the two div elements.

5.3.3 Calling a Function to Execute the Code

To call a function by clicking a button, you set the `onclick` attribute of the button to the function name, as shown here.

```
<button type="button" id="cost" onclick="ComputeCost()">Compute Cost</button>
```

The `onclick` attribute specifies that the function `ComputeCost()` is to be called when the `onclick` event of the button takes place—that is, when the user clicks the button.

A function, like `ComputeCost()`, or any code that is executed in response to an event is called an **event handler**. The attribute that specifies the event handler (e.g., `onclick`) is called the **event handler attribute** for the event.

You may specify more than one even handler function for an event, as in

```
onclick="ValidateInput()"; "ComputeCost()";>
```

However, there are special conditions under which the second function may not be called. Instead, you may call a single function that calls other functions.

5.3.4 HTML Events

You can use JavaScript to take actions in response to many other events, like an element getting focus or changing the value of an `<input>` element. Table 5-5 provides some examples of common events and brief descriptions of when they take place.

Table 5-5: HTML events

Event	When it takes place
onblur	When an element loses focus
onchange	For an input element, onchange is fired when the content of the element changes and the element loses focus. For a select element, onchange is fired when the selection changes.
onclick	When the element is clicked
onfocus	When an element gets focus
oninput	Immediately after the value of an `<input>` or `<textarea>` changes
onload	When the browser finishes loading a page
onmouseover	When the mouse is moved over an element

To respond to an event, you use the event handler attribute for that event to specify the event handler. For example, to change the background color of an `<input>` element to light gray when you move the insertion point into it, use the `onfocus` attribute as follows:

```
<input type="number" onfocus="this.style.backgroundColor='lightgrey'">
```

The keyword `this` refers to the current element. Thus the event handler could be a single statement or a function, like `ComputeCost()`.

Step 15: Modify the ComputeCost() function.

Modify the `ComputeCost()` function as shown in Figure 5-17 to get the input data from form elements and display the output in a form element.

Step 16: Call the ComputeCost() function when the Compute Cost button is clicked.

Set the `onclick` event of the button to `ComputeCost()` as discussed previously.

Display the page, enter the number of tickets, and click the button to test the code.

Make sure that the total cost is computed correctly using the number of tickets entered into the input field. Further, the instruction for the user should change after the total cost is computed.

Storing JavaScript in a Separate File

Currently, the Reservation-2.html file contains the JavaScript function in the head section. As the JavaScript code gets larger, it is more convenient to keep it in a separate file and specify the file name in the head section of HTML file.

Step 17: Store the JavaScript in a separate file.

Create a new file named Scripts.js.

Copy the `ComputeCost()` function (without the `<script>` element) from the head section into the Scripts.js file.

Delete the `ComputeCost()` function but, not the `<script>` element, from the head section.

Specify the `src` attribute for the `<script>` element, as shown in line 4 below.

```
1    <head>
2        <title>Reservation</title>
3        <link rel="stylesheet" href="styles.css">
4        <script src="Scripts.js"></script>
```

This code assumes that the Scripts.js file is in the workspace folder (html-forms). If not, you need to specify the path.

Display the page and test to make sure that clicking the button executes the function.

From now on, you will add all JavaScript to the Scripts.js file.

The current `ComputeCost()` function uses a fixed value of 17.5 for cost per ticket. Next, you will learn how to use **conditional statements** and **loops** to compute the **cost per ticket**, which is given by

Base price + Zone price − Member discount

Review Questions

14. Consider the following HTML element:

```
<h2 id="dept"></h2>
```

Which statement would display the heading "Info Tech" in the <h2> element?

    A. `document.getElementById("dept").value = "Info Tech";`
    B. `document.getElementById("dept").innterHtml = "Info Tech";`
    C. `document.getElementById("dept").content = "Info Tech";`

15. What is the difference between the *onchange* and *oninput* events?
16. How would you call a function named Discount() when the following button is clicked?

```
<button type="button" id="cost">Compute Discount</button>
```

5.4 Decision (Selection) Structures

Decision structures let you make the execution of a block of statements conditional. For example, you may want to execute a block statement that computes a student discount for an order—only if the customer is a student. JavaScript has different conditional statements that help you implement decision structures: **if, if-else, if-else-if, nested-if, if-of,** and **switch.**

5.4.1 *if Statement*

The simple `if` statement has the syntax shown in Figure 5-18.

```
1  if (condition){
2      code (one or more statements) to be executed, if condition is true.
3  }
```

Figure 5-18: The simple *if* statement

If there is only one statement to execute, the braces (curly brackets) are optional.

The `if` statement shown in Figure 5-19 computes a discount if the value of the variable `subTotal` is greater than or equal to 100.

```
1  var discount = 0;
2  if (subTotal >= 100){
3      discount = 10;
4  }
```

Figure 5-19: The simple *if* statement to compute discount

In the `if` statement presented in Figure 5-19, the **Boolean expression** (`subTotal>= 100`), which evaluates to true or false, is the *condition*. The expression uses the **relational operator** "`>=`" to compare the value of `subTotal` to the literal 100.

Table 5-6 presents additional relational operators.

Table 5-6: Relational operators

Relational operator	What it means	Example
>	Greater than	`if (roomCapacity>enrollment)`
<	Less than	`if (dateDue<DateTime.Today)`
>=	Greater than or equal to	`if (examScore>=90)`
<=	Less than or equal to	`if (dateOfBirth<=DateTime.Parse("12/31/1980")`

== (combination of two "=" signs)	Equal to	`if (city=="Oshkosh")        // city is string type`
		`if (inState==true)         // inState is Boolean type`
		`// The expression if (inState==true) may be replaced`
		`by if (inState), because inState itself is true or`
		`or false.`

!=	Not equal to	`if (middleInitial!="C")`

5.4.2 *if-else Statement*

The `if-else` statement lets you execute one of two different sets of statements based on the value of a Boolean expression. One set of statements will be executed if the expression evaluates to true, and the other set will be executed if the value is false. It has the syntax shown in Figure 5-20.

```
1   if (condition)
2   {
3       code to be executed, if condition is true
4   }
5   else
6   {
7       code to be executed, if condition is false
8   }
```

Figure 5-20: The *if-else* statement

The code that uses a simple `if` statement in Figure 5-19 may be written using the `if-else` statement as shown in Figure 5-21.

```
1   var discount;
2   if (subTotal >= 100)
3       discount = 10;
4   else
5       discount = 0;
```

Figure 5-21: The *if-else* statement to compute discount

Note that the braces are not used here because there is only a single statement in each block. If braces are not used, by default, the `else` clause includes only the first statement following the word `else`.

Comparing Strings

Just like in multiplication and division, if you mix a number and a string in comparison, the string is converted to a number. The output of the following code would be *true*:

```
1   if (3<"24")
2       document.write("true");
3   else
4       document.write("false");
```

However, if you compare them as strings, as in the next example, the output would be *false*:

```
1   if ("3"<"24")
2       document.write("true");
3   else
4   document.write("false");
```

This is because string comparison starts with comparing the first character of each string, and if they are equal, then compare the second character and so on. Here, 3 is not less than 2; thus the result is false.

Therefore, in order to compare numbers obtained from form fields and `prompt()` functions based on their numerical values, they must be converted from string to number before comparison and addition.

5.4.3 if-else-if Statement

The `if-else-if` statement nests a second `if` statement inside of the `else` portion of `if-else` statements. It has the syntax shown in Figure 5-22.

```
1   if (condition1)
2   {
3        code to be executed if condition1 is true
4   }
5   else if (condition2)
6   {
7        code to be executed if condition1 is false and condition2 is true
8   }
9   else
10  {
11       code to be executed if both condition1 and condition2 are false
12  }
```

Figure 5-22: The *if-else-if* statement

Though the indentation of statements is not essential for the proper execution of the code, it plays a major role in making the code easy to understand. This is particularly important when there are nested else and if blocks. You are strongly encouraged to use consistent indentation.

Let's apply the if-else-if statement to computing discount using the following decision rule.

Condition	Discount
Total cost is 100 or above	10
Total cost is less than 100 but is 50 or above	5
Total cost less than 50	0

The code in Figure 5-23 shows the if-else-if statement to compute discount using the above decision rule.

```
1   var discount;
2   if (subTotal >= 100)
3        discount = 10;
4   else if (subTotal >=50)
5        discount = 5;
6   else
7        discount = 0;
```

Figure 5-23: The *if-else-if* statement to compute discount

Note that the if statement in line 4 of Figure 5-23 doesn't have to check the condition subTotal<100 (in addition to subTotal>=50). This is because line 4 is executed only if the condition in line 2 (subTotal>=100) is false, which means subtotal is less than 100.

Essentially, the if-else-if statement shown above has the effect of nesting an if-else statement within the else clause in line 4.

You may have additional else-if clauses as shown in Figure 5-24, which gives a different discount if total sales are 25 or over but less than 50.

```
1   var discount;
2   if (subTotal >= 100)
3       discount = 10;
4   else if (subTotal >=50)
5       discount = 5;
6   else if (subTotal >=25)
7       discount = 1;
8   else
9       discount = 0;
```

Figure 5-24: The *if-else-if* statement with multiple *else-if* clauses

Having excessively nested `if-else` statements is a sign of poor programming structure and should lead to a restructuring of the code. In some cases, the `switch` statement (discussed next) can be used as a solution to this problem.

5.4.4 switch Statement

The `switch` statement is used to take action based on the value of a variable (test variable) or expression. Figure 5-25 shows the general syntax of the statement.

```
1   switch (test variable or expression)
2   {
3       case value1:
4           (code to be executed, if value1 matches the value of
                the test variable/expression)
5           break;
6       case value2:
7           (code to be executed, if value2 matches the value of
                the test variable/expression)
8           break;
9       case value3:
10          (code to be executed, if value3 matches the value of
                the test variable/expression)
11          break;
12      ...
13      default:
14          (code to be executed, if none of the values match the value of
                the test variable/expression)
15  }
```

Figure 5-25: The switch statement

If the value of the `test` variable/expression matches `value1`, the code associated with that case is executed and the `break` statement causes skipping the rest of the `switch` statement. If it doesn't match, the next case is executed, repeating the process.

A missing `break` statement may produce wrong results. For example, suppose the `break` statement is missing in the first case and the value of the variable matches `value1`. Oddly, the missing `break` statement will cause the code in the next (second) case to be executed, as if the value of the variable matches `value2`.

Let's use the switch statement to compute the base price based on the performance selected by the user from the drop-down list. Figure 5-26 shows the function named `BasePrice()`, which uses the `switch` statement to compute and display the base price. This function will be called from the `ComputeCost()` function to compute the total cost. The `ComputeCost()` function is discussed shortly.

```
1   function BasePrice() {  // Compute base price
2       var basePrice = 0;
3       var performance = document.getElementById("performance").value;
4       switch (performance) {
5           case "Young Irelanders":
6               basePrice = 65;
7               break;
8           case "Justin Hayward":
9               basePrice = 50;
10              break;
11          case "TedxOshkosh":
12              basePrice = 45;
13              break;
14          case "OSO Orchestral":
15              basePrice = 35;
16              break;
17          default:
18              alert("Invalid performance");
19      }
20      document.getElementById("basePrice").innerHTML = basePrice;
21  }
```

Figure 5-26: Code to compute the base price

In line 3, `document.getElementById("performance").value` gets the value of the `<select>` element, which gives the performance name selected by the user. The selected performance name is stored in the variable `performance`, whose value will determine which part of the switch statement is executed.

Lines 4–19 use the `switch` statement to compute the base price. If the value of performance matches the value specified in the first `case` ("Young Irelanders"), the code associated with that `case` is executed and the `break` statement skips the rest of the `switch` statement. If it doesn't match, the next `case` is executed and the process is repeated.

If the break statement is missing in the first *case* (line 7), for example, and you select `Young Irelanders`, the code in the next `case` will be executed and `basePrice` will be set to 50.

If none of the values matches the value of the test variable, then the statement in the `default` case is executed. This would be the case if the user doesn't select any performance.

Line 20 displays the base price in the `<output>` element for base price.

Step 18: Create the function BasePrice().

Add the code from Figure 5-26 to create the function `BasePrice()` below the `ComputeCost()` function within the Scripts.js file.

Calling the BasePrice() Function

You will invoke `BasePrice()` whenever the user selects a performance from the drop-down list. This is done by setting the `onchange` event handler attribute of the `<select>` element to `BasePrice()`, as shown here:

```
<select name="performance" id="performance" onchange="BasePrice()">
```

Step 19: Specify the onchange attribute for the <select> element.

Set the `onchange` attribute of the `<select>` element for the drop-down list to `BasePrice()` as shown above.

Display the reservation form and select a performance.

The `<output>` element should display the corresponding base price. In addition, the base price is stored in the global variable, `basePrice`, to be used in `ComputePrice()`.

To see the effect of a missing break statement, comment out the break statement in the first *case*. Display the page and select `Young Irelanders`. Instead of the correct base price of $65, the code displays the value $50 by executing the code in the second `case`.

For now, the base prices for different performances are hard-coded. Later, in Chapters 8 and 17, you will learn how to get the base prices from a database.

Next, you will create the `Discount()` function on your own to compute the discount when the user clicks the checkbox.

Do Steps 20 and 21.

Step 20: Create a function to compute the discount.

Create a function called `Discount()`. Use an `if` statement to compute a $10 discount if the `checked` property of the checkbox is true, and set it to $0 if it is false. Display it in the corresponding `<output>` field.

Step 21: Call the function.

Specify the `onclick` attribute of the checkbox to invoke the `Discount()` function when the user clicks the checkbox.

It's time to practice! Do Exercise 5-2 (see the end of the chapter).

Computing the Cost Per Ticket

The current `ComputeCost()` function, presented earlier in Figure 5-17, uses a fixed value of 17.5 for cost per ticket, as shown in Figure 5-27.

```
1   function ComputeCost() {
2       "use strict";
3       // Declare variables to hold data
4       var costPerTicket, tickets, totalCost;
5       // Get input data
6       costPerTicket = 17.5;
7       tickets = document.getElementById("tickets").value;
8       // Compute total cost
9       totalCost = costPerTicket * tickets
10      ...
```

Figure 5-27: The ComputeCost() function

Next, you will compute the cost per ticket using the expression

Base price + Zone price – Member discount

You already created the functions to compute the base price and discount. The zone price is computed using the zone selected by the user from the radio button group. Identifying the radio button selected by the user involves using an iterative process using a *loop* to check every button. Before we discuss the loop structure, let's update the ComputeCost() function so that it uses the base price and discount computed by the BasePrice() and Discount() functions.

To compute the cost per ticket, the ComputeCost function needs to access the prices and discount computed within the above three functions. Next, you will learn how to do it.

5.4.5 Sharing Data between Functions within a Page

It is often required to share the value of a variable between functions within the same page. For example, the base price will be computed in the function, BasePrice(), and used in ComputeCost(). A common method to share data between functions within the same page is using **global variables**, which are variables declared outside the functions.

Global variables, however, do not retain their values after a page refresh. In larger applications, it may be necessary to retain values when a page is reloaded. For example, you may want to restore the value of an element after a page reloads by storing the value in a variable. In such cases, you can store data in **sessionStorage**, **localStorage**, or **hidden field**, which survive page refreshes.

First, we will use global variables and then briefly describe the other methods.

Using Global Variables to Share Data

You will use global variables to store the base price, zone price, and discount, which are computed in three different functions, and share them with ComputeCost() function, where they are used to compute the total cost.

Step 22: Declare global variables to store the base price, zone price, and discount.

Declare global variables outside the ComputeCost() function as shown here:

```
1      // Declare global variables to hold prices
2      var basePrice = 0, zonePrice = 0, discount = 0;
3      function ComputeCost() {
```

Delete the statement var basePrice = 0 from the BasePrice() function (line 2, Figure 5-26).

Delete the statement that declares zonePrice within the Discount() function that you created in Step 20.

It should be noted that global variables will lose their values if the form is posted to the server. The use of JavaScript's sessionStorage, which survives a page refresh, is discussed shortly.

Next, you will update the ComputeCost() function to compute the cost per ticket using the base price, discount, and zone price computed in the functions.

Step 23: Modify the ComputeCost() function to compute cost per ticket using basePrice, zonePrice, and discount.

Make the change shown in line 10 of Figure 5-28. In addition, enable the Continue button, as shown in lines 19.

```
1   // Declare global variables to hold prices
2   var basePrice = 0, zonePrice = 0, discount = 0;
3   function ComputeCost() {
4       "use strict";
5       // Declare variables to hold data
6       var costPerTicket, tickets, totalCost;
7       // Get number of tickets
8       tickets = document.getElementById("tickets").valueAsNumber;
9       // Compute total cost
10      costPerTicket = basePrice + zonePrice - discount;
11      totalCost = costPerTicket * tickets;
12      // Display result
13      document.getElementById("totalCost").innerHTML = totalCost.toFixed(2);
14      // Modify instructions for the user
15      document.getElementById("msg1").innerHTML = "";
16      document.getElementById("msg2").innerHTML =
17                              "Plese click Continue button to proceed";
18      // Enable Continue button
19      document.getElementById("continue").disabled = false;
20  }
```

Figure 5-28: Modified ComputeCost() function

To test the code, display the reservation page, and select a performance, click the checkbox for member status, and enter the number of tickets.

Click the Compute Cost button. Make sure the correct total cost is displayed. Because you haven't created the function to compute the zone price, the zone price would be zero.

sessionStorage

A second way to share data within a page is using the `sessionStorage` object, which can be accessed using the `sessionStorage` property of the `window` object. It should be noted that the JavaScript `sessionStorage` available on the client is different from the ASP.NET Session State, which is available on the server.

The `sessionStorage` property lets you store data in the browser, which lasts for an entire session. Thus, unlike global variables, the value stored in the `sessionStorage` survives a page refresh. Data can be stored only as strings, making it necessary to convert other types of data.

The syntax for storing data in `sessionStorage` is:

```
sessionStorage.key = value
```

or,

```
sessionStorage.setItem("key", "value")   // Note that "s" in setItem is lower case.
```

For example, this is how you store a discount of 10 in a key named `Discount`:

```
sessionStorage.Discount = 10;
```

or,

```
sessionStorage.setItem("Discount", "10");
```

Similarly, you can store the value of the variable `discount` in a key named `Discount` as follows:

```
sessionStorage.Discount = discount;
```

or,

```
sessionStorage.setItem("Discount", discount);
```

The syntax to get the value from the `sessionStorage` is:

```
sessionStorage.getItem("key")
```

or,

```
sessionStorage.key
```

For example, this is how you get the value from the key, `Discount`, and store it in the variable `amt`:

```
var amt = (Number) sessionStorage.Discount;
```

or

```
var amt = (Number) sessionStorage.getItem("Discount");
```

Unlike cookies, the data stored in `sessionStorage` is not sent to the server with each request.

Let's look at how you can use `sessionStorage`, instead of global variable, to share the base price between the functions `BasePrice()` and `ComputeCost()`. First, you add a statement at the end of the `BasePrice()` function to store the value of the `basePrice` varible in `sessionStorage`, as follows:

```
sessionStorage.setItem("BasePrice", zonePrice);
```

Here, `BasePrice` is the key. Because you are using `sessionStorage`, this assumes that the variables are not declared as global variables.

You can get the base price from `sessionStorage` and store it in the local variable, `basePrice`, within the `ComputeCost()` function, as follows:

```
var zonePrice = Number(sessionStorage.getItem("ZonePrice"));
```

The same approach can be used to share the zone price and discount. Because there is no need to retain the values of the variables after a page refresh, you will not use `sessionState` in the current application.

localStorage

Another method of sharing data between client-side functions is using `localStorage` property, which has the same syntax as the `sessionStorage`, but the data persists until deleted.

Hidden Fields (<input type="hidden">)

Hidden fields, like all HTML fields and global variables, are reset when a page is refreshed.

5.4.6 Logical Operators

You may use multiple Boolean expressions for the condition in an `if` statement. Logical operators are used to combine Boolean expressions. Table 5-7 shows a list of logical operators. The current application doesn't use logical operators. Short examples of their use are provided in the table.

Table 5-7: Logical operators

Logical operator	Name	Example	What it means
&&	Conditional And	if((ACT>=20) && (GPA>=2.5)) The second Boolean expression is evaluated only if the first one is true.	If both expressions are true, "&&" returns true; if not, "&&" returns false. That is, true &&true yields true; true &&false yields false; false &&true yields false; false &&false yields false.
&	And	if((ACT>=20) & (GPA >=2.5)) Both expressions are always evaluated.	Same as "&" (i.e., if both expressions are true, "&" returns true; if not, "&" returns false).
||	Conditional Or	if ((ACT <20) || (GPA <2.5)) The second Boolean expression is evaluated only if the first one is false.	If either expression is true, "||" returns true; if not, "||" returns false. That is, true |false yields true; false |true yields true; true |true yields true; false |true yields false.
|	Or	if ((ACT <20) | (GPA <2.5)) Both expressions are always evaluated.	Same as "|" (i.e., if either expression is true, "|" returns true; if not, "|" returns false).
!	Not	if (!(ACT >= 20))	if (ACT >= 20), "!" returns false; if not, "!" returns true.

Review Questions

17. What would be the value of discount after executing the following program?

```
var totalCost = 120;
var discount;
if (totalCost >= 100)
    discount = 10;
else if (totalCost >= 50)
        discount = 5;
    else
        discount = 0;
```

18. What would be the output of the following code segment?

```
var zone = "Premium";
var price = 0;
switch (zone) {
    case "Suite":
        price = 25;
        break;
    case "Premium":
        price = 20;
        break;
    case "Circle":
        price = 10;
        break;
    default:
        alert("Invalid performance");
}
    document.write("Zone price: " + price);
```

5.5 Iteration Structures: Loops

Loops are used to process the same set of statements multiple times. For example, computing the payroll may involve reading and processing hundreds of similar records using the same set of statements. JavaScript uses the same iteration structures as many other languages: while, do-while, and the for loop. In the following section, you will learn how to use them.

Each of the three loop structures allows for repeating a set of statements a certain number of times, but each has a different mechanism for controlling the number of times the statements are repeated.

5.5.1 The while Loop

A commonly used form of loop is the while loop. This type of loop repeatedly executes a set of statements while a condition (Boolean expression) is true. It has the syntax shown in Figure 5-29.

```
1   while (condition) {
2       // statement(s) to be executed
3   }
```

Figure 5-29: The *while* loop

The code within the loop is executed as long as the condition is true.

The example in Figure 5-30 uses the while loop to prompt the user to enter a password, verifies it, and lets the user proceed only if the password is correct. In this example, the password is hardcoded to the value "guest," so the user will be prompted to enter a password until they enter "guest."

```
1   <script>
2       var password = null;
3       while (password != "guest") {
4           password = prompt("Please enter a valid password");
5       }
6       document.write("You may proceed");
7   </script>
```

Figure 5-30: Example of the *while* loop

The very first time through the code, the Boolean expression in line 3 (password != "guest") is *true*; therefore, the statement within the loop (line 4) is executed. If the user enters anything other than "guest," the Boolean expression (password != "guest") will be true and the loop will be executed again to ask for a password.

If the user enters the valid password (guest), the Boolean expression is false; therefore, the loop is not executed anymore. The statement following the loop (line 6) is executed to display the message.

TRY IT

Comment out the statements in the `<script>` element in Sandbox.html.

Type in the above code within the body element in the Sandbox.html file.

Display the page and test the code with invalid and valid passwords.

You also can limit the number of times a while loop is executed by using a counter variable to count the number of times the loop is processed, as shown in Figure 5-31.

```
1  <script>
2      var password;
3      var count = 0;
4      while (password != "guest") {
5          count = count + 1;
6          if (count > 3){
7              alert("sorry, only three attempts allowed");
8              exit;
9          }
10             password = prompt("Please enter password");
11     }
12     document.write("You may proceed");
13 </script>
```

Figure 5-31: The *while loop* with a counter

Line 5 increments the value of count by 1 each time the loop is executed. Line 6 checks the value of count, and if it exceeds 3, line 7 displays a message using the alert() function.

Figure 5-32 is another example that uses the while loop to ask the user to enter three numbers and calculate the average.

```
1  <script>
2      var count = 0;
3      var sum = 0;
4      while (count < 3){
5          count = count + 1;
6          sum = sum + Number(prompt("Enter number " + count + ":"));
7      }
8      document.write("Sum: " + sum);
9      document.write("<br>Average: " + sum/3);
10 </script>
```

Figure 5-32: A second example of the *while* loop with a counter

The first time through the loop, count will have the value 1, and line 6 displays the following prompt:

```
Enter number 1:
```

The prompt is formed by concatenating the string Enter number with the value of the variable count (1) and the colon (":").

The value entered by the user is converted to a number and added to the sum.

The second time, the value of count is 2, resulting in the prompt Enter number 2:, and the third time, the prompt would be Enter number: 3.

You may enter the code into Sandbox.html and try it.

5.5.2 Incrementing and Decrementing Variables

Line 5 in the previous example uses the following statement to increment the value of count by 1:

```
count = count + 1;
```

The ++ and -- Operators

In place of count = count + 1, you may use the shorter alternate forms using the ++ or += operators:

```
count++;
count += 1;
```

To decrement a variable by 1, replace the "+" sign with the "-" sign, as in

```
count = count - 1;
count--;
count -= 1;
```

You may replace the statement in line 5 with its other two versions in the sandbox file and test the code.

5.5.3 The do-while Loop

This is a variation of the while loop. The do-while loop checks the condition at the end of the loop so that the loop is always executed the first time, as shown here:

```
1  do {
2      // statements to be executed
3  }
4  while (condition)
```

Figure 5-33 uses the do-while loop to verify the password.

```
1  <script>
2      var password;
3      do{
4          password = prompt("Please enter password");
5      }
6      while (password != "guest");
7      document.write("You may proceed");
8  </script>
```

Figure 5-33: The *do-while* loop to verify the password

If the user enters the wrong password, the Boolean expression (password != "guest") will be true and the loop will be executed again to ask for a password.

5.5.4 The for Loop

The for loop is designed to perform a fixed number of iterations, determined by the value of a counter variable. The counter variable is initialized to a certain specified value and incremented/decremented at the end of each iteration by a specified value. It has the syntax shown in Figure 5-34.

```
1  for (initialization statement; Boolean expression, increment statement) {
2       // statements to be executed
3  }
```

Figure 5-34: The *for* loop

Figure 5-35 is an example that uses a for loop to compute the value of an investment at the end of each year for 30 years at a growth rate of 5% if a certain fixed amount is invested at the beginning of each year.

```
1  <script>
2      var yearlyInvestment = prompt("Enter yearly investment");
3      var yearlyInvestment = Number(yearlyInvestment);
4      var percentGrowth = 5, investmentValue = 0;
5      for (let year = 1; year <= 30; year++) {
6          investmentValue = (investmentValue + yearlyInvestment) *
7              (1 + percentGrowth/ 100);
8          document.write("Year: " + year +
9              " Investment value: " + investmentValue.toFixed(2) + "<br>");
10     }
11 </script>
```

Figure 5-35: Application of the *for* loop

The **initialization statement** (let year = 1) in line 5 that sets the counter variable to 1 is executed only once when the loop is processed.

Declaration using let: Note that the variable is declared using the keyword *let* because it is needed only within the loop. It is a good practice to limit the scope of a variable to where it is needed to reduce the chances of inadvertently changing its value from another place.

The **Boolean expression** (year <= 30) is evaluated at the beginning of each iteration through the loop. If it is true, the statements in the loop are executed; if not, processing of the loop stops and the statement following the loop is executed.

The **increment/decrement statement** (year++) increments the year by 1 at the end of each iteration through the loop.

Note that the prompt() function in line 2 returns the number that you type in as a string. Line 3 converts this string to a number using the Number() function. If you don't, the statement in line 6 will concatenate the string with the value of investmentValue.

Line 8 uses the toFixed() function to format the investment value to display 2 decimal places. The statement displays the year and investment value with the labels "Year:" and "Investment value:".

Because the document.write() statement is within the loop, it will display the year and investment value each time through the loop. If you want to display only the final values, you will place the statement outside the loop.

TRY IT

Type in the code to compute the investment value, and test the code as follows:

Comment out any JavaScript code in the Sandbox.html file and type in the above code.

Display the page and enter a value of 1000 for the yearly investment.

You should get an investment value of 69760.79 at the end of 30 years.

Next, test the effect of mixing strings and numbers:

Change the Boolean expression to `year <= 2`.

Comment out the statement in line 3 that converts the yearly investment to a number.

Display the page and enter a value of 1000 for yearly investment.

You should get an investment value of 11026050.00 at the end of two years, which is clearly wrong.

The yearly investment of 1000 was returned by the `prompt()` function as a string, which caused it to be concatenated to the investment value.

Test the effect of placing the `document.write()` statement outside the loop.

Uncomment line 3. Change the Boolean expression back to `year<=30`.

Remove the statement from line 8, and place it after the closing brace "}" of the loop.

Run the code. The output should consist of only the final values of year and investment value.

To summarize, `while` and `do-while`-loops are similar, except that the `while` loop checks the condition at the beginning of the loop, whereas the `do-while` loop checks the condition at the end of the loop. That means the while loop won't be executed even a single time if the condition is not met. The `do-while` loop is always executed the first time.

The `for` loop performs a fixed number of iterations, determined by the value of a counter variable that is incremented/decremented at the end of each iteration. The loop terminates when a condition involving the counter variable is met.

It's time to practice! Do Exercise 5-3 (see the end of the chapter).

Review Questions

19. What would be the output of the following program?

```
var totalSavings = 0, year = 0;
while (year < 3)
    {
        year = year + 1;
        totalSavings = totalSavings + 1000;
    }
document.write("Total Savings: " + totalSavings);
```

20. What would be the output of the following program?

```
 var population = 70000, year;
for (year = 2021; year <= 2023; year = year + 1)
{
        population = population + 1000;
}
document.write("Projected Population: " + population);
```

5.6 Using Arrays and Maps to Store Multiple Values

The variables we have used so far can store only one value at a time. Arrays and Maps let you store multiple values of the same type or different types, in a single variable. Typically, you store values in a database or file and read them into an array or Map in the memory for faster access. However, to keep it simple, here we will use hard-coded data.

5.6.1 Array

Create an array using the following syntax:

> var *array name* = [*value-1, value-2, value-3, …*];

or,

> var *array name* = new Array(*value-1, value-2, value-3, …*);

For example, you can store the names of all seating zones in an array as follows:

```
var zones = ["suite", "premium", "circle", "balcony"];
```

or,

```
var zones = new Array("suite", "premium", "circle", "balcony");
```

The first statement is a shorter version of the second statement, which creates the `Array` object more explicitly.

Each element in an array has an element number (index) associated with it. The index starts with 0 and increments by one for each successive element, as shown here:

Element	"suite"	"premium"	"circle"	"balcony"
Index	0	1	2	3

You can access an individual element using its index. This is how you access the third element (index 2) and store it in a variable:

```
var zone = zones[2];  // zone will have the value "circle"
```

Similarly, you can change the value of an element using its index. The following statement changes the value of the fourth element to "gallery" from "balcony."

```
zones[3] = "gallery";
```

You can access all elements of the array by changing the index from 0 to 3 using a `for` loop, as follows:

```
1  for(var i = 0; i < zones.length; i++)
2      document.write(zones[i] + "</br>");
```

The `for` loop here uses the `length` property of the array to get the number of elements (4) in the array. The output will appear as follows:

```
suite
premium
circle
balcony
```

You can use a pair of arrays together when you want to store two sets of related data, where one set consists of keys and the other set consists of corresponding values. For example, along with the zones array that stores the zone names, you may use a second array named `prices` to store the zone prices, as shown here:

```
var prices = [40,25,10,0];
```

In this array, the first element (40) is the price of the first element (suite) in the zones array, the second element (25) is the price of the second element (premium) in the zones array, and so on.

To find the price for a specific zone, you can use the indexOf method to find the index of the zone and use the index to find the price, as follows:

```
1  var index = zones.indexOf("circle");
2  var price = prices[index];
3  document.write("price for circle: " + price);
```

TRY IT

Use the code presented above to create the two arrays and find the price for a specific zone. You may use the Sandbox.html file to run the code.

The Array object has several methods to work with arrays. Examples of the methods are shown in Table 5-8.

Table 5-8: Methods of the Array object

Method	Example
Concat()	zones-1.concat(zones-2)—Concatenates the arrays, zones-1, and zones-2
sort()	zones.sort()—Sorts the array
toString()	zones.toString()—Returns a comma-separated string.

When you want to store a set of keys and corresponding values, it is more convenient to use a Map object, rather than a pair of arrays.

5.6.2 The Map Object

The Map object lets you store a set of key-value pairs. This is the syntax to create a Map object and store multiple key-value pairs in it:

var map name = new Map([[key-1, value-1], [key-2, value-2], [key-3, value-3], …])

The keys in a Map object could be of different types. Similarly, the values also could be of different types. This is how you create a Map object and store the zone names and prices in it.

```
var zonePrices = new Map([["suite", 40], ["premium", 25], ["circle", 15],["balcony", 0]]);
```

Each key-value pair is represented like an array within a pair of square brackets. Note that the entire set of key-value pairs is within another set of square brackets, thus creating the structure of an array of arrays.

Earlier, you saw how we first found the index of a value in one array and used that to look up a corresponding value in a second array. A Map makes those operations simpler by combining them into a single statement. When you use a Map, you can find the value corresponding to a key using the get method by specifying the key. The following statement gets and displays the price for the "premium" zone:

```
1  document.write(zonePrices.get("premium"));
```

TRY IT

Create a Map object named zonePrices as shown above and use it to find and display the price for a specific zone.

You can add a key-value pair to the Map object using its `set` method. This is how you add the price for a fifth zone named "gallery":

```
1  zonePrices.set("gallery", 5);
```

TRY IT

Add the key-value pair for the balcony and display the price for the balcony.

When all data in a `Map` object or array is to be accessed, it is more convenient to use the `for-of` loop, instead of the `for` loop, as explained next.

5.6.3 Iterating over Items in a Collection Using the for-of Loop

The `for-of` loop lets you iterate over items in a collection, including Maps, arrays, and strings. The following code gets each element from the `zones` array and displays it:

```
1  for (let zone of zones)
2      document.write(zone + "</br>");
```

Note that the variable zone is declared using the keyword `let` because the zone is used only locally within the `for-of` loop.

Similarly, you can use the `for-of` loop to iterate over the key-value pairs in a map, as follows

```
1  for (let [zone, price] of zonePrices)
2      document.write( zone + ": " + price + "<br>");
```

Each iteration of the `for-of` loop gets a key-value pair from the Map object and stores it in the variables `zone` and `price`.

It's time to practice! Do Exercises 5-4 and 5-5 (see the end of the chapter).

5.6.4 Objects

You have been using objects like `Map`, `Array`, and `document`, which are available in JavaScript. You also can create your own objects to make it easier to work with a group of data items and to make code reusable. Objects and their applications are discussed in detail in Appendix D* on object-oriented programming, which is an extensively used methodology in software development. This brief section gives you a peek at a JavaScript object.

An object has a set of properties representing multiple related data items (e.g., first name, last name, e-mail, phone). Each property has a name and a value (e.g., first name: Jane; phone: 920-424-9999).

An object also can have methods that perform actions on the object. For example, a customer object can have a method to get the full name by combining the first and last names.

You create an object by specifying the names and values of properties and methods, as shown in Figure 5-36.

```
1  var cust =  {id: 334455, fName: "Han", lName: "Solo",
2              email: "hansolo@starmail.com", phone: "920-424-9999",
3              fullName : function(){
4                  return this.fName + " " + this.lName; }
5              };
```

Figure 5-36: Creating an object

Here, id, fName, lName, email, and phone are names of properties, and fullName is a method (function) that combines the last and first names.

You may get the value of any property by specifying the name of the property, as in

---

* Appendix can be freely downloaded from https://www.prospectpressvt.com/textbooks/philip-web-development at "Student Resources."

```
document.write(cust.email);
```

The syntax to access a property is

Object name.property name

5.6.5 Using the for loop to Find the Selected Zone

Now that you understand how loops work, you will create the **ZonePrice()** function, which uses a loop to compute the zone price whenever the user clicks a radio button in the group. To find the zone price, you first need to identify which button was clicked and then use the value of the button to determine the price.

To identify the selected radio button, examine the checked property of each button. If the checked property of a button is true, then the value property of the button gives the selected zone. Recall that all radio buttons in the group have the name zone (but each button has a different id). You can get the entire set of radio buttons as a single collection using the getElementsByName() function.

The following statement would get all radio buttons that have the name zone and store the collection in a variable named radioList:

```
var radioList = document.getElementsByName("zone");
```

Each button in the radioList collection is identified by a different index, starting with the value zero. Thus the first button is referred to as radioList[0], the second button radioList [1], the third one radioList [2], and so on.

You can access each button in the radioList collection using a for-of loop, as shown in lines 5–7 in Figure 5-37.

The if statement in line 6 and 7 gets the value of the button, if it is checked.

```
1   function ZonePrice() {
2       // Find selected zone
3       var radioList, button, zone;
4       radioList = document.getElementsByName("zone");
5       for (button of radioList) {
6           if (button.checked)
7               zone = button.value;
8       }
9       // Compute zone price
10      switch (zone) {
11          ...
12          ...
13      }
14      // statement to display zone price in the output field, goes here
15  }
```

Figure 5-37: ZonePrice() function using a for--of loop

It's time to practice! Do Step 24.

Step 24: Compute the zone price.

Create a function named ZonePrice() in the Scripts.js file.

Add the code from Figure 5-37 to the function. Fill in the statements in the switch statement in the ZonePrice() function to compute zone price based on the zone selected. Use the following zone prices:

Zone	"suite"	"premium"	"circle"	"balcony"
Zone price	40	25	10	0

The function should save the zone price in the global variable `zonePrice`. If no zone is selected, display the alert "Invalid zone."

Step 25: Display the page and test the code.

The total cost should include the computed zone price.

5.6.6 Working with Dates

The `Date` type in JavaScript can be used to store a date along with a time. The `Date` type is an object. You can create a `Date` object from a date using the `new` keyword, as shown here.

```
var performDate = new Date("12/14/2022");
```

The statement converts the string "12/14/2022" to a date and stores the date in the variable `performDate`.

You can use a similar statement to get the performance date that you entered into the `<input type="date">` field on the form:

```
var performDate = new Date(document.getElementById("performDate").value);
```

Here you use the value property of the element to get the performance date as a string. The string is converted to a `Date` type object using `new Date()` and stored in the variable, `performDate`. You will use this statement to get the date and validate it when you click the Continue button.

If you don't specify a parameter value for the `Date()` function, then `new Date()` will return today's date. If you don't specify the time, zeros are used for the hour, minute, second, and millisecond. You may specify the time as shown here:

```
"12/14/2022 2:30"
```

Because the `Date` type is an object, it has its own built-in methods, like those shown in Table 5-9.

Table 5-9: Methods of the Date object

Method	Description
`getDate()`	Returns the day of the month as a number, 1–31
`getDay()`	Returns the day of the week as a number, 0–6
`getFullYear()`	Returns the year
`getMonth()`	Returns the month as a number, 0–11
`toDateString()`	Converts the date portion of a `Date` object into a string
`toString()`	Converts a `Date` object to a string

You can display a date using the `toDateString()` method as follows:

```
document.write(performDate.toDateString());
```

Validating the Performance Date Using the onblur Event

The performance date selected by the user must be the current date or a future date. You can use JavaScript to do the validation using the `onblur` event, which is fired when the field loses focus. First, specify the required code in a function, as shown in Figure 5-38.

```
1   function ValidateDate(){
2       var performDate = new Date(document.getElementById("performDate").value);
3       if (performDate < new Date())
4           alert("Date must be current or future date");
5   }
```

Figure 5-38: Function to validate performance data

Line 3 compares the performance date with today's date, given by `new Date()`.

Next, to call the function when the user enters a date and move focus away from the field, set the `onblur` attribute of the date field to the function, as follows:

```
1   <label for="performDate">Date:</label>
2       <input type="date" name="performDate" id="performDate" required
            onblur = "ValidateDate()"><br><br>
```

Step 26: Create the ValidateDate() function.

Add the function to the Scripts.js file.

Specify the `onblur` event handler for the date field.

Display the form and enter a past performance date, and press the tab key to move focus away from the field.

You should see the message "Date must be current or future date."

Note that the date field has the attribute `required`, which checks whether a date is entered at the time of submission of the form. You may wonder why you need this attribute if you validate the field using the `ValidateDate()` function. You still need the `required` attribute because validation using the `onblur` event will take place only if you move focus away from the field **after you enter or change data** in it. So, if you don't specify the `required` attribute and move focus away from the date field without entering a date, no client-side validation will take place because you haven't made any changes to the data.

TRY IT

Remove the *required* attribute from the `<input type="date">` element.

Display the reservation form, and click inside the date field. If you move focus away from the field, you won't get a validation message ("Date must be current or future date").

Thus you still need the `required` attribute, even with the validation for past dates.

Make sure you add the `required` attribute back.

Review Questions

21. What would be the output of the following program?

```
var quarterlySales = [125, 110, 120, 150];
document.write(quarterlySales[2]);
```

22. What would be the output of the following program?

```
var quarterlySales = [125, 110, 120, 150];
for (let sale of quarterlySales)
        document.write(sale + "</br>");
```

23. Consider the array created by the following statement:

    ```
    var examGrades = [90,82,94,88,77];
    ```

    Write the statement to change the last grade from 77 to 79.

24. Write the statement that would display which day of the week (as a number, 1–7) is the date stored in the performDate variable.

    ```
    var performDate = new Date("11/14/2022");
    ```

25. What would be the value of testDate after execution of the following statement?

    ```
    var testDate = new Date();
    ```

5.7 Validating Data: Checking for Missing Values

For input fields like `<input type="text">`, `<input type="number">`, and `<input type="date">`, where the user is expected to type in a value rather than select a value, you can specify the required attribute to make sure the user enters a value before submitting a form.

However, for a `<select>` element, the first item in the list is selected by default, which makes it difficult to check whether the user intended to select that item or just overlooked the field. To address this problem, you can add a prompt as the first one, which displays text such as "Select an option." With this option, if the selectedIndex of the `<select>` element is zero, that means the user didn't make a selection.

Lines 4–6 of the function shown in Figure 5-39, ValidData(), would check whether the user selected a performance from the drop-down list (line 3) and display a message if the user didn't.

```
1   function ValidData(){
2       let index,  radioList, isButtonSelected;
3       // Check whether a performance is selected
4       index = document.getElementById("performance").selectedIndex;
5       if (index==0)
6           alert("Please select a performance");
7       // Check whether a zone is selected
8       radioList = document.getElementsByName("zone");
9       isButtonSelected = false;
10      for (var i = 0; i < radioList.length; i++) {
11          if (radioList[i].checked)
12              isButtonSelected = true;
13      }
14      if (isButtonSelected==false){
15          alert("Please select a zone") ;
16      }
17      if ((index==0) || (isButtonSelected==false))
18          return false;
19      else
20          return true;
21 }
```

Figure 5-39: ValidData() function

Lines 8–15 check whether the user selected a zone. This code is similar to the one used in Figure 5-37 to identify the checked radio button. Line 8 gets the set of radio buttons and stores the collection in the variable radioList. The loop in line 10 checks whether any of the buttons is checked.

If no performance is selected (index==0) or if no zone is selected (isButtonSelected==false), lines 17–20 specify that the function returns false; if not, it returns true. You will use the value returned by the function (true or false) to cancel the submission of the form.

Step 27: Add the function ValidData() to the Scripts.js file.

Next, you will use the value returned by the function (`true` or `false`) in the `onsubmit` event of the form to cancel the submission of the form, if the value is false.

5.7.1 Canceling the Submission of a Form

Clicking the Continue button (`<button type="submit">`) would cause the form to be submitted. But you can cancel the submission of the form by specifying the event handler *return false* for the event attribute `onsubmit` of the form as follows:

```
onsubmit="return false"
```

In the current form, you will use the value returned by the `ValidData()` function in place of the literal value `false` to cancel submission only when `ValidDate()` returns false (i.e., no performance is selected). The `<form>` element with the `onsubmit` attribute is shown here.

```
<form action="Confirmation.html" id="reservation" onsubmit="return ValidData()" method="post">
```

Step 28: Modify the <form> element by specifying the onsubmit attribute.

Specify the `onsubmit` attribute as shown above.

Display the page.

To test the code, enter the date and select a zone, but don't select any performance.

Click the Compute Cost button.

Click the Continue button. The confirmation page won't be displayed, indicating the form was not submitted.

Validating Numeric Fields

The input field for the number of tickets has the attributes, min, max, and step that specify that if a value is entered into the field, it must be integers between 1 and 10. The field will allow only values between 1 and 10. However, if the user doesn't enter a value, these attributes have no effect. If the user clicks the Compute Cost button without entering a value for number of tickets, the output field for the total cost will display "NaN" because the value of the tickets variable will be "NaN."

Exercise 5-6 addresses this problem by adding the necessary code in the `ComputeCost()` function to make sure that the number of tickets is not "NaN."

It's time to practice! Do Exercise 5-6. (See the end of the chapter.)

Inconsistency between Input Data and Output

If a user computes the total cost and then changes the input data, the input and output would become inconsistent. One way to address this problem is to automatically compute and update the total cost whenever the user changes any of the input data. Exercises 5-7 and 5-8 at the end of the chapter use this approach.

Another approach is to disable the Continue button when the user changes any input data and enable it when the total cost is computed. Use jQuery in Exercises 5-10 and 5-11 for this approach.

5.7.2 Resetting a Form

Clicking the Reset button on the reservation form resets only the input fields, not the output fields. We will use a jQuery method to reset the output fields.

The complete code for the reservation page, including the jQuery from the next section, is provided at the end of the chapter.

It's time to practice! Do Exercise 5-7 or Exercise 5-8 (see the end of the chapter). You may also do E-commerce Assignment Part B or Auto Rental Assignment Part B.

5.8 Sharing Data with Another Web Page

Confirm reservation

First Name: Sam
Last Name: Adams
E-mail: null
Performance: Justin Hayward
Date: 2022-11-03
Zone: premium
Tickets: 2
Total Cost: 130.00

[Edit Reservation]   [Confirm]

Figure 5-40: The confirmation page

Sharing data between web pages is integral to most websites. On an e-commerce website, for example, information on products you select on different pages appears in a shopping cart. In a similar manner, the current application needs to share data between the reservation page and the confirmation page that displays the reservation data as shown in Figure 5-40.

There are multiple methods to share data. The more relevant ones are shown in Table 5-10.

It should be noted that the query string and the HTTP request send only input data, not the data displayed in `<output>` fields. To send data that is displayed in an `<output>` field (e.g., the total cost), you can store it in a hidden input field so that it will be sent with the query string.

Sending data by a query string using the `get` method and retrieving the data from the query string is relatively simple. However, as stated in Chapter 4, data shared through query strings can be tampered with by users. To demonstrate the use of query strings as a quick-and-dirty method, we will use the `get` method on the reservation form. Using the `post` method is discussed in Chapters 16 and 17.

Table 5-10: Approaches to sharing data between pages

Method	Description
`get` **method that uses a query string**	If you set the `method` attribute of a form element to `get`, the browser will send the input data to the page specified in the `action` attribute by appending the data to the URL as a query string. A drawback of using a query string is that it is not suitable for sending sensitive data.
`post` **method that uses an HTTP request**	If you set the `method` attribute to `post`, the input data is sent by adding it to the body of an HTTP request, making it more suitable for sensitive and larger sets of data. However, retrieving the values is not as simple as with a query string.
Session variables	Session variables store data in the server memory. Therefore, retrieving the data involves using server-side code. Session variables are discussed in more detail and used in Chapters 9 and 17.
`localStorage` **and** `sessionStorage` **objects**	You can use the `sessionStorage` object, which lasts only for a session, or the `localStorage` object, which spans multiple sessions, to save data in the browser as key-value pairs and retrieve them from any page. Similar to using a query string, these methods are not suitable for sharing sensitive data.
Hidden fields	The hidden field (`<input type="hidden">`) is an input field that is not visible to the user when the form is displayed in a browser.

An alternative to using query strings is to store each data item in the `sessionStorage`, described in Table 5-10, and access it from the confirmation page. The query string would be more convenient in the current example, because the data you enter into input fields on a form is automatically sent to the target page using a query string if the `method` attribute of the form is set to `get`.

Let's look at how to retrieve the data from a query string on the page that receives it.

5.8.1 Retrieving Data from a Query String and Displaying Them

The process of getting the variables (called *URL parameters* or *query string parameters*) and their values passed through a query string involves two steps:

1. Get the entire query string, including the initial "?" mark and the "&" that separates the variables, as a single string. The complete query string is given by the `search` property of the `location` object.
2. Get the values of individual variables using the `URLSearchParams` object.

Location Search Property

The `search` property of the `location` (`window.location`) object returns the entire query string as a single string. For example, the following statement would get the query string and store it in the variable `queryString`.

```
var queryString = location.search;
```

For ease of formatting, we will display the query string within a paragraph element using its `innerHTML` property. The code in Figure 5-41 specifies a `<p>` element and displays the query string within the paragraph.

```
1   <body>
2       <h2>Confirm reservation</h2>
3       <p id="qs"></p>
4       <script>
5         // Get the entire query string
6         var queryString = location.search;
7         document.getElementById("qs").innerHTML="Query String:" + "<br>"+queryString;
8       </script>
9   </body>
```

Figure 5-41: Code to display the query string

Step 29: Specify the style for a paragraph element.

Add the code shown in Figure 5-42 within the `<head>` element of the confirmation page.

```
1   <style>
2           P {
3                   margin-left: 40px; font-size: 15px;
4           }
5   </style>
```

Figure 5-42: The style element for a paragraph

Step 30: Make sure that the value of the *method* attribute on the reservation form is *get*, as shown here.

```
<form action="Confirmation.html" id="reservation" method="get">
```

Step 31: Create an HTML file named *Confirmation.html*.

Change the title of the page to *Confirmation*.

Step 32: Display the query string within the paragraph.

Type in the code from Figure 5-41 into the `<body>` element of the Confirmation.html page.

To test, display the reservation form and enter/select the input data.

Click the Continue button.

The confirmation page should display a query string that looks similar to the one shown here, except for the border.

Confirm reservation

Query String:
?fName=Sam&lName=Ada&emailAddress=&performance=Justin+Hayward&performDate=2022-02-22&zone=premium&memeberStatus=member&tickets=2&area=music&area=comedy&totalCost=130.00

Delete line 7, which is not needed for the rest of the code.

URLSearchParams Object

The URLSeachParams object has various methods for working with the query string. You can create the URL-SearchParams object using the syntax shown here:

```
var urlParams = new URLSearchParams(query string);
```

To get the value for any specific variable (key), use the get method as follows:

```
URLParams.get ('variable');
```

The code in Figure 5-43 creates the URLSearchParams object using the query string (line 5) and gets the value for the variable fName (line 7).

```
1   <script>
2       // Get the entire query string
3       var queryString = location.search;
4       // Create URLSearchParams object using the query string
5       var urlParams = new URLSearchParams(queryString);
6       // Get the first name
7       var firstName = urlParams.get('fName');
8       document.getElementById("qs").innerHTML = "First Name: " + firstName;
9   </script>
```

Figure 5-43: Creating the URLSearchParams object

Step 33: Get the first name from the query string and display it.

Type in lines 4–8 from Figure 5-43 into the `<script>` element, as shown in Figure 5-41.

Display the reservation page.

Enter/select the first name, date, and number of tickets, to start.

Click Continue. The confirmation page should display the first name.

Delete lines 6–8 of Figure 5-43.

In order to display all data items from the query string, we will concatenate the data items into a single string and display it within the paragraph element. Figure 5-44 shows how to concatenate the data items from the query string with line breaks between them.

```
1   // Get each data item and concatenate them together to form a single string
2   var resData = "";
3   resData += "First Name: " + urlParams.get('fName') + "<br>";
4   resData += "Last Name: " + urlParams.get('lName') + "<br>";
5   resData += "E-mail: " + urlParams.get('email') + "<br>";
6   resData += "Performance: " + urlParams.get('performance') + "<br>";
7   resData += "Date: " + urlParams.get('performDate') + "<br>";
8   resData += "Zone: " + urlParams.get('zone') + "<br>";
9   resData += "Tickets: " + urlParams.get('tickets') + "<br>";
10  resData += "Total Cost: " + urlParams.get('totalCost') + "<br>";
11  // Display the concatenated string in the <P> element
12  document.getElementById("qs").innerHTML = resData;
```

Figure 5-44: Code to display items from the query string

Starting with an empty string (restData) in line 2, the value of each data item and its caption is concatenated to the string in successive lines. Line 12 displays the string within the paragraph. This is a quick-and-dirty method for displaying the items. Later, in Chapters 16–18, you will display the items in a tabular format.

Step 34: Display the data items from the query string within the paragraph.

Add the code presented above to the existing code in the `<script>` element.

Display the reservation page and enter/select the input data.

Click the Compute Cost button.

Click the Continue button.

The complete set of data should be displayed as shown here, except for the border.

Confirm reservation

First Name: Sam
Last Name: Adams
E-mail: null
Performance: Justin Hayward
Date: 2022-02-22
Zone: premium
Tickets: 2
Total Cost: 130.00

It's time to practice! Do Exercise 5-9 (see the end of the chapter).

Review Questions

26. What is the drawback of using the query string to send data from one page to another?
27. What is the advantage of using the post method to send data from one page to another?

5.9 Displaying One Page from Another

In an HTML form, like the reservation form, the action attribute specifies the URL of the page to be displayed. However, pages like the confirmation page may not have a form. In such pages, you can use the window.history object or the window.location object to display another page.

5.9.1 Window.history Object

The `window.history` object contains information on the browser's history of pages visited. You can use this object to go back to the previous page or forward to the next page in the history of pages, using the `history.back()` function and `history.forward()` function. For example, the following statement on the confirmation page would let the user go back to the reservation page:

```
history.back();
```

With this method, the reservation page is not reloaded. The `history.back()` function has the same effect as hitting the back button on the browser. Therefore, the form fields will retain all data previously entered/selected by the user, allowing the user to make changes to the data. However, the dynamically generated contents like the content of the output fields and the message in the `<div id="msg2">` element are not retained. You will have to use other methods like `sessionStorage` to store that content and re-display it. You will learn more about `sessionStorage` later, in Chapter 16.

To call the `history.back()` function, create a button on the confirmation page as shown here:

```
1   <!-- Add a button to let the user go back and edit reservation-->
2   <button type="button" onclick="history.back()">Edit Reservation</button>    
```

Step 35: Create a button to let the user go back to the reservation page.

Add the code to create the button immediately below the end tag `</script>`.

Display the reservation page and select/enter the input data.

Click the Continue button to display the confirmation page.

Click the Edit Reservation button to go back to the reservation page.

The reservation page should show the selections that you made previously.

The `history.back()` function displays the reservation form, which becomes the previous page, because the confirmation page is always displayed from the reservation page. This won't work if the user has the option to open other pages from the confirmation page and come back to it. In such a scenario, `history.back()` will display whatever page was the previous page.

If you want to redirect the browser to a specific page, you can use the `window.location` object.

5.9.2 Redirect to a Page Using the Anchor (<a>) Element

You may use the anchor element on the confirmation page, as shown here, to let the user go back to the reservation form:

```
<a href="Reservation-2.html"> <input type="button" value="Edit Reservation"/>  </a>
```

Note that the anchor element's content is a button, instead of text, so that the link looks like a button. In this approach, the reservation page is reloaded and doesn't retain the data previously entered by the user. You can restore the data by storing it in a dictionary like `sessionStorage`, retrieving it when the reservation page is redisplayed.

TRY IT

Comment out the HTML code for the *Edit Reservation* button, and add the anchor element in its place. Display the confirmation page from the reservation page. Click the link to redisplay the reservation page. The page should be displayed in its original state, with no data displayed.

Anchor Element versus window.location Object to Redirect to a Page

An alternative to using the anchor element to redirect to a page is to use the `window.location` object. To do this, create a JavaScript function that sets the `href` property of the location object to the URL of the target page, as shown in Figure 5-45.

```
1   <script>
2       function GoToReservation() {
3           location.href = "reservation-2.html";
4       }
5   </script>
```

Figure 5-45: The window.location object

You can invoke this function using the `onclick()` event handler of a button, as shown here.

```
<button type="button" onclick= "GoToReservation()"> Edit Reservation </button>
```

In general, the anchor element should be used for links whenever possible because it provides several advantages over JavaScript code:

- It works even when JavaScript is disabled.
- It is easily followed by search engines.
- The destination page is displayed by moving the mouse over the link.
- Links are identified by screen readers.

The Confirm Button

To let the user confirm the reservation, add a second button to the confirmation page, which, in this project, will just display a message that the reservation is confirmed.

Step 36: Add a button and <div> element to confirm the reservation.

Add the code shown in Figure 5-46 immediately below the end tag `</script>`.

```
1   <!-- Add a button and a <div> element to display confirmation message -->
2   <button type="button" onclick= "DisplayMessage()"> Confirm </button> <br>
3   <div id="confirm"> </div>
```

Figure 5-46: The Confirm button

The button specified in line 2 would call a function named `DisplayMessage()`, which you will create in the next step.

Step 37: Create the function DisplayMessage().

Add the code shown in Figure 5-47 within the `head` element of the confirmation page.

```
1   <script>
2       function DisplayMessage(){
3           document.getElementById("confirm").innerHTML="Reservation confirmed"
4       }
5   </script>
```

Figure 5-47: The DisplayMessage() function

Test the code by displaying the confirmation page from the reservation page.

Click the Confirm button. The `<div>` element should display the message.

It's time to practice! Do E-Commerce Assignment Part C or Auto Rental Assignment Part C (see the end of the chapter).

Review Questions

28. Consider two different methods to go back to a page that the user visited: (1) use the `history.back()` method or (2) use the anchor element that specifies the page. What effects do the two methods have on the page displayed?

29. Consider two different methods to redirect to a page: (1) use the anchor element or (2) use JavaScript to set the `location.href` property to the page to be displayed. What are the advantages of using the anchor element?

5.10 A Brief Introduction to jQuery

jQuery is a library of JavaScript functions that perform many common tasks. By calling a jQuery function, you can avoid writing multiple lines of code or simplify a JavaScript statement. Before you can use jQuery, you have to either download the jQuery library into the project folder or use it from a content delivery network (CDN) like Google or Microsoft. To use it from Google, add the following code to the head section of the web page:

```
<script src="https://ajax.googleapis.com/ajax/libs/jquery/3.4.1/jquery.min.js"></script>
```

To use jQuery from Microsoft, add the following code:

```
<script src="https://ajax.aspnetcdn.com/ajax/jQuery/jquery-3.4.1.min.js"></script>
```

Essentially, jQuery methods help you perform an action (e.g., change the background color or content) on a group of selected elements (e.g., all input fields or headings). The general syntax is:

$(*selector*).*action*

The jQuery selector, similar to CSS selectors, selects one or more elements using the name, id, type, class, attribute, and so forth. The action is performed by jQuery methods. The types of actions performed by jQuery methods include:

1. Add/change HTML elements and attributes, like adding an `<output>` element or changing the value or background color of an `<input>` element.

2. Find HTML elements based on their position relative to other elements, like finding a child element or a sibling. A child element is an element within an element, like a `<p>` element within a `<div>` element. Two elements are siblings if both are within a third element and one is not a child of the other.

3. Produce effects, like animating, fading, or hiding.

4. Provide AJAX (Asynchronous JavaScript and XML) functionality. AJAX, as described in Chapter 14, lets you update parts of a web page without reloading it.

Let's look at three commonly used methods from the first group: `text()`, `html()`, and `val()`.

5.10.1 The text() Method

The text() method sets or returns the text content of the selected element(s). Consider the commented-out JavaScript statement in line 1 of Figure 5-48, which is used in the ComputeCost() function to change the content of the empty <div> element (id="msg2") to the string "Please click Continue…".

```
1   //document.getElementById("msg2").innerHTML="Please click Continue button to proceed";
2   $("#msg2").text("Please click Continue button to proceed");
```

Figure 5-48: The text() method

You can replace the statement in line 1 with the jQuery shown in line 2. In jQuery, "#msg2" selects the element that has the id "msg2", and text() is the action (jQuery method) that sets the content of the element to the string "Please click…".

Step 38: Use jQuery to add content to the empty <div> element (id="msg2").

First, add the <script> element within the head section to use jQuery from Google or Microsoft, as described earlier.

Next, comment out the statement shown in line 1 (from the ComputeCost() function), and add the statement shown in line 2, as shown here.

```
1   //document.getElementById("msg2").innerHTML= "Please click Continue button to proceed";
2   $("#msg2").text("Please click Continue button to proceed");
```

Display the reservation form and click the Compute Cost button.

The <div> element should display the message specified in jQuery.

5.10.2 The html() Method

This method is similar to the text() method, except that you can use HTML code instead of just text. For example, to add HTML code to display the message in bold, you can use the html() method as follows:

```
$("#msg2").html("<strong>Please click Coninue button</strong>");
```

This statement sets the content of the <div> element to the html element <strong>, to show the message in bold. The equivalent JavaScript would be:

```
document.getElementById("msg2").innerHTML =
            "<strong>Please click Continue button to proceed</strong>";
```

Step 39: Change the text() function you added in the previous step to an html() function.

Change the jQuery as shown here:

```
$("#msg2").html("<strong>Please click Continue button</strong>");
```

Display the reservation page and click the Compute Cost button.

You should see the message in bold.

5.10.3 The val() Function

The val() function sets or returns the value of form fields. Consider the following input field on the reservation form:

```
<input type="number" name="tickets" id="tickets" ... value="1" >
```

This is how you would change the value to zero:

```
$("#tickets").val("0");
```

5.10.4 Selecting Multiple Elements

You may use a jQuery method to perform an action on multiple elements. For example, you may select all <output> elements, all <input type="number"> elements, or all elements with the same value for the class or name attribute. This is how you reset the text content of every element whose class attribute is set to "output":

```
$("output").text("0");
```

Let's use this method to address the issue that the Reset button doesn't reset <output> elements. You may display the reservation form, enter some data, click the Compute Cost button, and then click the Reset button to observe that the output fields are not reset.

To address this issue, create a function named ResetOutput and call the function from the onreset event attribute of the form element. The function is shown in Figure 5-49.

```
1  <script>
2      function DisplayMessage(){
3          document.getElementById("confirm").innerHTML = "Reservation confirmed";
4      }
5  </script>
```

Figure 5-49: The ReserOutput() function

Step 40: Create the function ResetOutput.

Use the code presented above to create the `ResetOutput()` function within the Scripts.js file.

Step 41: Call ResetOutput when the form is reset.

Specify the `onreset` event attribute for the form element, as follows:

```
<form action="Confirmation.html" id="reservation" onsubmit="return ValidData()"
        method="get" onreset="ResetOutput()">
```

Display the form, select a performance, a zone, and click the Compute Cost button.

Click the Reset button. The output fields should be reset to zeros.

5.10.5 Adding Elements

In addition to changing the attributes of elements, you also can add new elements using jQuery methods like `after()` and `before()`.

For example, instead of creating an empty <div> element initially and adding content later, you can add the <div> element and the content when the user clicks the Compute Cost button. This is how you can use jQuery to add the <div> element immediately following the <output> element (#totalCost):

```
$("#totalCost").after("<br><br><div id='msg2'>Please click Continue button to proceed</div>")
```

TRY IT

Use jQuery to add a `<div>` element and its content.

Comment out the `<div>` element from the Reservation-2 file:

```
<!-- <div id="msg2"></div> -->
```

Comment out the jQuery you added in the previous step in the `ComputeCost()` function, and add a new statement as shown in line 3 of Figure 5-50.

```
1   //document.getElementById("msg2").innerHTML= "Please click Continue button to proceed";
2   //$("#msg2").append("Please click Continue button to proceed");
3   $("#totalCost").after("<br><br><div id='msg2'>Please click Continue button
4   to proceed</div>")
```

Figure 5-50: jQuery for the <div> element

Display the reservation form, and click the Compute Cost button.

The `<div>` element should display the message specified in the jQuery.

5.10.6 Event Methods

jQuery has event methods, like `click()`, `focus()`, `change()`, and `submit()`, that allow you to take action in response to events. Suppose you want to show a light-green background color for `<input>` fields whose values were changed by the user. You can use the `change()` event method to do it, as follows:

```
1   $("input").change(function () {
2       $(this).css("background-color", "lightgreen");
3   });
```

The statement in line 1 selects all `<input>` elements, and you can use the `change()` method to specify that the jQuery within the method (to change the color) is to be executed when the change event is fired. In line 2, the `css()` method is used to change the background color. The `css()` method can be used to specify styles, in general.

Note that what is passed to the event method, `change()`, is the function:

```
function () {
        $(this).css("background-color", "lightgreen");
    }
```

The syntax of event methods is as follows:

```
$(selector).eventmethod (function(){code})
```

Where do you place this jQuery code? This code can be placed in a `<script>` element in the head section. However, you want the code to be executed only after all elements have been loaded. To make sure all elements have been loaded, place the code inside the `document.ready()` event method, as shown in Figure 5-51.

```
1   <script>
2       $(document).ready(function(){
3           $("input").change(function () {
4               $(this).css("background-color", "lightgreen");
5           });
6       });
7   </script>
```

Figure 5-51: jQuery change() method

The statement `(document).ready` is optional. Thus you may replace line 1 by

$(function(){

Step 42: Use jQuery to change the background color when values of input fields are changed.

Add the code presented above to the `head` section.

Display the form and enter the input data. The background color of input fields should change when you move focus away after changing the values.

It's time to practice! Do Exercises 5-10 and 5-11. (See the end of the chapter.)

Figure 5-52 shows the complete HTML code for the Reservation-2 page.

```
1    !DOCTYPE html>
2    <html lang="en">
3    <head>
4        <title>Reservation-2</title>
5         <script src="Scripts.js"></script>
6        <link rel="stylesheet" href="styles.css">
7        <script src="https://ajax.aspnetcdn.com/ajax/jQuery/jquery-3.4.1.min.js">
         </script>
8        <script>
9            $(document).ready(function(){
10               $("input").change(function () {
11                   $(this).css("background-color", "lightgreen");
12               });
13           });
14       </script>
15
16   </head>
17
18   <body>
19       <h1>Reservation</h1>
20       <form action="confirmation.html" id="reservation"  method="get"
             onsubmit="return ValidData()" onreset="ResetOutput()">>
21           <!-- Input fields to enter personal information-->
22           <fieldset width="400px"><br>
23               <label for="fName">First Name</label>
24                   <input type="text" name="fName" id="fName"> <br><br>
25               <label for="lName">Last Name</label>
26                   <input type="text" name="lName" id="lName"><br><br>
27               <label for="emailAddress">e-mail:</label>
28                   <input type="email" name="emailAddress" id="emailAddress"
                         placeholder="username@mailserver.com"
                         a. pattern=".+@prospectpressvt.com"

                         title="The domain must be prospectpressvt.com"> <br><br>
29           </fieldset><br>
30
31           <fieldset><br>
32               <!-- Dropdown list to select a performance-->
33               <label for="performance">Performance:</label>
34               <select name="performance" id="performance" onchange="BasePrice()">
35                   <option >--Select an option--</option>
36                   <option value="Young Irelanders">Young Irelanders</option>
37                   <option value="Justin Hayward">Justin Hayward</option>
38                   <option value="TedxOshkosh">TedxOshkosh</option>
39                   <option value="OSO Orchestral">OSO Orchestral</option>
```

Figure 5-52: Complete HTML code for Reservation-2 (*continues*)

```
40              </select>
41              <!--Output element to dislay base price-->
42              <label for="basePrice" class="label"> Base Price: </label>
43              <output id="basePrice"for="performance"class="output">0</output><br><br>
44              <!--input field to enter date-->
45              <label for="performDate">Date:</label>
46              <input type="date" name="performDate" id="performDate" required> <br><br>
47              <!-- Radio buttons to select a zone-->
48              <input type="radio" name="zone" id="suite" value="suite"
                       onchange="ZonePrice()">
49                  <label for="suite">Suite</label>
50              <input type="radio" name="zone" id="premium" value="premium"
                       onclick="ZonePrice()">
51                  <label for="premium">Premium</label>
52              <input type="radio" name="zone" id="circle" value="circle"
                       onclick="ZonePrice()">
53                  <label for="circle">Circle</label>
54              <input type="radio" name="zone" id="balcony" value="balcony"
                       onclick="ZonePrice()">
55                  <label for="balcony">Balcony</label>
56              <!--Output element to dislay the zone price-->
57              <label for="zonePrice" class="label"> Zone Price: </label>
58              <output id="zonePrice" for="suite premium circle balcony"
                       class="output" >0</output><br> <br>
59
60              <!-- Checkbox to specify member status -->
61              <input type="checkbox" name="memeberStatus" id="status"
                       onclick="Discount()">
62                  <label for="status">Are you a member?</label>
63              <!--Output element to dislay the discount-->
64              <label for="discount" class="label"> Discount: </label>
65              <output id="discount" for="status" class="output" >0</output><br> <br>
66
67              <!-- Input field to enter number of tickets -->
68              <label for="tickets"># of tickets:</label>
69              <input type="number" name="tickets" id="tickets" min="1"max="10" step="1"
70                       onfocus="this.style.backgroundColor='lightgrey'"> <br><br>
71          </fieldset><br>
72          <!-- Checkboxes to enter preferences -->
73          <fieldset>
74              <h3>Choose one or more of your areas of interest</h3>
75              <input type="checkbox" name="area" id="drama" value="drama">
76              <label for="drama">Drama</label><br>
77              <input type="checkbox" name="area" id="music" value="music">
78              <label for="music">Music</label><br>
79              <input type="checkbox" name="area" id="comedy" value="comedy">
80              <label for="comedy">Comedy</label><br>
81              <input type="checkbox" name="area" id="symphony" value="symphony">
82              <label for="symphony">Symphony</label>
83              </fieldset><br>
84
85          <!-- Button to compute total cost -->
86          <div id="msg1">Please compute cost to continue</div> <br>
87          <button type="button" id="cost" onclick="ComputeCost()">Compute Cost</button>
88          <!-- Field to display total cost -->
89          <label for="totalCost" class="label"> Total Cost: </label>
90          <output id="totalCost" class="output" for="performance zone status tickets">0
                       </output> <br><br>
91          <input type="hidden" name="hiddenTotalCost" id="hiddenTotalCost" value="0">
92
93          <!-- Buttons to reset and continue -->
94          <div id="msg2"></div>
95          <button type="reset" id="reset"> Reset </button>
96          <button type="submit" id="continue" disabled=true> Contiune </button>
97      </form>
98  </body>
99  </html>
```

Figure 5-53 shows the complete set of JavaScript functions within the **Scripts.js** folder.

```
1    // Declare globale variables to hold data
2    var basePrice=0, zonePrice=0,  discount=0;
3
4    function ComputeCost() {
5        "use strict";
6        // Declare variables to hold data
7        var costPerTicket, tickets, totalCost;
8        // Get number of tickets
9        tickets = document.getElementById("tickets").valueAsNumber;
10       // Compute total cost
11       costPerTicket = basePrice + zonePrice - discount;
12       totalCost = costPerTicket * tickets;
13       // Display result
14       document.getElementById("totalCost").innerHTML = totalCost.toFixed(2);
15       // Modify instructions for the user
16       document.getElementById("msg1").innerHTML = "";
17       //document.getElementById("msg2").innerHTML = "Plese click Continue button to proceed";
18       $("#msg2").text("Please click Continue button to proceed"); // Alternative using jQuery
19       $("#msg2").html("<strong>Please click Continue button</strong>"); //Uses html() method to apply boldface
20       // Enable Continue button
21       document.getElementById("continue").disabled = false;
22       document.getElementById("hiddenTotalCost").value=totalCost.toFixed(2);
23   }
24
25   function BasePrice(){
26       // Compute base price
27       var performance = document.getElementById("performance").value;
28       switch (performance) {
29           case "Young Irelanders":
30               basePrice = 65;
31               break;
32           case "Justin Hayward":
33               basePrice = 50;
34               break;
35           case "TedxOshkosh":
36               basePrice = 45;
37               break;
38           case "Oshkosh Orchestral":
39               basePrice = 35;
40               break;
41           default:
42               alert("invalid performance");
43       }
44       document.getElementById("basePrice").innerHTML=basePrice;
45   }
46
47   function ZonePrice(){
48       // Find selected zone
49       var radioList, button, zone;
50       var radioList = document.getElementsByName("zone");
51       for (button of radioList){
52           if (button.checked)
53               zone = button.value;
54       }
55       // Compute zone price
56       switch (zone) {
57           case "suite":
```

Figure 5-53: JavaScript functions within the Scripts.js folder (*continues*)

```
58              zonePrice = 40;
59              break;
60          case "premium":
61              zonePrice = 25;
62              break;
63          case "circle":
64              zonePrice = 10
65              break;
66          case "balcony":
67              zonePrice = 0;
68              break;
69          default:
70              alert("invalid zone");
71      }
72      document.getElementById("zonePrice").innerHTML=zonePrice;
73 }
74
75 function Discount(){
76      // Compute member discount
77      var memberStatus = document.getElementById("status").checked;
78      if (memberStatus)
79          discount = 10;
80      else
81          discount = 0;
82      document.getElementById("discount").innerHTML = discount;
83 }
84
85 function ValidateDate(){
86      var performDate = new Date(document.getElementById("performDate").value);
87      if (performDate < new Date())
88          alert("Date must be current or future date");
89 }
90
91 function ValidData(){
92      let index,  radioList, isButtonSelected;
93      // Check whether a performance is selected
94      index = document.getElementById("performance").selectedIndex;
95      if (index==0)
96          alert("Please select a performance");
97      // Check whether a zone is selected
98      radioList = document.getElementsByName("zone");
99      isButtonSelected = false;
100     for (var i = 0; i < radioList.length; i++) {
101         if (radioList[i].checked)
102             isButtonSelected = true;
103     }
104     if (isButtonSelected==false){
105         alert("Please select a zone") ;
106     }
107     if ((index==0) || (isButtonSelected==false))
108         return false;
109     else
110         return true;
111 }
112
113 function ResetOutput(){
114     $(".output").text("0");
115 }
```

Exercises

Exercise 5-1: Sales Tax

Create a program in Sandbox.html that performs the following tasks:

Prompt the user to enter the total sales and the sales tax rate as a percent.

Compute the sales tax (=total sales*tax rate/100) and the amount due (=total sales+sales tax).

Display the total sales and amount due on separate lines, with proper labels.

Exercise 5-2: Room Rate

A hotel chain offers special discounts for different discount categories. The following discounts are applied to the standard rate to compute the discounted price:

Senior	20%
Govt	15%
AAA	10%

Use the Sandbox.html file to develop a form that lets the user enter the standard rate and the discount category into two different input fields. Provide a button to compute the discounted price using the `switch` statement and display it with a label. Display two decimal places.

Exercise 5-3: AEV Prices

All-electric vehicle (AEV) prices are expected to come down by 10% per year. Develop a form that lets the user enter the current price and a target price that is lower than the current price. Provide a button that uses a `while` loop to compute and display the number of years it takes for the price to fall below the target price. Provide two input fields to enter the current and target prices. Display the number of years in an output field with a label.

Provide a second button that uses a `for` loop to compute and display the expected price after three years. Use the current price to compute the expected price.

Exercise 5-4: Array and for-of Loop

Create an array and store the monthly sales for each of the 12 months, in an array named `monthlySales`. Use a `for-of` loop to compute and display the total sales for the year, and the average monthly sales (= total sales/12).

Exercise 5-5: Map and for-of Loop

Create a Map object named `productSales` that stores the product numbers and annual sales for seven different products.

Use a `for-of` loop to compute and display the combined total sales for all products and the average of sales over all products.

Use a second `for-of` loop to find and display the product numbers of products that have annual sales above the average.

Exercise 5-6: Validate Numeric Field

Add the necessary code in the `ComputeCost()` function to display an alert if the user clicks the Compute Cost button without entering a value for the number of tickets. You may do this by checking whether the value of the tickets variable is "NaN."

Exercise 5-7: oninput and onchange Events

When the user changes the number of tickets on the form, call the `ComputeCost()` function to automatically compute the total cost and display it. To do it, use the `oninput` event.

Next, use the `onchange` event to do the same. How do the two differ?

Exercise 5-8: Events of Select Element

Use an appropriate event of the `<select>` element to automatically compute and update the total cost whenever the user selects a performance.

Exercise 5-9

On the confirmation page, rather than concatenating the values together and displaying the values in a paragraph element, display the labels and values in a table. Display the performance, date, zone, member status, number of tickets, and total charge in a table with two columns, with the first column for the labels and the second for the values. Display each label-value pair in a separate row of the table.

Exercise 5-10

This exercise applies jQuery to the Reservation-2 form. In addition to changing the background color, it is desired to take the following actions when the user changes an **input field**:

1. Disable the Continue button to force the user to re-compute the total cost before submitting the form. Make sure you add the necessary statement in the `ComputeCost()` function to enable the Continue button.

2. Change the content of `<div id="msg1">` to "Please compute cost to continue," and remove the content of `<div id="msg2">`.

Modify the code in Figure 5-51 to perform the above actions.

Exercise 5-11

This exercise uses jQuery on the Reservation-2 form. In addition to changing the background color, it is desired to take the following actions when the user changes the selection in the **drop-down list**:

1. Disable the Continue button to force the user to re-compute the total cost before submitting the form. Make sure you add the necessary statement in the `ComputeCost()` function to enable the Continue button.

2. Change the content of `<div id="msg1">` to "Please compute cost to continue," and remove the content of `<div id="msg2">`.

Modify the code in Figure 5-51 to perform the above actions.

Assignments

E-Commerce Assignment: Part B

This assignment is a continuation of the E-Commerce Assignment: Part A, from Chapter 4. If you haven't done Part A, you need to do it first. Part A developed the user interface for the Laptop-selection page. This page is a part of the e-commerce system for P&I Laptops, which is a small business that assembles and sells a limited set of laptop models online. In this assignment, you will develop the JavaScript code to compute the costs and add additional functionality to the Laptop-selection form.

The Laptop-selection form allows the user to specify the laptop model, software, amount of extra RAM, whether a warranty is needed, and the quantity, as described in Part A. The following figure shows the form with sample input data and the corresponding output.

Laptop selection

Model:
S330
S380
S580

Hardware: $495

Software:
○ MS 365 Personal ● MS 365 Family ○ Office Home & Student

Software: $99

Extra RAM: ● 4 GB ○ 8GB

RAM: $45

☑ Ext. Warranty?

Warranty: $55

Quantity: 2

Compute Cost

Total Cost: $1398

Reset   Select

The Laptop-selection form should have the following features:
1. When the user selects a laptop model, compute and display the hardware cost (unit price). Here are the unit prices for each model:

Laptop model	Unit price
S330	$295
S380	$495
S580	$595
G780	$950
B780	$995

No laptop model should be set as the default selection.
2. When the user selects a software, compute and display the software cost:

Software	Price
MS 365 Personal	$65
MS 365 Family	$99
Office Home & Student	$95

3. When the user selects the RAM, compute and display the RAM price:
   4 GB: $45
   8 GB: $80
4. When the user selects an extended warranty, display the corresponding cost ($55).
5. When the user enters the number of tickets, change the background color of the input field and enable the Compute Cost button.
6. When the user clicks the Compute Cost button, verify that a laptop model is selected. If not, display a message. If a model is selected, compute and display the total cost (hardware + software + RAM + warranty) and enable the Select button.
7. When the user clicks the Reset button, reset the input data and the output.
8. When the user clicks the Select button, submit the form, display the Cart page, and pass the input data to the Cart page using a query string.

E-Commerce Assignment: Part C

This assignment develops the code to display the order data on the Cart page that you created in Part A. The Cart page should meet the following requirements:

Display the laptop model selected by the user on the Laptop-selection form, the software, the RAM, and the quantity. Provide proper labels for each data item displayed.

Display the total cost that was computed on the Laptop-selection form. Compute and display the sales tax (5% of total cost) and the amount due (total cost + sales tax).

Auto Rental Assignment: Part B

This assignment develops JavaScript to compute costs and provide additional functionality to the web form Auto-rental, developed in Auto Rental Assignment: Part A, in the previous chapter. If you haven't done Part A, you need to do it first.

Ace Auto Rentals is a small business that rents automobiles. The web form lets the user select the model of the vehicle, specify related information, and submit the form, as described in Part A.

The following figure shows the form with sample input data and the corresponding output.

Rental Reservation

Model:

Minivan:Chrysler Pacifica	▲
Compact car: Ford Escape	
Luxury Car: Tesla 3	▼

Daily Rate: 60

Start date: 12/15/2023

Special rate:
○ Senior   ○ Government   ● Corporate   ○ AAA

Discount: 6

☑ Do you want insurance?

Insurance: 10

Additonal drivers: 1

Drivers: 15

Cost/day: 91

# of days: 2

Total Cost: 182

Reset   Continue

Specifically, you should add the following functionality to the form:

1. When the user selects a model, compute and display the daily rental rate and the cost/day. These are the rental rates:

Model	Daily rental rate
SUV: Honda CR-V	50
Minivan: Chrysler Pacifica	60
Compact car: Ford Escape	25
Full-size car: Chevy Impala	35
Luxury Car: Tesla 3	60

Cost/day = Daily rate – Discount + Insurance + Charge for extra drivers

None of the models should be set as the default selection in the drop-down list.

2. When the user selects a special rate, display the discount amount and update the cost/day. No button should be selected as the default.

Discount: Senior = 25%, Government= 15%, Corporate = 10%, AAA = 20%

3. When the user selects insurance, display the corresponding insurance cost, and update the cost/day. The cost of insurance is $10/day.

4. When the user enters/changes the number of additional drivers, display the corresponding cost and update the cost/day. The cost per additional driver is $15.

5. When the user enters/changes the number of days, verify that a model is selected. If not, display a message. If a model is selected, compute and display the total cost, enable the Continue button, and display the message "Please click Continue to proceed." The message should not be displayed until the total cost is computed and displayed.

Total cost = Cost per day × Number of days

6. When the user clicks the Reset button, reset the input data and the output, and disable the Continue button.

7. When the user clicks the Continue button, verify that a start date is entered and that it is a current or future date. If not, display a message. If the date is valid, submit the form, display the confirmation page, and pass the input data to the confirmation form using a query string. The Continue button should be disabled until the total charge is computed and displayed.

Auto Rental Assignment: Part C

This assignment develops code to display the reservation data on the confirmation page that you created in Part A. The confirmation page should meet the following requirements:

Display the model selected by the user on the Auto-rental form, the start date, and the number of days. Provide proper labels for each data item displayed.

Display the total charge that was computed on the Auto-rental form. Compute and display the sales tax (5% of total charge) and the amount due (total charge + sales tax).

Provide a button with the caption "Reserve" that displays a web page named Payment. You should create an empty page named Payment.

Chapter 6

ASP.NET Web Forms

Chapters 4 and 5 presented HTML web forms that use client-side JavaScript to make web pages interactive. However, client-side scripts cannot and should not do everything. For example, it is not easy or safe to use a client-side script to access databases on the server. Similarly, you should not use client-side scripts for applications that require large amounts of memory and/or processing power, which would result in slow response. Further, client-side scripting generally makes applications less secure. Server-side scripts help address these issues.

This chapter introduces the use of the ASP.NET platform and ASP.NET Web Forms, which is a traditional approach to develop web applications involving server-side processing. You will develop the user interface for a simple web page that accepts user input, processes it, and displays results.

Learning Objectives

- After studying this chapter, you should be able to:
- Describe the characteristics of ASP.NET web forms.
- Describe the differences between stateless and stateful modes, ASP.NET and ASP.NET Core, ASP.NET code and HTML code, web server controls and HTML controls, AutoPostBack, and IsPostBack properties.
- Create an ASP.NET web application that includes web forms.
- Use common web server controls to build user interfaces.
- Create simple event handlers.
- Use validation controls to validate data.

6.1 Introduction

The ASP.NET web form approach develops web pages, called **web forms**, using drag-and-drop server-side controls, typically in the **Visual Studio environment**, making them particularly suitable for rapid application development of dynamic web pages. Those who are familiar with Windows forms, which use similar drag-and-drop controls, may find this method fairly easy to learn. As the name indicates, these controls can be directly accessed from server-side code, making it convenient to do server-side processing.

Newer approaches to developing dynamic web pages include Razor pages and MVC (model view controller), which use HTML forms but provide the framework to do server-side processing. These frameworks provide better support for the development and maintenance of larger high-traffic websites. However, they are more complex, with steeper learning curves. Chapter 16 introduces Razor pages, and Chapters 17 and 18 present MVC.

As you learned in previous chapters, dynamic web pages are created on-demand. So, they may look different each time they are displayed, depending on various factors like who the user is, the date/time, changes in the data read, and the parameter values passed to it. Dynamic web pages typically involve processing server-side **scripts** to do tasks like look up inventory levels from a database, compute order totals, validate data, and customize web pages.

When a web server receives a request for a web page that includes a server-side script, the server executes the script, generates the HTML code for the web page, and sends it to the browser. In addition to the web server that handles HTTP requests, a website may include an **application server** and a **database server**. The application server provides an environment to execute scripts and create dynamic web pages, while the database server executes the commands that access the database.

6.1.1 ASP.NET and ASP.NET Core

ASP.NET and ASP.NET Core are frameworks for developing web applications, performing the function of the application server. ASP.NET Core is relatively new. A major difference between the two frameworks is that ASP. NET Core is an open-source cross-platform framework, whereas ASP.NET is Windows-based.

Another significant difference is that ASP.NET supports the **web forms** approach, described earlier. ASP.NET Core, on the other hand, supports Razor pages and ASP.NET Core MVC (model view controller) frameworks, which generally are more complex, but makes it easier to scale and maintain large projects.

Because of its ease of use, first, we will use ASP.NET web forms to learn web development in Chapters 6–10. ASP.NET is based on the .NET Framework, so all .NET Framework features, like type safety, inheritance, and compatibility with .NET languages, are available for ASP.NET development.

Chapter 16 discusses Razor pages using the ASP.NET Core framework, and Chapters 17 and 18 present ASP. NET Core MVC.

In the ASP.NET environment, a single physical server may run the web server (which handles HTTP requests), as well as an application server that provides an environment to execute scripts and to create dynamic web pages. In addition, it can also serve the role of the database server by running SQL Server DBMS on it.

ASP.NET applications are typically developed in a network environment using an Internet or Intranet. However, you may use a stand-alone development environment where a single computer serves as the client and server. The process of developing websites is essentially the same in both environments. For the sake of convenience, you will use the **stand-alone environment**.

In the stand-alone environment, you may use the **IIS Express web server** that comes with Visual Studio as the web server and **SQL Server Express** as the DBMS for the database server.

You may use **Visual Studio 2019** or the free stand-alone version, **Visual Studio Community 2019**, to develop ASP.NET applications specified in the tutorials.

6.1.2 Stateless and Stateful Modes

HTTP is a stateless protocol. That means that as soon as a client receives a response for its request, the connection between the client and server is terminated and the server keeps no information about the client. So, if the same client makes another request, the server has no way to know whether it is the same client or another client. The stateless mode of communication between the server and the client is fine for applications involving only simple static pages. However, it does not meet the requirements of today's typical dynamic applications. For example, in an e-commerce application, the user may want to put an item in a shopping cart, then check out, and have the checkout page remember the items the user put in the cart. In a stateless mode, the server has no way to remember what is in the cart.

ASP.NET provides several features to maintain the state, including view state, session state, application state, and server-side caching. You will learn more about these features later.

Review Questions

1. How is the ASP.NET web form different from the HTML web form?
2. How does the server process the request for a dynamic web page?
3. How is ASP.NET Core different from ASP.NET?
4. What is the drawback of the stateless mode of communication between the server and the client?

6.2 Overview of the Website

In this chapter and the rest of this book, you will develop a website named Theater that makes reservations for performances at The Grand Oshkosh theater, described in Exercise 1-1 in Chapter 1. Chapters 6–10 use the ASP.NET framework to develop multiple web pages within the website. To help you get the big picture, a brief description of the web pages follows:

Reservation page: The reservation page, as the name indicates, lets the user make reservations. This page is shown in Figure 6-2 in the next section. You will develop this page in five stages, starting with the first version named Reservation-1; followed by the second version, Reservation-2; and so on, ending with Reservation-5. The current chapter develops the user interface for Reservation-1.

Confirmation page: The confirmation page, similar to a shopping cart, displays the selected options for a reservation and lets the user revisit the reservation page to change the reservation before confirming it.

Events page: The events page dynamically displays the name, description, image, and dates of each performance (event) and lets the user filter the performances and select one.

Different versions of the reservation page: As stated earlier, there are five different versions of the reservation page, each version adding to or improving the previous version.

The first version of the reservation page, **Reservation-1**, is developed in Chapters 6 and 7. It creates the user interface controls and the server-side code to compute and display the prices.

Reservation-2 is developed in Chapter 8. It uses a database to store data on performances, and to get prices and dates for the selected performance.

Reservation-3 is developed in Chapter 9. It uses session state to share data with the confirmation page.

Reservation-4 also is developed in Chapter 9. It uses objects and a list of objects to store multiple reservations so that the user can select multiple reservations, view them together, and edit a selected reservation.

Reservation-5 is developed in Chapter 10. It includes two additional features: (1) it interacts with the events page, and (2) it uses the application state to store global data and cookies to capture user data.

6.3 Creating ASP.NET Web Forms

In the rest of this chapter, you will learn the essentials of developing an ASP.NET web form in the **Visual Studio environment**. As stated earlier, you will first create a website named Theater. A website is also referred to as a web project or web application. To create a web project, you choose a *template* from a set of five different templates: **Empty, Web Forms, MVC, Web API**, and **Single Page Application**.

The Empty template and the Web Forms template use the **Web Forms** approach to develop web pages. This template creates a web project with no web pages, whereas a project created using the Web Forms template includes a default page and folders/files needed to provide functionality. You will use the Empty template in the current project because of its simplicity.

The **MVC** template, which is becoming increasingly popular, supports the ASP.NET MVC (model view controller) software design architecture presented in Chapters 17 and 18. This approach, which is more complex than Web Forms, divides an application into three components—model, view, and controller—providing better performance, maintenance, ease of testing, and better control over the rendered HTML code.

The **Web API** template lets you create Web API (application program interface) services. The **Single Page Application** (SPA) dynamically updates a web page as the user interacts with it, without having to reload the entire page, thus helping create highly interactive applications that include a single page.

Within the Theater website, you will develop a web page using the Web Forms approach. A web page developed using the Web Forms approach is also called a web form. A web form, which uses web server controls to build user interfaces, also may contain HTML markup, CSS, server-side scripts, and client-side scripts, though as you will learn, server-side scripts are typically placed in a separate code-behind page.

Before you get started with developing the Reservation-1 page, let's look at the key concepts/features covered in this chapter with the help of the tutorial:

- The current section shows how to create ASP.NET web applications and web forms, and examines the different components of a web form.

- Section 6.4 introduces the ASP.NET code and HTML.

- Section 6.5 describes the use of web server controls, event handlers, and automatic post-back of web forms. This section also provides a brief description of the life cycle of web pages.

- Section 6.6 applies CSS to improve its layout appearance.

- Section 6.7 discusses data validation using validation control.

- Section 6.8 describes the rendering of ASP.NET code to HTML.

- Section 6.9 provides a brief overview of HTML controls.

Tutorial: Developing a Simple Web Form—Reservation-1

This tutorial develops the first version of the reservation page named Reservation-1 within the **Theater** web project for The Grand Oshkosh theater, described in Exercise 1-1 in Chapter 1. **Reservation-1** lets the user select a performance (event), date, zone, and membership status, and enter the number of tickets, as shown in Figure 6-1.

When the user selects a performance, the corresponding **base price** (the lowest price, which doesn't include the extra charge for the zone) is displayed in the window on the right side. Similarly, when the user selects a zone, the corresponding **zone price** (extra charge, if any, for the zone, over and above the base price) is displayed. If the user checks the *Member* CheckBox, the discount is displayed. The Continue button is used in later versions of the page to let the user display the Confirmation page. The server-side code to compute the prices in the Reservation-1 form is developed in Chapter 7.

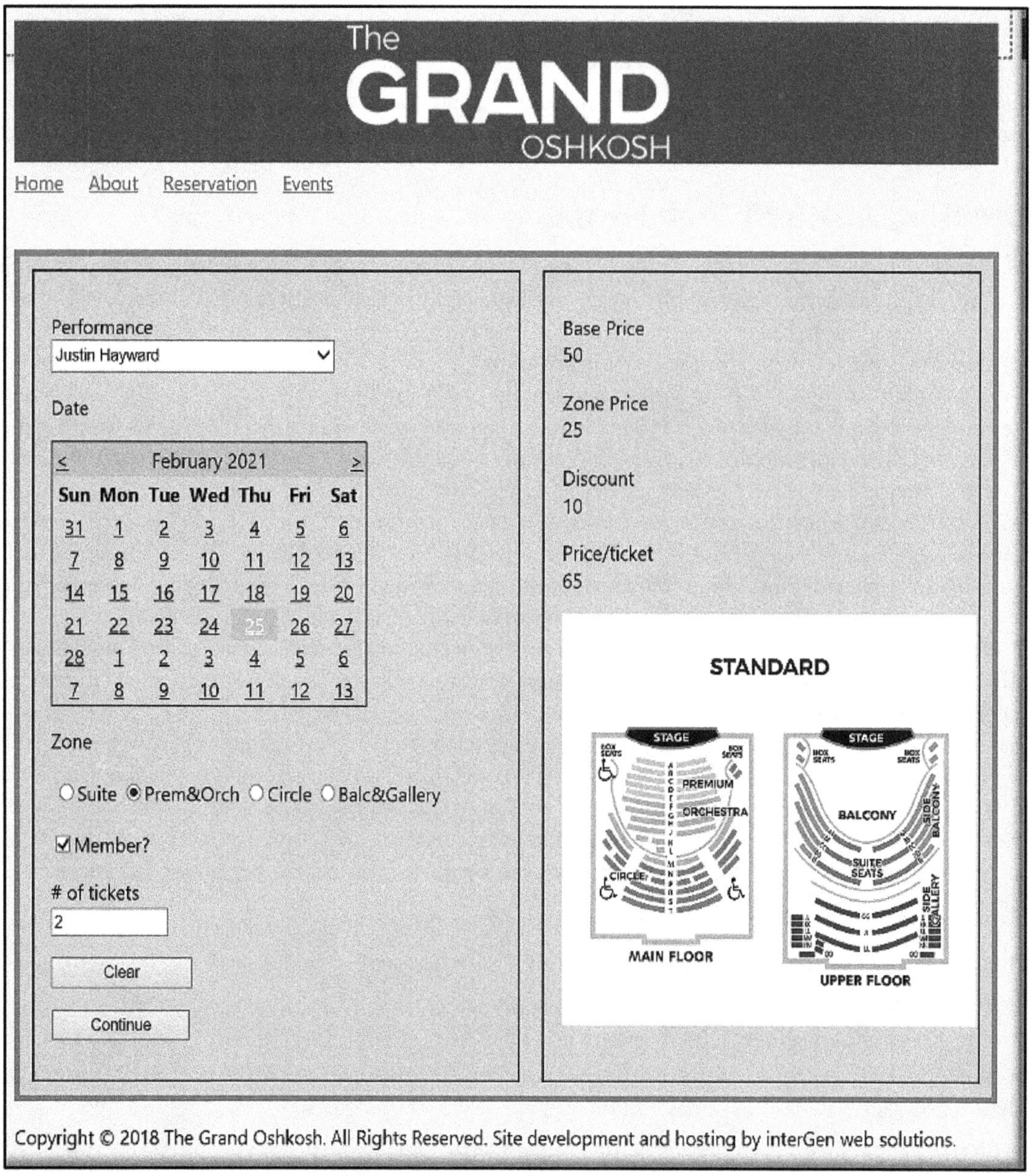

Figure 6-1: Reservation form—first version

6.3.1 Introduction to Visual Studio

The tutorials in the rest of this book use Visual Studio Community 2019 to develop the applications. You may download Visual Studio Community 2019 from the following *Downloads* page:

https://visualstudio.microsoft.com/downloads/

Click the link *How to install offline* on the Downloads page to find detailed information on installing the software and working in the Visual Studio environment.

It should be noted that the initial steps to create a project in Visual Studio for Mac are similar to the steps given here, but there are some differences that make it necessary for you to figure out certain steps yourself. Similarly, for database access, this book uses SQL Server Express LocalDB, which is a limited version of SQL Server. If you are using Visual Studio for Mac, you may use SQLite Database. Again, there are some differences between their use. Next, you will use Visual Studio to create the Theater web project and the Reservation-1 page within the project.

Step 1: Create a new project.

Open Visual Studio. You will see the *Get started* page, as shown in Figure 6-2.

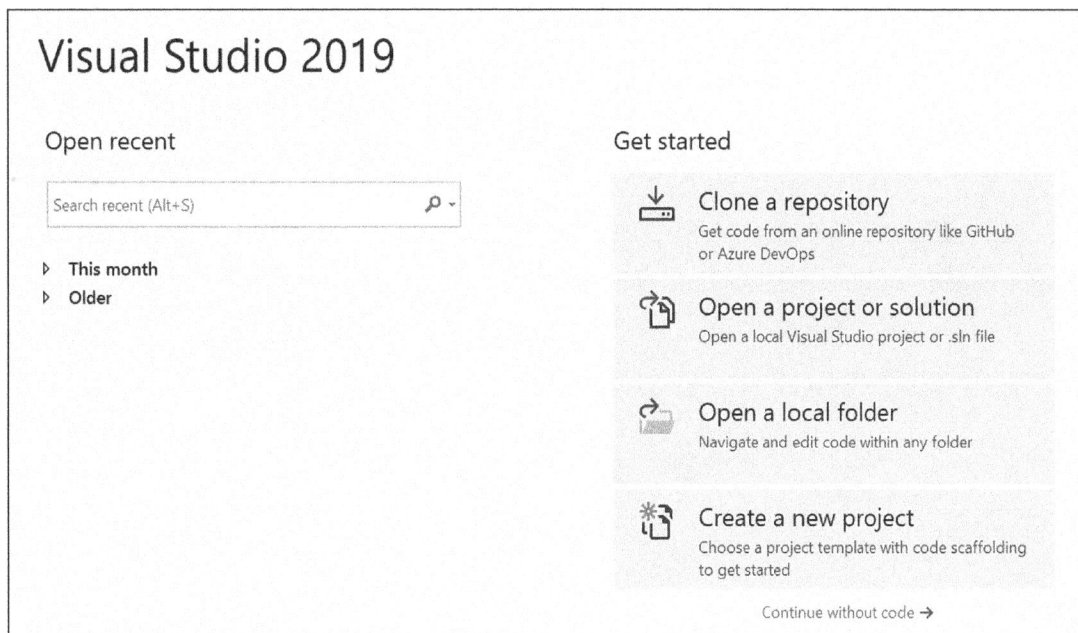

Figure 6-2: The start page

On the *start* page, choose *Create a new project.*

You will see the *Create a New Project* window.

Type *ASP.NET* in the search box to narrow down the choices, and select C# from the drop-down list for languages, as shown in. Figure 6-3.

Choose *ASP.NET Web Application (.NET Framework)* template for C#, and click *Next*.

You will see the *Configure your new project* window.

Enter **Theater** for *Project name*, **C:\WebProjects** for location, and uncheck the checkbox for *Place solution and project in the same directory*, as shown in Figure 6-4. If your project folder is different from C:\ WebProjects, enter the project folder that you use.

Caution! Using certain shared locations like Google Drive and Network drives can cause problems with websites not running properly.

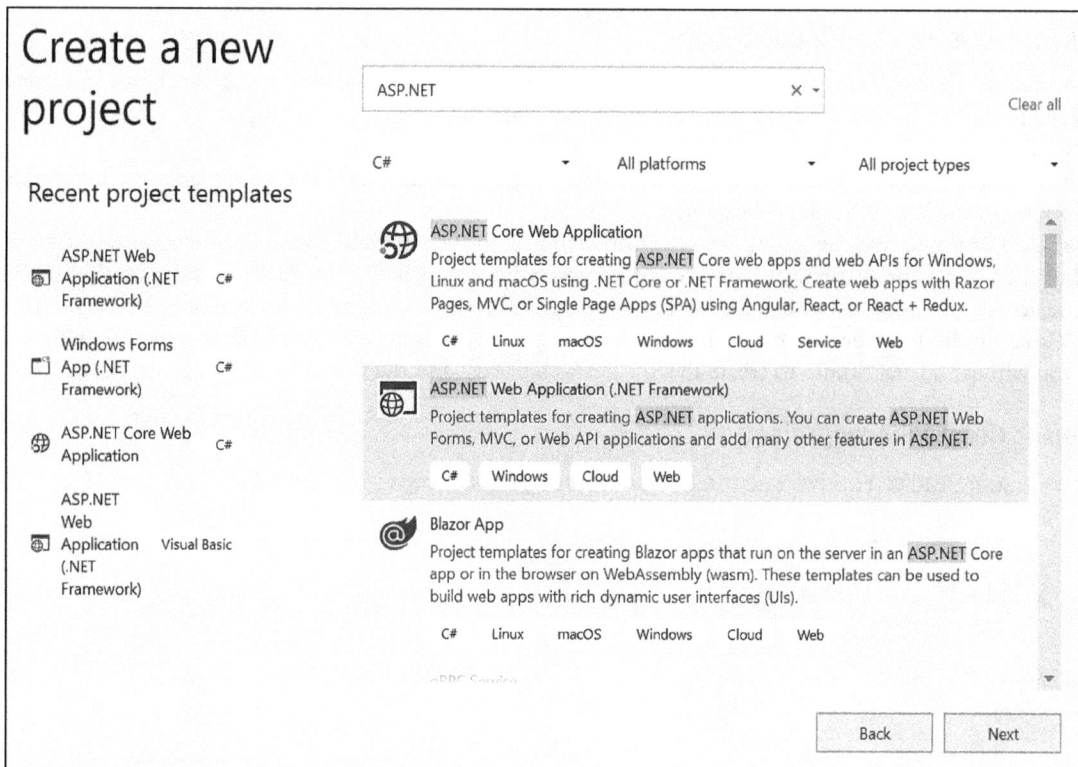

Figure 6-3: The Create a new project window

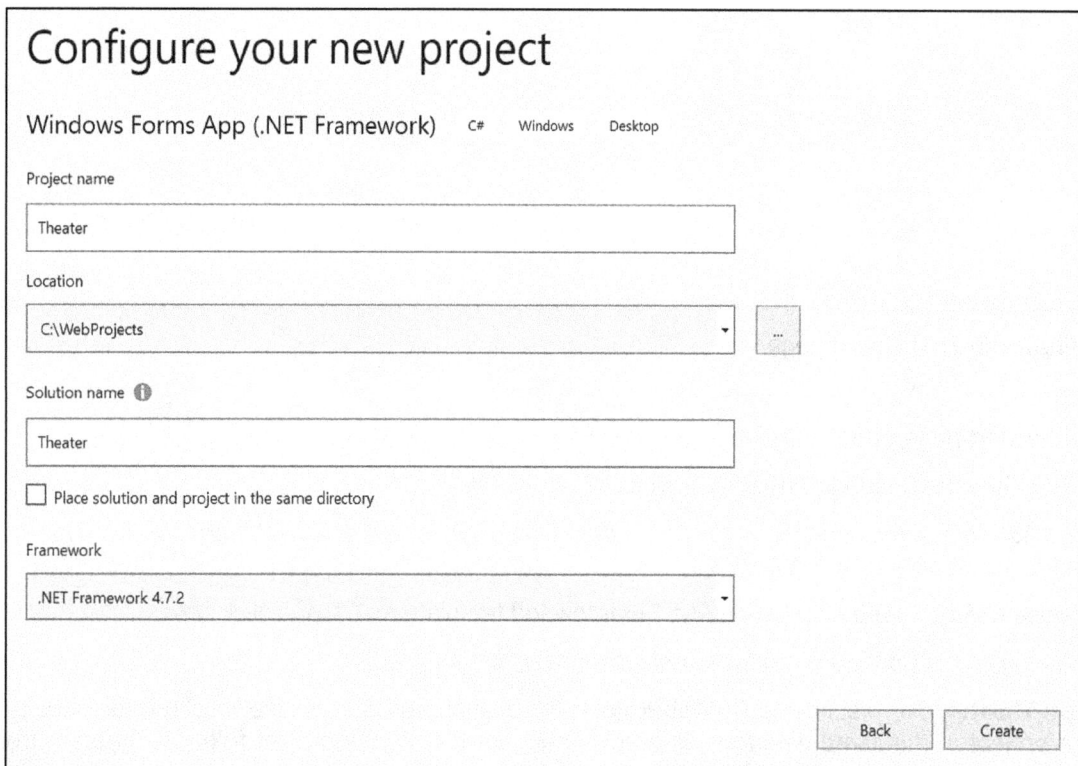

Figure 6-4: The Configure your new project window

Click **Create**.

You will see the **Create a new ASP.NET Web Application** window, as shown in Figure 6-5, which lets you select a template for your project.

Create a new ASP.NET Web Application

Empty

An empty project template for creating ASP.NET applications. This template does not have any content in it.

Web Forms

A project template for creating ASP.NET Web Forms applications. ASP.NET Web Forms lets you build dynamic websites using a familiar drag-and-drop, event-driven model. A design surface and hundreds of controls and components let you rapidly build sophisticated, powerful UI-driven sites with data access.

MVC

A project template for creating ASP.NET MVC applications. ASP.NET MVC allows you to build applications using the Model-View-Controller architecture. ASP.NET MVC includes many features that enable fast, test-driven development for creating applications that use the latest standards.

Web API

A project template for creating RESTful HTTP services that can reach a broad range of clients including browsers and mobile devices.

Single Page Application

A project template for creating rich client side JavaScript driven HTML5 applications using ASP.NET Web API. Single Page Applications provide a rich user experience which includes client-side interactions using HTML5, CSS3, and JavaScript.

Authentication

No Authentication

Change

Add folders & core references

☐ Web Forms
☐ MVC
☐ Web API

Advanced

☑ Configure for HTTPS

☐ Docker support

(Requires Docker Desktop)

☐ Also create a project for unit tests

Theater.Tests

Back   Create

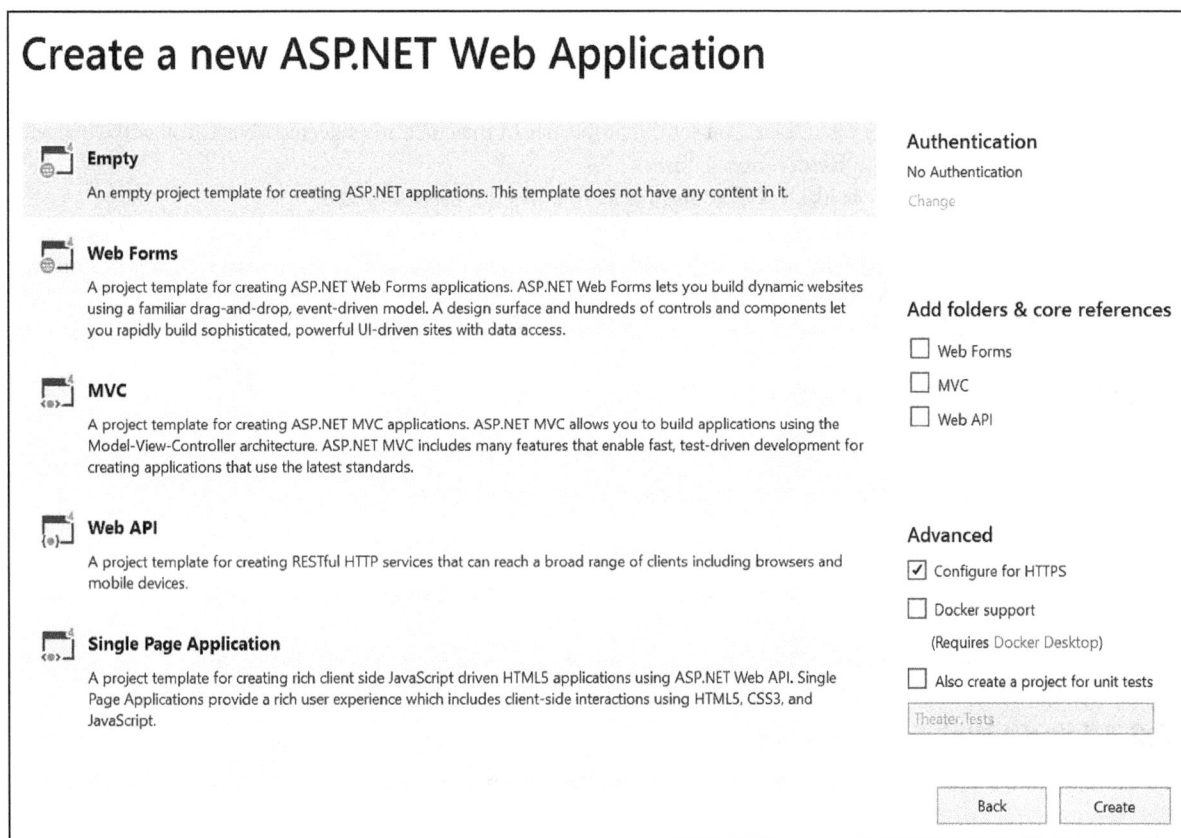

Figure 6-5: The Create a new ASP.NET Web Application window

Select the **Empty** template that was described earlier, and click *Create*.

You will see the Visual Studio integrated development environment (IDE), as shown in Figure 6-6.

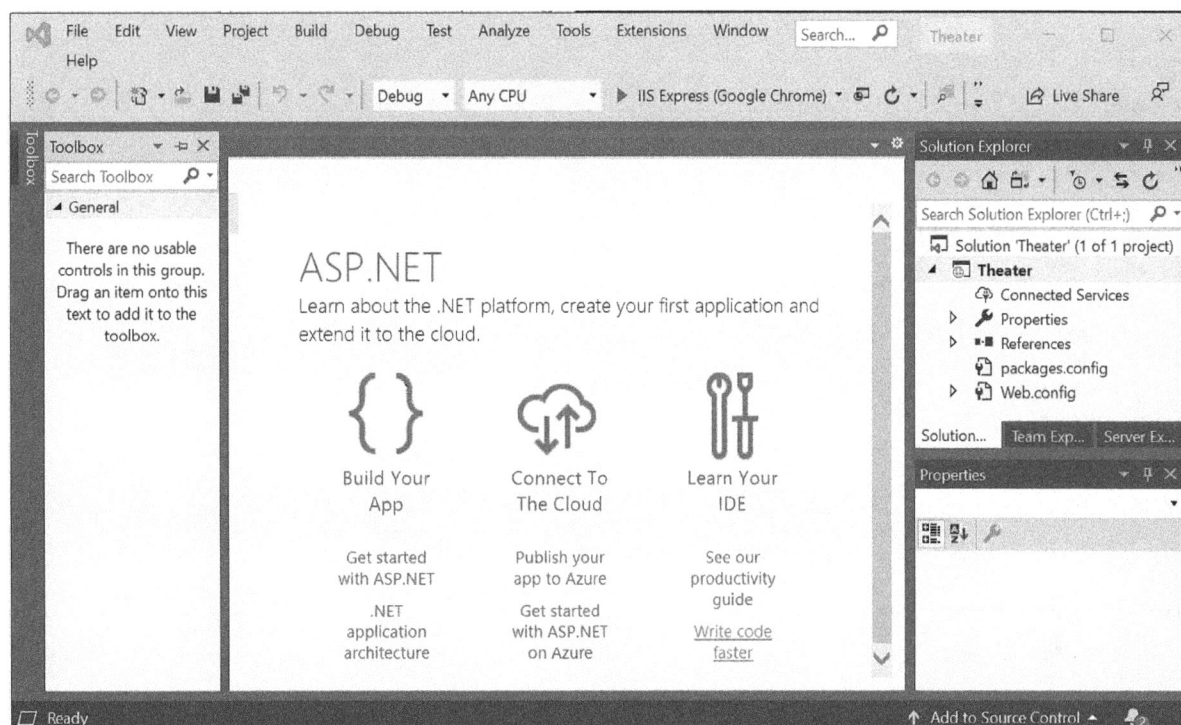

File   Edit   View   Project   Build   Debug   Test   Analyze   Tools   Extensions   Window   Search...   Theater

Help

Debug ▾   Any CPU ▾   ▶ IIS Express (Google Chrome) ▾   Live Share

Toolbox

Search Toolbox

◢ General

There are no usable controls in this group. Drag an item onto this text to add it to the toolbox.

ASP.NET

Learn about the .NET platform, create your first application and extend it to the cloud.

{ }
Build Your App

☁↕
Connect To The Cloud

Learn Your IDE

Get started with ASP.NET

.NET application architecture

Publish your app to Azure

Get started with ASP.NET on Azure

See our productivity guide

Write code faster

Solution Explorer

Search Solution Explorer (Ctrl+;)

Solution 'Theater' (1 of 1 project)
◢ Theater
    Connected Services
  ▷ Properties
  ▷ ▪▪ References
    packages.config
  ▷ Web.config

Solution...   Team Exp...   Server Ex...

Properties

Ready   ↑ Add to Source Control ▴

Figure 6-6: The Visual Studio IDE

Figure 6-6 shows the **Toolbox** on the left and the **Solution Explorer** window and the **Properties** window on the right. The **Solution Explorer** window shows the name of the Solution, which is a container of projects; the name of the project (or projects, if there are multiple projects); and the names of the files within each project.

The **Properties** window displays and lets you set the properties of the currently selected object. You will work with the properties after you create the Reservation-1 form.

The **Toolbox** contains controls like TextBox and Label to build the user interface.

Note that the WebProjects folder now has a new folder named Theater that holds the Solution file (Theater.sln) and the project folder, also named Theater, which holds configuration files and files that store information on web pages.

To familiarize you with specifying the settings within Visual Studio, you will use the Tools menu in the next step.

Step 2: Set Visual Studio settings to Web Development.

Open Visual studio and select Tools > *Import and export settings*.

On the *Welcome to the...* Wizard page, click *Reset all settings* > Next.

On the *Save Current Settings* page, select *Yes, save my current settings* > Next.

On the Choose a *Default Collection of Settings* page, select **Web Development** > Finish.

If you need to restore the saved settings, use the *Import selected environment settings* menu item.

6.3.2 Adding a Web Form to the Project

The next step shows how to add a web form (web page) to the project.

Step 3: Add a web form to the website.

Right-click the project name Theater (not the Solution, Theater) in the Solutions Explorer.

Select **Add > New Item > Web Form**.

The Add New Item window appears. Change the name of the web form to **Reservation-1.aspx**, as shown in Figure 6-7, and click **Add**.

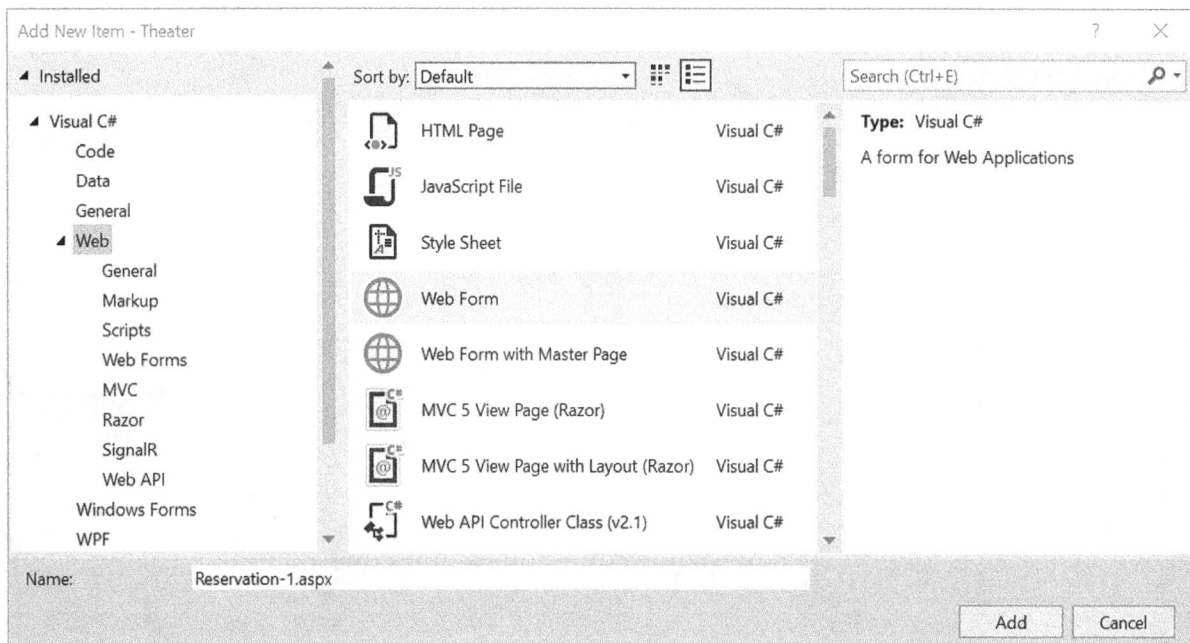

Figure 6-7: Add New Item window

An alternative is to select **Add > Add Webform**, which doesn't give you the option to add items other than web forms.

You will see the Reservation-1 form **Designer window** in **Source view**, as shown in Figure 6-8.

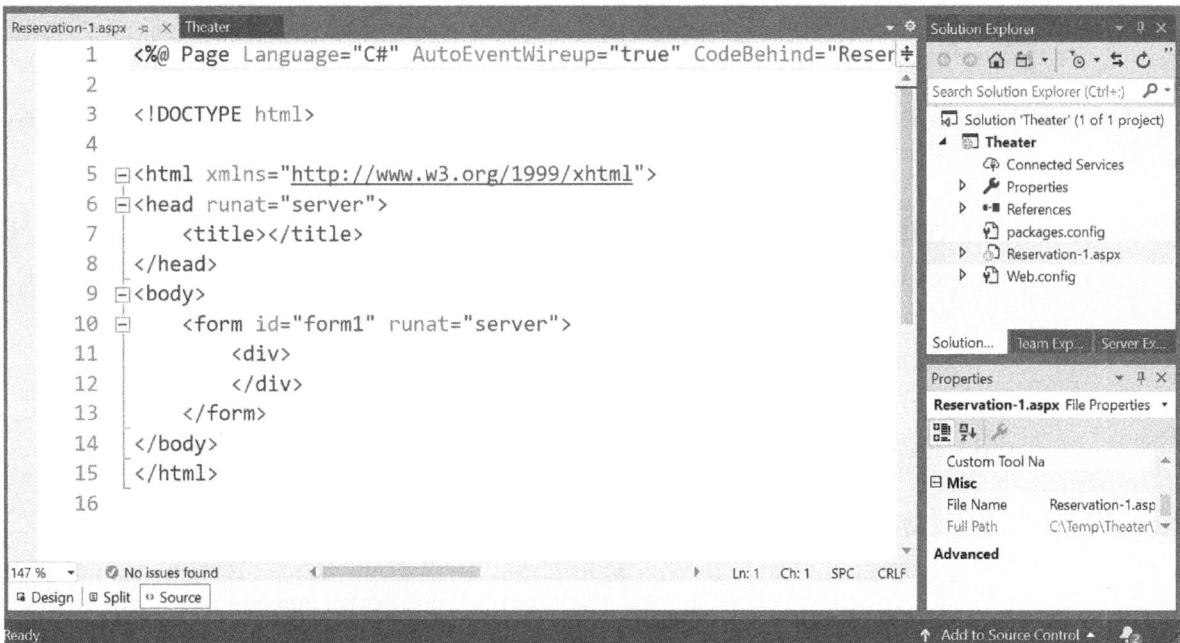

Figure 6-8: The Designer window in Source view

You may change the view to **Design** (or **Split**) by clicking the corresponding button at the bottom of the window. The Design view will show the empty Reservation-1 form.

6.3.3 Components of a Web Form

A web form typically includes three files with different types of code:

• The **actual web form** with an .aspx extension containing HTML code, ASP.NET code, and other codes like CSS, all of which deal with the user interface. The Solution Explorer in Figure 6-8 shows the Reservation-1. aspx file. You may click the triangle to the left of the file name to display the other two files, as shown here:

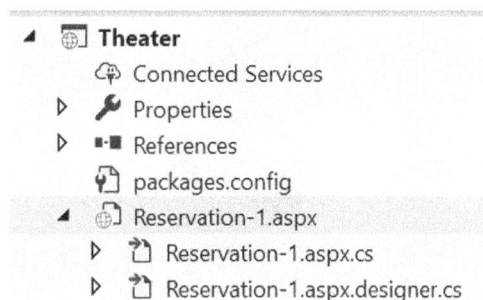

• A **code-behind file** with the extension .aspx.cs containing the C# for the application logic.
• A **design file** with the extension .aspx.designer.cs that contains the C# code for generating the user interface controls. This file is automatically generated and you shouldn't need to make changes to it.

You may display the C# code in the code-behind file by clicking the Reservation-1.aspx.cs file in the Solution Explorer.

The three-file system, which uses a separate code-behind file, provides some separation between the application logic and the user interface, making it easier to understand and maintain the code. An alternative to storing code in a code-behind file is to embed code in the .aspx file. This method is generally used with only relatively simple code. Code mixed in with HTML is harder to maintain.

Note that the WebProjects\Theater\Theater folder includes the additional files Reservation-1.aspx, Reservation-1.aspx.cs, and Reservation-1.aspx.designer.cs.

As you will recognize, the .aspx file shown in Figure 6-8 is essentially HTML code and additional elements that represent ASP.NET code, which we will discuss shortly.

Review Questions

5. What is the difference between the Web Forms approach and MVC approach to developing web pages?

6. How does code-behind differ from embedded code?

6.3.4 The Visual Studio Environment

In this section, we will look at some additional details on working in the Visual Studio environment.

Displaying a Closed Window

If any window, like the Toolbox, Solution Explorer, or Properties, is not displayed, select *View* from the menu, and then select the corresponding window. For practice, you may close the Properties window and then display it.

Changing the Docking Position of a Window

You may change the docking position of a window, such as the Properties window, by grabbing it with the pointer and then dropping it in one of the four positions (top, bottom, left, right), as shown in Figure 6-9.

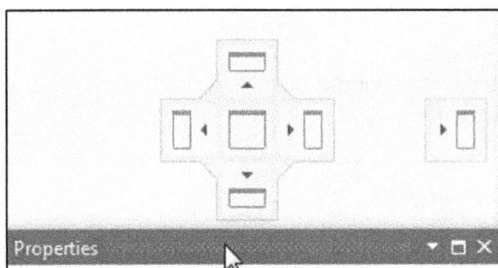

Figure 6-9: Docking positions

Undocking (Floating) a Window

By default, all windows are attached to a side of the Visual Studio window. To undock a window, right-click its title bar and select float. To dock a floating window, grab it with the pointer and drop it in one of the positions shown in Figure 6-9.

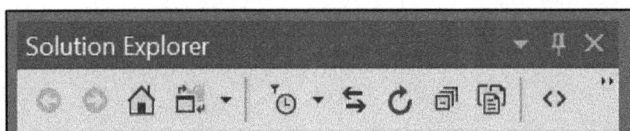

Figure 6-10: The pushpin on the Solution Explorer

Autohide

Click the pushpin at the top of the window to turn autohide on and reduce the window to a tab (Figure 6-10).

When you click the pushpin on the Solution Explorer, the window becomes a vertical bar to the right of the Properties window, as shown in Figure 6-11.

To turn autohide off, click on the tab and click the pushpin.

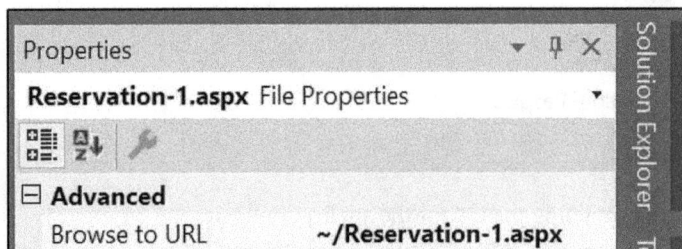

Figure 6-11: Solution Explorer when autohide is on

Menu and Toolbar Items

The menu and toolbar, shown in Figure 6-12, provide access to an extensive set of functions.

The function of each toolbar item is displayed in a **ToolTip** box when you hover the mouse pointer over the item. For a general understanding of the functions, you may view the descriptions of the Toolbar items by displaying the Tooltips. You will become more familiar with them through the projects in this book.

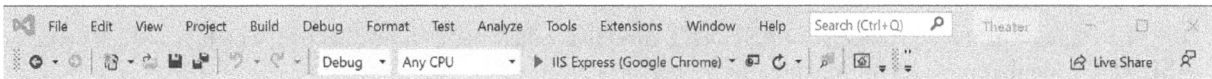

Figure 6-12: Menu and toolbar

Deleting a Form

To delete a form, right-click the name of the file (.aspx) in the Solution Explorer window, and select *Delete*. Deleting a file deletes the form and all files associated with the form permanently.

Excluding a Form from the Project

You may exclude a form from the Solution Explorer but still keep the files associated with the form, in the project folder, as follows:

Right-click the file name (.aspx) in the *Solution Explorer*, and select *Exclude from Project*.
To bring back a form that was excluded from the project,
Select *Project, Add Existing Item*.
Select the file (.aspx) from the project folder.
The form will be added to the Solution Explorer.

6.4 Aspx: HTML and ASP.NET codes

Figure 6-13 presents the initial source code, referred to as aspx code, which was generated by ASP.NET and stored in the Reservation-1.aspx file when you added the web form to the website.

```
1   <%@ Page Language="C#" AutoEventWireup="true" CodeBehind="Reservation1.aspx.cs"
    Inherits="Theater.Reservation_1" %>
2
3   <!DOCTYPE html>
4
5   <html xmlns="http://www.w3.org/1999/xhtml">
6   <head runat="server">
7       <title></title>
8   </head>
9   <body>
10      <form id="form1" runat="server">
11          <div>
12          </div>
13      </form>
14  </body>
15  </html>
```

Figure 6-13: Initial aspx code

All files with the .aspx extension, like the Reservation-1.aspx file, are processed on the server when the page is loaded.

The code within the .aspx file (aspx code) contains two types of code:

1. Code that is processed on the server, called ASP.NET code, which the browser doesn't understand. The first line of code, which starts with `<%@ Page,` is an example of such code, as described shortly. The form element `<form id="form1" runat="server">` is another ASP.NET code. ASP.NET code is typically translated (rendered) by the server to produce HTML markup that the browser understands.

2. Code that is processed by the browser. This code consists of HTML markup, CSS, and any client-side script that the browser understands. The browser can process this code directly without having the server translate it.

The following sections will describe the elements of the initial web form code.

@Page Directive

The first element in the aspx code is an **ASP.NET directive** called the `@Page` directive, which is a special type of ASP.NET code that provides information for the server:

```
<%@ Page Language="C#" AutoEventWireup="true" CodeFile="Reservation-1.aspx.cs"
Inherits="_Reservation-1" %>
```

The `Language` attribute says C# is the language, `AutoEventWireup` enables the event-driven feature, `Code-File` says Reservation-1.aspx.cs is the **code-behind file**, and `Inherits` specifies the _Reservation-1 class as the parent of the page. Because directives are information for the server, no corresponding markup is sent to the browser.

In general, the set of tags `<% %>` are used to insert embedded code blocks within the markup in an aspx page. There are different types of embedded code blocks, denoted by symbols like "@" following the start tag. You will learn about other types of code blocks in Chapter 8.

The runat="server" Attribute

Other than the initial `@Page` directive, the aspx code consists of an **HTML document.** However, the HTML document in the .aspx file contains both the familiar HTML elements and additional ASP.NET code, which consists of elements with the `runat="server"` attribute, as in the head and form elements.

```
<head runat="server">
<form id="form1" runat="server
```

The `runat` attribute specifies that the code is run on the server. All web server controls have the `runat` attribute, as you will see in the body element.

<head> Element

Specifying the `runat` attribute for the head element makes the head accessible by code on the server so that changes can be made to the head using server-side code (code that is run on the server).

<form> Element

The `runat` attribute for the form element specifies that the form is processed on the server so you can refer to the form and the controls on the page using server-side code. The form element in the initial code doesn't have any content other than the `div` element.

A `form` element is required if a page contains web server controls. The server controls must be placed inside the form element. Though other HTML markup may be placed outside the `form` element, it is recommended that all content displayed on the page be placed inside.

You may delete the `div` element, which you won't be using for now.

Step 4: Add the title.

Insert the title, "Reservation-1," in the `title` element, and delete the `div` element so that the code appears as follows:

```
1   <head runat="server">
2       <title>Reservation-1</title>
3   </head>
4   <body>
5       <form id="form1" runat="server">
6
7       </form>
8   </body>
```

Review Questions

7. What is the difference between ASP.NET code and standard HTML markup that is part of the HTML language?

8. What is the effect of specifying the `runat="sever"` attribute for the form element?

6.5 Working with Web Server Controls

ASP.NET provides a variety of web server controls (also called **server controls** and **Web Form controls**), including TextBox, Label, Button, RadioButtonList, DropDownList, ListView and GridView, to help you create the user interface without having to write HTML code. An important characteristic of web server controls is that their properties are accessible on the server.

In the .Net environment, code is often attached to specific events like clicking a button, loading a page, or pressing a key to take action when those events occur. For example, TextBoxes, which let uers enter input data on a form, have the TextChanged event that you can use to take an action when the value of a TextBox changes. You will use events to run code, as described later.

If you are familiar with Windows forms, you will find that Web server controls have similar functionality. But they differ from Windows Form controls because of their special features to work with web pages. For example, server controls have additional properties, including **AutoPostBack** and **Runat**, which are discussed later. Server controls generally have a limited set of *events* compared with Windows Forms controls. The TextBox server control, for example, doesn't have several events, like Click, KeyPress, Enter, Leave, and Validating, which are available for the Windows Forms Text Box.

6.5.1 Adding Web Server Controls to a Web Form

You can add web server controls to a web form either by adding code to the .aspx file in Source view or by adding controls from the Toolbox in Design view. Working directly with the aspx code to add controls will give you more control and flexibility. You will use this approach in some cases, so understanding the generated code is important. However, to help focus on ASP.NET and C# coding, we will generally add controls from the Toolbox, and let Visual Studio generate the code for the controls you added. In Design view, you can drag-and-drop controls or double-click the controls. After you drop it on the form, you can select a control and set its properties.

Let's look at how to develop the Reservation-1 page that was presented in Figure 6-1, using web server controls.

The Initial Layout of the Web Form

Initially, you will place controls and their captions on the form in a simple single-column layout, as shown in Figure 6-14, and later use CSS to split the content and present it in two boxes, side by side, as shown earlier in Figure 6-1. Note that positioning text and controls using spaces and tabs will not be reliable across different client environments.

The form uses the web server control **DropDownList** to display a list of performances and let the user select a performance. When the user selects a performance, the corresponding **base price** is displayed in the Label, lblBasePrice.

Figure 6-14: Initial layout of controls on the form

The Calendar control is used to let the user select the date of the performance. You should be aware that the appearance of the calendar control in the browser may vary with the browser.

The RadioButtonList, which is described shortly, lets the user select a zone (section) of the theater. When the user selects a zone, the corresponding **zone price** (extra charge, if any, for the zone, over and above the base price) is displayed in the Label, lblBasePrice

Next, you will add the controls to the form and set their properties.

Step 5: Add the caption and DropDownList control to the form.

Make sure the form is displayed in *Design* view.

Type in the caption, *Performance*, in the first line of the form.

Press the Enter key to go to the next line.

Double-click the DropDownList in the Toolbar to add it to the form.

Step 6: Set properties of DropDownList.

Make sure the DropDownList is selected.

In the Properties window, set the **ID** property of the DropDownList to *ddlPerformance*.

Set the AutoPostBack property to True.

Step 7: Continue adding controls.

Add the rest of the captions and controls, except the RadioButtonList, to the form as shown in Figure 6-14.

Adjust the width of the controls as necessary.

Step 8: Set properties of controls.

Set properties of the controls as specified in Table 6-1.

To specify a scheme for the calendar, click the arrow at the top right of the Calendar control and choose *Auto Format* (or click *Auto Format* at the bottom of the Properties window). Select a formatting scheme.

Table 6-1: Controls and their properties

Purpose	Type of control	Property	Setting
Select performance	DropDownList	ID	ddlPerformances
		AutoPostBack	True
Select date	Calendar	ID	calPerformDate
Select zone	RadioButtonList	ID	rblZones
		AutoPostBack	True
		RepeatDirection	Horizontal
Specify member status	CheckBox	ID	chkMember
		Text	Member?
		AutoPostBack	True
Enter number of tickets	TextBox	ID	txtNoOfTickets
Clear controls	Button	ID	btnClear
		CauseValidation	False
Redirect to Confirmation page	Button	ID	btnContinue
Display base price	Label	ID	lblBasePrice

Table 6-1: Controls and their properties (continued)

Purpose	Type of control	Property	Setting
Display zone price	Label	ID	lblZonePrice
Display discount	Label	ID	lblDiscount
Display unit cost	Label	ID	lblUnitPrice
Display seat map	Image	ID	imgSeatMap
		ImageUrl	/img/Seat-map.jpg
		Width	360px

This property is explained later in this section. When this property is set to True, the web page will be posted back to the server whenever you interact with the control.

RadioButtonList

A RadioButtonList consists of a group of RadioButtons that let the user select a single item from a group. In Chapter 5, you created a set of radio buttons (`<input type="radio">`) with a common name to do the same function. But the RadioButtonList is a single control that includes multiple buttons. An advantage is that you can use the **SelectedValue** property of the RadioButtonList to identify the button selected by the user.

You use the **Items** collection of the RadioButtonList to create a group of RadioButtons. The **RepeatDirection** property of the control determines whether the RadioButtons are arranged horizontally or vertically.

Similar to RadioButtonList, ASP.NET also provides a **CheckBoxList** that lets you add a group of checkboxes to the control so the user can select multiple items from a list.

Step 9: Add a RadioButtonList and add four buttons:

Drag and drop the RadioButtonList from the ToolBox into a new line right below the caption, "*Zones.*"

Set the ID property of the control to rblZones.

Select RepeatDirection property, and set it to Horizontal.

Select the **Items** property of rblZones, and click on the ellipses (...) on the right side.

On the *ListItem Collection Editor*, click the **Add** button to add a RadioButton. Type in "Suite" for Text and Value properties, as shown in Figure 6-15.

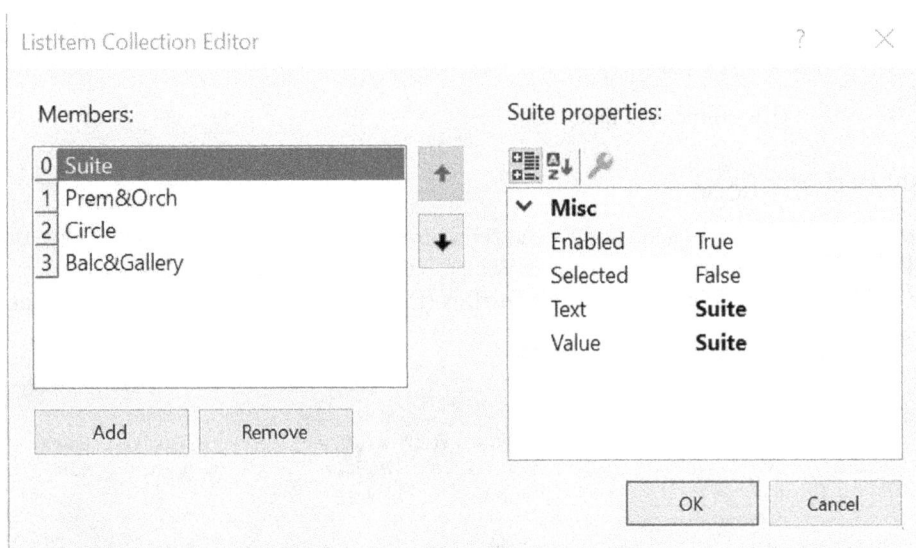

Figure 6-15: ListItem collection

Repeat the previous step to add RadioButtons for Prem&Orch (Premium and Orchestra zones) and Circle. **Do not** add the RadioButton for "Balc&Gallery"; you will add it using code. Click OK.

Switch to Source view. The Source view displays the ASP.NET code for the controls you added. Web server controls are signified by the "**asp:**" prefix. For example, the code for the TextBox appears as follows:

```
<asp:TextBox ID="txtNoOfTickets" runat="server" Width="99px"></asp:TextBox>
```

Similarly, DropDownList is represented by the following ASP.NET code:

```
<asp:DropDownList ID="ddlPerformance" runat="server" AutoPostBack="True" Height="26px"  Width="249px">
```

DropDownList vs. RadioButtonList

Often, there could be more than one control that can be used to let the user make a selection on a form. Choosing between a RadioButtonList, DropDownList, and CheckBox may not be an obvious decision. We will look at some guidelines to make decisions that are consistent with commonly accepted practices.

When you want to let the user select a single item from a group of items, how do you decide between a Drop-DownList and a RadioButtonList? A DropDownList, as opposed to a RadioButtonList, is recommended when the list is large (more than about seven items) or dynamic. The list is dynamic when the content of the list or the number of items in the list is likely to undergo changes.

To select a month from a list of the 12 months, for example, a DropDownList is recommended over a RadioButtonList because the list is large, though it is static. In the Reservation form, we use a RadioButtonList for zones because the list is static and it is not large. Similarly, we use a DropDownList for the list of performances because the list is dynamic (it is likely to change frequently).

You also may consider other factors like the importance of visibility, ease of selection, and space limitations on the form. For example, if visibility and ease of selection are not important, and space is limited, one may use a Drop-DownList for a static list of five items.

CheckBox vs. RadioButtons

When the selection is true/false or yes/no, a single CheckBox, **not** two RadioButtons, is recommended. For example, the following use of RadioButton is **not** recommended:

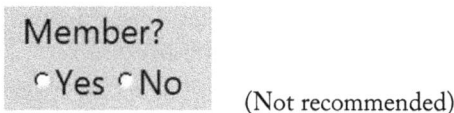

Member?
ᵒYes ᵒNo   (Not recommended)

Use a CheckBox instead:

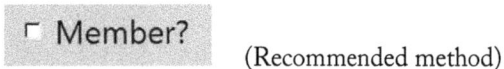

☐ Member?   (Recommended method)

Working with ASP.NET code

Instead of working in Design view to add controls and set their properties, you can add controls by adding the corresponding ASP.NET code in Source view and specify the properties using code. For example, you may add a new TextBox named txtDemo to the right of the existing TextBox (for the number of tickets), by adding the code shown in line 4 in the Source view.

```
1  <p>
2      # of tickets <br />
3      <asp:TextBox ID="txtNoOfTickets" runat="server" Width="99px"> </asp:TextBox>
4      <asp:TextBox ID="txtDemo"        runat="server" Width="80px"> </asp:TextBox>
5  </p>
```

The Design view would show the newly added TextBox. If you add the TextBox, txtDemo, make sure you delete it either in the Source view or in the Design view.

It's time to practice! Do Step 1-9. This step needs to be completed before proceeding further.

Step 10: Add a fourth RadioButton for "Balc&Gallery".

Open the Source view of the page and view the ASP.NET code for the RadioButtonList.

Note that each RadioButton is represented by a ListItem element.

Add a fourth RadioButton for "Balc&Gallery" by adding a new element to the group using the syntax for the "Prem&Orch" button.

Because "&" is a reserved character (like "<" and ">"), you need to use the name of the character entity with an "&" preceding it and a ";" following it, as in "&".

6.5.2 Event Handlers of the Web Page and Web Server Controls

An event handler is a function that runs in response to an event, like clicking a button, loading a page, or selecting an item from a drop-down list.

In an HTML form, you would use the event handler attributes of HTML elements to specify event handler functions, as discussed in Chapter 5. For example, to invoke a function named ComputeCost() when the user clicks a button, you would set ComputeCost() as the value of the onclick() attribute of the button, as shown here:

```
<button type="button" … onclick="ComputeCost()">Compute Cost</button>
```

In an ASP.NET form, you just have to double-click the button to create the method that will be invoked when you click the button. For a web server control like the *Continue* button on the Reservation-1 form, you can double-click the control to create the default event handler, which is the click event handler for a button. The click event handler of the button, which is fired when the user clicks the *Continue* button, looks as follows:

```
1  protected void btnContinue_Click(object sender, EventArgs e)
2  {
3      (add code here)
4  }
```

You will use event handlers extensively in Chapter 7.

For a web page, the default event handler is the **Page_Load** event handler, which is fired when the page is loaded. Loading is one of the multiple stages involved in displaying a page, as described next.

Life Cycle of a Web Page

The process of displaying a web page involves multiple stages, in sequence:

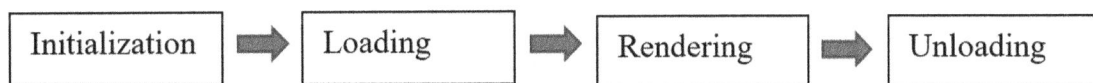

Typically, there are multiple events associated with each stage. You can create event handlers, like the Page_Load event handler, to take action when these events occur. For example, later in Chapter 8, you will use the DayRender event handler of the Calendar control to change the appearance of certain dates when they are rendered. Because the scope of this book doesn't require creating event handlers for other events, we won't get into the details of page life-cycle events beyond Page_Load.

Initialization: This stage is used to determine whether the request is a new request or a postback, and to create the page and the control objects. Values of controls are **not** restored in this stage. This stage involves three events: PreInit, Init, and InitComplete.

Loading: If it is a postback, loading involves restoring the values of controls (from the view state, explained later) and making connections to databases. The events in this stage are PreLoad, Load, and LoadComplete. The Page_Load event handler associated with the Load event is commonly used to do initial tasks when the page is loaded. It should be noted that event handlers of controls are processed **after** the Page-Load event.

Rendering: Rendering generates the client-side HTML and script that are sent to the browser to display the page and its controls. PreRender, Render, and RenderComplete events take place at this stage.

Unloading: Essentially, unloading involves clean-up work like closing the files and database connections.

Using Page_Load and Click Event Handlers

You create default event handlers, like the Click event handler of a button, by double-clicking the control. The event handler will be created in the .aspx.cs file. For a page, the Page_Load event handler is already created in the .aspx.cs file.

To help you understand how event handlers are processed, you will use the Page_Load event handler to add items to the DropDownList and use the Click event handler of Clear button to clear the TextBox. If you don't fully understand the code, don't worry. Here, the intent is to get familiar with event handlers.

Currently, when you display the page in the browser, the DropDownList is empty. You will add items (performance names) to the list using C# code in the Page_Load event handler of the web page, as shown in Figure 6-16.

```
1  protected void Page_Load(object sender, EventArgs e)
2  {
3    string[] performances = {"--Select one--", "Young Irelanders", "Justin Hayward",
                              "TedxOshkosh", "Oshkosh     Orchestral" };
4    foreach (string performance in performances)
5        ddlPerformance.Items.Add(performance);
6  }
```

Figure 6-16: Page-Load event handler of Reservation-1

Line 3 creates a string array that contains the performance names and stores it in the variable, performances. In this first version, the performance names are hard-coded. In later versions of the page, you will get the names from a database.

Lines 4 and 5 use a foreach loop to get each performance name from the array and add it to the DropDownList.

Step 11: Create the Page_Load event handler.

Select Reservation-1.aspx.cs file in Solution Explorer to display the code-behind file.

Visual Studio opens the aspx.cs file and displays the Page_Load event handler.

Type in the C# code from Figure 6-16 into the Page_Load event handler.

Step 12: Specify the start page and run the website.

In a website with multiple web pages, you can display a specific page like Reservation-1 by setting it as the Start Page. To specify the **start page**, right-click the .aspx file for the page in the Solution Explorer and select *Set as Start Page*.

If no start page is specified for a multipage project, Visual Studio runs the currently selected web form, if any. If no web form is selected, the form named Default.aspx, if any, is displayed. Running a project with no start page specified, no form selected, and no file with the name Default.apsx results in an error. To view the different start options, right-click the project name Theater and select *Start Options*.

To specify Reservation-1 as the start page, right-click Reservation-1.aspx and select,

Set as Start Page.

Display the page and make sure that the DropDownList displays the performance names when you click it.

Step 13: Create the Click event handler of btnClear.

Double-click btnClear to create the Click event handler.

Enter the statement shown in line 3 of Figure 6-17 to the event handler.

```
1   protected void btnClear_Click(object sender, EventArgs e)
2   {
3       txtNoOfTickets.Text = String.Empty;
4   }
```

Figure 6-17: Click event handler of btnClear

Line 3 clears the TextBox, txtNoOfTickets, by displaying an empty string in it.

Display the Reservation-1 page, enter a number into the TextBox, and click the Clear button.

The code in the Click event handler of the button is executed to clear the TextBox.

The next section discusses how the AutoPostBack property of controls affect the processing of event handlers.

6.5.3 Postback of Forms and AutoPostBack Property of Server Controls

An important aspect of running event handler code associated with web server controls is that the code is run on the server. For example, when you click the Clear button on the reservation button, the C# code in the Click event handler in Figure 6-17 is run on the server. This is different from the client-side processing of event handlers in HTML forms that you created in Chapters 4 and 5.

Server-side processing of the code in the event handler is done by automatically posting back (submitting) the form when you click the button. That is, the client sends a message to the server, which includes the selections the user made and the input data like the number of tickets entered into the TextBox. The server executes the C# code in all event handlers in the form and sends the HTML markup for the refreshed page. In the reservation page, the refreshed page includes the changes like the added list of performances in the DropDownList and the null value for the TextBox set by the `btnClear` event handler.

The automatic post back is a default feature of all Buttons. This feature is available also for other server controls that accept user input, like TextBox and DropDownList. To enable this feature, you set the *AutoPostBack* property of the control to *True*, as you did when you added the controls to the form.

If the AutoPostBack property of a web server control is set to True, the web page will be posted back to the server whenever you interact with the control, like select an item from the DropDownList, check the CheckBox, or click a RadioButton. Similarly, if you enter a value into a TextBox, the page will be posted back when the TextBox loses focus. So, it is important to remember to set the AutoPostBack property of a control to True, if you want to take an action when a certain event is raised.

Every time the page is posted back, the page is refreshed and loaded. So, the Page_Load event is fired whenever you interact with any control, including the RadioButtonList and CheckBox, which have the AutoPostBack property set to True. To see the effect of the repeated firing of Page_Load events, do the following experiment.

Step 14: Explore the effects of the AutoPostBack property.

Display the page in the browser, click the DropDownList, and select a performance.

Click the DropDownList again. You will see the same set of performances listed two times. This is because selecting a performance the first time caused a page postback and loading of the refreshed page a second time. Of course, you can avoid this problem by setting the AutoPostBack property of the controls to False. But you won't be able to execute the server-side code in the event handlers because the page won't be posted back to the server.

Click outside the DropDownList (to avoid another postback), and click the Clear button. You will see that the DropDownList will have the performances listed three times because clicking any button causes a postback and loading of the page.

IsPostBack Property of ASP.NET

To avoid adding the performances when the page is posted back, you will add an `if` statement, as shown in Figure 6-18, to check whether the page is loaded for the first time or it is loaded in response to a postback.

```
1   protected void Page_Load(object sender, EventArgs e)
2   {
3       if (!IsPostBack) // if it is not a postback
4       {
5        string[] performances = {"--Select one--",
             "Young Irelanders", "Justin Hayward", "TedxOshkosh", "Oshkosh Orchestral"
6        foreach (string performance in performances)
7            ddlPerformance.Items.Add(performance);
8       }
9   }
```

Figure 6-18: The updated Page_Load event handler

The code in line 3 uses the `IsPostBack` property of ASP.NET page to check whether the page is being loaded for the first time. A value of `true` indicates that the page is being loaded (rendered) in response to a postback. A value of `false` means that the page is being loaded the first time.

Step 15: Control the Page Load and AutoPostBack.

Add the `if` statement in the Page_Load event handler to make sure that the performances are added to the DropDownList only when the page is loaded the first time.

Display the page, and select a performance and make sure the list is not duplicated.

6.5.4 Structuring Contents

This section applies what you learned about structuring the contents of the page. So far you have created the main content of the page. Next, you will mark up the main content and add the header and footer.

Step 16: Mark up the main content and add the header and footer as shown in Figure 6-19.

Add a `<main>` element that contains all elements within the body.

Copy the `<header>` element from Default.html that you may have created in Exercise 1-1 and 1-2 in Chapter 1, paste it above the `<main>` element and modify it as shown in Figure 6-19. If you didn't create Default.html, type in the header element as shown in Figure 6-19.

Copy the `<footer>` element from Default.html and paste it after the end tag `</main>`, or type it in.

The page should have the structure shown in Figure 6-19.

```
1  <body style="height: 16px; width: 800px" >
2      <header>
3          <img src="img/Grand_Logo.png" style="background-
           color: darkmagenta; width: 100%" alt="The Gran Oshkosh" />
4          <nav>
5              <a href="default.html">Home</a>
6              <a href="about.html">About</a>
7              <a href="Reservation-1.html">Plan Your Visit</a>
8              <a href="Events.html">Support</a>
9          </nav>
10     </header>
11
12     <main>
13         <form id="form1" runat="server">
14             ...
15             ...
16         </form>
17     </main>
18
19     <footer>
20         <p>Copyright &copy; 2018 The Grand Oshkosh. All Rights Reserved.
21         Site development and hosting by interGen web solutions. </p>
22     </footer>
23 </body>
```

Figure 6-19: Main, header, and footer elements of Reservation-1.

Review Questions

9. How does a single RadioButtonList control differ from a set of RadioButton controls?

10. What is the effect of setting the AutoPostBack property of a web server control to True?

11. What is the difference between AutoPostBack and IsPostBack properties?

6.6 CSS Styling

This section is just an application of CSS that you learned in Chapter 2, to improve the layout of web server controls and the overall appearance. Because the section doesn't deal with the main content of this chapter, you may skip it or just do Step 17.

Currently, when you display the page, the controls appear in a single column based on the order they are listed in the code. So, the three prices are displayed at the bottom of the page. Instead, you will display the prices in a second column on the right side of the page using the Flexbox layout described in Chapter 2.

In order to use the Flexbox, you will add a `<section>` element that includes all controls on the form. Within the section, you will add two `<div>` elements: the first `<div>` containing the controls to be displayed in the first column and the second `<div>` containing the controls to be displayed in the second column, as shown in Figure 6-20.

```
1   <form id="form1" runat="server">
2       <section class="flex-box">
3       <div id="select-performance">
4           <p>
5               Performance <br />
6               <asp:DropDownList ID="ddlPerformance" runat="server" AutoPostBack="True"
7               ...
8           </p>
9           ...
10          <p>
11              <asp:Button ID="btnContinue" runat="server" Text="Continue" Width="110px"
12              style="margin-left: 3px" />
13          </p>
14      </div>
15
16      <div id="display-price">
17          <p>
18              Base Price <br />
19              <asp:Label ID="lblBasePrice" runat="server"></asp:Label>
20          </p>
21          ...
22      </div>
23      </section>
24  </form>
```

Figure 6-20: HTML form with CSS

The class attribute of the section element is set to "flex-box" (it can be any meaningful name) so that you can refer to it to apply CSS style. The id values of the div elements are used to refer to the elements in the CSS.

Step 17: Add <section> and <div> elements to help apply the CSS styling.

Add the section and div elements as shown in Figure 6-20.

Add the paragraph elements to separate the controls into different paragraphs.

Step 18: Specify the CSS styling.

Open Styles.css and add the CSS code shown in Figure 6-21, which also includes the code to display the navigation links as a block. You may copy the code from Tutorial-starts/Data-files/Ch6-CSS.txt.

Make sure to specify the link to the styles.css file as shown here:

```
<link rel="stylesheet" type="text/css" href="stylesheets/styles.css" />
```

```
1   body {
2       background-color: honeydew;
3       font-family: 'Segoe UI', Tahoma, Geneva, Verdana, sans-serif;
4       border: 2px dashed green;
5       padding: 10px;
6   }
7   h1 {color: blue;}
8   h2 {color: lightblue;}
9   .flex-box{
10      display: flex;
11      flex-direction: row;
12      border: 5px solid gray;
13      width: 850px;
14      background-color: lightgray;
15  }
16      #select-performance {
17          border: 2px solid black;
18          width: 400px;
19          margin: 10px;
20          padding: 15px;
21          background-color: lightblue;
22      }
23      #display-price {
24          border: 2px solid black;
25          width: 350px;
26          margin: 10px;
27          padding: 20px;
28          background-color: lightblue;
29      }
30
31  nav {
32      overflow: hidden;
33      background-color: cadetblue;
34  }
35      nav a {
36          float: left;
37          border-right: 1px solid black;
38          padding: 16px;
39          display: block;
40          text-decoration: none;
41          color: aliceblue;
42      }
43      nav a:hover {
44          background-color: lightblue;
45      }
```

Figure 6-21: The CSS code

To refresh your memory, line 10 specifies that Flexbox layout is to be used for the elements (the two div elements) within the class, outer-box.

Line 16 specifies that the element whose id is "select-performance" (div) within the "flex-box" class will have the style specified in lines 17–21. (Because id's are unique, the class name "flex-box" is not required.) Similarly, lines 23–28 specify the style for the element whose id is "display-price."

Lines 31–42 display the hyperlinks as a block to give the following appearance:

Lines 43–45 change the back color of the hyperlink when the pointer hovers over it.

Display the page and make sure the three prices are displayed in a separate box on the right side of the page, and that the borders and colors are applied as specified.

6.7 Validating Data Using Validation Controls

ASP.NET provides several easy-to-use and powerful controls to validate input data using client-side validation and display error messages, if any. Table 6-2 describes the ASP.NET validation controls.

Table 6-2: ASP.NET validation controls

Validation control	Description
RequiredFieldValidator	Makes sure that the user does not skip an entry
RangeValidator	Checks whether an entry is between specified limits
CompareValidator	Compares an entry to a specified value, including the value of another control. You may also compare the datatype of an entry to a specified data type.
RegularExpressionValidator	Checks whether an entry matches a specified pattern, like a phone number, zip code, or e-mail address
CustomValidator	Validates entries using custom code in the ServerValidate event handler of the control

Note that if there is no data in a control, none of the validators, except the RequiredFieldValidator, are evaluated.

The available validation controls can be found in the **Validation** tab in the Toolbox. This group also includes the **ValidationSummary** control, which doesn't do any validation but is used to display the messages from all validation controls on the page together.

First, let's use the RequiredFieldValidator and CompareValidator to make sure that the user enters the number of Tickets into txtNoOfTickets and that it is a valid integer.

Step 19: Add the RequiredFieldValidator and CompareValidator to the form.

In Design view, move the insertion point to the right of the txtNoOfTickets TextBox, and press Enter to create a new line.

Add the RequiredFieldValidator control from the Validation tab, as shown in Figure 6-22, to the new line.

The validator appears as follows:

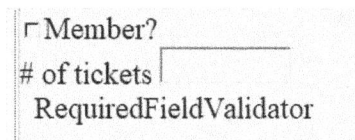

```
┌ Member?
# of tickets
   RequiredFieldValidator
```

Set the ID property of the validator to **reqTickets**.

Add the CompareValidator to the right of reqTickets. Set its ID to **cmpTickets**.

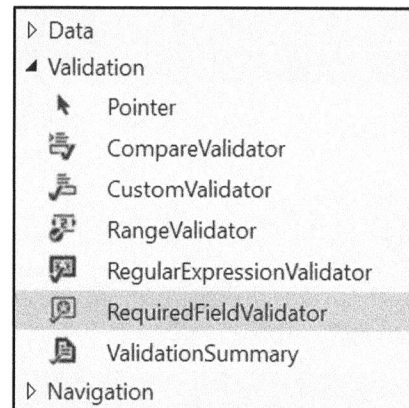

Figure 6-22: The RequiredFieldValidator

To validate a control, set the additional properties of the validation control as shown in Table 6-3.

Table 6-3: Properties of validation controls

Property	Description
ControlToValidate	The ID of the control to be validated; for example, txtNoOfTickets
Display	Specifies how the error message is displayed. There are three settings: Static: Screen space for the error message is taken even when there is no error. Dynamic: Space for the error message is taken only when there is an error. None: Display the error in a ValidationSummary control.
ErrorMessage	The message to be displayed in the Validation Summary, if there is one
Text	The message to be displayed in the validation control. If you don't enter a value for Text, it is set to the value of the ErrorMessage.

Step 20: Set the validation control properties.

Set the properties of the reqTickets validator like this:

Property	Value
ControlToValidate	txtNoOfTickets
Display	Dynamic
ErrorMessage	Please enter the number of tickets
Text	** Required **
ForeColor	(choose an attention-catching color)

We haven't added the ValidationSummary control yet. So, the ErrorMessage won't be used until the control is added.

Similarly, set the properties of the cmpTickets validator:

Property	Value
ControlToValidate	txtNoOfTickets
Display	Dynamic
ErrorMessage	Please enter an integer for NoOfTickets
Text	** Invalid number **
ForeColor	(choose a color)
Operator	DataTypeCheck
Type	integer

Note that you can have multiple validators for the same control.

6.7.1 Unobtrusive Validation

In addition to setting properties of validation controls, you need to choose from one of two different methods of client-side validation. As stated earlier, validation controls use client-side validation.

One option is to choose the traditional approach that was used prior to the release of .Net Framework 4.5. In this approach, the validation control generates a lot of JavaScript that is sent to the browser in-line with the HTML to be used for client-side validation. The generated in-line JavaScript significantly increases the size of the code.

The second option is to use a relatively new method called **unobtrusive validation**, which is the default method in .NET Framework 4.5 and later releases. This method uses the JavaScript from the jQuery library without having to add a significant amount of JavaScript to the page.

To use unobtrusive validation, first you need to install the jQuery library because websites created from *Empty Web Site* template do not have jQuery installed. An easy way to install and configure jQuery is to use the Visual Studio extension manager, NuGet, which can be used to install and configure many different packages that improve the functionality of web projects. It is fairly easy to install jQuery, as detailed in the tutorial step below.

Both methods require that the browser supports JavaScript (all major browsers do). First, you will use the traditional method. To use the traditional method, you need to disable unobtrusive validation for the web page using the following code in the Load event handler of the page:

```
UnobtrusiveValidationMode = UnobtrusiveValidationMode.None;
```

Step 21: Validation without unobtrusive validation.

Disable unobtrusive validation and set initial focus to txtNoOfTickets (to make it easier to test validation of the TextBox) by inserting lines 9 and 10 into the Load event handler of the page in Figure 6-23.

```
1   protected void Page_Load(object sender, EventArgs e)
2   {
3       if (!IsPostBack) // if it is not a postback
4       {
5           string[] performances = { "Young Irelanders", "Justin Hayward", "TedxOshkosh",
6                                      "OSO Orchestral" };
7           foreach (string performance in performances)
8               ddlPerformance.Items.Add(performance);
9       }
10      UnobtrusiveValidationMode = UnobtrusiveValidationMode.None;
11      txtNoOfTickets.Focus();
12  }
```

Figure 6-23: Validation without unobtrusive validation

Set the TabIndex property of ddlPerformance, calPerformDate, rblZone, chkMember, txtNoOfTickets, and btnContinue to 0, 1, 2, 3, 4, and 5, respectively.

Step 22: Run the web page.

Click the Continue button. (Currently, it doesn't do anything; but clicking it causes a page postback, which triggers validation.) You should see the message "** Required **" below the TextBox.

Enter a non-integer value for number of tickets and click the Continue button. You should see the message "** Invalid number."

6.7.2 Timing of Validation

In general, validation of data takes place both on the client and the server. Client-side validation takes place in an input control that has a validator associated with it, when you move focus away from the control after entering data (or make a change in the data). Thus, if you enter a non-integer into txtNoOfTickets and Tab to the next field, the Compare validation will take place, resulting in displaying the error message.

An advantage of client-side validation is that it helps give immediate feedback to the user without a postback of the page that causes a roundtrip to the server. However, client-side validation won't identify missing data, unless a value was entered and then deleted before moving the focus away, because validation will take place only if you move focus away from the control after you enter or change data in it. So, if you move focus away from txtNoOfTickets without entering any data, no client-side validation will take place because you haven't made any changes to the data. Then how did you get the error message in the last step when you clicked the Continue button without entering the number of tickets? It was due to server-side validation that takes place when the page is posted back.

Server-side validation rechecks the data using the same validation criteria as the client-side validation, but it is done on the server. This ensures that browser incompatibilities with the client-side validation scripts don't cause the submission of invalid data. Server-side validation takes place in every control that has a validator associated with it, whenever you click a button whose *CauseValidation* property is set to *True*. Hence, if you click the Continue button without entering the number of tickets, the RequiredField validation will take place, and you will get the message to enter the number of tickets.

Server-side validation takes place before processing the code in the event handlers. If you do not want validation to take place when you click a button like the Clear button, you can set the *CauseValidation* property of the button to *False*.

Setting the **EnableClientScript** property of the validator to *False* would turn off client-side validation. Let's check it out.

Step 23: Test the behavior of validators.

Click the Stop button if the web page is running, and then run it again.

Make sure that txtNoOfTickets has the focus.

Press Tab to move focus away from txtNoOfTickets without entering any data.

No message is displayed, indicating that the data in txtNoOfTickets was not validated because you haven't yet changed the value of txtNoOfTickets.

Click inside txtNoOfTickets so that it has the focus. Enter a decimal number like 3.5, and press Tab. You will see the error message "Invalid number."

Next, delete the invalid number, and press Tab. You will see the error message "Required." The RequiredFieldValidator was evaluated this time because the data in txtNoOfTickets was changed.

Step 24: Validation with unobtrusive validation.

Comment out the following statement that disables unobtrusive validation:

```
UnobtrusiveValidationMode = UnobtrusiveValidationMode.None;
```

Right-click the project Theater in the Solution Explorer.

Select Manage **NuGet Packages.** You will see the NuGet Theater window.

Select **Browse**.

Type "**aspnet.scriptmanager.jquery**" in the search box. The name of the package will appear in the bottom-left pane.

Select the package and click **Install**. Click OK.

Run the web page, and repeat the test as before with no value for the number of NoOfTickets, and also with non-integer values. You should see the same results as before.

If you have a problem with installing, you may uncomment the statement that enables unobtrusive validation.

Next, you will use the RequiredField validator to check whether the user selected an item from the DropDownList and from the RadioButtonList.

Step 25: Add a RequiredFieldValidator to check whether the user selected a performance from the DropDownList.

Add a new line below the DropDownList, and add the RequiredFieldValidator to the new line. Set its properties as follows:

Property	Value
Id	reqPerformance
ControlToValidate	ddlPerformance
Display	Dynamic
ErrorMessage	Select performance
Text	** Required **
ForeColor	(choose a color)

Step 26: Add a RequiredFieldValidator to check whether the user selected a zone from the RadioButtonList.

Comment out the statement that sets focus to txtNoOfTickets.

Display the web form and click the Continue button to test the Validator controls. You should see the messages you specified in the Text property.

The RequiredFieldValidator is not appropriate for the CheckBox because it always will have the value True or False. The CheckBox web control doesn't have a CheckState property with an Indeterminate value, as in the Windows form control, to help you check whether the user made a selection.

The RequiredFieldValidator doesn't work with Calendar control. We will use JavaScript later in Chapter 9 to validate the Calendar.

6.7.3 ValidationSummary Control

The ValidationSummary control displays the ErrorMessages from all validation controls together.

You may set the *ShowMessageBox* property to *True* to display the messages in a pop-up box. By default, the messages are displayed in a bullet list, but the *DisplayMode* property lets you select a simple *List* or *SingleParagraph*. The *HeaderText* property gives a heading for the summary. You may add it to a place in the form that catches the user's attention.

Step 27: Add the ValidationSummary control at the top of the form.

Insert a new line above the DropDownList, and add the ValidationSummary control.

Set the properties as follows:

Property	Value
Id	valReservation
HeadText	Oops!
ForeColor	(*choose a color*)

Display the page and click the Continue button without making any selections.

You should see the summary of error messages at the top and the individual messages where you placed the validation controls.

You may set the ShowMessageBox property to True to display the summary in a pop-up box.

Additional information on validators and examples of their use can be found at https://msdn.microsoft.com/en-us/library/ms972961.aspx.

It's time to practice! Do Exercise 6-1 (see the end of the chapter).

Review Questions

12. When does client-side validation take place in the ASP.NET web form? When does server-side validation take place?

13. What are the advantages and disadvantages of client-side validation compared with server-side validation?

14. How do you prevent validation when you click a button?

15. Which validator would you use to check the data type of the data you enter into a TextBox?

6.8 Rendering ASP.NET Code to HTML Elements

When the user requests a web page, the ASP.NET code for the page is rendered (converted) to HTML markup by the server so that the browser can understand it. The page that is sent to the client contains only HTML markup and client-side scripts, if any.

To view the HTML that the browser receives, run the website, right-click the page in the browser, and select **View page source**.[1] In general, an understanding of the rendered HTML is not essential for basic web development. However, it is beneficial to understand the relationship between web server controls and the rendered HTML.

The ASP.NET code (in the aspx file) for the TextBox, txtNoOfTickets, and the corresponding HTML element rendered by the server (as it appears in the browser) are shown here:

ASP.NET code: `<asp:TextBox ID="txtNoOfTickets" runat="server" Width="99px"> </asp:TextBox>`

HTML element: `input name="txtNoOfTickets" type="text" id="txtNoOfTickets" style="width:99px;"/>`

The web server control TextBox is rendered by the server to the HTML element `<input>`. Note that the HTML element doesn't have the `runat` attribute because it is processed in the browser.

The ASP.NET code and the corresponding HTML element for the Button, btnClear, are:

ASP.NET code: `<asp:Button ID="btnClear" runat="server" Text="Clear" Width= "111px" />`

HTML element: `<input name="btnClear" type="submit" value="Clear" id="btnClear" style="width:111px;" />`

The form element, which is not a web server control, is represented by the following ASP.NET code and the corresponding HTML element:

ASP.NET code: `<form id="form1" runat="server">`

HTML element: `<form method="post" action="./Reservation-1.aspx" id="form1">`

The ASP.NET code for the form doesn't have the "`asp:`" prefix because the form is not a web server control.

In the previous examples, each server control, like TextBox and Button, is represented by a single HTML `<input>` element (discussed next). However, because web server controls are not part of the HTML language, some web server controls use multiple HTML elements to expand the functionality of individual HTML controls. For example, the **DropDownList** control is represented by two HTML elements, `<select>` and `<option>`, as follows:

```
<select name="ddlPlays" style="width:184px;" >
   <option value="3"> Young Irelanders </option>
   <option value="1"> Justin Hayward </option>
   <option value="2"> TedxOshkosh </option>
 ...
</select>
```

---

1. Displaying source code may vary by browser, so you may have to research how to do it in your browser.

Categories of Web Server Controls

You can find other commonly used server controls on the **standard** tab of the Toolbox. In addition to the *standard* tab, there are several other Toolbox tabs that contain different groups of controls, including:

- **Login:** Controls that enable you to create a login page and build secure web applications
- **Data:** Controls that help access data from databases and display the data
- **Validation:** Controls that validate input data
- **Navigation:** Controls that enable you to create navigation aids, including menus
- **WebParts:** Controls to create web pages that let users personalize web pages from the browser
- **Ajax Extensions:** Controls that help manage client-side scripts

Review Question

16. Consider the following ASP.NET code:

```
<asp:TextBox ID="txtNoOfTickets" runat="server" </asp:TextBox>
```

The corresponding HTML markup is

```
<input name="txtNoOfTickets" type="text" id="txtNoOfTickets" />
```

Explain why the `runat="server"` attribute is missing in the HTML markup.

6.9 HTML Controls

The HTML tag of the Toolbox provides a group of controls called **HTML controls**, as shown in Figure 6-24, which you can drag and drop on the form to generate the HTML elements.

These controls are part of the HTML language. Each control is represented by a single HTML element.

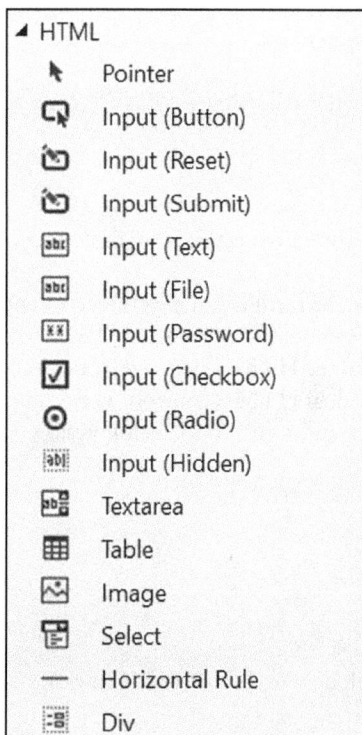

As an example, let's look at the **Input (Text)** control, which is the HTML control corresponding to TextBox. Adding this control to the form adds an `<input>` element to the aspx code that looks like this:

```
<input id="Text1" type="text"/>
```

This is identical to the HTML code rendered by the server when the web server control, TextBox, is processed on the server. It is tempting to use HTML controls because there is no need to render them to HTML, and they may improve the performance of the web page compared with web server controls.

However, there is a significant difference that makes HTML controls generally unsuited in web form applications. The HTML controls are **client-side** controls, which means they are not accessible by programs running on the server. So, you cannot access the data entered into an HTML **Input (Text)** control (also called **Text Field**) to do calculations using server-side code. Further, the data you enter into a Text Field is lost when the page is refreshed as a result of a page postback.

HTML controls may be used in special cases where you need only client-side functionality for the control. For example, to do animations with images using client-side JavaScript, you may use the HTML control Image.

HTML
- Pointer
- Input (Button)
- Input (Reset)
- Input (Submit)
- Input (Text)
- Input (File)
- Input (Password)
- Input (Checkbox)
- Input (Radio)
- Input (Hidden)
- Textarea
- Table
- Image
- Select
- Horizontal Rule
- Div

Figure 6-24: HTML controls in the Toolbox

6.9.1 HTML Server Controls

You can easily convert an HTML control to an HTML server control to address the limitations of HTML controls by adding the **runat** attribute to the HTML element as in this code:

```
<input id="Text1" type="text" runat="server"/>
```

HTML server controls, like web server controls, are processed on the server. The data you enter into them can be accessed by server-side code, and the data is not lost when the page is refreshed.

However, HTML server controls have less functionality than web server controls. You also cannot double-click an HTML button (or submit) control to generate the click event handler header in the code file. You typically will have to add a function to the aspx file within a script element.

Web server controls have several other advantages over HTML server controls:

- The ability to detect the capabilities of the browser and render HTML markup accordingly. Thus a control may look different in a mobile browser than in a laptop.
- A richer object model
- The ability to define a custom layout for the controls using templates
- The ability to define a consistent look for controls using themes

If the content of a control needs to be accessed by code on the server, **the web server control is preferable** because of its advantages over HTML server controls. However, understanding HTML server controls is beneficial because they are suitable under certain conditions. For example, developers may use an HTML server control when they want to write their own HTML markup and/or when they want every browser to see the same set of HTML markups.

Step 28: Verify that HTML controls cannot be used in server-side code.

In Design view, add the HTML control Input (Text) to the right of `txtNoOfTickets`.

Set the ID property of the control to `txtDemo`.

Double-click the Clear button to open the Click event handler, and type in the following statement:

```
txtDemo.Text = null;          //invalid
```

You will see a red line under txtDemo, indicating that there is an error.

Move the cursor over the red line to view the error message,

```
"The name txtDemo does not exist in the current context."
```

What causes the error, even though `txtDemo` is a valid name? The C# code is server-side code, which is processed on the server, but the HTML control, `txtDemo`, is a client-side control, which cannot be accessed on the server.

Step 29: Verify that the value you enter into an HTML control is lost when the page is refreshed as a result of a page postback.

Comment out the following statement:

```
txtDemo.Text = null;
```

Display the page in the browser.

Enter a number into each of the two TextBoxes, `txtNoOfTickets` and `txtDemo`.

Select a performance from the DropDownList.

Note that the number you entered into txtDemo disappears but the number entered into `txtNoOfTickets` is preserved. Why?

Selecting an item from the DropDownList causes a page postback that refreshes the page. A web server control retains its value when a page is refreshed. But an HTML control like `txtDemo` doesn't retain its value.

Step 30: Verify that an HTML server control can be accessed on the server.

Change the HTML control, txtDemo, to an HTML server control by adding the `runat= "server"` attribute to the Input element. The Input element looks like this:

```
<input id="txtDemo" runat="server" type="text" />
```

Uncomment the statement, `txtDemo.Text = null;`.

There is no red line under the name, txtDemo. But there is a red line under the Text property.

Move the cursor on the red line to view the error message:

```
"htmlInputText does not contain a definition for 'Text' …"
```

As the message says, the `Input(Text)` control doesn't have a Text property. Instead, it has a Value property.

Change the Text property of `txtUnitrPrice` to Value, like this:

```
txtDemo.Value = null;
```

There is no more error because `txtDemo` now is an HTML server control that can be accessed from the server-side code.

Display the page in the browser, enter a number into txtDemo, and click the Clear button.

The number in txtDemo is cleared, indicating that the server-side code in the Click event handler of the Clear button was able to access txtDemo.

Step 31: Verify that an HTML server control retains its value when the page is refreshed as a result of postback.

Display the page in the browser; enter a number into `txtNoOfTickets` and `txtDemo`.

Select a performance from the DropDownList, which causes a page postback.

The HTML server control, txtDemo, doesn't lose its value when the page is refreshed, just like the web server control `txtNoOfTickets`.

Thus, for this application, the HTML server control works well. However, because of the advantages of web server control discussed earlier, we will use web server controls for user input.

Delete the statement `txtDemo.Value = null;` from the click event handler of the Clear button, and delete the HTML server control, txtDemo.

Now the page is back to where it was before our experimentation with HTML control.

In general, stay away from the HTML controls in the Toolbox while developing web forms in this chapter.

Review Questions

17. List two ways HTML controls differ from web server controls.

18. What are the advantages of web server controls over HTML controls?

19. How do you convert an HTML control like Input(Text) control to an HTML server control?

20. List three advantages of web server controls over HTML server controls.

Exercises

Exercise 6-1

Use the appropriate Validator control to check whether the number of tickets entered by the user is between 1 and 5. (Reservations for more than five tickets are considered group reservations, which are treated differently.) If the number is not within the range, give immediate feedback with an appropriate message.

Assignments

Auto Rental Assignment: Part A (ASP.NET)—UI Controls

This assignment is a modified version of the Auto Rental Assignment from Chapter 4. The assignment in Chapter 4 used the HTML web form, whereas this assignment uses the ASP.NET web form. "Ace Auto Rentals" is a small business that rents automobiles. Ace would like you to develop a web page that lets the user select a type of rental vehicle and specify related information. This exercise develops the user interface.

Develop a web form named **Auto-selection**, within an ASP.NET web application named **AutoRental-ASP-WebForm.** You may develop the application within the WebProjects folder. The auto-selection web form lets the user select an auto type, a discount category, and insurance, and enter the number of rental days, as shown below. Choose an appropriate layout. What is shown here is just an example.

The first phase of the web form developed in this exercise should let the user do the following:

1. Select a type of auto from a drop-down list that displays a list of auto types; for example,

    Standard SUV, Full-size SUV, Minivan, 12 passenger van

2. Select one of the discount categories (Best rate, Govt. Employees, Business or Favorite Customer) from a RadioButtonList.

3. Specify whether the customer needs insurance by checking a CheckBox.

4. Enter the number of days of the rental into a TextBox.

Create the captions and controls, including the buttons, shown in the figure. You don't have to write the code to compute the costs and discount.

E-Commerce Assignment: Part A (ASP.NET)—UI Controls

This assignment is a modified version of the E-Commerce Assignment from Chapter 4. The assignment in Chapter 4 used the HTML Web Form, whereas this assignment uses the ASP.NET Web Form. P&I Laptops is a small business that assembles and sells a limited set of laptop models online. In this exercise, you will develop a segment of an e-commerce web form for P&I.

Create an ASP.NET Web Application named **PI-eComm-ASPWebForm** and develop a web form named **laptop-selection**. You may develop the application within the WebProjects folder. The **web page** should allow the user to select a laptop model and other options, as shown below. Choose an appropriate layout. What is shown here is just an example.

The form should allow the user to do the following:

1 Select a model from a drop-down list that displays a list of models. The current models include S330, S380, S580, G780, and B780.

2 Select an optional productivity software from a radio button list.

3 Specify whether the user wants an extended warranty.

4 Enter the quantity into a TextBox.

Create the captions and controls, including the buttons, as shown in the figure. You don't have to write the code to compute the costs.

Chapter 7

C# Server-Side Scripts

This chapter describes the development of server-side code in ASP.NET web forms to do tasks in response to events such as selecting an item from a DropDownList or CheckBox. You will continue the development of the Reservation-1 form by adding the code to process the input data entered by the user. You also will learn how the value of the AutoPostBack property of a control affects the processing of event handlers, and how to preserve data when a web page is posted back.

Learning Objectives

After studying this chapter, you should be able to:

- Use server-side code in event procedures.
- Describe the effect of the AutoPostBack property on controls.
- Explain the limitation of class level variables in web forms.
- Use View State to preserve the data between page postbacks.

7.1 C# Code in a Web Form

C# server-side code is used in a web form to specify the actions to be taken in response to events. The events could be those associated with controls (e.g., `Click`, `Validating`, `SelectedIndexChanged`) or with the page (e.g., `Page_Load`, `Init`). You have already used the `Page_Load` event handler in Chapter 6 to add performance names to the drop-down list in the reservation form. In this chapter, you will work with the reservation form to develop additional event handlers that compute prices and discounts. Typically, input data from a web form is submitted to the server and processed on the server.

Through the tutorial in this chapter, you will learn how to use C# code to do the following tasks in the Reservation-1 form:

1. Compute and display the base price when the user selects a performance.
2. Compute and display the zone price when the user selects a zone.
3. Display the discount when the user selects the member status.
4. Display the total price per ticket (unit price) when the user makes a selection.
5. Reset all controls when the user clicks the Clear button.

Figure 7-1 shows the reservation form.

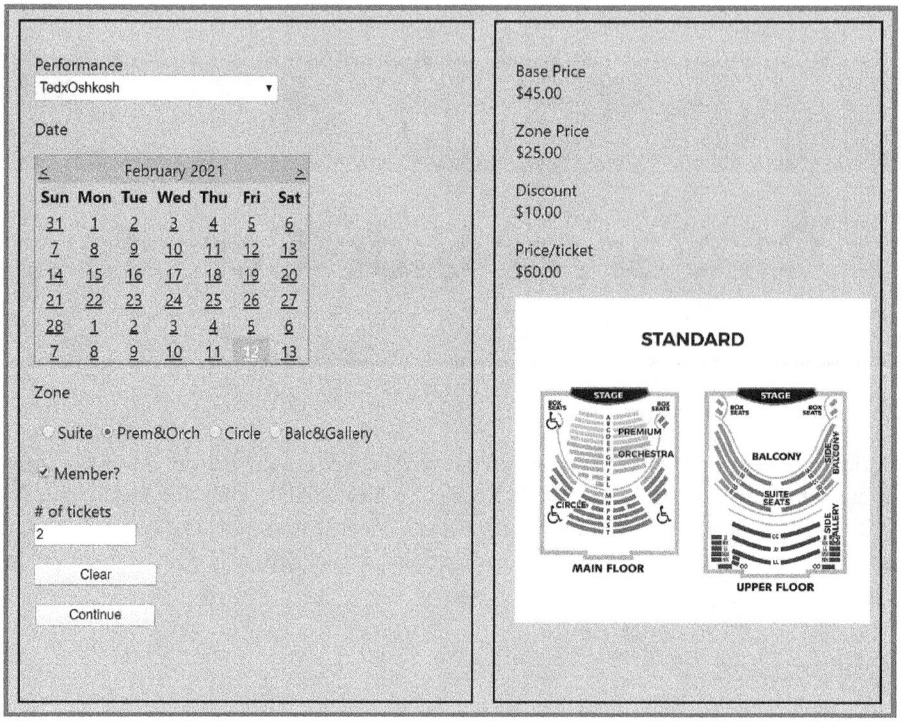

Figure 7-1: The Reservation-1 form

Computing the Base Price

The base price is computed and displayed when the user selects a performance from a drop-down list. You do this using the `SelectedIndexChanged` event handler, which is invoked when the user selects an item from the drop-down list. The event handler is shown in Figure 7-2.

```
1  protected void ddlPerformance_SelectedIndexChanged(object sender, EventArgs e)
2  {
3      string selectedPerformance = ddlPerformance.SelectedItem.ToString();
4      switch (selectedPerformance)
5          {
6          case "Young Irelanders":
7              basePrice = 65;
8              break;
9          case "Justin Hayward":
10             basePrice = 50;
11             break;
12         case "TedxOshkosh":
13             basePrice = 45;
14             break;
15         case "OSO Orchestral":
16             basePrice = 35;
17             break;
18         default:
19             Response.Write("<script> alert('Invalid performance!'); </script>");
20             break;
21
22         }
23      lblBasePrice.Text = basePrice.ToString("C"); // Display base price
24  }
```

Figure 7-2: SelectedIndexChanged event handler of ddlPerformance

Note that the variable basePrice used in this event procedure is not declared within the event procedure. Because basePrice is to be shared between this event procedure and similar event handlers associated with the RadioButtoList and the CheckBox, as explained later, basePrice is declared at the class level (outside of all event procedures), as shown in Step 1.

Line 3 gets the selected performance using the `SelectedItem` property of the drop-down list.

The Response.Write() function used in line 19 displays the message "Invalid performance" using the JavaScript function *alert()*. In general, the Response.Write() writes the specified string (e.g., `"<script> alert('Invalid performance!') </script>"`) into the HTTP response sent to the client. This function is described in more detail in Chapter 9. Note that the *default* clause will be executed only if the performances specified in the *switch* statement is inconsistent with the performances added to the drop-down list.

Lines 4–22 use the `switch` statement to find the base price for the selected performance. The prices are hard-coded in this version of the page. In the next version, you will look up the price from a database.

Line 19 displays the base price in a Label.

Step 1: Declare class-level variables to be shared between event procedures:

Declare variables basePrice, zonePrice, and discount at the class level, as shown in line 3:

```
1    public partial class Reservation_1 : System.Web.UI.Page
2        {
3            decimal basePrice, zonePrice, discount, unitPrice;
4            protected void Page_Load(object sender, EventArgs e)
```

Step 2: Create the SelectedIndexChanged event handler for the drop-down list.

Double-click the DropDownList in the Design view.

Visual Studio opens the Reservation-1.aspx.cs file and displays the `SelectedIndexChanged` event handler.

Insert the code from Figure 7-2 to compute and display the base price.

Display the page and make sure the right price is displayed when you select a performance.

AutoPostBack Property of the DropDownList

It is important to note that the above code that computes the base price is executed on the server when the page is posted back in response to the user clicking the DropDownList. The Postback happens because the AutoPostBack property of the DropDownList is set to *True*.

You may set the AuoPostBack to *False* and display the page. The base price won't be displayed because the server-side code in the event handler is not executed without a postback.

The following step shows what happens if the AutoPostBack property is set to False, which is the default for all controls except Buttons,

Step 3: Demonstrate the effect of AutoPostBack property.

Set the AutoPostBack property of the DropDownList to False.

Display the page and select a performance.

Note that the base price is not displayed in the Label. Though the `SelectedIndexChanged` event is raised on the client, the server-side C# code in the event handler is not executed because the page is not posted back to the server.

Set the AutoPostBack property back to True.

Computing the Zone Price

The process of finding the zone price when the user selects a zone is identical to finding the base price. You use the SelectedIndexChanged event handler of the RadioButtonList to execute the code, as shown in Figure 7-3, to find the zone price.

```
1   protected void rblZones_SelectedIndexChanged(object sender, EventArgs e)
2   {
3           string selectedZone = rblZones.SelectedValue.ToString();
4           switch (selectedZone)
5           {
6               case "Suite":
7                   zonePrice = 40;
8                   break;
9               case "Prem&Orch":
10                  zonePrice = 25;
11                  break;
12              case "Circle":
13                  zonePrice = 10;
14                  break;
15              case "Balc&Gallery":
16                  zonePrice = 0;
17                  break;
18          default:
19              Response.Write("<script> alert('Invalid zone!') </script>");
20              break;
21          }
22          lblZonePrice.Text = zonePrice.ToString("C"); // Display zone price
23  }
```

Figure 7-3: SelectedIndexChanged event handler of rblZones

Line 3 uses gets the selected zone using the SelectedValue property of the RadioButtonList. Lines 4–21 use a switch statement to find the zone price.

Step 4: Get the zone price.

Repeat the process from Steps 1–2 to create the SelectedIndexChanged event handler for the RadioButtonList, and insert the code from Figure 7-3 to compute and display the zone price.

Display the page and verify that the correct price is displayed when you select a zone.

Computing the Discount

To compute and display the discount when the user checks the CheckBox, use the CheckedChanged event to execute the code shown in Figure 7-3.

```
1   protected void chkMember_CheckedChanged(object sender, EventArgs e)
2   {
3       // Compute and display discount for members
4       if (chkMember.Checked)
5           discount = 10;
6       lblDiscount.Text = discount.ToString("C"); // Display discount
7   }
```

Figure 7-4: CheckedChanged event handler of chkMember

Double-click the CheckBox to create the event handler, and add the code from Figure 7-4 to display discount, if any.

Display the page and test the code.

Clearing the Controls

The following click event handler of btnClear clears all controls, as described below.

```
1   protected void btnClear_Click(object sender, EventArgs e)
2   {
3       ddlPerformance.SelectedIndex = -1;
4       calPerformDate.SelectedDates.Clear();
5       rblZones.SelectedIndex = -1;
6       chkMember.Checked = false;
7       txtNoOfTickets.Text = String.Empty;
8       foreach (Control control in form1.Controls)
9       {
10          if (control.GetType() == typeof(Label))
11          {
12              Label lbl = (Label)control;
13              lbl.Text = String.Empty;
14          }
15      }
16  }
```

Lines 3 and 5 set the `SelectedIndex` property of the two controls to -1, which makes no item selected in each control.

Line 4 unselects the currently selected date(s), if any. An alternative is to select today's date and make the corresponding month visible on the form using the following code:

```
calPerformDate.SelectedDate = DateTime.Today;
calPerformDate.VisibleDate = calPerformDate.SelectedDate;
```

In line 8, each iteration through the loop stores the reference to a member of the Controls collection into the variable *control*. Form1 is the value of the id attribute of the form element.

Line 10 checks whether the control is a Label. The *GetType()* method gets the type of a control. Essentially, *typeof(Label)* gets the (compile time) type of *Label*.

If it is, line 12 casts the variable control to Label.

Line 13 sets the Text property of the control to an empty string.

Insert the code presented above into the click event handlers of btnClear to clear the controls.

Display the page and test the code.

The Complete Code

Figure 7-5 shows the complete C# code to display the base price, zone price, and discount.

```
1  decimal basePrice, zonePrice, discount;
2  protected void Page_Load(object sender, EventArgs e)
3  {
4      if (!IsPostBack)
5      {
6          string[] performances = { "—Select one—",Young Irelanders", "Justin Hayward",
                                    "TedxOshkosh", "OSO Orchestral" };
7          foreach (string performance in performances)
8              ddlPerformance.Items.Add(performance);
9          UnobtrusiveValidationMode = UnobtrusiveValidationMode.None;
10         txtNoOfTickets.Focus();
11     }
12 }
13
14 protected void ddlPerformance_SelectedIndexChanged(object sender, EventArgs e)
15 {
16     string selectedPerformance = ddlPerformance.SelectedItem.ToString();
17     switch (selectedPerformance)
18     {
19         case "Young Irelanders":
20             basePrice = 65;
21             break;
22         case "Justin Hayward":
23             basePrice = 50;
24             break;
25         case "TedxOshkosh":
26             basePrice = 45;
27             break;
28         case "OSO Orchestral":
29             basePrice = 35;
30             break;
31         default:
32             Response.Write("<script> alert('Invalid performance!') </script>");
33             break;
34     }
35     lblBasePrice.Text = basePrice. ToString("C"); // Display base price
36 }
37
38 protected void rblZones_SelectedIndexChanged(object sender, EventArgs e)
39 {
40     string selectedZone = rblZones.SelectedValue.ToString();
41     switch (selectedZone)
42     {
43         case "Suite":
44             zonePrice = 40;
45             break;
46         case "Prem&Orch":
47             zonePrice = 25;
48             break;
49         case "Circle":
50             zonePrice = 10;
51             break;
52         case "Balc&Gallery":
53             zonePrice = 0;
54             break;
55         default:
56             Response.Write("<script> alert('Invalid zone!') </script>");
57             break;
58     }
59     lblZonePrice.Text = zonePrice.ToString("C") // Display zone price
60 }
61
62 protected void chkMember_CheckedChanged(object sender, EventArgs e)
63 {
64     // Compute and display discount for members
65     if (chkMember.Checked)
66         discount = 10;
```

Figure 7-5: The complete code to display prices (*continues*)

```
67     lblDiscount.Text = discount.ToString("C"); // Display discount
68 }
69
70 protected void btnClear_Click(object sender, EventArgs e)
71 {
72     ddlPerformance.SelectedIndex = -1;
73     calPerformDate.SelectedDates.Clear();
74     rblZones.SelectedIndex = -1;
75     chkMember.Checked = false;
76     txtNoOfTickets.Text = String.Empty;
77     foreach (Control control in this.Controls)
78     {
79         if (control is Label)
80         {
81             Label lbl = (Label)control;
82             lbl.Text = String.Empty;
83         }
84     }
85 }
```

Figure 7-5: The complete code to display prices (*continued*)

7.1.1 Preserving Values of Web Server Controls during Postback

When a page is refreshed in response to a postback, the values and properties of all web server controls are automatically preserved as you will see in the next step. These values are retained in hidden fields in the form.

Step 7: Verify that the values of web server controls are preserved during postback.

Display the page, and enter a value for the number of tickets. (You don't use this value in the current version of the form.)

Select a performance, which results in a postback of the page.

The C# code in the `SelectedIndexChanged` event of the DropDownList runs and displays the base price, as expected.

Notice that the value you entered into the TextBox is preserved when the page is displayed in the browser after postback.

However, values of variables are not automatically preserved during a postback. Special care should be taken to preserve them. We will discuss this important aspect when we develop the code to compute and display the total price per ticket in the next section.

Review Questions

1. Why is the code in the SelectedIndexChanged event handler not processed when the user selects a performance from the drop-down list if the AutoPostBack property of the DropDownList is set to False?

2. How is the value you enter into a TextBox preserved when the page is posted back?

7.2 Preserving Values of Variables on Postback

The event handlers developed in previous sections to compute prices used variables that are accessed only within a single event handler. For example, as Figure 7-5 shows, the variable `basePrice` is used within the `SelectedIndexChanged` event handler to temporarily store the base price when the event handler is processed on the server. The variable is not accessed from any other event handler when the page is posted back.

But oftentimes it is necessary to assign a value to a variable in one event handler and use it in another one. This would be the case when we update the current form to compute the total price per ticket (Base price + Zone price − Discount), whenever the user selects a performance, zone, or member status, and display the price per ticket (unit price), as shown in Figure 7-6.

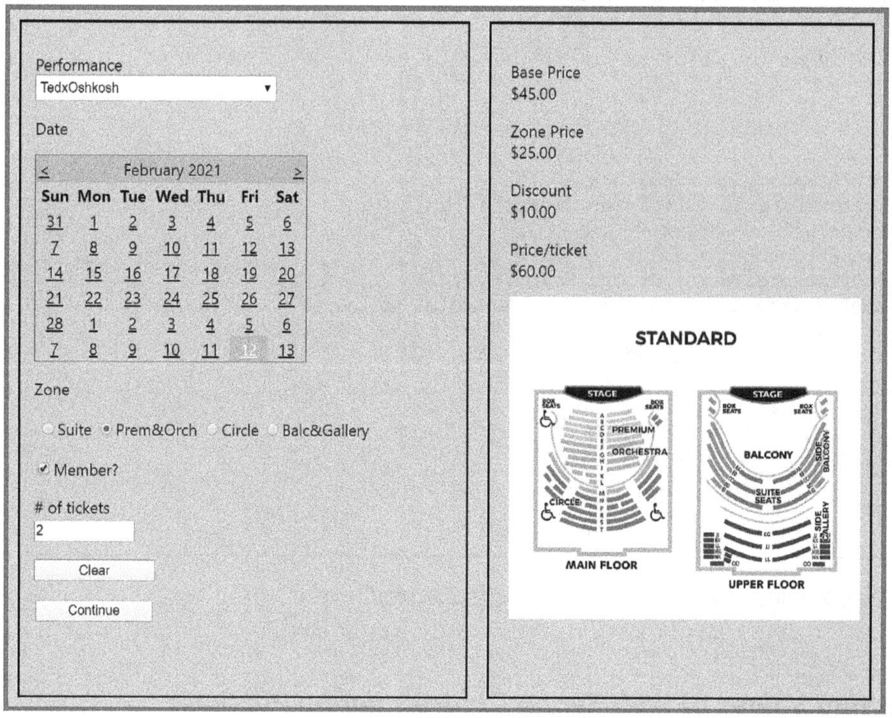

Figure 7-6: Reservation form with price per ticket

This would require assigning a value for basePrice, for example, in the `SelectedIndexChanged` event handler of the DropDownList control and using it in other event handlers to compute the total price per ticket.

7.2.1 The Problem with Using Class (Field) Level Variables to Preserve Data in a Web Form

The variables basePrice, zonePrice, and discount are declared in Step 7-1 as class-level variables (outside all methods) so that you can use them in multiple methods, as shown here:

```
1  public partial class Reservation_1 : System.Web.UI.Page
2      {
3      decimal basePrice, zonePrice, discount;
4      protected void Page_Load(object sender, EventArgs e)
5      ...
```

However, in a web form, the values stored in class level variables are **not** preserved when the page is refreshed after a postback. This is because every time the user selects an item, the page is posted back to the server and a new page object is created.

So, if you compute the base price in one event hander, how do you access it from another event handler after the page is posted back? Computing the price per ticket using the prices and discount displayed in the Labels may look like an obvious choice. However, programs that are tightly linked to the user interface tend to be difficult to reuse and maintain. For example, a program that uses the data from the Labels won't work if the user interface is changed in such a way that a ListBox is used in place of the Labels to display data or the input data items are not displayed at all.

Let's see what happens if we use the class level variables `basePrice`, `zonePrice`, and `discount` to compute the unit price whenever the user selects a performance, zone, or member status.

Step 8: Compute and display the unit price.

Insert the comment and code shown in lines 12–14 of Figure 7-7 to compute and display the unit price, at the end of the `SelectedIndexChanged` event handler of the DropDownList.

```
1   protected void ddlPerformance_SelectedIndexChanged(object sender, ...)
2   {
3       string selectedPerformance = ddlPerformance.SelectedItem.ToString();
4       switch (selectedPerformance)
5           {
6           case "Young Irelanders":
7               basePrice = 65;
8               ...
9               ...
10          }
11      lblBasePrice.Text = basePrice.ToString(); // Display base price
12      // Compute and display price per ticket
13      unitPrice = basePrice + zonePrice - discount;
14      lblUnitPrice.Text = unitPrice.ToString("C");
15  }
```

Figure 7-7: Updated event handler for ddlPerformance

Step 9: Update zone price.

Insert the lines of code from lines 12–14 of Figure 7-8 at the end of the SelectedIndexChanged event handler of the RadioButtonList, and put a debugger **breakpoint** at the closing brace of the event handler.

```
1   protected void rblZones_SelectedIndexChanged(object sender, EventArgs e)
2   {
3       string selectedZone = rblZones.SelectedValue.ToString();
4       switch (selectedZone)
5       {
6           case "Suite":
7               zonePrice = 40;
8           ...
9           ...
10      }
11      lblZonePrice.Text = zonePrice.ToString(); // Display zone price
12      // Compute and display price per ticket
13      unitPrice = basePrice + zonePrice - discount;
14      lblUnitPrice.Text = unitPrice.ToString("C");
15  }
```

Figure 7-8: SelectedIndexChanged event handler of the RadioButtonList

Insert the same code at the end of the CheckedChanged event handler of the CheckBox, and **put a break at the closing brace**.

Note that the two lines of code to compute and display the unit price is repeated in three different methods. In general, when a group of statements that do a well-defined task is repeated in multiple locations, it is recommended that you create a separate method to do the task and call the method from each location. This would make maintenance and reuse easier.

Step 10: Display the page in the browser.

Select TedxOshkosh for performance. You should see the following values for base price and price/ticket (unit price):

Base Price
45

Zone Price

Discount

Price/ticket
45

The base price and unitPrice (basePrice + zonePrice − discount) are correctly computed as 45; basePrice has the value 45, and the variables zonePrice and discount still have their default value of zero. So, everything is fine.

The code in the `SelectedIndexChanged` event handler of the DropDownList was run on the server to find the value of the base price. The event handler also computed the unit price and displayed it in the Label. Because everything was done within the same method, there was no need to persist data during a postback.

Next, select the zone, *Suite*, from the RadioButtonList. The program breaks at the closing brace of the `SelectedIndexChanged` event handler of the RadioButtonList. Hover the pointer over the four variables.

The `zonePrice` has the value 40, as expected. But notice that the class level variable `basePrice` **shows a value of 0**, resulting in the wrong `unitPrice` value of 40, instead of the correct value of 85 (45 + 40 − 0).

Remove the break. Click Continue on the Toolbar to display the values as shown:

Base Price

Zone Price
40

Discount

Price/ticket
40

Why did the value of `basePrice` change from 45 to 0 when you selected a zone? When you select a zone, the page is posted back to the server so that ASP.NET can run the code in the `SelectedIndexChanged` event handler of the RadioButtonList. After running the code, the server generates the HTML markup for the page that displays the unit cost and preserves the properties of the server controls. However, the values of variables declared in the code are not preserved.

Click Continue on the Toolbar, and check the CheckBox, indicating the client is a member. The program breaks at the closing brace. The values of variables basePrice and zonePrice are zero, indicating that the values of these class-level variables are not preserved when the page is posted back. Click Continue to display the values. Remove the break.

So, how do you preserve the prices and discount between postbacks of the page? One method to preserve the values of variables within a page between postbacks is to use what is called **view state**.

It's time to practice! Do Exercise 7-1 (see the end of the chapter).

7.3 Preserving Values of Variables Using View State

Because HTTP is a stateless protocol, each HTTP request for a page is treated as an independent request. The server does not retain information from previous requests from a browser by a user.

ASP.NET features that help address this limitation include **view state**, **session state**, and **application state**. Storing data in view state allows you to access the data from a single page during a session, whereas storing data in session state lets you access the data from multiple pages during a session. Application state preserves data across all sessions of an application, for all users. Next, you will learn how to use view state to preserve the values of variables when the page is posted back.

7.3.1 View State

View state stores data within a page (in hidden fields) to be accessed only from the same page. Storing data within the page makes view state less secure than session state, which stores data on the server.

As stated earlier, when a web page is refreshed due to a postback, ASP.NET automatically preserves the state of server controls and the data entered by the user into TextBoxes. ASP.NET also preserves values of page properties using view state. So, you don't have to do any coding to persist the values of page properties or server controls when a page is posted back.

However, the values of variables are not preserved when a page is posted back. You can preserve a data item like base price by storing it in a view state item, which is also referred to as a **view state variable**.

You store data in view state items by adding items to the *view state object*. The items you store in the view state object are of the key/value type (consists of a key and a corresponding value). You access the view state object using the ViewState property of the page. So, you store an item in the view state object using the pattern:

```
ViewState.Add (name, value);
```

or

```
ViewState["name"] = value;
```

For instance, the following statement adds the value of *basePrice* to the collection using the item name (view state variable) *BasePrice*:

```
ViewState.Add("BasePrice", basePrice);
```

or

```
ViewState["BasePrice"] = basePrice;
```

Here BasePrice is the key, and basePrice is the value. An item could be of any .Net Framework type, including value types, arrays, and certain types of collections and objects.

Note that you don't have to declare the view state variable, BasePrice. However, when you retrieve its value, you need to cast it to the appropriate type, determined by the type of the value assigned to it.

To retrieve the value from the view state object, use the name of the item, like this:

```
double basePrice = (double) ViewState["BasePrice"];
```

The item should be cast to the appropriate type.

Because view state is stored within the page, data stored in view state variables is transferred between the browser and the server every time the page is posted back. So, storing large amounts of data in ViewState may affect the performance of the page. Table 7-1 describes the properties/methods of ViewState.

Table 7-1: Properties/methods of ViewState objects

Properties/methods	Description
`Count`	Number of items in the collection: e.g., `int count = ViewState.Count;`
`Add (name, value)`	Adds an item to the collection, or replaces it if it exists: e.g., `ViewState.Add("BasePrice", basePrice);`
`Clear()`	Removes all items from the collection: e.g., `ViewState.Clear();`
`Remove (name)`	Removes the specified item from the collection: e.g., `ViewState.Remove("BasePrice");`

Using View State to Preserve Data in the Reservation-1 Form

Because the prices and discount data are needed only on the reservation page, you will use view state to store the prices and discounts in this initial version of the website. When the website is expanded to include a confirmation page that displays the data from the reservation page, you will use the session state instead of the view state.

As you learned in Section 7.2.1, using class-level variables as shown in Figure 7-9 to share data between methods doesn't work.

```
1   public partial class Reservation-1 : System.Web.UI.Page
2       {
3       decimal basePrice, zonePrice, discount;
4       decimal unitPrice;
5       ...
6       protected void ddlPerformance_SelectedIndexChanged(object sender, EventArgs e)
7       {
8           switch (selectedPerformance)
9               {
10              case "Young Irelanders":
11                  basePrice = 65;
12                  ...
13          // Compute and display price per ticket
14          unitPrice = basePrice + zonePrice - discount;  // Doesn't work
15          lblUnitPrice.Text = unitPrice.ToString("C");
16      }
```

Figure 7-9: Global variables to store prices

The variables `zonePrice` and `discount` do not retain their values assigned to them in other event handlers because the web page was posted back and refreshed after they were assigned the values. So, the values of these variables need to be stored in ViewState when they are computed within the event handlers so that the view state variables can be used to compute the unit price.

First, consider the modified event handler for the DropDownList shown in the next listing. The statement in line 11 of Figure 7-10 that computes the unit price using class-level variables is commented out and replaced by the two statements shown in lines 12 and 13.

```
 1  protected void ddlPerformance_SelectedIndexChanged(object sender, EventArgs e)
 2  {
 3      string selectedPerformance = ddlPerformance.SelectedItem.ToString();
 4      switch (selectedPerformance)
 5          {
 6          case "Young Irelanders":
 7              basePrice = 65;
 8          ...
 9      lblBasePrice.Text = basePrice.ToString("C"); // Display base price
10      // Compute and display price per ticket
11      // unitPrice = basePrice + zonePrice - discount; // Doesn't work
12      ViewState["BasePrice"] = basePrice;
13      unitPrice = (decimal)ViewState["BasePrice"] +
                  (decimal)ViewState["ZonePrice"] - (decimal)ViewState["Discount"];
14      lblUnitPrice.Text = unitPrice.ToString("C");
15 }
```

Figure 7-10: ViewState to store base price

The first statement (line 12) stores the value of basePrice in the view state variable BasePrice so that it can be used in other event handlers after page postbacks. To be precise, the statement adds an item named BasePrice to the view state and stores the value of basePrice in the item.

The second statement (line 13) computes the unit price using the zone price and discount that are stored in view state variables in the event handlers of the RadioButtonList and CheckBox, respectively. Unit price doesn't have to be stored in view state because the unit price computed in this event handler is not needed in any other event handlers; it is computed in each event handler.

You need to make similar changes to store zone price and discount in view states in the corresponding event handlers. Lines 2 and 3 in Figure 7-11 store the zone price in view state and use it to compute and display unit price in the event handler of the RadioButtonList.

```
 1  // unitPrice = basePrice + zonePrice - discount; // -- Doesn't work
 2  ViewState["ZonePrice"] = zonePrice;
 3  unitPrice = (decimal)ViewState["BasePrice"] +
             (decimal)ViewState["ZonePrice"] - (decimal)ViewState["Discount"];
 4  lblUnitPrice.Text = unitPrice.ToString("C");
```

Figure 7-11: ViewState to store zone price

For the CheckBox, use the corresponding code that appears in Figure 7-12.

```
 1  // unitPrice = basePrice + zonePrice - discount;  // -- Doesn't work
 2  ViewState["Discount"] = discount;
 3  unitPrice = (decimal)ViewState["BasePrice"] +
             (decimal)ViewState["ZonePrice"] - (decimal)ViewState["Discount"];
 4  lblUnitPrice.Text = unitPrice.ToString("C");
```

Figure 7-12: ViewState to store discount

Note that when the user selects a Performance, the user might not have selected the zone or the member status. So, the ZonePrice and Discount items may not exist in the view state when the event handler for the DropDownList is executed. Because the user may make selections in any order, or may reselect an item, it is necessary to create view state items for base price, zone price, and balance in the load event handler of the page, when the page is loaded the first time in a session. The code in Figure 7-13 creates the view state variables.

```
1  ViewState["BasePrice"] = (decimal)0;
2  ViewState["ZonePrice"] = (decimal)0;
3  ViewState["Discount"] = (decimal)0;
```

Figure 7-13: Creating view state variables

Figure 7-14 shows the complete code for the Reservation-1 page using view state.

```
1   public partial class Reservation_1 : System.Web.UI.Page
2       {
3       decimal basePrice, zonePrice, discount, unitPrice;
4       protected void Page_Load(object sender, EventArgs e)
5       {
6           if (!IsPostBack) // if it is not a postback
7           {
8               string[] performances = { "Young Irelanders", "Justin Hayward",
                                          "TedxOshkosh", "OSO Orchestral" };
9               foreach (string performance in performances)
10                  ddlPerformance.Items.Add(performance);
11              ViewState["BasePrice"] = (decimal)0;
12              ViewState["ZonePrice"] = (decimal)0;
13              ViewState["Discount"] = (decimal)0;
14          }
15          UnobtrusiveValidationMode = UnobtrusiveValidationMode.None;
16          txtNoOfTickets.Focus();
17      }
18
19      protected void ddlPerformance_SelectedIndexChanged(object sender,EventArgs e)
20      {
21          string selectedPerformance = ddlPerformance.SelectedItem.ToString();
22          switch (selectedPerformance)
23              {
24              case "Young Irelanders":
25                  basePrice = 65;
26                  break;
27              case "Justin Hayward":
28                  basePrice = 50;
29                  break;
30              case "TedxOshkosh":
31                  basePrice = 45;
32                  break;
33              case "Oshkosh Orchestral":
34                  basePrice = 35;
35                  break;
36              default:
37                  Response.Write("<script>alert('Invalid performance')</script>");
38                  break;
39              }
40          lblBasePrice.Text = basePrice.ToString("C"); // Display base price
41          // Compute and display price per ticket
42          // unitPrice = basePrice + zonePrice - discount; // -- Doesn't work
43          ViewState["BasePrice"] = basePrice;
44          unitPrice = (decimal)ViewState["BasePrice"] +
                      (decimal)ViewState["ZonePrice"] - (decimal)ViewState["Discount"]
45          lblUnitPrice.Text = unitPrice.ToString("C");
46      }
47
48      protected void rblZones_SelectedIndexChanged(object sender, EventArgs e)
49      {
50          string selectedZone = rblZones.SelectedValue.ToString();
51          switch (selectedZone)
52              {
```

Figure 7-14: The complete code for Reservation-1 page (*continues*)

```
53              case "Suite":
54                  zonePrice = 40;
55                  break;
56              case "Prem&Orch":
57                  zonePrice = 25;
58                  break;
59              case "Circle":
60                  zonePrice = 10;
61                  break;
62              case "Balc&Gallery":
63                  zonePrice = 0;
64                  break;
65              default:
66                  Response.Write("<script> alert('Invalid zone!') </script>");
67                  break;
68
69          }
70          lblZonePrice.Text = zonePrice.ToString("C"); // Display zone price
71          // Compute and display price per ticket
72          // unitPrice = basePrice + zonePrice - discount; // -- Doesn't work
73          ViewState["ZonePrice"] = zonePrice;
74          unitPrice = (decimal)ViewState["BasePrice"] +
75                  (decimal)ViewState["ZonePrice"] - (decimal)ViewState["Discount"]
75          lblUnitPrice.Text = unitPrice.ToString("C");
76      }
77
78      protected void chkMember_CheckedChanged(object sender, EventArgs e)
79      {
80          // Compute and display discount for members
81          if (chkMember.Checked)
82              discount = 10;
83          lblDiscount.Text = discount.ToString("C"); // Display discount
84          // Compute and display price per ticket
85          // unitPrice = basePrice + zonePrice - discount;  // -- Doesn't work
86          ViewState["Discount"] = discount;
87          unitPrice = (decimal)ViewState["BasePrice"] +
                    (decimal)ViewState["ZonePrice"] - (decimal)ViewState["Discount"];
88          lblUnitPrice.Text = unitPrice.ToString("C");
89      }
90
91      protected void btnClear_Click(object sender, EventArgs e)
92      {
93          ddlPerformance.SelectedIndex = -1;
94          calPerformDate.SelectedDates.Clear();
95          rblZones.SelectedIndex = -1;
96          chkMember.Checked = false;
97          txtNoOfTickets.Text = String.Empty;
98          foreach (Control control in form1.Controls)
99          {
100             if (control.GetType() == typeof(Label))
101             {
102                 Lable lbl = (Label)control;
103                 control.Text = String.Empty;
104             }
105         }
106     }
107}
```

Figure 7-14: The complete code for Reservation-1 page (*continued*)

Step 11: Preserve data in the view state.

To preserve the data in the view state, you will need to make the changes in the code as shown in Figure 7-14. To do this, start by adding the statements shown in lines 11–13 to the Load event handler of the page.

Make the changes to the SelectedIndexChanged event handler of ddlPlays, as shown in lines 42–44, to use view state variables in the place of class-level variables.

Similarly, make the changes to the event handler of `rblZones`, as shown in lines 72–74, and make changes to the event handler `chkMember`, as shown in lines 85–87.

Run the website and test the code. Make sure the unit price (Base price + Zone price – Discount) is computed correctly.

Review Questions

3. What is the limitation of using a class-level variable to share data within a page?
4. Describe the characteristics of the view state.
5. What is the effect of executing this statement?

```
ViewState.Add("BasePrice", basePrice);
```

Exercises

Exercise 7.1

This exercise adds a CheckBox to let the user select the option to have snacks served at the Theater.

Add a new CheckBox to the Reservation-1 form to let the user select snacks. If the user selects snacks, then display $5 in a Label, as shown in the figure below. In addition, display the unit price that includes the snacks. Making other selections like performance, zone, and discount must display the updated unit price.

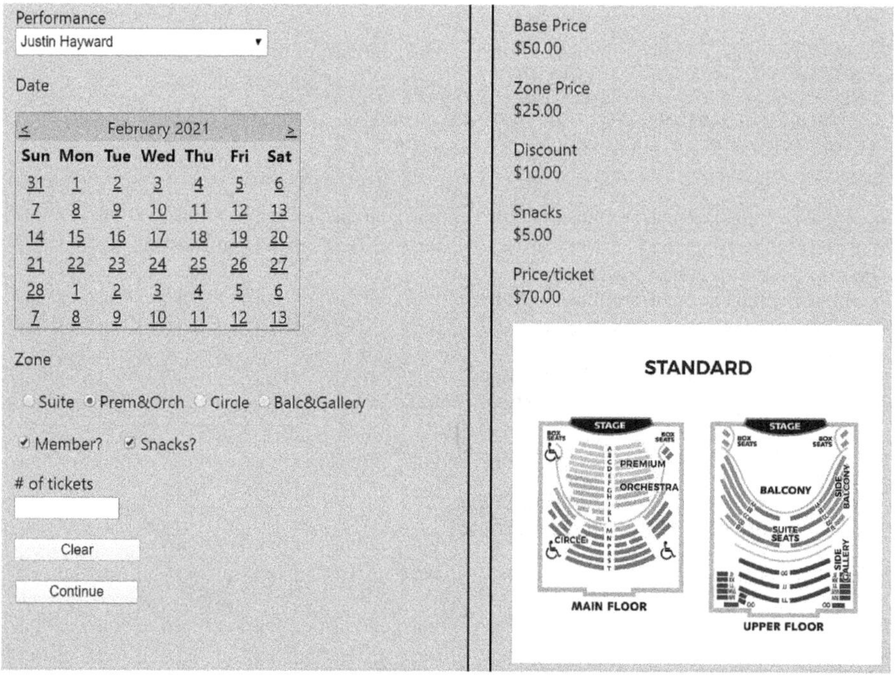

Assignments

Auto Rental Assignment: Part B (ASP.NET)—Server-Side Code

This assignment is an expanded version of the assignment, *Auto Rental: Part A (ASP.NET)—UI Controls*, from Chapter 6, which develops the user interface controls for a web page. This assignment develops the server-side code to process data. If you have already done *Part A*, you may add the code to the project; if not create a new project, as described here.

Ace Auto Rentals is a small business that rents automobiles. Ace would like you to develop a web page that lets the user select a type of rental vehicle and specify related information.

Develop a web form named **Auto-selection**, within the ASP.NEGT Web application named **AutoRental** that lets the user select an auto type, a discount category, and insurance, and enter the number of rental days as shown below. Choose an appropriate layout. What is shown here is just an example of the layout.

Auto Type: Standard SUV ▾	Standard Rate: $45.00
Discount Category:	Discount: $4.50
○ Best Rate ○ Govt. Emp. ○ Business ○ Favorite	Insurance: $10.00
☐ Insurance?	Cost/day: $50.50
Number of days: 4	Total Cost: $202.00
[Compute Cost] [Clear]	

The web form should let the user do the following:

1. Select a type of auto from a drop-down list that displays a list of auto types.

   When the user selects a model, display the corresponding standard (daily) rate.

   The auto types and rates are as follows:

   Compact car: $30, Full-size car: $40, Standard SUV: $45, Minivan: $60

   In addition, compute and display the cost/day (Standard rate + Insurance – Discount).

2. Select one of the discount categories from a RadioButtonList that displays a set of categories. When the user selects a category, display the corresponding discount amount. The categories and corresponding discount percent are as follows:

   Best Rate: 0%, Govt. Emp: 15%, Business: 10%, Favorite: 20%

   In addition, compute and display the cost/day.

3. Specify whether the customer needs insurance by checking a CheckBox. When the user selects or unselects insurance, display the cost. The cost of insurance is $10/day. In addition, compute and display the cost/day.

4. Enter the number of days of the rental into a TextBox.

5. Compute and display the total cost (Number of days × Cost/day) by clicking a button.

6. Clear the data by clicking a button.

Develop the user interface, including the captions, controls, and the server-side code to compute the results as specified above.

E-Commerce Assignment: Part B (ASP.NET)—Server-Side Code

This assignment is an expanded version of the assignment, *E-Commerce Assignment: Part A (ASP.NET)—UI Controls*, from Chapter 6, which develops the user interface controls for a web page. This assignment develops the server-side code to process data. If you have already done Part A, you may add the code to the project; if not, create a new project as described here.

P&I Laptops is a small business that assembles and sells a limited set of laptop models online. In this exercise, you will develop an e-commerce web form for P&I.

Create an ASP.NET Web Application named **PI-eComm** and develop a web form named **laptop-selection**. The web page should allow the user to select a laptop model and other options as shown below. Choose an appropriate layout. What is shown here is just an example of the layout.

The form should allow the user to do the following:

1. Select a model from a drop-down list that displays a list of models. When the user selects a model, display the corresponding price and the unit price (Laptop + Software + Warranty). The current models and their prices are:

   S330: $295, S380: $450, G780: $950, and B780: $995.

2. Select an optional productivity software from a radio button list. When the user selects a software, display the corresponding price and the unit price. Here are the software and their prices:

   Office (Pro): $119.95, Office (H): $149.95, Office 365 (H): $69.95, Office 365 (P): $69.95

3. Specify whether the user wants an extended warranty. If the user selects the warranty, display the cost ($99.50/unit) and the unit price.

4. Enter the quantity into a TextBox.

5. Display the total cost (Quantity × Unit price) by clicking a button.

6. Clear the data by clicking a button.

Develop the user interface, including the captions, controls, and the server-side code, to compute the results as specified above.

## Chapter 8

Database Applications

A typical characteristic of dynamic web pages is accessing databases to look up information that determines the contents of a web page. This chapter describes how to access a Microsoft SQL Server database to look up and display dynamic data on a web page. In addition, Appendix B* provides a brief introduction to organizing data and to the Structured Query Language (SQL), which is the standard language used to access databases. In the tutorial, you will create a new version of the reservation form named **Reservation-2** that gets the performance data from a database.

Learning Objectives

After studying this chapter, you should be able to:

- Describe how data is organized in a relational database.
- Create queries using the SQL language to retrieve data from one or more tables.
- Configure an SQL Server DataSource and bind it to a control.
- Look up information from an SQL Server database table using the DataSouce control.
- Filter data obtained from a database.

8.1 Introduction

The current version of the reservation form uses hard-coded data to display performances in the drop-down list and compute prices. For example, say you used a *switch* statement to set the price of *Young Irelanders* to $65 and *Justin Hayward* to $50, thus embedding the performance names and prices within the code itself. While this method of using data makes coding relatively easy, it makes programs difficult to maintain and reuse.

What would happen if *Justin Hayward* is no longer featured at The Grand Theater or the price for a play changes? Someone has to go through the tedious task of identifying the place(s) in the code where that data is used and changing them. Every time data is deleted, added, or changed, the code will have to be changed.

You can address the above problem with data maintenance by storing the data in a database that is accessed from the code but separate from the code. There are other types of data storage, including spreadsheets, text files, and XML (Extensible Markup Language) files. However, databases provide several advantages because they are managed by specialized software known as database management systems (DBMS), which incorporate features that make it easy to access data, reduce redundancy, and ensure security.

Some of the most commonly used DBMS are Oracle, Microsoft SQL Server, MySQL, DB2, and Microsoft Access. This book uses SQL Server. It should be noted that SQL (Structured Query Language) is the common language used in all these systems to access databases. Though there is some variation between the implementation of SQL in different systems, they are very similar. Here you will use SQL Server, which is available in Visual Studio.

To help you learn how to access databases using ASP.NET, you will create a second version of the reservation form named Reservation-2 that includes two additional features. Table 8-1 shows the differences between Reservation-1 and Reservation-2.

---

* Appendix can be freely downloaded from https://www.prospectpressvt.com/textbooks/philip-web-development at "Student Resources."

Table 8-1: Reservation-1 versus Resservation-2

Reservation-1	Reservation-2
The list of performances displayed in the drop-down list and the corresponding prices are hard-coded.	The list of performances displayed in the drop-down list and their prices are obtained from the database Theater.mdb.
The user selects the desired date of performance from a calendar that doesn't indicate the dates for which the performance is scheduled.	The user selects the desired date of performance from a drop-down list that gets and displays the dates for which the performance is scheduled.

The user interface for Reservation-2 is the same as in Reservation-1, except that the user selects the date from a second drop-down list that shows only applicable dates, instead of selecting from a calendar that shows all dates. Figure 8-1 shows the drop-down list, which replaces the Calendar control, to select the date.

Figure 8-1: Dropdown list for date

The tutorial in this chapter focuses on helping you learn how to use an SqlDataSource control to select data from a database and bind the DataSource to a drop-down list to display selected data in the drop-down list. You will create two separate DataSource controls—one bound to ddlPerformance and the other bound to ddlDate.

Theater Database, Data Organization, and SQL

In this chapter, you will work with the **Theater.mdb** database that is available in the Tutorial-starts/Data-files folder. If you are not familiar with how data is organized in relational databases, you should review the section on data organization in Appendix B* before proceeding further. There is only limited use of the SQL (Structured Query Language) in this chapter. SQL is the language used to select and retrieve data from databases. For the most part, you will use the SqlDataSource control to generate the SQL. However, if you are unfamiliar with the use of SQL, you are encouraged to review the section on SQL in Appendix B*.

You don't need to know how to create a database to understand the materials presented in this book. However, if you would like to learn how to create an SQL Server database, please see the steps given in Appendix C*.

The Theater database contains four tables: Performance, PerformDate, Category, and Zone. Only Performance and PerformDate tables are used in the tutorials. These two tables are shown here with subsets of the records:

1. **Performance** (PerformanceId, PerformanceName, BasePrice, Description, Presenter, Image)
   The Performance table stores the name, base price, description, presenter, and image for each performance.

PerformanceId	PerformanceName	BasePrice	Description	Presenter	image
11	Young Irelanders	35	Each and every year the Irish Cultu...	ICA	*NULL*
12	Justin Hayward	30	Justin David Hayward is an English...	ACG	*NULL*
13	TedxOshkosh	59	We believe in the power of a great...	Tedx	*NULL*
14	Orchestral Presents	20	Join the Oshkosh Symphony Orch...	OSO	0xFFD8F...
15	Ladies of Laughter	30	Since 2012 the Ladies of Laughter...	*NULL*	*NULL*
16	Steve Earl & The Dukes	25	If you ever had any doubt about ...	ACG	*NULL*
17	Harmonious Wail	35	Join us for a night of Wail n' good...	Wail	*NULL*

---

* Appendix can be freely downloaded from https://www.prospectpressvt.com/textbooks/philip-web-development at "Student Resources."

2. **PerformDate** (PeformDateId, PerformanceId, PerformDate)
   The PerformDate table stores the scheduled dates for each performance. PerformDateId is a unique id for each record.

PerformDateId	PerformanceId	PerformDate
101	11	1/3/2021 7:30:00 PM
102	11	1/4/2021 7:30:00 PM
103	11	1/5/2021 7:30:00 PM
104	11	1/10/2021 7:30:00 PM
105	12	1/11/2021 7:30:00 PM
106	12	1/6/2021 7:30:00 PM
107	13	1/7/2021 8:00:00 PM
108	13	1/7/2021 7:30:00 AM
109	14	1/8/2021 8:00:00 PM

Note that the PerformanceId field, which is the foreign key, links each PerformDate record with the corresponding Performance record, as represented below.

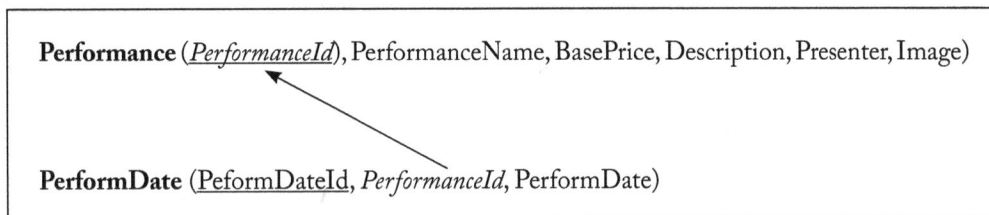

Performance (*PerformanceId*), PerformanceName, BasePrice, Description, Presenter, Image)

PerformDate (PeformDateId, *PerformanceId*, PerformDate)

The first four records in PerformDate, for example, have the PerformanceId value, 11, which is the PerformanceID value for Young Islanders. That is, Young Islanders is scheduled for the dates shown in the first four records. Thus there is a one-to-many relationship (described in Appendix B*) between Performance and PerformDate. You will use this relationship to display the dates in the second drop-down list when the user selects a performance from the first drop-down list.

8.2 Configuring a DataSource

To dynamically display the performance names in the drop-down list on the reservation form, you will get the performance names from the Performance table and add them to the drop-down list. When the user selects a performance, the corresponding price is looked up from the same table.

A relatively easy way to access a database table from a web form is using the SqlDataSource control, which includes the following three initial steps:

1. Create an ASP.NET folder named App_Data and copy the database to the App_Data folder. It is important to note that because the database is accessed by ASP.NET running on a server, the web forms cannot access databases stored on the client. You need to create an **ASP.NET folder** named **App_Data** that becomes a part of the application so that it can be accessed by the server.

2. Add the sqlDataSource control from the Data tab of the Toolbox to the reservation form.

3. Configure the DataSource to specify the database, table, fields, and selection criteria, if any.

---

* Appendix can be freely downloaded from https://www.prospectpressvt.com/textbooks/philip-web-development at "Student Resources."

Tutorial: Accessing the Theater Database

Before you do the above steps, you will make a copy of the Reservation-1 form named Reservation-2 that you will use in this chapter.

Step 1: Copy and paste the Reservation-1 form and rename it to Reservation-2 as follows:

Make a backup copy of the project folder, Theater, and store it in another location, just in case you damage the current web page (Right-click the folder; select copy. Right-click the folder where you want to store the backup and select paste.)

Open the web project, Theater.

Right-click Reservation-1.aspx in the Solution Explorer and select copy.

Right-click the project name, Theater, in Solution Explorer and select paste.

Right-click the newly created file, Reservation-1 copy.aspx, and select rename.

Change the name of the new file to Reservation-2.aspx.

Before you can use the Reservation-2 page, you need to make a few important changes. Be careful to follow the steps exactly.

Open Reservation-2.aspx.designer.cs file from the Solution Explorer.

Change the class name from Reservation_1 to Reservation_2 as shown here.

```
1    namespace Theater {
2
3
4        public partial class Reservation_2 {
```

Make sure you use an **underscore** (_), not a hyphen, in Reservation_2.

DO NOT use "Show potential fixes" to make changes, which will change other forms too.

Save and close Reservation-2.aspx.designer.cs.

Open Reservation-2.aspx.cs file.

Change the class name from Reservation_1 to Reservation_2 as shown here.

```
1    namespace Theater
2    {
3        public partial class Reservation_2 : System.Web.UI.Page
```

Save and close Reservation-2.aspx.cs.

Open Reservation-2.aspx file from the Solution Explorer.

In the first statement, change the form name from Theater.Reservation_1 to Theater.Reservation_2, as shown here.

```
<%@ Page Language="C#" AutoEventWireup="true" CodeBehind=
            "Reservation-2.aspx.cs" Inherits="Theater.Reservation_2" %>
```

In addition, change the Title to Reservation-2.

```
<title>Reservation-2</title>
```

Close Reservation-1 form.

Save Reservation-2 form.

You will work with Reservation-2 form in the rest of this chapter.

Step 2: Copy the database.

Create an ASP.NET folder named App_Data and copy Theater.mdf into the folder, as follows:

Right-click the project name Theater in the Solution Explorer.

Select Add > Add ASP.NET Folder > App_Data.

Caution: Creating a standard folder and naming it to App_Data won't work.

Copy Theater.mdf from Project_starts/DataFiles to the App_Data folder.

Step 3: Add the DataSource control.

Make sure the Reservation-2 form is displayed in Design view.

Add the SqlDataSource control from the Data tab of the ToolBox into a new line below the validation controls. Change the ID of the data source to SqlDataSource_Performance.

Step 4: Configure the DataSource control.

Configuring the DataSource consists of specifying the database, table, fields, and selection criteria, if any. Configure the DataSource as follows:

Select **Configure DataSource** from the smart tag menu of the DataSource.

Click the **New Connection** button.

On the *Choose DataSource* window, make sure that the DataSource is a Microsoft SQL Server Database File.

Click Continue.

Click Browse, and select **Theater.mdf** from the **App_Data folder**.

Caution: It is important that you select the database file from the App_Data folder; otherwise, your application won't work if moved to a different location.

You may click the **Test Connection** button to test the connection, and click **OK.**

Click **OK** on the *Add Connection* window.

On the Configure DataSource window, click **Next**.

Leave the **Yes** box checked to save the connection to the application configuration file.

Click **Next**.

Make selections as indicated in Figure 8-2 to specify that the id, name, and base price from the Performance table are selected. If all columns of the table are to be selected, you would check the first checkbox with the "*" next to it.

To sort the Performances by Performance name, select **ORDER BY** button, and select **PerformanceName** from the first drop-down list. Click **OK**. Note that the SQL now includes the ORDER BY clause.

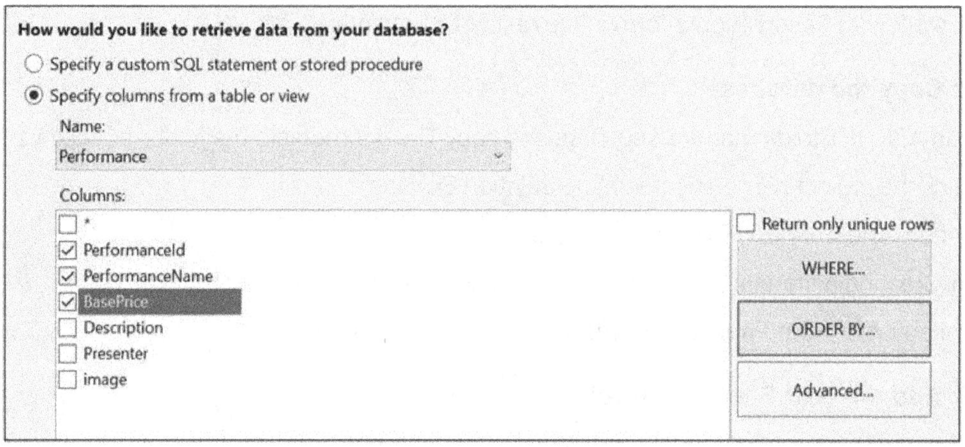

Figure 8-2: Configuring the DataSource: Selecting the fields

Click **Next**, and click **Test Query** button. You should see the records from the Performances table.

Click **Finish**.

Open Web.Config file from the Solution Explorer to view the connection string:

```
"DataSource=(LocalDB)\MSSQLLocalDB;AttachDbFilename=|DataDirectory|\Theater.mdf;
Integrated Security=True;Connect Timeout=30"
```

Note that the directory for the database file name is |**DataDirectory**|, not the local folder. Again, if you connect to a database in a local folder, the application won't work correctly if moved to a different directory or computer.

8.3 Binding a Drop-Down List to a DataSource

After the DataSource is configured, you can display any field from the DataSource in a drop-down list by binding the control to the DataSource. Binding a drop-down list to the DataSource involves setting three properties of the drop-down list as specified here:

DataSourceId: SqlDataSource_Performance
DataTextField: PerformanceName
DataValueField: PerformanceId

DataSourceId is the name of the DataSource.

DataTextField specifies the name of the field that is **displayed** in the drop-down list. This field is set to PerformanceName so that the drop-down list displays the Performance names.

DataValueField specifies the name of the field whose value is returned by the SelectedValue property. We set the DataValueField to PerformanceId so that we can use the PerformanceId to find the base price for the selected Performance, though we let the user select the Performance name.

DataValueField could be set to the same field as the DataTextField. Thus both properties could be set to PerformanceName, if you want to use the Performance name rather than the id.

These three properties could be set in the Properties window of the drop-down list, in the wizard or using code. You will use the wizard and then view the property settings in the Property window.

Step 5: Bind the drop-down list control to SqlDataSource_Performance.

Select **ChooseDataSource** from the smart tag menu of the drop-down list.

Select SqlDataSource_Performance for DataSource.

Select *PerformanceName* for the field to display, and *PerformanceId* for the value of the drop-down list, like this:

> Select a data source:
>
> SqlDataSource_Performance ⌄
>
> Select a data field to display in the DropDownList:
>
> PerformanceName ⌄
>
> Select a data field for the value of the DropDownList:
>
> PerformanceId ⌄

Click OK.

Note that the Properties window displays these settings:

DataSourceId:	SqlDataSource_Performance
DataTextField:	PerformanceName
DataValueField:	PerformanceId

Because the performance names are obtained from the database, you don't need to hard code the names. The next step removes the statements that adds hard-coded data to the drop-down list.

Step 6: Delete the statements that add static performances to the drop-down list.

Delete the statements shown in lines 5–7 of Figure 8-3, which are from the previous version of the web page.

```
1   protected void Page_Load(object sender, EventArgs e)
2   {
3       if (!IsPostBack) // if it is not a postback
4       {
5           //string[] performances = {"Young Irelanders", "Justin Hayward",
            "TedxOshkosh","OSO Orchestral"}
6           //foreach (string performance in performances)
7           //    ddlPerformance.Items.Add(performance);
```

Figure 8-3: The Page-Load event handler

Step 7: Set Reservation-2 as the start page.

Right-click Reservation-2.apsx and select Set as Start Page.

Run the Web Form and verify that the drop-down list displays performance names.
  Next, you will use the PerformanceId of the selected Performance to find the base price from the database.

Review Question

1. How does the DataTextField property of the DataSource control differ from the DataValueField?

8.4 Filtering Retrieved Records to Look Up a Value from a Table

There are different ways to select records and look up information from a table. Three commonly used methods are described below. You will use the third method to find the base price for the selected performance.

1. **Use a data table in combination with the table adapter and the binding source.**

   The table adapter uses SQL to select data from the database table and load the data into the data table, which is maintained in the client's memory.

   The binding source allows the user to navigate the data table and filter records.

   The use of data tables provides a convenient way to navigate through the records, add, update, and delete records, and look at different groups of data.

2. **Use the SqlCommand object in combination with a DataReader.**

   The SQL associated with the SqlCommand object is used to select records, and the DataReader is used to sequentially access the data retrieved by the SQL. Unlike the data tables used in the previous method, this method doesn't store records in the client's memory. It is used later in Chapter 16.

3. **Retrieve records using the Select method of the SqlDataSource**

   To look up information, like the price for a performance, this method uses the Select method of the SqlDataSource, which involves four steps:

   1. Use the Select method to retrieve the records specified in the DataSource, and store the data in a **DataView** object.
   2. Apply a filter to select the row for the specific Performance selected by the user.
   3. Store the selected row of the DataView in a **DataRowView** object.
   4. Get the base price from the DataRowView and store it in a variable.

8.4.1 Retrieving Record(S) Using the Select Method of SqlDataSource

The Select method has the following pattern:

DataSourceName.Select(*ParameterValues)*

Here is an example that uses the Select method of SqlDataSource_Performance to retrieve the records specified when the DataSource was configured:

```
SqlDataSource_Performance.Select(DataSourceSelectArguments.Empty);
```

When the SQL specified in the DataSource doesn't include any parameters, the Empty argument, `DataSourceSelectArguments.Empty`, is specified.

The above Select method would execute the SQL statement,

```
SELECT    PerformanceId, PerformanceName, BasePrice
FROM      [Performance]
ORDER BY  [PerformanceName];
```

and return the result set.

Finding the Base Price for the Selected Performance

To find the specific record for the selected Performance, the result set returned by the above Select method is cast to a **DataView** as shown on the right side of the equal sign in the statement shown below. A `DataView` consists of a set of rows of type `DataRowView`. The result set is then stored in a variable of `DataView` type, named `dvPerformanceTable`.

```
DataView dvPerformanceTable =
(DataView) SqlDataSource_Performance.Select(DataSourceSelectArguments.Empty);
```

After you store the data in a `DataView`, you can specify the criterion to select a row (or multiple rows) using the `RowFilter` property of the `DataView`. The following code uses the `PerformanceId` of the selected performance to filter the rows:

```
dvPerformanceTable.RowFilter = "PerformanceId = " + ddlPerformance.SelectedValue.ToString();
```

Note that the `SelectedValue` property gives the `PerformanceId` because the `DataFieldValue` of the `DataSource` is set to the `PerformanceId`. The right-hand side of the equal sign concatenates two strings into a single string that will appear as follows:

```
"PerformanceId = 101"
```

An alternative is to use the `PerformanceName` to filter the rows. The `PerformanceName` is given by the `SelectedItem` property of the drop-down list because the `DataTextValue` of the `DataSource` is set to `PerformanceName`.

```
dvPerformanceTable.RowFilter =
"PerformanceName = " + "'" + ddlPerformance.SelectedItem.ToString() + "'";
```

Because `PerformanceName` is a text field (not numeric), the value must be enclosed in apostrophes ('). The right-hand side of the equal sign concatenates four strings into a single string, like

```
"PerformanceName = 'TedxOshkosh'"
```

In general, it is more efficient to use the primary key when a single record is to be selected.

To get the base price from the selected row, first you store the selected row in a DataRowView named `drvPerformanceRow`:

```
DataRowView drvPerformanceRow = (DataRowView)dvPerformanceTable[0];
```

The index 0 is used to select the first (and only) row from the DataView.

Next, you get the base price from the `BasePrice` column of the `DataRowView` and store it in the variable `basePrice`:

```
decimal basePrice = (decimal) drvPerformanceRow["BasePrice"];
```

The code shown in Figure 8-4 combines the above set of statements to find the base price.

```
1   DataView dvPerformanceTable   =
        (DataView)SqlDataSource_Performance.Select(DataSourceSelectArguments.Empty);
2   dvPerformanceTable.RowFilter =
        "PerformanceId = " + ddlPerformance.SelectedValue.ToString();
3   DataRowView drvPerformanceRow = (DataRowView)dvPerformanceTable[0];
4   double decimal basePrice      = (decimalouble)drvPerformanceRow["BasePrice"];
```

Figure 8-4: Code to find the base price

Now let's look at how to put the above code to work.

Event Handler to Execute the Code

Which event handler would you use to execute the above code to find the price for the Performance selected in the drop-down list? The drop-down list control has the `SelectedIndexChanged` event that is fired when the user selects an item. So, you will use the `SelectedIndexChanged` event handler.

Step 8: Display the base price.

Delete the switch statement from the SelectedIndexChanged event handler of `ddlPerformance` and add the code that looks up the base price from the database, as shown in lines 3–8 of Figure 8-5.

```
1   protected void ddlPerformance_SelectedIndexChanged(object sender, EventArgs e)
2   {
3       DataView dvPerformanceTable =
4           (DataView)SqlDataSource_Performance.Select(DataSourceSelectArguments.Empty);
5       dvPerformanceTable.RowFilter =
6           "PerformanceId = " + ddlPerformance.SelectedValue.ToString();
7       DataRowView drvPerformanceRow = (DataRowView)dvPerformanceTable[0];
8       decimal basePrice = (decimal)drvPerformanceRow["BasePrice"];
9
10      lblBasePrice.Text = basePrice.ToString(); // Display base price
11      // Compute and display price per ticket
12      // unitPrice = basePrice + zonePrice - discount; // -- Doesn't work
13      ViewState["BasePrice"] = basePrice;
14      unitPrice = (decimal)ViewState["BasePrice"] +
15              (decimal)ViewState["ZonePrice"] - (decimal)ViewState["Discount"];
16      lblUnitPrice.Text = unitPrice.ToString();
17  }
```

Figure 8-5: SelectedIndexChanged event of ddlPerformance

Add the following **directive** that specifies the namespace for the DataView class:

```
using System.Data;
```

Display the form in the browser and verify that the right base price is displayed when you select a performance.

There is, however, a minor problem that needs to be addressed. The SelectedIndexChanged event is not fired the first time the page is loaded. So, the page won't automatically display the price for the performance initially displayed in the drop-down list (the first performance in the list). That is, the drop-down list will display a performance, but its price won't be displayed in the Label. Further, if you click the first performance in the list to select it, the corresponding base price is not displayed because you haven't changed the selected item.

Adding a Prompt to the Drop-Down List

To address the issue described previously, add a prompt as the first item in the list of performances, when the page is loaded for the first time. You need to bind the drop-down list to the DataSource before adding the prompt to the list because databinding of the control hasn't taken place at this stage of the page life cycle.

The statements in lines 5 and 7 of Figure 8-6 implement the above steps.

```
1   protected void Page_Load(object sender, EventArgs e)
2   {
3       if (!IsPostBack) // if it is not a postback
4       {
5           ddlPerformance.DataBind(); // bind the dropdown list to the DataSource
6           // Add a prompt
7           ddlPerformance.Items.
8               Insert(0, new ListItem("--Select performance--", ""));
8           ViewState["BasePrice"] = (decimal)0;
9           ViewState["ZonePrice"] = (decimal)0;
10          ViewState["Discount"] = (decimal)0;
11      }
```

Figure 8-6: Adding a prompt to the drop-down list

Line 5 uses the DataBind() method of the drop-down list to bind the control to the DataSource, which is specified by the value of the DataSourceId of the control. This step is necessary because when the Load event is fired, the drop-down list is not yet bound to the DataSource.

Line 7 inserts the prompt into the drop-down list. To insert an item into the list at a specified position, you use the Insert method of the Items class:

```
ddlPerformance.Items.Insert(0, new ListItem("--Select performance--", ""));
```

The first parameter of `ListItem()` represents the string returned by the `SelectedItem` property of `Items` collection, and the second parameter is the value returned by the `SelectedValue` property.

Step 9: Add a prompt to the drop-down list.

Add the code from lines 5–7 of Figure 8-6 to the Load event handler to bind the drop-down list to the DataSource and to add a prompt.

Test to make sure that the prompt is displayed and that the base price for the selected performance is automatically displayed when the page is displayed.

If you click on the prompt, after selecting a performance, you will get an error because the prompt is not found in the database table. To avoid this error, you need to add an if statement to the ddlPerformance_SelectedIndexChanged() event procedure to make sure the code is executed only if the index of the selected item is greater than zero.

Add the if statement as shown in Figure 8-7.

```
1   protected void ddlPerformance_SelectedIndexChanged(object sender,EventArgs e)
2   {
3       if (ddlPerformance.SelectedIndex > 0) {
4           ...
5       }
6   }
```

Figure 8-7: Updated SelectedIndexChanged event handler

Make selections for performance, zone, and member status and make sure the prices, discount, and unit price are displayed as in Reservation-1 form.

Other than the changes in the `Page_Load` event handler and the `SelectedIndexChanged` event handler of the drop-down list, the rest of the code in the Reservation-2 form is the same as the code in Reservation-1 form.

It's time to practice! Do Exercise 8-1 (see the end of the chapter).

Review Question

2. What is the difference between a DataView and DataRowView?

8.5 Retrieving Selected Records from a Table

In the previous version of the reservation form (Reservation-1), the user first selects a performance and then selects a date from the Calendar control. However, the Calendar doesn't identify the scheduled dates for the performance, making it hard to choose an applicable date. In Reservation-2, the user will select the performance date from a drop-down list (ddlDate) that shows only those dates for which the performance is scheduled.

In the previous section, you used the RowFilter property of the DataView table to specify criteria that select a record from the Performance table. This approach extracts **all records** from the database table, stores them in a DataView table, and then uses the RowFilter property to select a specific record. When working with large tables, this approach of retrieving the entire table and storing it in memory may negatively impact network traffic and client memory.

An alternative is to specify the selection criterion in the DataSource so that only the required records are retrieved and stored on the client. You will use this method in this section to select dates for the selected performance and display the dates in a drop-down list.

8.5.1 Displaying Dates for the Selected Performance in a Drop-Down List

Just like you configured the `SqlDataSource_Performance` to get all records from the Performance table, you will configure another DataSource that gets only those dates for the performance selected by the user from the first drop-down list, ddlPerformance. Then you can use this DataSource property to display the dates in the second drop-down list, ddlDate.

Sample records from the PerformDate table are shown here.

PerformDateId	PerformanceId	PerformDate
101	11	1/3/2021 7:30:00 PM
102	11	1/4/2021 7:30:00 PM
103	11	1/5/2021 7:30:00 PM

Step 10: Add an SqlDataSource.

Add the SqlDataSource control to the end of Reservation-2 form in Design view.

Change its name to `SqlDataSource_Dates`.

Step 11: Configure the DataSource.

Select **Configure DataSource** from the smart tag menu of the DataSource.

Because you created a ConnectionString when you created the `SqlDataSource_Performance`, you can use it here without creating a new one. Make sure **ConnectionString** is selected in the drop-down list. You may check the CheckBox to display the ConnectionString.

Click **Next**.

Make the selections indicated in Figure 8-8 to specify that all columns of the PerformDate table are selected.

To sort records by date, click the **ORDER BY** button, select PerformDate, and click OK.

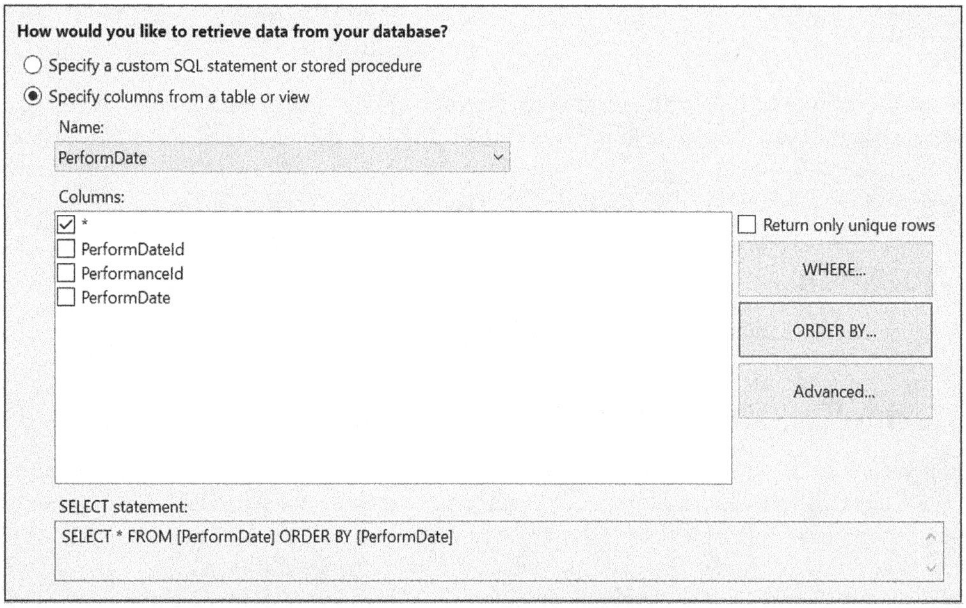

Figure 8-8: Configuring SqlDataSource_PerformDates

Specifying the Selection Criterion

You can specify the criterion (to select scheduled dates) during the configuration process by clicking the WHERE button. The criterion is that the value of the PerformanceId should be equal to the id of the performance selected in the drop-down list, ddlPerformance. The next step describes how to specify the criterion.

Step 12: Specify the criterion to select scheduled dates for the selected performance.

Click the WHERE button in the configuration window. The Add *WHERE Clause* window appears.

Select Column, Operator, Source, and Control ID as shown in Figure 8-9.

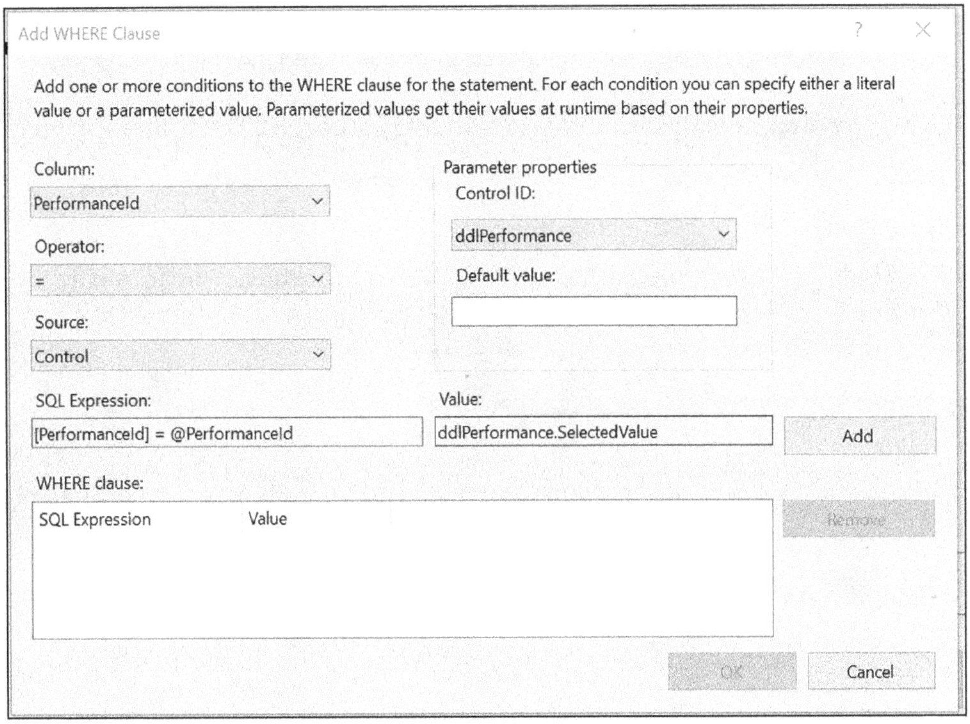

Figure 8-9: The *Add WHERE Clause* window

Click *Add*. Click *OK*.

The SELECT statement on the Configure SqlDataSource window changes to

```
SELECT * FROM [PerformDate] WHERE ([PerformanceId] = @PerformanceId) ORDER BY [PerformDate]
```

Click *Next* and then *Finish*.

It might be of interest to you to note that the information you provided to create and configure the DataSource is used to generate the ASP.NET code from Figure 8-10 in the Reservation-2.asp file.

```
1  <asp:SqlDataSource ID="SqlDataSource_Dates" runat="server"
2      ConnectionString="<%$ ConnectionStrings:ConnectionString %>"
3      SelectCommand="SELECT * FROM [PerformDate]
4                  WHERE ([PerformanceId] = @PerformanceId) ORDER BY [PerformDate]">
5      <SelectParameters>
6          <asp:ControlParameter Name="PerformanceId" ControlID="ddlPerformance"
7              PropertyName="SelectedValue" Type="Int32" />
8      </SelectParameters>
9  </asp:SqlDataSource>
```

Figure 8-10: Generated ASP.NET code

Lines 1–3 show the start tag of the SqlDataSource element, which includes the `ID`, `runat`, `Connection-String`, and `SelectCommand` parameters.

Line 3 shows the SELECT command. The parameter `@PerformanceId` represents the `SelectedValue` of `ddlPerformance`, as specified in the `<SelectParameter>` element in lines 4–6.

As you get more familiar with the ASP.NET elements, you may write the code without having to use the wizard to generate it.

Displaying Scheduled Dates

Step 13: Add a DropDown list named ddlDate to display scheduled dates.

Add a DropDown list named ddlDate between the caption "Date:" and the Calendar control.

Step 14: Bind the drop-down list ddlDate to the DataSource, SqlDataSource_Date.

Choose DataSource from the smart tag menu of the drop-down list, ddlDate.

Select SqlDataSource_Date for DataSource.

Select PerformDate for the field to display and for the value of the drop-down list, like this:

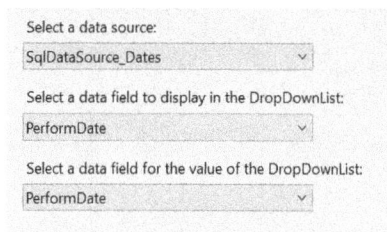

```
Select a data source:
SqlDataSource_Dates                    ∨

Select a data field to display in the DropDownList:
PerformDate                            ∨

Select a data field for the value of the DropDownList:
PerformDate                            ∨
```

Step 15: Display date with the day name.

Set the DataTextFormatString property of the drop-down list to {0:dddd, MMMM dd, yyyy}.

Step 16: Test the code.

Display the Reservation-2 page.

Select a performance.

The second drop-down list should display all scheduled dates for the selected performance.

Now you have the option to select the date from the drop-down list or the calendar.

Figure 8-11: The Calendar control with scheduled dates

8.5.2 Identifying Selected Dates in a Calendar Control

An alternative to displaying scheduled dates for a performance in a drop-down list is to identify the scheduled dates in a Calendar control, using color, as shown in Figure 8-11. The user will be able to select one of the dates marked in color.

Identifying dates in a calendar makes it easier for the user because the calendar shows other information about the date, like the day of the week. However, working with the Calendar control is more complex. Using this method is not a part of the tutorial; it is presented here for your information.

Though you can get the dates for a selected performance using the DataSource, **you cannot directly bind a Calendar control to the DataSource**, as you did for the drop-down list, in order to identify those dates in the Calendar. To do this, compare each scheduled date for a performance with each date displayed on the Calendar, and if they match, mark the date on the Calendar with a different color. This can be done within the DayRender event handler of the Calendar.

DayRender Event of the Calendar

The Calendar control displays each day in a table cell. The DayRender event of the Calendar is fired when the table cell for each day is generated. So, you can use this event to take an action for each date.

The DayRender event handler has an argument "e" of type `DayRenderEventArgs`, as shown in Figure 8-12, which has two properties: Cell and Day.

```
1   protected void calPerformDate_DayRender(object sender, DayRenderEventArgs e)
2   {
3       if (e.Day.IsWeekend)
4           e.Cell.BackColor = System.Drawing.Color.Yellow;
5   }
```

Figure 8-12: Arguments of the DayRender event handler

The `Cell` property may be used to set the style of display, like the `BackColor`. The `Day` property lets you set or get the properties of the date and other contents, if any. You will use these properties to set the `BackColor` for the days when a performance is scheduled.

Looping through Each Row Returned by the DataSource

As before, use a DataView table to store the records returned by the DataSource:

```
DataView dvDatesTable = (DataView)SqlDataSource_Dates.Select(DataSourceSelectArguments.Empty);
```

Next, use a loop to cycle through each row in the DataView, store the row in a DataRowView, and compare the date from the row to the date from the current Calendar cell. The complete code is shown in Figure 8-13.

```
1   protected void calPerformDate_DayRender(object sender, DayRenderEventArgs e)
2   {
3       if (IsPostBack && ddlPerformance.SelectedIndex > 0)
4       {
5           SqlDataSource_Dates.SelectCommand = "Select * From PerformDate
            WHERE PerformanceId = " + ddlPerformance.SelectedValue.ToString();
6           DataView dvDatesTable =
            (DataView)SqlDataSource_Dates.Select(DataSourceSelectArguments.Empty);
7           int NoOfDates = dvDatesTable.Count;
8           for (int count = 0; count < NoOfDates; count = count + 1)
9           {
10              DataRowView drvDateRow = dvDatesTable[count];
11              if (((DateTime)drvDateRow["PerformDate"]).ToShortDateString() ==
                                        e.Day.Date.ToShortDateString())
12                  e.Cell.BackColor = Color.Green;
13          }
14      }
15  }
```

Figure 8-13: DayRender event handler of calPerformDate

The `DayRender` event handler is run for each day of the calendar to do the following:
Line 3 checks whether the event is raised due to a page postback, and a performance is selected.
Line 7 gets the number of rows in the DataView table, which is used in the `for` loop in line 8.
Line 10 stores the current row in a `DataRowView`, each time through the loop.
Line 11 compares the value of `PerformDate` field with the date in the current Calendar cell.
Line 12 sets the `BackColor` of the current cell to green. The statement in this line assumes that the namespace for *color* is specified using the directive *using System.Drawing*.

TRY IT

Create the DayRender event handler for calPerformDate, as follows:

Select the Calendar in Design view.

In the Properties window, click the *lightning bolt* icon for Events and double-click DayRender event to create the event handler.

Switch to Source view (Reservation-2.aspx.cs) and add the Directive,

```
using System.Drawing.
```

Enter the code from Figure 8-13. Add the directive *using System.Drawing*.

To help test the code set the VisibleDate property of the Calendar to 2021-02-01 (or any date in December 2020, January 2021, or February 2021 – these are the months that have performances in the table).

Display the page, and select a performance.

To view the scheduled dates, change the month displayed on the Calendar to December 2020, January 2021, and February 2021 (which are the months when the current performances are scheduled). You should see the scheduled dates in green color.

Select a date from the set of scheduled dates.

Note that now you can select the performance date from the drop-down list or the calendar. But the later versions of the reservation form will save the date only from the dropdown list.

8.6 Using Grid View Control to Display Multiple Fields from a Table

The GridView control is a powerful tool that provides a simple way to display multiple fields from a DataSource as a table, with various options to customize the presentation of data and add functionality. Typically, you customize the control by modifying the aspx code generated by Visual Studio.

You display data in a GridView by linking it to a DataSource using the DataSource property. To let the user select individual rows, you add a button to each row by setting the AutoGenerateSelfButton to True.

To demonstrate, you will display the performance name and base price of all performances in a GridView and let the user select a performance.

Step 17: Add the GridView control from the Data tab of the Toolbox, below the drop-down list.

Set the properties of the GridView as shown here:

ID:	gvPerform
DataSourceId:	SqlDataSource_Performance
AutoGenerateSelectButton:	True

Step 18: Display the web page.

Display the web page in the browser. When you display the web page, you should see the GridView with the data as shown in Figure 8-14.

	PerformanceId	PerformanceName	BasePrice
Select	23	A Grand Night	15
Select	21	Charlie Berens	40
Select	26	Christmas Sing	15
Select	24	Cocktail Hour	15
Select	20	Hamlet	30
Select	17	Harmonious Wail	35
Select	12	Justin Hayward	30
Select	15	Ladies of Laughter	30

Figure 8-14: The GridView control

The user can select a row by clicking the Select button in the row.

To get the value in the selected row, use the `SelectedRow.Cells[]` property.

The code in Figure 8-15 shows the `SelectedIndexChanged` event handler of the GridView to change the `BackColor` of the selected row and display the base price in the Label.

```
1  protected void gvPerform_SelectedIndexChanged(object sender, EventArgs e)
2  {
3      gvPerform.SelectedRow.BackColor = System.Drawing.Color.LightGray;
4      lblBasePrice.Text = gvPerform.SelectedRow.Cells[3].Text;
5  }
```

Figure 8-15: The gvPerform_SelectedIndexChanged event handler

Line 4 uses the `Cells` property to get the base price from column 3, which has the index 2.

Step 19: Create the SelectedIndexChanged event handler of the GridView.

Double-click the GridView control to create the event handler and type in the code shown in Figure 8-15 to set the `BackColor` and display the base price.

Display the web page and select a button to select a row. The base price from the row should be displayed in the Label `lblBasePrice`. The unit price won't be displayed because you haven't included the code to compute it.

Step 20: Reset the web page to its original state as follows:

Comment out the `SelectedIndexChanged` event handler of the GridView.

Comment out the aspx code for the GridView in Source view, as follows:

```
1  <%--
2  <asp:GridView ID="gvPerform" runat="server" AutoGenerateColumns="False"
3  ...
4  </asp:GridView>
5  --%>
```

In addition to adding the Select button to select rows, GridView provides you with the option to display several other controls like CheckBoxes, hyperlinks, and images within the GridView. Other options include the ability to sort the data in one or more fields, and to format the data into pages and let the user choose the page to display.

Exercises

Exercise 8.1

In the current reservation form, the zone prices are hardcoded within a switch statement that finds the zone price for the zone selected by the user. Instead, you will create code to look up the zone price from the table, Zone (ZoneId, ZoneName, ZonePrice), which contains the zone prices for six different zones: Suite, Premium, Orchestra, Circle, Balcony, and Gallery.

To help the user select an individual zone, you need to replace the RadioButton "Prem&Orch" with two RadioButtons: "Premium" and "Orchestra." Similarly, replace the button "Balc&Gallery" with "Balcony" and "Gallery." So, you need to add two more radio buttons and change the names of the current buttons that combine zones.

Hint: The method to find the zone price is similar to the one you used to find the base price when the user selects a performance. Create a new DataSource named SqlDataSource_Zone and bind it to the RadioButtonList. Use the Select method of the sqlDataSource to retrieve the Zone records. Use the DataView and DataRowView objects to find the zone price for the selected zone.

Assignments

Auto Rental Assignment: Part C (ASP.NET)—Using Databases

This assignment is a modified version of the assignment, *Auto Rental Assignment: Part B (Variant)—Server-Side Code*, from Chapter 7, which develops the code for the web form named Auto-selection. The primary difference is that this assignment uses data from a database. If you have already done Part B from Chapter 7, you may modify it; if not, create a new project, as described here.

This assignment develops a web project for "Ace Auto Rentals," which is a small business that rents automobiles. Ace would like you to develop a web page that lets the user select a vehicle and specify the related information.

Ace has a database named AutoRental.mdf that contains two tables, AutoTypes and Models. The database can be found in the folder Tutorial-starts/Data-files. A subset of the records in the AutoTypes table is shown here.

AutoTypes

AutoTypeId	AutoType	Capacity	DailyRate	WeeklyRate
11	Standard SUV	5	45	250
12	Full-size SUV	7	65	400
13	Ludury SUV	5	55	350
14	Standard Minivan	7	60	400
15	12 Passenger Van	12	75	500

AutoType represents the type of auto, such as standard SUV, economy car, and small pickup. AutoTypeId is a unique number that identifies each record in the RentalRate table. Capacity represents the passenger (seating) capacity.

The table below shows a subset of the records in the Models table.

Models

ModelId	ModelName	AutoTypeId	CargoSpace
101	Honda CRV	11	39
102	Ford Escape	11	34
103	Chevy Suburban	12	39
104	Ford Expedition MAX	12	38
105	BMW X3	13	28
106	Dodge Caravan	14	15
107	Chrysler Pacifica	14	32
108	GMC Savana	15	92

ModelName includes both the make and model (e.g., Tesla Model 3). ModelId is a number that identifies each record in the Model table. Cargo space is in cubic feet.

Develop a web form named **Auto-selection**, within the web project named **AutoRental** that lets the user select an auto type, the model, the discount category, and insurance as shown below. Choose an appropriate layout. What is shown here is just an example.

The major differences from the previous version of the assignment are:

i) The drop-down list gets the list of auto types from the AutoTypes table. Similarly, the standard (daily) rate is obtained from the AutoTypes table.

ii) When the user selects an auto type, the corresponding set of models are obtained from the Models table and displayed on the form in a GridView so that the user can select a model. There is no difference in price between the models of the same auto type.

The form should let the user do the following:

1. Select the type of auto from a drop-down list, which automatically displays the standard daily rental rate. Your application must look up the daily rental rate from the RentalRate table. In addition, compute and display the cost/day (Standard rate + Insurance − Discount).

2. Select one of the discount categories from a RadioButtonList that displays a set of categories. When the user selects a category, display the corresponding discount amount. The categories and corresponding discount percent are as follows:

   Best Rate: 0%, Govt. Emp: 15%, Business: 10%, Favorite: 20%

   In addition, compute and display the cost/day.

3. Specify whether the customer needs insurance by checking a CheckBox. When the user selects or unselects insurance, display the cost. The cost of insurance is $10/day. In addition, compute and display the cost/day.

4. Enter the number of days of the rental into a TextBox.

5. Compute and display the total cost (Number of days × Cost/day) by clicking a button.

6. Clear the data by clicking a button.

You may use the daily rate, irrespective of the number of days. Store the rental rate, discount, and insurance in view state variables when the user makes the selections, and use the view state variables to compute the unit price.

E-Commerce Assignment: Part C (ASP.NET)—Using Databases

This assignment is a modified version of the assignment, *E-Commerce Assignment: Part C (Variant)—Server-Side Code*, from Chapter 7, which develops the code for a web page named Laptop-selection. The primary difference is that this assignment uses data from a database. If you have already done *Part B* from Chapter 7, you may modify it; if not, create a new project, as described here.

P&I Laptops is a small business that assembles and sells a limited set of laptop models online, each with multiple configurations. In this assignment, you will develop a web page that is part of an e-commerce system.

Information on the different models and configurations of laptops sold by P&I and the software bundled with the laptops are maintained in the database **PI-Laptops** that can be found in the folder Tutorial-starts/DataFiles. The database contains the following three tables:

Model

ModelId	ModelName	Description	CPU	OS	Touch	StartingPrice
101	S330	15 inch lap top with essential perf...	A6-9225	*NULL*	*NULL*	295.00
102	S380	Durable, ready for school, 14 inch ...	i3-8130U	Windows...	*NULL*	450.00
103	S580	S580 is powered by Intel Core™ i...	i5-8250U	Windows...	True	595.00
104	G780	14 inch touch-screen gaming lapt...	i7-8750H	Windows...	True	950.00
105	B780	A light-weight 2-in-one 15-inch to...	C i7-875...	Windows...	True	995.00

The StartingPrice of a model is the price for the default configuration. ConfigPrice is the price for a specific configuration.

Configuration

ModelConfigId	ModelId	RAM	SSD	HardDrive	ConfigPrice
1001	101	4 GB	0	500 GB	295.00
1002	101	4 GB	0	1 TB	350.00
1003	102	4 GB	0	500 tGB	450.00
1004	102	8 GB	256 GB	0	595.00
1005	102	8GB	256 GB	500 GB	650.00
1006	103	8 GB	128 GB	0	750.00
1007	103	8 GB	0	1 TB	595.00
1008	103	8GB	256 GB	0	795.00

Software

SoftwareId	SoftwareName	Price
11	Office Home	119.95
12	Office Professional	149.95
13	Office 365 Home	69.95
14	Office 365 Personal	69.95

Create a web project named PI-eComm, and develop a web page named **laptop-selection.** The web page allows the user to select a laptop configuration and other options, as shown below. Choose an appropriate layout. What is shown here is just an example.

Note that the major differences from the previous version of the assignment are in steps 1–3.
The form should allow the user to do the following:

1. Select a model from a drop-down list that displays a list of current models. The set of models should be obtained from the Models table.
2. When the user selects a model, display the description of the model. The description should be obtained from the Models table. In addition, the set of configurations for the selected model are obtained from the Configuration table and displayed on the form in a GridView so that the user can select a configuration.
3. Select a configuration from the GridView. When the user selects a configuration, display the corresponding price (ConfigPrice) and update the unit price (Laptop + Software + Warranty).
4. Select optional productivity software from a RadioButtonlist and automatically display the software price and update the unit price. Look up the software price from the Software table.
5. Specify whether the user wants an extended warranty. If the user selects a warranty, display the cost ($99.50) and update the unit price.
6. Enter the number of days of the rental into a TextBox.
7. Compute and display the total cost (Quantity × Unit price) by clicking a button.
8. Clear the data by clicking a button

You may store the prices in the view state variables and use them to compute the unit price.

## Chapter 9

Multipage Websites

Developing multipage websites involves additional features of ASP.NET to enable interaction between pages. In this chapter, you will learn different methods to move from one page to another page, how to share data between pages using session state and query string, and how to set a page to its previous state.

Learning Objectives

After studying this chapter, you should be able to:

- Create multipage websites.
- Describe methods to transfer from one page to another: Redirect, Transfer, and Cross-page PostBack.
- Validate input data using JavaScript.
- Share data between pages using session state.
- Describe the use of query strings to transfer data.
- Set a page to its previous state using PreviousPage property and session state.
- Use objects and lists to store multiple reservations.

9.1 Introduction

In order to learn the concepts that help you develop multiple-page projects, you will expand the current Theater website to include a new web page named **Confirmation**.

Figure 9-1 shows the *Confirmation* page. This page displays the selections made by the user on the reservation page, computes and displays the total cost (= Number of tickets × Unit price), and provides the option to change the reservation or confirm it. You will create a third version of the reservation form named **Reservation-3** that incorporates the changes necessary to interact with the page. *Reservation-3* and *Confirmation* pages let the user work with a single reservation at a time.

You will also create another version of the reservation and confirmation pages named **Reservation-4** and **Confirmation-4** that let the user work with multiple reservations at a time using objects and lists.

Tutorial: Moving between Reservation and Confirmation Pages

First, you will develop a new version of the reservation form named Reservation-3 and develop the *Confirmation page*.

Confirmation

Your selection

Performance: Young Irelanders
Date: 1/5/2021
Zone: Orchestra
Member? Yes
# of tickets: 2
Total Cost:90

Change Reservation
Confirm

Figure 9-1: The Confirmation page

Step 1: Copy and paste the Reservation-2 form and rename Reservation-2 copy to Reservation-3.

Open the web project, Theater.

Right-click Reservation-2.aspx in Solution Explorer and select copy.

Right-click the project name, Theater, in Solution Explorer and select paste.

Right-click the newly created file, Reservation-2 copy.aspx, and select Rename.

Change the name of the new file to Reservation-3.aspx.

Close the Reservation-2 form, if it is open.

Before you can use the Reservation-3, you need to make a few important changes.

The process is the same as the one used earlier to make a copy of Reservation-1. The instructions are repeated here. Be careful to follow the steps exactly.

Open Reservation-3.aspx.designer.cs file from the Solution Explorer.

Change the class name from Reservation_2 to Reservation_3 as shown here:

```
namespace Theater {

        public partial class Reservation_3 {
```

Make sure you use an underscore (_) not a hyphen in Reservation_3.

Do not use "Show potential fixes" to make changes, which will change other forms too.

Save and close Reservation-3.aspx.designer.cs.

Open Reservation-3.aspx.cs file.

Change the class name from Reservation_2 to Reservation_3, as shown here.

```
namespace Theater
{
        public partial class Reservation_3 : System.Web.UI.Page
```

Save and close Reservation-3.aspx.cs.

Open Reservation-3.aspx file from the Solution Explorer.

In the first statement, change the form name from Theater.Reservation_2 to Theater.Reservation_3, as shown here.

```
<%@ Page Language="C#" AutoEventWireup="true" CodeBehind=
        "Reservation-3.aspx.cs" Inherits="Theater.Reservation_3" %>
```

In addition, change the Title to Reservation-3.

```
<title>Reservation-3</title>
```

Save Reservation-3 form.

Next, you will create the *Confirmation page*.

Step 2: Add a new web form named Confirmation.

Right-click the project name, Theater.

Select **Add > Web Form**.

Change the Item Name to **Confirmation**. Click **OK**.

Figure 9-2 presents the initial layout of the *Confirmation page* in the design view.

The Label *lblReservation* will display multiple lines of reservation information.

The Change Reservation button takes the user back to the reservation to make changes.

The Confirm button displays the confirmation message.

The Label lblConfirm will display the confirmation message.

Step 3: Add captions and controls to the *Confirmation* page as shown in Figure 9-2.

Enter the heading "Confirmation" into the first line.

Highlight "Confirmation" and select Heading 1 from the Block Format drop-down list on the Toolbar:

(None)   ▾

Move to the next line, and type in "Your selection:".

Highlight the text "Your selection:" and select Heading 2 from the Block Format drop-down list.

Add the Label, lblReservation.

Add the two buttons and set their ID property to btnChange and btnConfirm, respectively.

Set their Text property as shown in Figure 9-2.

Add the Label, lblConfirm, in the next line.

Figure 9-2: The Confirmation page in design view

9.2 Displaying One Page from Another

When the user clicks the *Continue* (submit) button on the reservations page, we want that click to cause the browser to request a whole different URL from the server, namely the *Confirmation page*. We also want the values of form fields to be passed along with that new request to the server. We will look at four commonly used methods to request a page (target page) from another page (source page).

9.2.1 Methods to Redirect to Another Page

The first three methods are server-based, while the fourth method is significantly different because it is client-based. These methods also vary in their features. The selection of a method for a specific application would depend on the requirements.

Redirect

This method uses the Redirect method of the HttpResponse class. You invoke the method using this syntax:

```
Response.Redirect(URL)
```

The URL could be that of a page within the same web site or a page in another website.

To redirect to the *Confirmation page* (target page) from the Reservation page (source page), you would code the Redirect method in the Reservation page like this:

```
Response.Redirect("Confirmation.aspx");
```

The way the Redirect method works may seem odd. When this method is executed on the server, it sends an HTTP request redirect message to the browser, which sends a request (HTTP GET) for the target page, just like when you click a hyperlink. In response, the server sends the *Confirmation* page. This method lets you use the *back* button of the browser to go back to the previous page.

You may ask, Why not have the server send the *Confirmation* page the first time? If the server sends a page without an HTTP request, the browser wouldn't know that it is a new page and hence would display the original URL (URL of the Reservation page) when it displays the *Confirmation* page. The Transfer method, described shortly, uses this approach.

The Redirect method also lets you pass values to the target page through a query string (GET parameters) using the following pattern:

```
Response.Redirect (URL?variableName=value&variableName=value& …)
```

The "?" indicates the start of the query string, and variableName and value indicate the name of the parameter and its value to be passed to the page. You may pass multiple values using "&" between parameters, but the size of the query string is limited.

You will learn more about the query string in Section 9.3 and use it to pass the selected performance and zone to the *Confirmation* page.

RedirectPermanent

When a page is permanently moved to a different location or is renamed, you use a different method of the HttpResponse class:

```
Response.RedirectPermanent (URL)
```

The `RedirectPermanent` works similar to Redirect, except that the server sends a different HTTP status code to the browser.

Transfer

A third option is to use the `Transfer` method of the `HttpServerUtility` class using this syntax:

```
Server.Transfer (URL)
```

To pass values from the Reservation page to the *Confirmation* page, you would code the `Transfer` method in the Reservation page, like this:

```
Server.Transfer ("Confirmation.aspx");
```

Similar to the `Redirect` method, the `Transfer` method allows you to programmatically redirect to another page, giving you the ability to dynamically specify the target URL and to include a query string.

This method, which is specific to ASP.NET, works differently from the `Redirect` method. When the `Transfer` method is executed on the server, it sends a request directly to the web server, which loads the *Confirmation* page and sends it to the browser, while keeping the previous page in server memory. Since the page was sent by the server without an HTTP request, the browser is unaware of the transfer, and it would continue to display the previous URL—that is, the target page (Confirmation) would display the URL of the source page (Reservation-3). Further, the target page must be on the same website as the source page.

The `Transfer` method, however, allows the target page to reference the source page using the **PreviousPage** property exposed by the `Page` class because the previous page is still in the server memory. Thus the target page can access the controls and public properties on the source page. You will use the `PreviousPage` property later in this chapter. With the `Redirect` method, you cannot access the previous page because the previous page is not in the server memory.

Cross-page posting

By default, clicking a button on a page causes a postback of the page. However, you can use the PostBackUrl property of a button to redirect to another page and post the form data to that page, without reloading the current page. To do this, you set the PostBackUrl property of a button on the source page to the URL of the target page, as shown in Figure 9-3.

The tilde (~) operator is automatically inserted when you enter the URL. You may also specify a query string with the URL.

Properties ▾ ⬛ ✕

btnReserve System.Web.UI.WebControls.Button ▾

⬛ ⬛ ⬛ ⚡ 🔧

OnClientClick ▲

PostBackUrl **~/Confirmation.aspx**

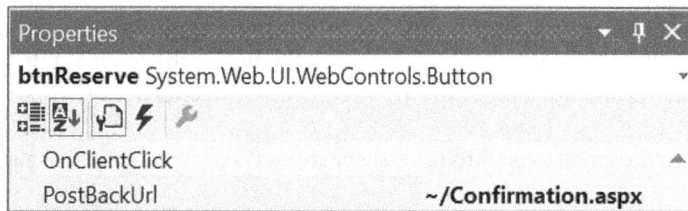

Figure 9-3: Cross-page posting

When you click the button, the **client** posts the page to the server (by sending an HTTP POST command), which sends the values of controls on the source page to the target page. It is important to note that it is the server that handles the click event. The browser displays the target page and its URL. This method is particularly important because of its similarity to how non-ASP forms work.

The cross-page posting, similar to the Transfer method, allows the target page to reference the source page using the `PreviousPage` property of the Page class. For example, if you use a cross-page posting or Transfer method to display the *Confirmation* page, you may access the Reservation page control `txtNoOfTickets` from the *Confirmation* page and display the number of tickets in the Label `lblReservation` using the following code:

```
1   TextBox txtNumTickets = (TextBox) PreviousPage.FindControl("txtNoOfTickets");
2   lblReservation.Text = txtNumTickets.Text;
```

The first statement uses the `PreviousPage` property to get the control, `txtNoOfTickets`; cast it to a TextBox; and assign it to the variable, `txtNumTickets`. The second statement uses the Text property of the TextBox to display the number of tickets.

Thus, `Redirect`, `Transfer`, and cross-page posting can access information stored in session state and also use the query string to pass data between two pages. But only the `Transfer` method and cross-page posting can use the `PreviousPage` property to access the source page.

9.2.2 Redirecting from the Reservation Page to the Confirmation Page

Which of the four methods discussed in the previous section should you use to display the *Confirmation* page? The `RedirectPermanent` method doesn't apply to the current example. The `Transfer` method has the drawback that it doesn't update the URL after a transfer. Both the `Redirect` method and cross-page posting would work.

To help familiarize you with both methods, you will use the `Redirect` method to display the *Confirmation* page, and use cross-page posting to go back to the Reservation page. Session variables will be used to make reservation information available to the *Confirmation* page and also to the Reservation page when the user revisits the page.

Step 4: Use the Redirect method in the Reservation page to display the Confirmation page.

Insert the following code into the click event handler of btnContinue on the Reservation page:

```
protected void btnContinue_Click(object sender, EventArgs e)
{
        Response.Redirect("Confirmation.aspx");
}
```

Step 5: Specify the start page and run the website.

To specify Reservation-3 page as the start page, right-click Reservation-3.aspx and select *Set as Start Page*.

Run the website and click the *Continue* button to display the *Confirmation* page. The *Confirmation* page should open with no data in the Label, *lblReservation*.

Next, you will learn how to share data between the Reservation page and the *Confirmation* page. Recall that the Redirect method doesn't allow you to access the previous page using the `PreviousPage` property of the Page class. You will use the session state to share data between pages.

Review Questions

1. From the user's perspective, what would be a drawback of using the Transfer method to move to another page?
2. What are two differences in functionality between the Redirect and Transfer methods?

9.3 Sharing Data between Pages

The session state stores information that needs to be shared between multiple pages during a specific session of a user. For example, the session state may be used to store a user's shopping cart as the user navigates the web pages. Unlike the view state, the session state stores the data in the server memory, and the data is available until the user closes the browser or the session is timed out.

How does ASP.NET maintain sessions? To identify a session, ASP.NET uses the value of the *sessionID* property of the Session object. If the request from the browser doesn't include a *sessionID* (e.g., in the first request for the page), ASP.NET starts a new session and sends the *sessionID* to the browser with the response. The browser returns the *sessionID* value with each request. If the time between requests exceeds a time-out value, a new session is started.

By default, the session state is stored on the server in what is called a non-persistent cookie, or a session cookie, which is deleted when the session ends.

Alternatively, the application can be configured for a cookieless session that stores the *sessionID* in the URL of the page. This can be done by using the following setting in the web.config file:

```
<sessionState cookieless="true" />
```

9.3.1 Storing and Retrieving Data Using Session Variables

Just like adding items to the ViewState object, you can add items to the collection contained in the Session object, which is accessed using the Session property of the Page class. You use the Session property to add an item (a key/value pair) using a syntax similar to the one used for view state:

```
Session.Add("name", value);
```

or

```
Session["name"] = value
```

Here is an example of each syntax:

```
Session.Add("BasePrice", basePrice);
```

or

```
Session["BasePrice"] = basePrice;
```

Both statements use an item named `BasePrice` to store the value of the variable `basePrice`.

An item that is added to the session state is often referred to as a **session variable**. In the previous example, `BasePrice` is the session variable.

Again, you don't have to declare the session variable. When you retrieve it, you cast it to the appropriate type, determined by the type of the value assigned to it. To retrieve the value of the session variable, BasePrice, you may use its name or its index in the collection:

```
double price = (double)Session["BasePrice"];
```

You may access the Session object from a class other than the Page class (or a class that inherits the Page class), using the Session property of the HttpContext object:

```
HttpContext.Current.Session["BasePrice"] = basePrice;
```

Table 9-1 describes the properties/methods of Session objects.

Table 9-1: Properties/methods of Session objects

Properties/methods	Description
`Count`	Number of items in the collection: e.g., `int count = Session.Count;`
`Add (name, value)`	Adds an item to the collection or replaces it if it exists: e.g., `Session.Add("BasePrice", basePrice);`
`Clear()`	Removes all items from the collection: e.g., `Session.Clear();`
`Remove (name)`	Removes the specified item from the collection: e.g., `Session.Remove("BasePrice");`

Next, you will learn how to move from one page to another, and use session state to share data between pages.

Storing Reservation Information in Session Variables

The following reservation information is needed for display and computation on the *Confirmation* page:

- Selected performance, selected date, selected zone, and member status – these are displayed on the *Confirmation* page.
- Unit price and number of tickets – these are needed to compute the total cost on the *Confirmation* page.

In addition, if the user comes back to the Reservation page from the *Confirmation* page, it is necessary to re-display the current values of controls on the Reservation page. Though the values of controls persist during a postback, they are not automatically preserved when you leave the page and reload the page. These are the additional data needed to set the reservation page to its current state:

- Base price, zone price, and discount to re-display the data in the Labels on the Reservation page
- Index of selected Performance and index of the selected zone, which are to be used to redisplay the selections made by the user from the DropDownList and RadioButtonList.

We can store all these values in session state so that they can be used on the *Confirmation* page or on the reservation page if the user returns to it. However, selected performance and selected zone are needed only on the *Confirmation* page. We will handle these two values differently from all the rest, just to illustrate how the query string can be used to transfer information between pages as well, though this is not a typical use of query strings. Session state will be used to share the rest of the data.

Currently, the base price, zone price, and discount are stored in view state variables. These three view state variables can be easily changed to session state variables by doing a search and replace for all occurrences of "ViewState" by "Session" in the entire C# code.

> **Step 6: Replace all occurrences of "ViewState" by "Session" in the C# code in Reservation-3.aspx.cs.**
>
> There should be fifteen replacements.

The unit price is currently stored in the class-level variable unitPrice within each of the three event handlers where it is computed. Unit Price also needs to be stored in a session variable within each event handler.

> **Step 7: Store the unit price in a session state.**
>
> Add the following statement to the end of the SelectedIndexChanged event handlers of ddlPerformance and rblZones, and the CheckedChange event handler of chkDiscount:

```
Session["UnitPrice"] = unitPrice;
```

The remaining data needs to be stored in session variables using the code in the click event handler of btnContinue, as shown in Figure 9-4. The event handler also includes the code (lines 3–7) to verify that the user made the required selections, as explained shortly.

```
1  protected void btnContinue_Click(object sender, EventArgs e)
2  {
3      if (calPerformDate.SelectedDate == DateTime.MinValue)
4      {   // display a message
5          Response.Write("<script> alert('Please select a date') </script>");
6          return;
7      }
8      // Store selected date, member status and number of tickets in session variables
9      Session["SelectedDate"] = calPerformDate.SelectedDate;
10     Session["IsMember"] = chkMember.Checked;
11     Session["NoOfTickets"] = int.Parse(txtNoOfTickets.Text);
12     // Store indices of selected play and selected zone in session variables
13     Session["PerformanceIndex"] = ddlPerformance.SelectedIndex;
14     Session["ZoneIndex"] = rblZones.SelectedIndex;
15
16     //Move to confirmation page. Pass selected performance and zone using query
17       string
18     Response.Redirect("Confirmation.aspx?performance=" +
19                       ddlPerformance.SelectedItem.ToString()+
20                         "&zone=" +rblZones.SelectedItem.ToString())
21 }
```

Figure 9-4: btnContinue_Click event handler

Line 5 displays an error message using the server-side Response.Write() function to insert the <script> element (that contains the alert() JavaScript function) into the HTML document. An alternative to using the alert() function to display an error message is to write a custom server-side function that inserts the message into the HTML document.

Lines 9 to 11 store the selected date, member status, and the number of tickets in session variables. These data items are needed on the *Confirmation* page and on the reservation page if the user revisits it.

Lines 13 and 14 store the indices of the selected Performance and zone in session variables to help redisplay the selections if the user revisits the Reservation page.

Line 17 uses the Redirect method to display the *Confirmation* page, as described herein.

9.3.2 Using the Query String to Pass Data

As stated earlier, the Redirect, Transfer, and Cross-Page Posting methods let you pass values to the target page through a **query string.** This is how you use a query string in the Redirect method:

```
Response.Redirect (URL?variableName=value&variableName=value& …)
```

The "?" indicates the start of the query string, and variableName and value indicates the name of the parameter and its value to be passed to the page. You may pass multiple values using "&" between parameters, but the size of the query string is limited.

For example, the following statement would pass the values "Hamlet" and "Suite" using the parameters performance and zone in a query string:

```
Response.Redirect("Confirmation.aspx? performance=Hamlet&zone=Suite")
```

Instead of passing literals like "Hamlet" and "Suite," you may pass the selected Performance and selected zone using the following query string that concatenates four different strings together:

```
Response.Redirect("Confirmation.aspx?performance=" +
ddlPerformance.SelectedItem.ToString() + "&zone=" + rblZones.SelectedItem.ToString());
```

So, this method allows you to programmatically redirect to another page using code, by dynamically specifying the target URL and the query string.

On the target page (Confirmation), you may get the value of the variable, Performance, from the query string collection using the syntax

```
Request.QueryString["performance"]
```

Query strings are an alternative to using session variables to share data. Session states share data by storing data on the server, whereas query strings store the data in the URL of the page. When you pass variables using query strings, the variables and their values are visible on the target page and the links can be shared and bookmarked by users. So, this method is not appropriate for passing sensitive information. Further, if the same data is needed in multiple pages, you need to pass the data from page to page, whereas session variables can be stored once and then accessed from any page. However, unlike session variables, query strings do not take up room in the server memory.

Query strings may be used for a variety of purposes, including passing data that determines the contents of a dynamic web page (e.g., a product number to look up product information from a database), passing data that specifies how the page is displayed (e.g., sort data in a specific order), and sharing data that identifies the source from where the user requested the URL.

The statement in line 18 in Figure 9-4, copied here, uses the Redirect method to display the *Confirmation* page and pass the selected performance and zone to the *Confirmation* page using the query string parameters *confirmation* and *zone*.

```
Response.Redirect("Confirmation.aspx?performance=" +
ddlPerformance.SelectedItem.ToString()+ "&zone=" +rblZones.SelectedItem.ToString());
```

Realize that this is not a typical application of query string, but we use it to help you learn how it is used.

Next, let's develop the code to verify that the user selected a date. The Validating controls that you added earlier check whether the user selected the performance and zone, and entered the number of tickets. Recall that the RequiredFieldValidator doesn't work with Calendar control.

9.4 Using JavaScript to validate the Calendar

In Chapter 6, you used Validation controls to do client-side and server-side validation of some input data. Client-side validation using JavaScript is a commonly used flexible way to validate input data. To familiarize you with this method, you will use simple JavaScript code that invokes the method alert() to display a pop-up message when no date is selected. The alert method has the syntax alert('message'), as in

```
alert('Please select a date')
<script> Element
```

The script element is commonly used to specify client-side JavaScript code within the aspx file using the following pattern:

```
<script>
      JavaScript code
</script>
```

For example, you can include the alert() method within the aspx code using the statement

```
<script>  alert('Please select a date');  </script>
```

You also may use the script element with the runat="server" attribute to specify simple server-side code (e.g., a simple event handler, a function, or code to set a control's property), though it is preferable to place server-side code in a separate code-behind file. Chapter 10 presents more applications of the script element and also shows how to embed server-side code within the aspx file.

Next, you will insert the above script element into the HTML markup using server-side C# code that checks whether a date was selected. You can insert code into HTML using the *Response.Write()* method.

Response.Write() Method

In general, the `Response.Write()` method inserts a specified string into the HTTP response sent by the server to the client. The string could be any HTML element like the script element that contains JavaScript. To insert a `<script>` element into HTML when the server renders the HTML for a page, include the `Response.Write()` method in the server-side C# code, as shown here:

```
Response.Write("<script> alert('Please select a date'); </script>");
```

The *Response.Write()* method in combination with the `<script>` element can be used to insert any JavaScript code that needs to be run on the client.

The following code (included in lines 3–7 in Figure 9-4) uses this statement to display a pop-up alert if the user didn't select a date.

```
1  if (calPerformDate.SelectedDate == DateTime.MinValue)
2  {   // display a message
3      Response.Write("<script> alert('Please select a date'); </script>");
4      return;
5  }
```

Step 8: Work with JavaScript.

Insert the code from Figure 9-4 to the click event handler of btnContinue on the Reservation page.

To test validation of the Calendar, display the page, select values for all controls except the Calendar, and click Continue.

The form should display the pop-up message "Please select a date."

The `Response.Write()` method, which is processed on the server when the page is posted back, inserts the script element outside (before) the HTML document. So, the alert is displayed before the page is redisplayed after postback, which may have some negative effects on the processing of the page in certain situations. Further, the script won't be able to access the controls on the form.

To test the location of the `script` element, you may insert the following statement after line 3:

```
Response.Write("<h1> Testing </h1>");
```

When you display the page, you will see the word "Testing" displayed outside the page.

Make sure you delete the added statement.

Additional applications of *Response.Write* are presented in Chapter 8.

9.4.1 RegisterStartupScript() and RegisterClientScriptBlock() versus Response.Write

`RegisterStartupScript()` and `RegisterClientScriptBlock()` are methods that insert scripts into the HTML document. They differ from the `Response.Write()` method in that they insert a script inside the HTML document, whereas Response.Write() inserts the script outside (before) the HTML document.

The `RegisterStartupScript()` inserts the script at the end of the form, immediately before the end tag of the form, `</form>`. So the script is run after the controls are loaded on the form and therefore the script can access the controls.

The `RegisterClientScriptBlock()` inserts the script at the beginning of the form, immediately after the start tag of the form, `<form>`; so, the script won't have access to the controls. Because the alert method doesn't need to access any controls on the form, it won't make any difference in the current application. Both methods register the script with the page that contains the script, using the type of the script and the key (name) to identify each script, using the following syntax:

```
ClientScript.RegisterStartupScript(type, key, script)
```

Here is an example:

```
ClientScript.RegisterStartupScript(this.GetType(), "MissingDate",
                        "<script>alert('Select date');</script>");
```

ClientScript is the class that contains the method. The first parameter is typically set to this.GetType(), which gets the type of the page. The second parameter could be any name to serve as a key, and the third parameter is the script.

Step 9: Use RegisterStartupScript and RegisterClientScriptBlock.

Add the RegisterStartupScript() and RegisterClientScriptBlock() methods in the btnContinue_Click event handler as shown in Figure 9-5.

```
1   protected void btnContinue_Click(object sender, EventArgs e)
2   {
3       if (calPerformDate.SelectedDate == DateTime.MinValue)
4       {   // display a message
5           //Response.Write("<script> alert('Please select a date') </script>")
6           // Alternative methods:
7           //ClientScript.RegisterStartupScript(this.GetType(), "MissingDate",
            //    "<script>alert('Please select a date'); </script> ");
8           ClientScript.RegisterClientScriptBlock(this.GetType(), "MissingDate",
                "<script>alert('Please select a date'); </script> ");
9           lblDate.Text = "Please select a date"; // Alternative to an alert
10          return;
11      }
```

Figure 9-5: The RegisterClientScriptBlock() method

Display the page, select values for all controls except the Calendar, and click Continue. The pop-up message is displayed before the page is redisplayed following the postback.

Uncomment line 7 and comment out line 8. Display the page, select values for all controls except the Calendar, and click Continue. The pop-up message is displayed after the controls on the form are redisplayed following the postback.

If a pop-up message is not needed, you may display the message in a Label as shown in Line 12. You also may create a JavaScript function (method) and call it from a CustomValidator control.

9.4.2 Validation Using Code versus Using Validation Control

As you learned earlier, when you use a validation control to validate the value of a control, the validation takes place immediately if the user enters (or changes) the value of a control and moves focus away from it. This helps give immediate feedback to the user before the page is posted back to the server, thus avoiding a round trip. (Recall that validation also takes place on the server because the client-side validation may not take place under certain conditions like the user skipping a field or browser incompatibility.)

When you use JavaScript code to validate the Calendar as shown above, the C# code is executed on the server when you click the Continue button that causes a page postback. Thus a page postback is necessary to detect that the date was not selected. The Response.Write() method (and the RegisterStartupScript and RegisterClientScriptBlock methods) would insert the string "<script> alert('Please select a date') </script>" into the HTML document that is sent to the browser. The browser will execute the JavaScript (the alert() method) to display the message.

9.5 Using Data from Query Strings and Session Variables

The confirmation page gets the reservation data through session variables and query strings and displays them in the Label, lblReservation. Figure 9-6 presents the code in the Page_Load event handler of the *Confirmation* page.

```
1   protected void Page_Load(object sender, EventArgs e)
2   {
3       if (!IsPostBack)      // If it is not a postback of the page,
4       {
5           // Get selected performance and zone from query string
6           string selectedPerformance = Request.QueryString["performance"];
7           string selectedZone = Request.QueryString["zone"];
8           // Get date, member status, # of tickets & unit price from session variables
9           DateTime selectedDate = DateTime.Parse(Session["SelectedDate"].ToString());
10          bool isMember = (bool)Session["IsMember"];
11          int noOfTickets = (int)Session["NoOfTickets"];
12          decimal unitPrice = (decimal)Session["UnitPrice"];
13          decimal totalCost = noOfTickets * unitPrice;
14
15          // Concatenate the reservation data into a single string with line breaks
16          string reservation =
17              "performance:      " + selectedPerformance.ToString() + "<br />" +
18              "Date:         " + selectedDate.ToShortDateString() + "<br />" +
19              "Zone:         " + selectedZone.ToString() + "<br />";
20          reservation = reservation + "Member?   " + (isMember ? "Yes": "No ") +
                            " <br/>";
21          reservation = reservation + "# of tickets: " + noOfTickets.ToString()
22                            + "<br />" + "Total Cost:" + totalCost.ToString();
23          // Display the string in the Label, lblReservation
24          lblReservation.Text = reservation;
25      }
26  }
```

Figure 9-6: The Page_Load event handler of the Confirmation page

Lines 6 and 7 get the selected performance and zone from query string parameters, performance and zone.

Line 9 casts the value of the session state item (session variable), SelectedDate, to DateTime type and stores it in the DateTime variable selectedDate.

Lines 10 to 12 get the values of the remaining session variables.

Lines 16 to 22 concatenate the reservation information into a single string with line feeds inserted using the HTML tag <br />. Line 24 displays the reservation data.

Line 20 uses the ternary conditional operator (?:), which evaluates the Boolean expression, isMember, and returns the value "Yes" if it is true and "No" if it is false.

Note that the Label expands automatically, depending on the number of lines displayed. So, unlike viewing information displayed in a ListBox, the user doesn't have to scroll when the size of the display exceeds the size of the control.

Step 10: Confirm the selection.

Insert the code from Figure 9-6 into the Load event handler of the *Confirmation* page.

Clicking the Confirm button on the *Confirmation* page should display a message in the Label, lblConfirm, and also disable the btnChange button. The following click event handler in Figure 9-7 of btnConfirm shows the code for this.

```
1   protected void btnConfirm_Click(object sender, EventArgs e)
2   {
3       lblConfirm.Text = "Your reservation is confirmed!";
4       // Display pop-up message:
5       Response.Write("<script> alert('Your reservation is confirmed') </script>");
6       //ClientScript.RegisterStartupScript(this.GetType(), "Confirm",
7       //    "<script>alert('Your reservation is confirmed'); </script> ");
8       btnChange.Enabled = false;
9   }
```

Figure 9-7: btnConfirm event handler

To demonstrate the difference between *Response.Write* and *RegisterStartupScript* methods, the code also includes these methods to display a pop-up message.

Step 11: Create the click event handler of btnConfirm to display a confirmation message.

Insert the code from Figure 9-7 into the click event handler of btnConfirm on the **Confirmation** page.

Run the website. Make selections on the Reservation page and click the Continue button. Make sure the **Confirmation** page displays the data, as shown in Figure 9-8.

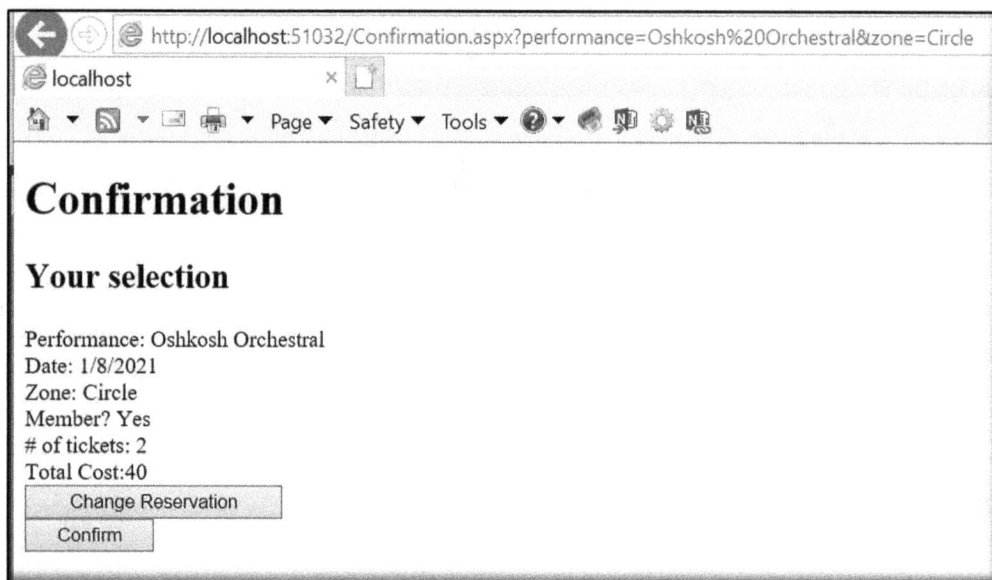

Figure 9-8: The Confirmation page with reservation data

Note that the values ("Oshkosh Orchestral" and "Circle") that are passed through the query string are visible on the page with the URL. Again, this shows why it is not desirable to pass a data item like the unit price using a query string.

It should be noted that if you access the *Confirmation* page directly, or from a page other than the Reservation page, you will get an error because the session state items that are used in the *Confirmation* page are created in the Reservation page. Here we assume that the *Confirmation* page is accessed only from the Reservation page. In the next section, you will learn how to check what the source page is.

Step 12: Click the Confirm button to display the confirmation message.

Note that the pop-up message is displayed before the message appears in the Label, because the script is inserted by *Response.Write* before controls are loaded into the page.

Now, comment out the code in line 5 and uncomment the code in lines 6 and 7, and re-display the **Confirmation** page.

Click the Confirm button.

The message already appears in the Label when the pop-up message is displayed, because the RegisterStartupScript method inserts the script after the controls are loaded on the page.

Review Questions

3. Describe the Session state.

4. Explain the effect of executing each of the following statements:

    (a)   `Session.Add("BasePrice", basePrice);`

    (b)   `Session["BasePrice"] = basePrice;`

5. How would you have your server-side C# code insert JavaScript code into the HTML document, which will be run by the client?

6. What is the difference between validating input using a validation control and validating using server-side C# code?

7. In Section 9.3, what mechanism did we use to have the user presented with the *Confirmation* page when the user clicks the *Confirm* button on the Reservation page?

    (a) Transfer method

    (b) Redirect method

    (c) Cross-page posting

8. How did we share the data about the number of tickets and the unit price between those two pages?

    (a) Session variables

    (b) Query-string parameters

    (c) Accessing the fields through the PreviousPage property

9. How did we share the data about the performance and zone between those two pages?

    (a) Session variables

    (b) Query-string parameters

    (c) Accessing the fields through the PreviousPage property

9.6 Setting a Page to Its Previous State

Users may want to go back to a page that was visited earlier and view and change data previously entered or selected by the user. For example, if the user goes back to the reservation page from the *Confirmation* page, the controls on the reservation page should display the values previously set by the user, so that the user can make changes. However, if the user displays the reservation page from a page other than the confirmation page, or displays it directly, there are no previously set values to display. So, the reservation page should identify the source page and set the properties of the controls only if the source page is *Confirmation*.

Both the Transfer method and cross-page posting provides you the `PreviousPage` property of the Page class to get access to the previous page from the Reservation page. But, because the Transfer method doesn't update the URL, we will use **cross-page posting** to go back to the Reservation page. This involves setting the `PostBackUrl` property of `btnChange`.

Step 13: Return to the Reservation page.

Set the PostBackUrl property of btnChange on the **Confirmation** page to Reservation-3.aspx, as illustrated here.

You may type in **Reservation-3.aspx** or click the ellipses (…) on the right side and select Reservation.aspx. The characters "~/" are automatically inserted.

OnClientClick	
PostBackUrl	**~/Reservation-3.aspx**
SkinID	

Next, we will look at the code required in the Reservation page to check whether the previous page, if any, is the *Confirmation* page and, if so, to set the controls to their previous state.

9.6.1 Getting the URL of the Previous Page

To check whether there is a previous page (source page), use the `PreviousPage` property of the current page. The `PreviousPage` property returns the previous page if there is one, and returns null if there is none. For instance, you may use the following `if` statement to take action based on the existence of a previous page:

```
if (PreviousPage != null)
```

To get the name of the previous page, use the `UrlReferrer` property of the `HttpRequest` class. The `UrlReferrer` returns the URL of the previous page. To get the previous page name from the URL, use the `AbsolutePath` property of the `UrlReferrer`:

```
Request.UrlReferrer.AbsolutePath
```

The above statement uses the `Request` property of the current page to access the `HttpRequest` class. The `AbsolutePath` returns the previous page name with a slash (/) preceding the name, as in "/Confirmation.aspx."

The following code would get the name of the previous page, if it exists:

```
1   string previousPageName;
2   if (PreviousPage != null)
3       previousPageName = Request.UrlReferrer.AbsolutePath;
```

The value of `previousPageName` could be null, "Confirmation," or some other page name. If the name of the previous page is *Confirmation*, you need to display previously selected values in the controls on the Reservation page, as explained next.

9.6.2 Setting Control Properties to Previously Selected Values

To help set the properties of controls on the Reservation page to their previously set values, you have already stored the values in session variables. Now you will include the code in the Load event handler of the reservation page to use those values to set the controls to their previous state. Figure 9-9 presents the code.

The `if` statement in line 6 ensures that the code in the Load event handler is executed only when the Reservation page is loaded the first time, either directly or from another page.

As discussed earlier, the `if` statement in line 12 gets the name of the previous page, if there is one.

The if statement starting on line 16 specifies that if it is a transfer from *Confirmation* page, the control properties are set to previously selected values using the statements in lines 18–28. It is important to note that it could be a transfer from a *Confirmation* page from an entirely different host, which could lead to error in the statements in lines 18–28. To address this problem, you may check whether the hostname of the previous page, given by Request.UrlReferrer. Host property, is the same as the current hostname given by Request.Url.Host. You may do this check as an exercise.

If the reservation page is not displayed from the *Confirmation* page (i.e., if the reservation page is displayed directly or from a different page), the session variables are created and set to zero in lines 32–34.

```
1   public partial class Reservation_3 : System.Web.UI.Page
2       {
3       decimal basePrice, zonePrice, discount, unitPrice;
4       protected void Page_Load(object sender, EventArgs e)
5       {
6           if (!IsPostBack)      // If it is not a postback of the page,
7           {
8               ddlPerformance.DataBind();    // bind the DropDownList
9               ddlPerformance.Items.Insert(0, new ListItem("", "")); //add an empty item
10              // if transferring from another page, get the name of that page
11              string previousPageName = "";
12              if (PreviousPage != null)
13                  previousPageName = Request.UrlReferrer.AbsolutePath;
14
15              // If transferring from Confirmation page
16              if (previousPageName == "/Confirmation.aspx")
17              {   // set control properties to previously selected values
18                  ddlPerformance.SelectedIndex = (int)Session["PerformanceIndex"];
19                  ddlDate.SelectedValue = Session["SelectedDate"].ToString();
20                  rblZones.SelectedIndex = (int)Session["ZoneIndex"];
21                  chkMember.Checked = (bool)Session["IsMember"];
22                  txtNoOfTickets.Text = Session["NoOfTickets"].ToString();
23
24                  // Re-display prices and discount
25                  lblBasePrice.Text = Session["BasePrice"].ToString();
```

Figure 9-9: Complete code for the Reservation-3 page (*continues*)

```
26              lblZonePrice.Text = Session["ZonePrice"].ToString();
27              lblDiscount.Text = Session["Discount"].ToString();
28              lblUnitPrice.Text = Session["UnitPrice"].ToString();
29          }
30          else//If not transferring from Confirmation page, create session variable
31          {
32              Session["BasePrice"] = (decimal)0;
33              Session["ZonePrice"] = (decimal)0;
34              Session["Discount"] = (decimal)0;
35          }
36      }
37      UnobtrusiveValidationMode = UnobtrusiveValidationMode.None;
38  }
39
40  protected void ddlPerformance_SelectedIndexChanged(object sender, EventArgs e)
41  {
42      DataView dvPerformanceTable =
                (DataView)SqlDataSourceTheater.Select(DataSourceSelectArguments.Empty);
43      dvPerformanceTable.RowFilter =
                "PerformanceId = " + ddlPerformance.SelectedValue.ToString();
44      // Or,
45      // dvPerformanceTable.RowFilter =
                "PerformanceName = " + "'" + ddlPerformance.SelectedItem.ToString() + "'";
46      DataRowView drvPerformanceRow = (DataRowView)dvPerformanceTable[0];
47      basePrice = (decimal)drvPerformanceRow["BasePrice"];
48
49      lblBasePrice.Text = basePrice.ToString(); // Display base price
50      // Compute and display price per ticket
51      Session["BasePrice"] = basePrice;
52      unitPrice = (decimal)Session["BasePrice"] +
53                  (decimal)Session["ZonePrice"] - (decimal)Session["Discount"];
54      lblUnitPrice.Text = unitPrice.ToString();
55      Session["UnitPrice"] = unitPrice;
56  }
57
58  protected void rblZones_SelectedIndexChanged(object sender, EventArgs e)
59  {
60      string selectedZone = rblZones.SelectedValue.ToString();
61      switch (selectedZone)
62      {
63          case "Suite":
64              zonePrice = 40;
65              break;
66          case "Prem&Orch":
67              zonePrice = 25;
68              break;
69          case "Circle":
70              zonePrice = 10;
71              break;
72          case "Balc&Gallery":
73              zonePrice = 0;
74              break;
75      }
76      lblZonePrice.Text = zonePrice.ToString(); // Display zone price
77      // Compute and display price per ticket
78      Session["ZonePrice"] = zonePrice;
79      unitPrice = (decimal)Session["BasePrice"] +
80                  (decimal)Session["ZonePrice"] - (decimal)Session["Discount"];
81      lblUnitPrice.Text = unitPrice.ToString();
82      Session["UnitPrice"] = unitPrice;
83  }
84
85  protected void chkMember_CheckedChanged(object sender, EventArgs e)
86  {
87      // Compute and display discount for members
88      if (chkMember.Checked)
89          discount = 10;
```

Figure 9-9: Complete code for the Reservation-3 page (*continued*)

```
90          lblDiscount.Text = discount.ToString(); // Display discount
91          // Compute and display price per ticket
92      Session["Discount"] = discount;
93      unitPrice = (decimal)Session["BasePrice"] +
94                  (decimal)Session["ZonePrice"] - (decimal)Session["Discount"];
95      lblUnitPrice.Text = unitPrice.ToString();
96      Session["UnitPrice"] = unitPrice;
97      }
98
99      protected void btnContinue_Click(object sender, EventArgs e)
100     {
101         if (calPerformDate.SelectedDate == DateTime.MinValue)
102         {   // display a message
103             Response.Write("<script> alert('Please select a date') </script>");
104             // Alternative methods:
105             ClientScript.RegisterStartupScript(this.GetType(), "MissingDate",
106             "<script>alert('Please select a date'); </script> ");
107         //ClientScript.RegisterClientScriptBlock(this.GetType(), "MissingDate",
108         //    "<script>alert('Please select a date'); </script> ");
109             lblDate.Text = "Please select a date"; // Alternative to an alert
110             return;
111         }
112         // Store date, member status and number of tickets in session variables
113         Session["SelectedDate"] = calPerformDate.SelectedDate;
114         Session["IsMember"] = chkMember.Checked;
115         Session["NoOfTickets"] = int.Parse(txtNoOfTickets.Text);
116
117         // Store indices of selected play and zone in session variables
118         Session["PerformanceIndex"] = ddlPerformance.SelectedIndex;
119         Session["ZoneIndex"] = rblZones.SelectedIndex;
120
121         //Move to confirmation page. Pass performance and zone using QueryStrin
122         Response.Redirect("Confirmation.aspx?performance=" +
123                         ddlPerformance.SelectedItem.ToString() + "&zone=" +
124                         rblZones.SelectedItem.ToString()
123     }
124
125     protected void btnClear_Click(object sender, EventArgs e)
126     {
127         ddlPerformance.SelectedIndex = -1;
128         calPerformDate.SelectedDates.Clear();
129         rblZones.SelectedIndex = -1;
130         chkMember.Checked = false;
131         txtNoOfTickets.Text = String.Empty;
132         foreach (Control control in form1.Controls)
133         {
134             if (control.GetType() == typeof(Label))
135             {
136                 Label lbl = (Label)control;
137                 lbl.Text = String.Empty;
138             }
139         }
140     }
141 }
```

Figure 9-9: Complete code for the Reservation-3 page (*continued*)

Step 14: Set the control properties to the previously selected values.

Make changes shown in Figure 9-9 to the Load event handler of the Reservation-3 page.

Run the website. Test the code by specifying the reservation information and transferring it to the *Confirmation* page.

Next, click the *Change Reservation* button to go back to the Reservation page. Make sure the selections you made previously on the Reservation page are displayed on the page.

Change the selections. Make sure the prices/discount are updated when you change the selections.

9.7 Session Variables of array/class/struct Type

In the current project, you used session variables that store a single data item. A session variable could be of any *.NET Framework* type, including value types, arrays, collections, structures, and objects. Rather than storing individual data items in separate session variables, you can store a group of related data items in an array and store the array in a single session variable. The following code illustrates the use of a session variable of array type, which stores several reservation data items:

```
1  string[] reservationInfo = {calDate.SelectedDate.ToString(),
       chkMember.Checked.ToString(), txtNoOfTickets.Text,
       ddlPerformance.SelectedIndex.ToString(),rblZones.SelectedIndex.ToString()};
2  Session["ReservationInfo"] = reservationInfo;
```

The first statement in the above code stores the seven data items (selected date, member status, …) from the click event handler of btnContinue in a string array, reservationInfo. The second statement stores the array in the session variable, ReservationInfo. The above two statements could replace the statements in lines 113–118 in the btnContinue_Click event handler shown in Figure 9-9.

To display the data on the *Confirmation* page, you may retrieve the data from the session variable and store them in a string array:

```
string[] reservationInfo = (string[]) Session["ReservationInfo"];
```

The data stored in the string array may be displayed in the Label *lblReservation* on the *Confirmation* page using Figure 9-10.

```
1  string selection =        "Date:      " + reservationInfo[0] + "<br />";
2  if (reservationInfo[1] == "True")
3         selection = selection + "Member?   " + "Yes";
4  else
5         selection = selection + "Member?   " + "No";
6  selection = selection + "# of tickets: " + reservationInfo[2];
7
8  lblReservation.Text = selection;
```

Figure 9-10: Displaying data from a string array

The code in Figure 9-10 uses the index of the array element to access each individual data item.

You also may use a structure or class in place of the array so that you can refer to individual data items by a name rather than by the index. The next section shows the use of a class to store reservations.

It's time to practice! Do Exercise 9-1. (See the end of the chapter.)

Review Questions

10. A website has three pages. The user enters an order using the first two pages in sequence, and the third page displays the data entered on the first two pages for confirmation. What is the drawback of sharing data between the pages using cross-page posting and PreviousPage property (without using session variables or query string) compared to the Redirect method with session variables?

11. A website has three pages. The user enters an order using the first two pages in sequence, and the third page displays the data entered on the first two pages for confirmation. What is the drawback of sharing data between the pages using the Transfer method with the query string compared to the Redirect method with session variables?

9.8 Storing Multiple Reservations Using Classes

Typical e-commerce sites allow users to place an order that includes multiple products. To help you understand how to work with multiple sets of data and share them between pages, you will update the reservation and confirmation forms to let the user make multiple reservations at a time, view them together, edit, and confirm them. The updated confirmation page named *Confirmation-4* is shown in Figure 9-11.

Confirmation

Your selections

Justin Hayward, Date: 1/11/2021, Zone: Orchestra, Member?: True, Tickets: 2
Cocktail Hour, Date: 3/1/2021, Zone: Premium, Member?: False, Tickets: 1
TedxOshkosh, Date: 1/20/2021, Zone: Suite , Member?: True, Tickets: 3

| Remove Reservation | Change Reservation | Add New Reservation |

| Confirm |

Figure 9-11: The Confirmation-4 page

To allow users to make multiple reservations at a time, you will store each reservation in an object and store multiple objects in a List of objects.

If you are not familiar with creating and working with classes and objects, you are encouraged to review Appendix D*, which provides an introduction to object-oriented programming.

You will create a class named **Reservation** to store all data that are needed on the confirmation page or on the reservation page (to restore its state if the user revisits the page). This class will be used (in place of the individual session variables that are currently used) to represent the following data items:

Performance Id, performance name, performance date,

      zone Id, zone name, member status, tickets,

      base price, zone price, discount and unit price

Note that the performance id and zone id are used here to help restore their states. (The previous version of the reservation page used the `SelectedIndex` of the `ComboBox` and `RadioButtonList`.) In the real world, you would typically save the reservations into a database using an ID field to identify each reservation. Chapter 18 discusses saving reservations into a database.

You will create a new version of the reservation form named Reservation-4.aspx that uses the Reservation class, along with a new version of the confirmation page named Confirmation-4 that works with Reservation-4.

9.8.1 Creating the Reservation Class

Typically, you will add a class to a folder named App_Code. To add a new class to the App_Code folder, right-click the folder and select the following:

Add > New Item > Visual C# > class

Step 15: Add the App_Code folder and add a class named Reservation to the folder.

In Solution Explorer, right-click the project name Theater, and add a folder named App_Code.

Right-click the App_Code folder and select

    Add > New Item > Visual C# > class

Change the class name to Reservation.cs.

---

* Appendix can be freely downloaded from https://www.prospectpressvt.com/textbooks/philip-web-development at "Student Resources."

Properties of the Reservation Class

You create a property corresponding to each data item of a reservation so you can store and retrieve the data item. Thus you specify eleven properties for the Reservation class corresponding to the eleven data items, as shown in Figure 9-12.

```
1   public string PerformanceId { get; set; }
2   public string Performance { get; set; }
3   public DateTime PerformDate { get; set; }
4   public string ZoneId { get; set; }
5   public string Zone { get; set; }
6   public bool IsMember { get; set; }
7   public int Tickets { get; set; }
8   public decimal BasePrice { get; set; }
9   public decimal ZonePrice { get; set; }
10  public decimal Discount { get; set; }
11  public decimal UnitPrice { get; set; }
```

Figure 9-12: Properties of the Reservation class

Because we need to use properties only to store and retrieve values, here we used automatic properties, which are simpler to specify. The Performance property represents the performance name, and Zone represents the zone name.

The Constructor of the Reservation Class

The constructor is a special method that is automatically called when you create an object from the class. In addition to creating the object, the constructor sets initial values for the properties. Figure 9-13 shows the constructor.

```
1   public Reservation(string performanceId, string performance, DateTime performDate,
2       string zoneId, string zone, bool isMember, int tickets,
3       decimal basePrice, decimal zonePrice, decimal discount, decimal unitPrice)
4   {
5       this.PerformanceId = performanceId;
6       this.Performance = performance;
7       this.PerformDate = performDate;
8       this.Zone = zone;
9       this.ZoneId = zoneId;
10      this.IsMember = isMember;
11      this.Tickets = tickets;
12      this.BasePrice = basePrice;
13      this.ZonePrice = zonePrice;
14      this.Discount = discount;
15      this.UnitPrice = unitPrice;
16  }
```

Figure 9-13: Constructor of the Reservation class

When you create a Reservation object, this method is called and the values passed to the parameters (performanceId, performance, ...) are assigned to the corresponding properties (this.PerformanceId, this.Performance, ...)

This is how you would create an object named reservation:

```
1. Reservation reservation = new Reservation(performanceId, performance, performDate,
2.      zoneId, zone, isMember, tickets, basePrice, zonePrice, discount, unitPrice);
```

Here, performanceId, performance, and so forth are variables whose values are passed to the constructor to initialize the properties.

The ToString() Method of the Reservation Class

The ToString() method is used to display the values of specified properties. The ToString() method shown in Figure 9-14 would display the performance name, date, zone, member status, and number of tickets.

```
1   public override string ToString()
2   {
3   return string.Format("{0}, Date: {1:d}, Zone: {2}, Member?: {3}, Tickets: {4}",
4                          Performance, PerformDate, Zone, IsMember, Tickets);
5   }
```

Figure 9-14: The ToString() method of the Reservation class

Step 16: Insert the code for the Reservation class.

Insert the code from Figures 9-12, 9-13, and 9-14 within the starting and ending braces of the Reservation class, shown here:

```
1   public class Reservation
2   {
3       <<Insert above code for properties, constructor and the Tostring() method here>>
4   }
```

9.8.2 Using the Reservation Class in the Reservation Form

The **current version** of the reservation form shares data by storing it in session variables and query strings specified within the **btnContinue_Click** event handler, as shown in Figure 9-15.

```
1   protected void btnContinue_Click(object sender, EventArgs e)
2   {
3       if (calPerformDate.SelectedDate == DateTime.MinValue)
4       {   // display a message
5           // Response.Write("<script> alert('Please select a date') </script>");
6           // return;
7       }
8       //Store selected date, member status and number of tickets in session variables
9       Session["SelectedDate"] = ddlDate.SelectedValue;
10      Session["IsMember"] = chkMember.Checked;
11      Session["NoOfTickets"] = int.Parse(txtNoOfTickets.Text);
12      // Store indices of selected play and selected zone in session variables
13      Session["PerformanceIndex"] = ddlPerformance.SelectedIndex;
14      Session["ZoneIndex"] = rblZones.SelectedIndex;
15      //Move to confirmation page. Pass selected performance & zone using QueryString
16      Response.Redirect("Confirmation.aspx?performance="
17          + ddlPerformance.SelectedItem.ToString() +"&zone="
18          + rblZones.SelectedItem.ToString());
19  }
```

Figure 9-15: brnContinure_Click event handler using session variables

In addition, four other session variables are used in other event procedures to store the prices and discounts.

Storing Data in Properties

Instead of storing data in session variables and query strings, you will create a Reservation object and store the data in the properties when you create the object. Figure 9-16 shows the revised code.

Lines 8–20 store the eleven data items in simple variables that are passed to the constructor.

Lines 22 and 23 create the Reservation object, which in turn initializes the eleven properties to the values of the eleven variables.

The rest of the code deals with storing the `Reservation` object in a `List` of `Reservation` objects, as explained next.

```
1   protected void btnContinue_Click(object sender, EventArgs e)
2   {
3       if (calPerformDate.SelectedDate == DateTime.MinValue)
4       {   // display a message
5           Response.Write("<script> alert('Please select a date') </script>");
6           return;
7       }
8       // Get reservation data from controls
9       string performanceId = ddlPerformance.SelectedValue;
10      string performance = ddlPerformance.SelectedItem.ToString();
11      DateTime performDate = calPerformDate.SelectedDate;
12      string zoneId = rblZones.SelectedValue;
13      string zone = rblZones.SelectedItem.ToString();
14      bool isMember = chkMember.Checked;
15      int tickets = int.Parse(txtNoOfTickets.Text);
16      // Get prices from session variables
17      basePrice = (decimal)Session["BasePrice"];
18      zonePrice = (decimal)Session["ZonePrice"];
19      discount = (decimal)Session["Discount"];
20      unitPrice = (decimal)Session["UnitPrice"];
21      // Create the Reservation object
22      Reservation reservation= new Reservation(performanceId, performance, performDate,
23          zoneId, zone, isMember, tickets, basePrice, zonePrice, discount, unitPrice);
24      // If it is the first reservation, create a List; if not get the List
25      List<Reservation> myReservations;
26      myReservations = (List<Reservation>)Session["MyReservations"];
27      if (myReservations == null)
28      {
29          myReservations = new List<Reservation>();
30      }
31      // If it is an edit, delete the reservation from the List before adding
32          the updated reservation.
32      string previousPageName = (string)ViewState["PreviousPageName"];
33      if ((previousPageName == "/Confirmation-4.aspx") &&
                            (Request.QueryString["action"] == "edit"))
34      {
35          int selectedResIndex = (int)Session["SelectedResIndex"];
36          myReservations.RemoveAt(selectedResIndex);
37      }
38      myReservations.Add(reservation);  // add the reservation
39      Session["MyReservations"] = myReservations;
40      // Move to confirmation page.
41      Response.Redirect("Confirmation-4.aspx?performance=");
42  }
```

Figure 9-16: btnContinue_Click event handler using an object

List<T> Collection

A `List` is a typed collection of items, similar to an array. The items could be of any data type, like a simple number, string, an array or an object; but within any one `List`, all items must be of the same type. You may create a `List` using the following syntax:

List<*Type*> *listName* = new List<*Type*>();

or

List<*Type*> *listName*
listName = new List<*Type*>();

Here are some examples:

```
List<int> monthlySales = new List<int>();
                    // creates a List of int type named monthlySales.
List<Customer> customers = new List<Customer>(); // creates a List of Customer objects
```

You add an item to the end of the List using the Add() method of the List. Some examples are as follows:

```
monthlySales.Add(355); // adds 355 to the end of the monthlySales List.
customers.Add(JohnSmith); // adds a Customer object named JohnSmith to the List.
```

Let's look at the rest of the code in lines 25–43 in Figure 9-16. You may first review the code and the internal comments that give a brief description of each section of the code. A more detailed description follows.

Lines 25–30 get the List of reservations, if it exists. If not (i.e., if it is the first reservation), a new List is created, as explained next.

Line 25 declares the reference variable **myReservations** of type List<Reservation>, which is later stored in a session variable **MyReservations** in line 39, so that it can be shared with the confirmation page.

Line 26 gets the List of reservations from the session variable MyReservations, if it exists, casts it to the List, and stores it (a reference to it) in myReservations. For the first reservation, the session variable is not yet created when line 26 is processed. So, myReservations would be null in line 27. Line 29 would create the List when the first reservation is made, and line 38 would add the reservation object to the List. Line 39 would store the List in the session variable, MyReservations.

For successive reservations, line 26 would get the List from the session variable, and line 29 would be skipped.

Lines 32 to 37 delete the reservation, if it is being edited, before adding the updated reservation to the List. Line 32 gets the name of the previous page from the view state variable created in the Page_Load event handler, as explained later. Line 33 checks whether it is an edit of an existing reservation. If so, line 35 would get the index of the selected reservation, and line 36 would remove the item from the List.

Creating the Reservation-4 Form That Uses the Reservation Class

Step 17: Create a copy of the Reservation-3 form and name it Reservation-4.

The process is the same as the one used earlier to make copies of Reservation-1 and Reservation-2. The instructions are repeated here.

Copy and paste the Reservation-3 form and rename Reservation-3 copy as Reservation-4.

Before you can use the Reservation-4 page, you need to make a few important changes. Be careful to follow the steps exactly.

Open the Reservation-4.aspx.designer.cs file from the Solution Explorer.

Change the class name from *Reservation_3* to *Reservation_4*, as shown here:

```
1    namespace Theater {
2        public partial class Reservation_4 {
```

Save and close Reservation-4.aspx.designer.cs.

Open the Reservation-4.aspx.cs file.

Change the class name from *Reservation_3* to *Reservation_4*, as shown here.

```
1    namespace Theater
2    {
3        public partial class Reservation_4 : System.Web.UI.Page
```

Save and close *Reservation-4.aspx.cs*.

Open the *Reservation-4.aspx* file from the Solution Explorer.

In the first statement, change the form name from *Theater.Reservation_3* to *Theater.Reservation_4*, as shown here:

```
<%@ Page Language="C#" AutoEventWireup="true" CodeBehind=
                "Reservation-4.aspx.cs" Inherits="Theater.Reservation_4" %>
```

In addition, change the `Title` to `Reservation-4`.

```
<title>Reservation-4</title>
```

Save the Reservation-4 form. Close the Reservation-3 form.

Step 18: Modify the btnContinue_Click event handler in Reservation-4 form to use the Reservation class.

To be able to use the `Reservation` class without specifying the directory where it is stored, add the following directive to the Reservation-4 page:

```
using Theater.App_Code;
```

Delete the current code in the btnContinue_Click event handler and add the code from Figure 9-16.

Next, you will make a copy of the Confirmation page and modify it to work with the revised reservation form.

Create the Confirmation-4 Form That Uses the Reservation Class

The confirmation page that displays the reservation data should be updated to work with the list of reservations that were created in the Reservation-4 form. You will create a copy of the Confirmation page named Confirmation-4. (The new version is named Confirmation-4, to indicate that it works with Reservation-4.)

Step 19: Make a copy of the Confirmation page and name it Confirmation-4.

Carefully follow the steps that you performed to make a copy of the Reservation-3 page.

Change **Confirmation** to **Confirmation_4** in the **.aspx.designer.cs**, **.aspx.cs**, and **.aspx** files.

Add the following directive to the Confirmation-4 page:

```
using Theater.App_Code;
```

The Confirmation page was designed to display a single reservation in the Label, *lblReservation*. The Confirmation-4 page will use a `ListBox` to display multiple reservations and let the user add a reservation, edit a reservation, or delete it, as shown in Figure 9-17.

Figure 9-17: Confirmation-4 form

Step 20: Modify the user interface on the Confirmation-4 page as shown in Figure 9-17.

Delete *lblReservation* and add a ListBox named lstMyReservations, in its place.

Add a button named btnRemove and another button named btnAdd.

Provide captions for the buttons.

Next, you will update the code on the Confirmation-4 page. The current code in the Page_Load event handler gets data from the session variables and query strings. You will modify this code to get reservations from the List of reservations, MyReservations, and add them to the ListBox.

Adding Items to the ListBox

You cannot directly add a Reservation object to a server control ListBox. You can add any string as an item of the ListBox using the Add method:

```
lstMyReservations.Items.Add(string)
```

To add a reservation to the ListBox, you can use the ToString() method of the Reservation class, which combines the reservation data into a single string:

```
lstMyReservations.Items.Add(reservation.ToString())
```

It should be noted that the items of a ListBox are of type ListItem, which has a **Text** property and an optional **Value** property. The above statement stores the specified string to the Text property. Typically, the optional Value property is used to store the identifier for the data stored in the Text property.

You will use the above statement in the Page_Load event handler of the confirmation page to get the data for each reservation from the List, MyReservations, and add it to the ListBox. The event handler is shown in Figure 9-18.

```
1  List<Reservation> myReservations;
2  protected void Page_Load(object sender, EventArgs e)
3  {
4      if (!IsPostBack)      // If it is not a postback of the page,
5      {
6          myReservations = (List<Reservation>)Session["MyReservations"];
7          lstMyReservations.Items.Clear();
8          foreach (Reservation reservation in myReservations)
9          {
10             lstMyReservations.Items.Add(reservation.ToString());
11         }
12         btnChange.Enabled = false;
13         btnRemove.Enabled = false;
14     }
15 }
```

Figure 9-18: Code to display reservations on the Confirmation-4 page

Line 1 declares the reference variable, myReservations, as of type List(Reservation).

Line 6 gets the List of Reservations from the session variable, MyReservations, and stores a reference to it in myReservations.

Line 8 uses a foreach loop to get each reservation. Line 10 adds the reservation data as an item of the ListBox.

Lines 12 and 13 disables btnChange and btnRemove, respectively. These buttons are enabled when the user selects an item from the ListBox.

Step 21: Add code from Figure 9-18 to the Confirmation-4 page to display reservations.

To test the code, display the Reservation-4 page, make selections for a reservation, and click the Continue button. The Confirmation-4 page should display the reservation data.

Adding, Changing, and Deleting Reservations

To **add** a new reservation when the user clicks the Add button, you can display the reservation page using the cross-page posting method by setting the `PostBackUrl` property of `btnAdd` to *Reservation-4.aspx*

Recall that the cross-page posting lets you access the previous page.

To **change** a reservation, the user selects a reservation from the `ListBox`, which enables the Change button. Clicking the change button displays the reservation form, which displays the selected reservation so that the user can edit it.

To display the reservation page when the user clicks the Change button, set the `PostBackUrl` property of btnChange to Reservation-4.aspx.

To let the reservation page know that the reservation is to be displayed for editing, pass a query string with the URL, like

```
Reservation-4.aspx?action=edit
```

Here, the query string *action* is set to the value *edit* and passed to the reservation page to indicate the selected reservation is to be edited.

How do you code the reservation form to get the Reservation object corresponding to the reservation selected in the `ListBox` on the confirmation page? Because the items in the `ListBox` are kept in the same order as those in the `List`, `MyReservations`, you can use the index of the item selected in the `ListBox` to find the corresponding Reservation object from the `List`.

You can share the index of the selected `reservation` with the reservation page by saving the index in a session variable whenever the user selects an item from the ListBox. This can be done using the `SelectedIndexChanged` event handler, as shown in Figure 9-19.

```
1   protected void lstMyReservations_SelectedIndexChanged(object sender, EventArgs e)
2   {
3       Session["SelectedResIndex"] = lstMyReservations.SelectedIndex;
4       btnChange.Enabled = true;
5       btnRemove.Enabled = true;
6   }
```

Figure 9-19: The SelectedIndexChanged event handler of lstMyReservations

Note that the following statement in line 35 in Figure 9-16 gets the index from the session state to remove the corresponding reservation:

```
int selectedResIndex = (int)Session["SelectedResIndex"];
```

The event handler also enables `btnChange` and `btnDelete`, which are initially disabled.

To **delete** the selected reservation when the user clicks `btnRemove`, it is necessary to delete the reservation from the `ListBox` and also from the `List`, `MyReservations`.

You remove an item from the `ListBox` using the `Remove()` method by specifying the index of the item to be removed, as follows:

```
lstMyReservations.Items.Remove(index)
```

To remove an item from the `List`, use the `RemoveAt()` method by specifying the index of the item, as in

```
myReservations.RemoveAt(index);
```

Recall that you used the `RemoveAt()` method earlier in line 36 in Figure 9-16 to remove a reservation from the List.

Figure 9-20 shows the click event handler of btnRemove, which removes the selected item from the `ListBox` and from the `List` MyReservations.

```
1   protected void btnRemove_Click(object sender, EventArgs e)
2   {
3       int selectedIndex = lstMyReservations.SelectedIndex;
4       lstMyReservations.Items.Remove(selectedIndex);
5
6       List<Reservation> myReservations= (List<Reservation>)Session["MyReservations"];
7       myReservations.RemoveAt(selectedIndex);
8       Session["MyReservations"] = myReservations;
9   }
```

Figure 9-20: The btnRemove_Click() event handler to remove a reservation

Line 4 removes the selected item from the ListBox by specifying its index.

Lines 6–8 get the List from the session variable MyReservations (line 6), remove the item at the specified index (line 7), and store the updated List back in MyReservations (line 8).

Step 22: Add code to add, change, and delete a reservation.

Set the PostBackUrl property of btnAdd to:

> Reservation-4.aspx

Set the PostBackUrl property of btnChange to:

> Reservation-4.aspx?action=edit

Create the SelectedIndexChanged event handler of the ListBox, as shown in Figure 9-19, to store the index in a session variable.

Create the click event handler of btnRemove, as shown in Figure 9-20.

Updating the Reservation-4 Page to Display the Reservation Object for Editing

You will update the code in the Page_Load event handler of the Reservation-4 form to display the selected reservation by accessing it from the List using its index. Figure 9-21 shows the code.

Lines 1–12 are the same as those in Reservation-3.

Line 13 stores the name of the previous page in the view state to be used later in the btnContinue_Click event handler.

Line 16 checks whether the page was displayed from the confirmation page for editing.

Line 18 gets the List of reservations from the session variable. Line 19 gets the index of the selected reservation, and line 20 gets the reservation object from the List.

Lines 22–32 set the controls to their previously set values for editing.

Lines 36–38 initialize session variables to share the prices and discount between event handlers within the same page. Note that because this is to be shared within the same page, you could use view state variables instead.

```
1   decimal basePrice, zonePrice, discount, unitPrice;
2   protected void Page_Load(object sender, EventArgs e)
3   {
4       if (!IsPostBack)      // If it is not a postback of the page,
5       {
6           ddlPerformance.DataBind();     // bind the DropDownList
7           ddlPerformance.Items.Insert(0, new ListItem("", "")); // add an empty item
8           // if transferring from another page, get the name of that page
9           string previousPageName = "";
10          if (PreviousPage != null)
11          {
12              previousPageName = Request.UrlReferrer.AbsolutePath;
13              ViewState["PreviousPageName"] = previousPageName;
14          }
15          // If transferring from Confirmation page for editing
16          if ((previousPageName == "/Confirmation-4.aspx") &&
                                    (Request.QueryString["action"] == "edit"))
17          {   // get reservation data from the List
18              List<Reservation> myReservations =
                                    (List<Reservation>)Session["MyReservations"];
19              int selectedResIndex = (int)Session["SelectedResIndex"];
20              Reservation res = myReservations[selectedResIndex];
21              // set control properties
22              ddlPerformance.SelectedValue = res.PerformanceId;
23              ddlDate.SelectedValue = res.PerformDate.ToString();
24              rblZones.SelectedValue = res.ZoneId;
25              chkMember.Checked = res.IsMember;
26              txtNoOfTickets.Text = res.Tickets.ToString();
27
28              // Display prices and discount
29              lblBasePrice.Text = res.BasePrice.ToString();
30              lblZonePrice.Text = res.ZonePrice.ToString();
31              lblDiscount.Text = res.Discount.ToString();
32              lblUnitPrice.Text = res.UnitPrice.ToString();
33          }
34          else    // If not transferring from Confirmation page for editing,
                        create session variables
35          {
36              Session["BasePrice"] = (decimal)0;
37              Session["ZonePrice"] = (decimal)0;
38              Session["Discount"] = (decimal)0;
39          }
40      }
41      UnobtrusiveValidationMode = UnobtrusiveValidationMode.None;
42  }
```

Figure 9-21: The `Page_Load` event handler of the Reservation-4 page

Step 23: Update Reservation-4 to display a Reservation object for editing.

Update the Page_Load event handler of the Reservation-4 page as shown in Figure 9-21.

Step 24: Test the code in the Reservation-4 and Confirmation-4 pages.

Display Reservation-4, enter data for a new reservation, and display the data on the Confirmation-4 page.

Click the Add button and add another reservation. The confirmation page should display both reservations.

Select the first reservation and click the Change button to edit it. Make one or more changes to the reservation, and display the confirmation page. The first reservation should show the changes.

Select the first reservation and click the Remove button. The ListBox should display only the second reservation.

It's time to practice! Do Exercise 9.2 (see the end of the chapter).

Review Questions

12. A class named `Zone` is specified as follows:

```
1    public class Zone
2    {
3        public string Id { get; set; }
4        public string Name { get; set; }
5        public decimal Price { get; set; }
6        public Zone(string id, string name, decimal price)
7        {
8            this.Id = id;
9            this.Name = name;
10           this.Price = price;
11       }
12   }
```

Show the code to create three objects named suite, circle, and balcony that have the following property values for `Id`, `Name`, and `Price`, respectively:

"11", "Suite", 40

"14", "Circle", 10

"15", "Balcony", 5

13. Refer to the class and objects specified in question 12. Show the code to create a `List` of `Zone` objects named `listOfZones`, and add the objects *suite*, *circle*, and *balcony* to the `List`.

14. Refer to the `List` specified in question 13. Show the code to create a session variable named `ListOfZones` and store the `List` in the session variable.

15. Refer to the `List` specified in question 14. Show the code to display the value of the `Name` attribute from each `Zone` object stored in the List. Display the names in a `ListBox` named `lstZoneNames`.

Exercises

Exercise 9.1: Snacks

Make sure you do Exercise 7.1, if you haven't already. Exercise 7.1 adds a `CheckBox` to the Reservation form for the user to specify whether or not snacks are needed. In the current exercise, the `Label` on the Confirmation page should display the snack selection made by the user. Other information displayed on the Confirmation page remains the same. In addition, if the user goes back to the Reservation page from the Confirmation page, the Reservation page should show the snack selection, if any.

Exercise 9.2: Update the Reservation Class

Add a twelfth property named `CancelBy` to the `Reservation` class. Compute the *cancel by* date (one day from today's date) within the click event handler of `btnContinue` on the Reservation page and store it in the `CancelBy` property of the `Reservation` object. Display the cancel by date along with the five data items currently displayed in the `ListBox` on the Confirmation page.

Assignments

E-Commerce Assignment: Part D (ASP.NET)—Multipage

This assignment expands E-Commerce Assignment Part C (ASP.NET)—Using Databases (see Chapter 8) to develop a shopping cart page named cart. Make sure you first develop the laptop-selection page as specified in Part C, if you haven't already done so.

Provide a button on the laptop-selection page to display the cart page to be created.

Add a new web form named **cart**. When the user displays the cart page, display the following selections made by the user on the laptop-selection page:

Model, RAM, SSD, hard drive, software, warranty, and total cost

Provide a button with the caption "Checkout." When the user clicks the Checkout button, for now, display the message "Order confirmed" on the page.

Provide three additional buttons on the cart page to add a model to the cart, delete a model from the cart, or change the selections for a model. The Add button should let the user go to the laptop-selection page and select another model and add it to the cart. The Delete button should delete the selected model from the cart.

The Change button lets the user go back to the laptop-selection page and make changes in the specifications for a selected model. When the user returns to the laptop-selection page, the page should display the model, software, and warranty that were previously selected by the user. For now, the user will re-select the configuration. The page should also display the prices.

Share data between pages using an object stored with the session by storing data items in the properties of the object. Use a List of objects to store information on multiple models.

Auto Rental Assignment: Part D (ASP.NET)—Multipage

This assignment expands Auto Rental Assignment: Part C (ASP.NET)—Using Databases (see Chapter 8) to add a new Confirmation page named **Rental-confirmation**. Make sure you develop the web page **Auto-selection** as specified in Part C, except that the page doesn't have the button to compute the total cost (step 5).

Add a new page named Rental-confirmation. Provide a button on the Auto-selection page to display the Rental-confirmation page. When you display the Confirmation page, it should display the selections made by the user on the **Auto-selection** page, the number of days, and the total cost, as follows:

Auto type, model, discount category, insurance, number of days, total cost

Provide a button with the caption "Confirm." When the user clicks the Confirm button, display the message "Rental confirmed" on the page.

Provide three additional buttons on the Confirmation page to let the user add, delete, or change a rental.

The Add button should let the user go back to the Auto-rental page and make selections for another rental. The Delete button should delete the rental selected by the user.

The Change button lets the user go back to **Auto-selection** page and make changes to the rental information. When the user returns to the **Auto-selection** page, the page should display the selections previously made by the user and the data displayed in the controls (daily rental rate, discount amount, insurance cost, and number of days).

Share data between pages using an object stored with the session, by storing the data items in the properties of the object. Use a List of objects to store information on multiple rentals.

Chapter 10

Data-Bound Lists

The previous two chapters presented several aspects of developing dynamic web pages, including linking web pages to databases and sharing data when moving between pages. Another important aspect of creating web pages is to dynamically generate code that presents information on a group of items like a list of products or services, and to provide the user the ability to search for information on specific items. For example, visitors to The Grand Theater website would benefit from a page that presents information on each performance (e.g., name, image, description, dates, prices) and provide the option to view performances for a certain month or a specific type. Typically, such information is obtained from a database table or file. In this chapter, you will learn how to use a database to generate, present, and search for information on items from a list.

Learning Objectives

After studying this chapter, you should be able to:

- Use a Repeater control to present a list of items from a database table.
- Upload images to a database table and display images from a database on a web page.
- Use the script element and embedded code blocks to insert code within the markup.
- Use nested Repeaters to display a list within another list.
- Use the Command event of buttons to pass values from buttons to the event handler.
- Scroll to a selected item and filter items from a list.
- Use client-side JavaScript and server-side code to display selected item(s) on another page.

10.1 Introduction

To help you understand how to generate HTML and ASP.NET code to present information on multiple performances, you will create a new page named **Events.aspx.** (The term *Events* is used here to represent performances.) You also will create an updated final version of the reservation form named **Reservation-5,** which incorporates changes to interact with the Events form. Figure 10-1 shows a segment of the Events page.

The main panel on the right side shows the name, image, description, scheduled dates, and a link button that takes the user to the Reservation page.

The set of hyperlinks in the left panel lets the user scroll to a specific performance in the right panel.

The *Filter by Month* button filters the performances to display only those that are scheduled for the selected month.

The *Reserve* button for each performance displays the reservation page and automatically selects the performance in the DropDownList on the reservation page.

A key characteristic of this page is that it displays information on a set of performances that are retrieved from a database table. A commonly used control that makes it easy to generate such a list is the **Repeater** control.

10.2 Generating a List Using the Repeater Control

A `Repeater` control lets you retrieve data from the rows in a data source and display them on a web page as a list using the layout you specify in a template. The `Repeater` loops through each row and generates the controls and other content specified in the template. The `Repeater` control may be bound to data sources like the `SqlDataSource`, `DataSet`, `SqlDataReader`, and most collections.

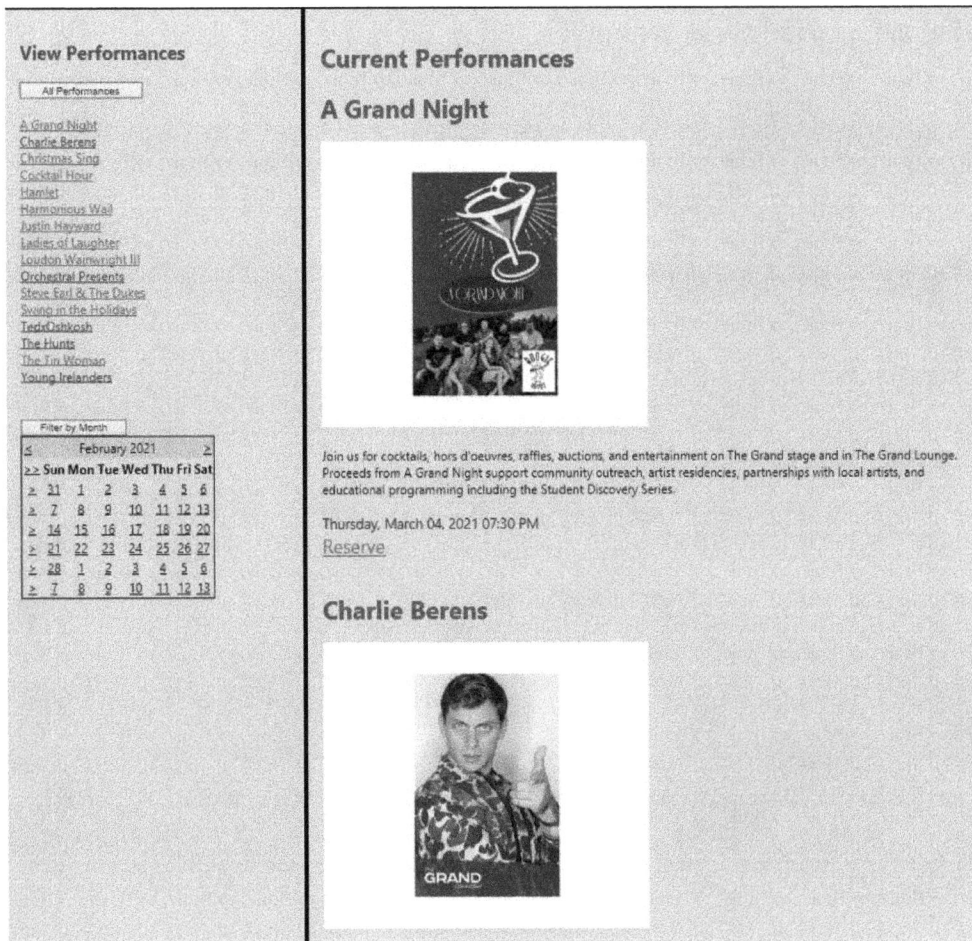

Figure 10-1: The Events page

There are five different types of templates. The most important template is the `ItemTemplate`, which specifies the content and layout of each item. The `ItemTemplate` contains markup that applies to each row of the dataset returned by the data source. Data that is to be displayed once at the beginning and end of the list is specified in a `HeaderTemplate` and `FooterTemplate`, respectively.

In addition, the `AlternatingItemTemplate` applies to every other row in the data source to provide a different look for alternating rows, and `SeparatorTemplate` specifies contents between items, like a line.

First, you will use the `Repeater` control to display a list of performance names, descriptions, and images of performances from the Performance table. Next, you will use another `Repeater` nested inside the first one to display multiple dates for each performance, from the PerformDate table.

Tutorial: The Events Page

This tutorial creates the Events page shown in Figure 10-1. Initially you will create the list consisting of performance names and descriptions. The `Repeater` will be bound to a data source configured to retrieve the data from the Performance table.

Step 1: Add a new web form.

Add a new web form named Events.aspx to the Theater project.

In Source view, copy the header element from the Reservation-4 page to the Events page immediately below the start tag of the body element, `<body>`.

Step 2: Set up the datasource control.

Add the `SqlDataSource` control to the web form after the header within the form element.

Set its `ID` to `SqlDataSource_Performance` and configure it using the existing connection string to select all fields from the Performance table. (See Sections 6.3 and 6.6 in Chapter 6.)

As you add more controls, place them above the DataSource to keep it at the bottom of the form element.

Step 3: Set up the Repeater control.

Add a `Repeater` control to the web form from the Data tab of the Toolbox.

Set its `ID` to `rptEvents`.

Bind the `Repeater` to the DataSource by setting the `DataSourceID` property of the `Repeater` to `SqlDataSource_Performance`.

Switch to the Source view and type in the start and end tags of the `<ItemTemplate>` element as shown in lines 2 and 4 of the following code:

```
1    <asp:Repeater ID="rptEvents" runat="server"
                    DataSourceID="SqlDataSource_Performance">
2        <ItemTemplate>
3
4        </ItemTemplate>
5    </asp:Repeater>
```

The next step is to specify the performance name and description as the content of the `<ItemTemplate>` element. Because the `Repeater` is bound to the data source, the field names `PerformanceName` and `Description` are accessible within the Repeater. You will use embedded code blocks within the ItemTemplate to get data from the data source and display it.

10.2.1 Embedded Code Block

Embedded code blocks are generally used to embed relatively simple server-side code within the HTML document. The code must be written in the default language of the page, which is C#, as specified in the directive `<%@ Page Language="C#"` in the aspx file. The embedded code is enclosed in the server tags `<% %>`. It should be noted that the Razor pages framework presented in Chapter 16 uses a different syntax called the Razor syntax to embed code blocks. The rest of this section will introduce some common uses of code blocks.

One use is to execute multiple lines of code within the HTML document. For example, the following code block contains code that displays a greeting based on the time of the day:

```
1   <body>
2   <form id="form1" runat="server">
3   <h2>
4       <%
5           if (DateTime.Now <= DateTime.Parse("12:00 PM"))
6               Response.Write("<h3>Good Morning</h3>");
7           else
8               Response.Write("<h3>Good Afternoon</h3>");
9       %>
10  </h2>
```

As you learned in the previous chapter, anything passed to the `Response.Write` method will be inserted into the HTTP response sent by the server to the client. Thus the above code will display the greeting "Good Morning" or "Good Afternoon" using the Heading 3 style.

The "<%=" Tag

You use the equal sign (=) in the tag to specify code that gets the value of a variable, or an expression or the value returned by a function. For example, consider a C# server-side function specified using the script element within the head element, as shown here:

```
1        <script runat="server">
2    string Greeting()
3    {
4        if (DateTime.Now <= DateTime.Parse("12:00 PM"))
5            return "Good Morning";
6        else
7            return "Good Afternoon";
8    }
9 </script>
```

You can call the above function and display the greeting using the following code block placed within the HTML document.

```
<%= Greeting() %>
```

The equal sign is a shortcut for Response.Write. So, you also may write the above statement as follows:

```
<% Response.Write(Greeting()); %>
```

Though it can be a useful shorthand to add code directly to the HTML page, it is preferable to define server-side functions in the code-behind file.

The "<%#" tag and Eval() Function

In database applications, you may have to access data that is bound to a control. For example, on the Events page, you need to get the values of performance names and descriptions from the DataSource that is bound to the Repeater. To access data bound to a control, use the "#" sign, instead of the "=" sign, along with the Eval() function. This is one of the more common uses of code blocks.

Eval() is a function in JavaScript (and other languages like Perl, PHP, and Python). It is commonly used in the context of a data-bound control (like the Repeater) to get the value from a data source field. The argument of the Eval function must be represented as a string (enclosed in quotes). For example, this is how you would use Eval() within the Repeater to get the value of PerformanceName from the DataSource that is bound to the Repeater control:

```
<%# Eval("PerformanceName"); %>
```

Because PerformanceName is not a simple variable, but a data item bound to the Repeater, use the "#" sign, rather than the "=" sign.

You also may use Eval() to evaluate an expression as follows:

```
<%# Eval("BasePrice - 0.1 * BasePrice"); %>
```

The function will return the computed value.

Another use of <%# tag is to set properties of server-side controls. For example, you may display the greeting in a Label by setting it as the value of the Text property, as follows:

```
<asp:Label ID="Performance" runat="server" Text="<%# Greeting() %>"
BorderStyle="Solid"></asp:Label>
```

It is invalid to set properties using the <%= tag.

In general, embedded code blocks, which are mixed in with the markup, are harder to maintain. Further, the code block is executed on the server when the page is rendered, giving the developer less flexibility regarding the stage when the code is processed. So, its use should be limited to relatively simpler code. The preferred place for server-side code is in a separate code-behind file.

Let's use an embedded code block to display the PerformanceName and Description of each performance within the ItemTemplate of the Repeater control.

Step 4: Set up the ItemTemplate.

Insert the following code in the ItemTemplate element to display the PerformanceName and Description:

```
1    <asp:Repeater ID="rptEvents" runat="server"
2                  DataSourceID="SqlDataSource_Performance">
3      <ItemTemplate>
4        <h1> <%# Eval("PerformanceName") %> </h1>
5        <%# Eval( "Description") %>
6      </ItemTemplate>
7    </asp:Repeater>
```

Line 4 uses the embedded code along with the `Eval()` function to get the value of `PerfomanceName` from the data source bound to the Repeater. Line 5 gets the `Description`.

Once you are finished, display the page.

The web page should list the name and description of each performance, as shown in Figure 10-2.

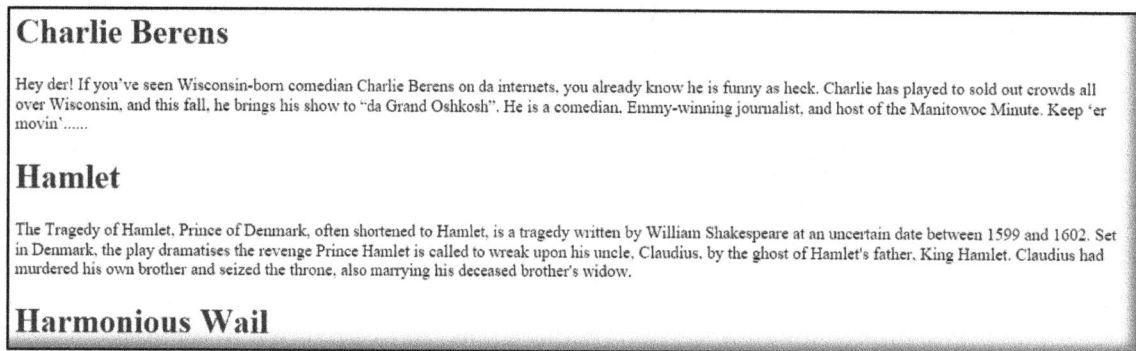

Charlie Berens

Hey der! If you've seen Wisconsin-born comedian Charlie Berens on da internets, you already know he is funny as heck. Charlie has played to sold out crowds all over Wisconsin, and this fall, he brings his show to "da Grand Oshkosh". He is a comedian, Emmy-winning journalist, and host of the Manitowoc Minute. Keep 'er movin'......

Hamlet

The Tragedy of Hamlet, Prince of Denmark, often shortened to Hamlet, is a tragedy written by William Shakespeare at an uncertain date between 1599 and 1602. Set in Denmark, the play dramatises the revenge Prince Hamlet is called to wreak upon his uncle, Claudius, by the ghost of Hamlet's father, King Hamlet. Claudius had murdered his own brother and seized the throne, also marrying his deceased brother's widow.

Harmonious Wail

Figure 10-2: Displaying performances using the ItemTemplate

In the next section, you will display the images on the page.

Review Questions

1. What is the difference between the `HeaderTemplate` and `ItemTemplate` of the `Repeater` control?
2. When do you use embedded code as opposed to code within a `<script>` element?
3. What is a common use of the `Eval()` function?

10.3 Displaying Images on Web Pages

A convenient way to dynamically display images on web pages is to store them in a database table and display them using code.

10.3.1 Uploading Images to Database Tables

The performance table has a field named `Image` of type `varbinary(MAX)`, which is the recommended type to store images. Several images are already loaded into the table.

It should be noted that storing images in a table significantly increases the size of the table, which in turn could affect performance and backup operations. In larger systems, it is recommended that images be kept separate from the rest of the data, especially if the images are not accessed frequently. For example, you can store each image as a separate file on the server and only keep the image file name in the database table. Or the images could be kept in a table separate from the Performance table, which requires joining the two tables when you want to combine other data with the images.

You can load an image into a table using the `UPDATE` command in the SQL statement. The statement shown in Figure 10-3 would upload the image Hamlet.jpg into the `Image` field of the `Performance` record that has the value 20 for `PerformanceId`.

```
1   UPDATE Performance
    SET Image = (SELECT * FROM OPENROWSET(BULK 'C:\img\Hamlet.jpg', SINGLE_BLOB) AS img)
      WHERE PerformanceID = 20;
```

Figure 10-3: Code to upload an image

Line 2 uses the OPENROWSET() method to get the single image (BLOB), CocktailHour.jpg, from the file C:\img\Hamlet.jpg. Though OPENROWSET is a method, it can be used like a table in an SQL statement to access a file. The alias, img, is required, but it is not used in this application.

As you learned in Chapter 8, you can type in and execute SQL statements in Visual Studio by creating a query.

Step 5: Start a SQL query.

Right-click *Tables* in Server Explorer and select *New Query*.

A new window opens with a caption like SQLQuery1.sql.

You can type any SQL statement into the window and execute it by clicking the Execute button (the first button on the Toolbar at the top of the window).

Step 6: Upload an image of the performance *The Cocktail Hour*.

Copy the file CocktailHourl.jpg from the Tutorial-starts/img folder to a local drive on your computer.

Refer to the code in Figure 10-3 and type in the SQL statement necessary to upload the image CocktailHour.jpg to the Performance record that has the value 24 for PerformanceId.

Make sure the SQL specifies the folder where you stored the image on your computer.

Click the Execute button.

You should see the message "(1 row(s) affected)" in the lower window.

10.3.2 Displaying Images from Database Tables

You can display images dynamically using the server control, Image. The imageUrl property of the Image control specifies the path to the image to be displayed.

To display the image, if any, of each performance from the Performance table, you will add an Image control within the ItemTemplate of the Repeater, without specifying the imageUrl. The imageUrl will be set in the server-side code. Line 3 in Figure 10-4 shows the code to add the Image control.

```
1   <div id="<%# Eval("PerformanceId")  %>" >
2       <h1> <%# Eval("PerformanceName") %>   </h1>
3       <asp:Image style="width: 50%" ID="imgPerformance" runat="server" /> <br /><br />
4       <%# Eval( "Description") %> <br /><br />
5   </div>
```

Figure 10-4: Code to insert the Image control

Step 7: Insert the image control.

Add the code from line 3 in Figure 10-4 to insert the Image control.

To set the imageURL property of the Image control to display the images from the table, use the **Item-DataBound** event handler of the Repeater. The ItemDataBound event of the Repeater takes place before each item is rendered on the page but after the item is bound to the DataSource. So, you can code this event handler to do tasks involving each row of the DataSource.

The code in the `ItemDataBound` event handler shown in Figure 10-5 gets the image from the `Image` field of the current row, converts the bytes that represent the image to a string, and assigns the string to the `imageUrl` property of the control to display the image, as explained next.

```
1   protected void rptEvents_ItemDataBound(object sender, RepeaterItemEventArgs e)
2   {
3       if (e.Item.DataItem != null)
4       {
5           DataRowView dr = (DataRowView)e.Item.DataItem;
6           if (dr["Image"] != DBNull.Value)
7           {
8               string imageUrl = "data:image/jpg;base64," +
                                  Convert.ToBase64String((byte  [])dr["Image"]);
9               (e.Item.FindControl("imgPerformance") as Image).ImageUrl = imageUrl;
10          }
11          else
12              (e.Item.FindControl("imgPerformance") as Image).Visible = false;
13      }
14  }
```

Figure 10-5: ItemDataBound event handler of rptEvents

In line 3, `e.Item.DataItem` is an object that represents the current row that is bound to the Repeater item. Because this event is raised also for the header, line 3 checks whether the `DataItem` is not null.

Line 5 converts the `DataItem` object to a `DataRowView`.

Line 6 checks whether the `Image` field of the `DataRowView` has an image in the current row.

Line 8 converts the image into a string consisting of a byte array.

Line 9 assigns the string to the `ImageUrl` property of the `Image` control to display the image.

Line 12 makes the `Image` control not visible, if there is no image in the current row, to avoid leaving a large blank space.

Step 8: Create the ItemDataBound event handler of the Repeater.

In Design view, click anywhere within `rptEvents`.

Select the Events button in the Properties window. Double-click `ItemDataBound` event.

Add the code from Figure 10-5.

Display the page and make sure the image of each performance that has an image in the table is displayed on the web page.

The next task is to add the scheduled dates for each performance to the list.

It's time to practice! Do Exercise 10.1 (see the end of the chapter).

10.4 A List within Another List Using Nested Repeater Controls

Currently, the web page displays the name, image, and description of the performances using a `Repeater` control. Each performance also has a set of scheduled dates, which are stored in the Perform-Date table shown in Figure 10-6. For example, there are four scheduled dates for PerformanceId 11.

As shown in Figure 10-7, the list of performance dates is to be displayed underneath each event. This means we need a list (of dates) nested inside another list (of events).

PerformDateId	PerformanceId	PerformDate
101	11	1/3/2021 7:30:00 PM
102	11	1/4/2021 7:30:00 PM
103	11	1/5/2021 7:30:00 PM
104	11	1/10/2021 7:30:00 PM
105	12	1/11/2021 7:30:00 PM
106	12	1/6/2021 7:30:00 PM
107	13	1/7/2021 8:00:00 PM
108	13	1/7/2021 7:30:00 AM

Figure 10-6: PerfomDate table

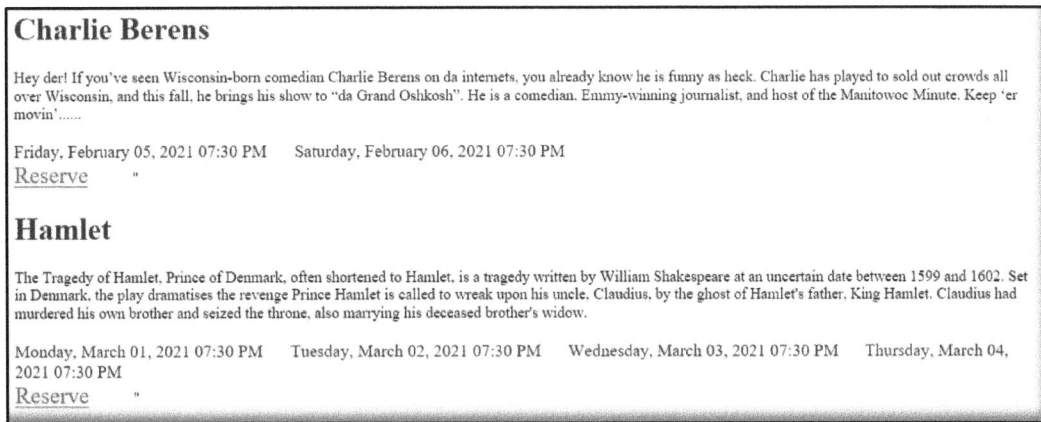

Figure 10-7: Nested Repeater output

To display a list within another list, use nested `Repeaters`. You will add a second `Repeater` that is bound to the PerformDate table and nest it within the existing `Repeater` (rptEvents) that is bound to the Performance table. So, you will have a parent `Repeater`, rptPerfomance, and a child `Repeater` named rptDates.

Step 9: Add a Repeater within the ItemTemplate of the parent Repeater.

In **Source view**, add a new line within the `ItemTemplate` below the code block, `<%# Eval( "Description") %>`.

Add the `Repeater` control in the new line. (Keep the insertion point in the line and double click the control.)

Change its `id` to `rptDates`. Line 2 shows the code for the nested Repeater.

```
1    <%# Eval( "Description") %> <br /><br />
2    <asp:Repeater id="rptDates" runat="server">  </asp:Repeater>
```

To get the dates for each performance from the PerformDate table, you will create a new data source, SqlData-Source_Dates, and bind it to rptDates. As the parent Repeater cycles through each performance in its data source, you will store the PerformanceId in a hidden field and use it in the WHERE clause in DataSource_Dates to select only the dates for the specific performance.

The `HiddenField` control, which is not visible by default, is commonly used in the aspx page to store values. To access PerformanceId from the current performance, the hidden field must be placed in the `ItemTemplate` of rptPerformance, as shown in line 10 in Figure 10-8. The figure also shows the SqlDataSource_Dates Data-Source to be created and bound to rptDates.

```
1    <asp:Repeater ID="rptPerformance" ... DataSourceID="SqlDataSource_Performance">
2        <ItemTemplate>
3            ...
4            <asp:Repeater id="rptDates" ... DataSourceId="SqlDataSource_Dates" >
5                <ItemTemplate>
6                    <span ... > <%# Eval("PerformDate") %> </span>
7                </ItemTemplate>
8            </asp:Repeater>
9
10           <asp:HiddenField ID="hidPerformId" Value='<%# Eval("PerformanceId")%>... />
11
12           <asp:SqlDataSource ID="SqlDataSource_Dates" ...>
13               <SelectParameters>
14                   ...
15               </SelectParameters>
16           </asp:SqlDataSource>
17           ...
18       </ItemTemplate>
19   </asp:Repeater>
20   <asp:SqlDataSource ID="SqlDataSource_Performance" runat="server"
```

Figure 10-8: Structure of nested Repeaters

Step 10: Add the HiddenField control to store the current PerformanceId.

Add the `HiddenField` control from the Standard tag of the Toolbox to a new line below the `Repeater`, `rptDates`, as shown in Figure 10-8.

Set the `ID` property to `hidPerformId`, and add the Value attribute as shown here:

```
<asp:HiddenField Value='<%# Eval("PerformanceId") %>' ID="hidPerformId" runat="server"
                                Visible="false"/>
```

Note the code block, `<%# Eval("PerformanceId") %>`, gets the value of PerformanceId from the current row of the dataset returned by the DataSource, *SqlDataSource_Performance*.

Next, you will add the data source for `rptDates` and configure it. The data source must be within the `Item-Template` of the parent `Repeater` that contains the hidden field so the data source can use the current `Perfor-manceId` from the hidden field to select the dates. However, for convenience in configuring the data source, initially you will add the data source outside the parent `Repeater` in Design view, configure it, and then move it inside the `ItemTemplate`.

Step 11: Add a DataSource to get the dates for each performance.

In Design view, add a `SqlDataSource` at the bottom of the page right below `SqlDataSource_Performance`.

Change its `ID` to `SqlDataSource_Dates`.

Step 12: Configure the SqlDataSource_Dates.

Configure the data source using the existing connection string. Select the `PerformDate` field from the `PerformDate` table, sorted by `PerformDate`.

In addition, on the *Configure the Select command* screen, click the *WHERE* button and make the following selections:

Column: PerformanceId
Operator: =
Source: None

Click Add, then OK, then Finish.

You will specify the `Source` (value) for `PerformanceId` shortly.

Switch to the Source view. The generated markup for the data source should look as shown in Figure 10-9.

```
1   <asp:SqlDataSource ID="SqlDataSource_Dates" runat="server"
2       ConnectionString="<%$ ConnectionStrings:ConnectionString %>"
3       SelectCommand="SELECT [PerformDate] FROM [PerformDate]
4               WHERE ([PerformanceId] = @PerformId) ORDER BY [PerformDate]">
5       <SelectParameters>
6           <asp:Parameter Name="PerformanceId" Type="Int32" />
7       </SelectParameters>
8   </asp:SqlDataSource>
```

Figure 10-9: Markup for SqlDataSource_Dates

Line 4 shows the `WHERE` clause that includes the parameter represented by `@PerformId`, whose value is not specified, yet. You will set its value to the value of `PerformanceId` stored in the hidden field. This can be done using the **ControlParameter** that will replace the `Parameter` in line 6, as described in the next section.

10.4.1 Using a ControlParameter to Specify the Value of a Parameter

The `ControlParameter` object sets the value of a parameter to the value of a control's property. To help understand it, let us use the `ControlParameter` to set the value of the parameter, `PerformId`, used in the SQL.

Step 13: Set up the ControlParameter.

Change the content of the `SelectParameter` element (line 6 in Figure 10-9) as shown here:

```
<asp:ControlParameter Name="PerformId" ControlID="hidPerformId"
                      PropertyName="Value" Type="Int32" />
```

The `Name` property ("`PerformId`") specifies the name of the SQL parameter whose value is set. This name must match the parameter name specified in the `WHERE` clause (`@PerformId`).

The `ControlID` ("`hidPerformId`") specifies the `ID` of the control that provides value for the parameter.

The `PropertyName` ("`Value`") specifies which property of the control provides the value of the parameter.

Thus the `ControlParameter` sets the value of parameter `PerformId` to the Value property of the hidden field, `hidPerformId`.

Step 14: Move the code for SqlDataSource_Dates to the ItemTemplate of the parent Repeater.

Add a new line immediately below the `HiddenField`.

Cut the modified code for `SqlDataSource_Dates` and paste it in the new line below `HiddenField`, as shown in Figure 10-8.

`SqlDataSource_Dates` is now configured to get the dates for the current `PerformanceId` stored in the hidden field. To display the dates, bind the `Repeater` to the data source and specify the field(s) to be displayed.

Step 15: Bind rptDates to SqlDataSource_Dates, and specify PerformDate as the data to be displayed.

Set the `DataSourceId` to `SqlDataSource_Dates`.

Add the `ItemTemplate` in `rptDates` and use a code block to display the dates as shown here.

```
1   <asp:Repeater id="rptDates" runat="server" DataSourceId="SqlDataSource_Dates"
2       <ItemTemplate>
```

```
3              <span style="color: darkblue; fontsize: 18px" >
               <%# Eval("PerformDate","{0:dddd, MMMM dd, yyyy hh:mm tt}")  %> </span>
4          </ItemTemplate>
5      </asp:Repeater>
```

The code block in line 3 uses the `Eval()` function to get the `PerformDate` from the DataSource.

The format code for the date, "`dddd, MMMM dd, yyyy hh:mm tt`," consists of the code for day name, month name, day number, year, hour minutes, and am/pm, respectively.

For spacing, add line breaks after the code block preceding `rptDates`:

```
<%# Eval( "Description") %> <br /><br />
```

Figure 10-10 shows the complete code of the outer (`rptEvents`) and nested (`rptDates`) Repeaters, as well as the two data sources.

```
1  <asp:Repeater ID="rptEvents" runat="server" DataSourceID="SqlDataSource_Performance"
2              OnItemDataBound="rptEvents_ItemDataBound" DataMember ="DefaultView" >
3    <ItemTemplate>
4        <h1> <%# Eval("PerformanceName") %>   </h1>
5        <asp:Image style="width: 50%" ID="imgPerformance" runat="server"/> <br /><br
6        <%# Eval( "Description") %> <br /><br />
7
8        <asp:Repeater id="rptDates" runat="server" DataSourceId="SqlDataSource_Dates"
9            <ItemTemplate>
10              <span style="color: darkblue; fontsize: 18px" > <%# Eval("PerformDate",
11                        "{0:dddd, MMMM dd, yyyy hh:mm tt}")  %> </span>
12          </ItemTemplate>
13        </asp:Repeater>
14
15        asp:HiddenField Value='<%# Eval("PerformanceId") %>' ID="hidPerformId" .../>
16
17        <asp:SqlDataSource ID="SqlDataSource_Dates" runat="server" ConnectionString=
18            "<%$ ConnectionStrings:ConnectionString %>" SelectCommand=
              "SELECT PerformDate FROM [PerformDate]
19            WHERE PerformanceId = @PerformId ORDER BY [PerformDate]">
20          <selectparameter>
21            <asp:ControlParameter Name="PerformId" ControlID="hidPerformId"
                            PropertyName="Value" ... />
22          </selectparameters>
23        </asp:SqlDataSource> <br />
24
25    </ItemTemplate>
26 </asp:Repeater>
27
28 <asp:SqlDataSource ID="SqlDataSource_Performance" runat="server" ConnectionString=
29        "<%$ ConnectionStrings:ConnectionString %>" SelectCommand=
          "SELECT * FROM [Performance] ORDER BY [PerformanceName]">
30 </asp:SqlDataSource>
```

Figure 10-10: Code within the nested Repeaters

Display the page to test the code.

The top part of the page should appear as in Figure 10-11, except for the image.

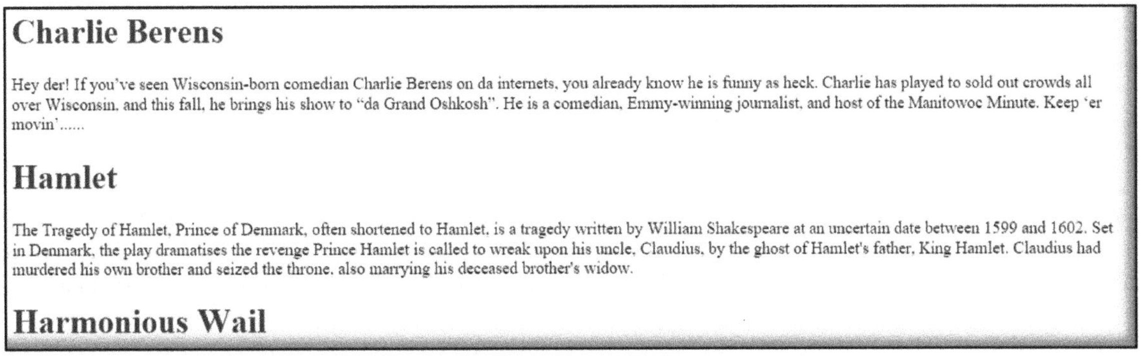

Figure 10-11: Top of the reservation page

Next, you will add a button for each performance to take the user to the Reservation page and to have the performance automatically selected in the `DropDownList` on the page. The next section discusses the **Command** event of buttons and how it helps to use a single event handler for multiple buttons associated with the items in a list.

It's time to practice! Do Exercise 10-2 (see the end of the chapter).

10.5 Taking Action on Items in a Repeater

When a web page displays a list of items using a `Repeater`, you may provide a button for each item so that the user can take an action related to that item. For example, a web page displaying a set of products may have a button associated with each product to let the user add the item to a shopping cart. In such cases, it would be convenient to use a common event handler for all buttons in the list to display the target page. However, the event handler would need the product number associated with the specific button that was clicked, so that the information can be sent to the proper target page. The Command event allows you to pass such information from a button to the event handler.

To help you understand the Command event, let's look at how you will use this event for the Reserve button for each performance, as shown in Figure 10-12.

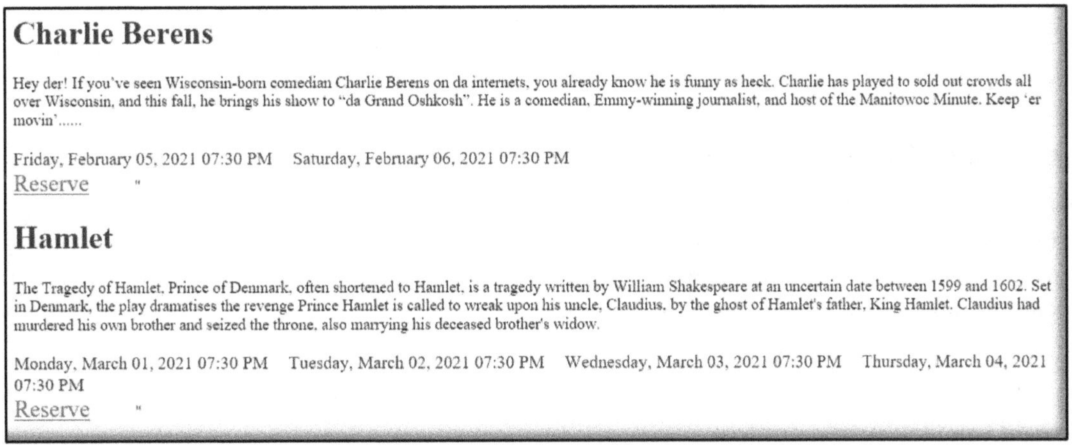

Figure 10-12: Events page with the Reserve button

You will use the **Command** event of the button to send the `PerformanceId` from the button to the event handler, which in turn will pass it to the Reservation page.

The command event is similar to the click event. Both events are raised when the user clicks a control. But the command event is raised also when the user interacts with the control using keys and other means. The more significant difference, especially for this application, is that the command event lets you send two additional pieces of data to the event handler by storing them in the **CommandName** and **CommandArgument** properties of the control.

The `CommandName` property is typically used to specify the action to be taken by the event handler, and the `CommandArgument` is used to send additional information related to the action. The following code for the Reserve button shows the `CommandName` set to "`ReDirect`" and `CommandArgument` set to "18", which is the Performance Id for *The Hunt*.

```
1   <asp:Button name="btnReserve" runat="server" OnCommand="btnReserve_Command"
2       Text="Reserve" CommandName="ReDirect" CommandArgument="18" />"
```

The `OnCommand` property specifies that the name of the event handler for the Command event is "`btnReserve_Command`." (In contrast, to create an event handler for the Click event, set the `OnClick` property of the button to an event handler name, as in `OnClick="btnReserve_Click`.) You can give any meaningful name to the event handler. What is important is which property, `OnCommand` or `OnClick`, you use.

The Command event handler for the button, shown in Figure 10-13, gets the value of `PerformanceId` from the `CommandArgument` property and sends it to Reservation-5 page using a query string. Reservation-5 is the final version of the reservation form, which you will create shortly.

```
1   protected void btnReserve_Command(object sender, CommandEventArgs e)
2   {
3       string PerformId = e.CommandArgument.ToString();
4       Response.Redirect("Reservation-5.aspx?performanceId=" + PerformId);
5   }
```

Figure 10-13: Command event handler of btnReserve

Note that the event handler for the Command event is passed an argument (e) of type `CommandEventArgs`, which has two properties, `CommandName` and `CommandArgument`. In comparison, the argument for the Click event is passed an argument of type `EventArgs`, which doesn't have these two properties. These two properties give the values of the corresponding properties of the button.

Thus line 3 gets the `PerformanceId` stored in the `CommandArgument` property of the button.

Line 4 displays the Reservation-5 page and passes the Performance Id to the page using a query string.

We do not use the `CommandName` property in this example. The `CommandName` property is typically used when there are multiple buttons representing different actions that can be taken for each item. By having a different `CommandName` for each button, you can distinguish between the commands and take appropriate action.

10.5.1 Passing a Different Value from Each Button

In the previous example, the CommandArgument was set to a fixed value of 18 (`CommandArgument="18"`). When you create the buttons using a Repeater, the `CommandArgument` of each button needs to be set to the ID of the performance associated with the current performance so that the value can be sent to the common event handler. This can be done by using the `Eval()` function in an embedded code block:

```
CommandArgument='<%# Eval("PerformanceId") %>'
```

Step 16: Add the LinkButton control.

Insert the following code between lines 23 and 25 in Figure 10-10 (inside the `ItemTemplate` of `rptEvents`).

```
<asp:LinkButton name="btnReserve" runat="server" OnCommand="btnReserve_Command"
  Text="Reserve" CommandName="ReDirect" CommandArgument='<%# Eval("PerformanceId") %>' />
```

Note that the Reserve buttons in the Events page are LinkButtons, which are essentially regular buttons with the appearance of a hyperlink (different color and underline).

Next, create the Command event handler that receives the value of PerformanceId from the `CommandName` and sends it to the Reservation-5 page.

Step 17: Create the Command event handler for the Reserve button.

Type in the following code in the code-behind file:

```
1    protected void btnReserve_Click(object sender, CommandEventArgs e)
2    {
3        string PerformId = e.CommandArgument.ToString();
4        Session["SourcePage"] = "Events";
5        Response.Redirect("Reservation-5.aspx?performanceId=" + PerformId);
6    }
```

This is the same code presented in Figure 10-13, except that line 4 is added to store the name of the page in a session variable, so that the Reservation page can identify the source page. Recall that, unlike the Transfer and Cross-page Posting methods, the Redirect method cannot use the `PreviousPage` property to get the name of the source page.

Display the Events page and click the Reserve button for any performance. The Reservation-5 page should be displayed.

The next step is to have the performance associated with the button displayed in the drop-down list on the Reservation page and the corresponding price displayed in the Label.

Step 18: Make a copy of the Reservation-4 page and name it Reservation-5.

Copy and paste Reservation-4 form and rename Reservation-4 copy as Reservation-5.

The process is the same as the one used earlier to make copies of earlier versions of the reservation form. The instructions are repeated here.

Open the Reservation-5.aspx.designer.cs file from the Solution Explorer.

Change the class name from `Reservation_4` to `Reservation_5`, as shown here.

```
1    namespace Theater {
2
3        public partial class Reservation_5 {
```

Save and close Reservation-5.aspc.designer.cs.

Open the Reservation-5.aspx.cs file.

Change the class name from `Reservation_4` to `Reservation_5`, as shown here.

```
1    namespace Theater
2    {
3        public partial class Reservation_5 : System.Web.UI.Page
```

Save and close Reservation-5.aspc.cs.

Open Reservation-5.aspx file from the Solution Explorer.

In the first statement, change the form name from `Theater.Reservation_4` to `Theater.Reservation_5`, as shown here.

```
<%@ Page Language="C#" AutoEventWireup="true" CodeBehind=
                "Reservation-5.aspx.cs" Inherits="Theater.Reservation_5" %>
```

In addition, change the Title to `Reservation-5`.

```
<title>Reservation-5</title>
```

Save the Reservation-5 form. Close the Reservation-4 form.

Step 19: Update the navigation links.

Change the hyperlinks within the `<nav>` element of each page from Reservation-4 to Reservation-5.aspx, as shown here:

```
a href="reservation-4.aspx">Reservation</a>
```

Note you will work with the Reservation-5 form in the rest of this chapter.

Modifying the Page_Load Event Handler of Reservation-5 Page

The following is a segment of the current code in the Load event handler. The code uses an `if` statement to check whether the previous page is the Confirmation page and, if so, resets the controls to their previous values.

```
1   // If transferring from Confirmation page
2   if (previousPageName == "/Confirmation.aspx")
3                       && (Request.QueryString["action"] == "edit"))
4   {   // set control properties to previously selected values
5       ...
6       // Re-display prices and discount
7   }
8   else   // If not transferring from Confirmation page for editing,
9   {       // create session variables
10      ...
11  }
```

You will add an `else if` clause in the if statement as shown below, to check whether the reservation page was opened from the Events page and the `performanceId` variable was passed by the query string.

```
1   else if ((Session["SourcePage"] != null) && (Session["SourcePage"].ToString() ==
            "Events") && (Request.QueryString["performanceId"] != null))
2   {
3       string PerformanceId = Request.QueryString["performanceId"];
4       ddlPerformance.Items.FindByValue(PerformanceId).Selected = true;
5       Session["BasePrice"] = (decimal)0;
6       Session["ZonePrice"] = (decimal)0;
7       Session["Discount"] = (decimal)0;
8       ComputeBasePrice();
9   }
```

Line 3 gets the value of `performanceId`.

Line 4 selects the corresponding performance in the DropDownList.

Lines 5–7 initialize the session variables.

Line 8 invokes the `ComputeBasePrice()` method that computes the base price and the total price, and displays them. `ComputeBasePrice()` is a new method to be created with the same functionality as the `ddlPerformance_SelectedIndexChanged` event handler.

Step 20: Create the ComputeBasePrice() method and invoke it.

Create a new method named `ComputeBasePrice()`.

Copy the code from the `ddlPerformance_SelectedIndexChanged` event handler into `ComputeBasePrice()`.

To avoid duplication of code, replace the entire code in the `ddlPerformance_SelectedIndexChanged` event handler with a single statement that invokes the `ComputeBasePrice()`, as shown here:

```
1   protected void ddlPerformance_SelectedIndexChanged(object sender, EventArgs e)
2   {
3       ComputeBasePrice();
4   }
```

Step 21: Select the specified performance.

Next, you modify the `Page_Load` event handler to add the `else if` clause that selects the specified performance in the DropDownList and find base price. To do this, insert the `else if` statement within the `if` statement, as shown in lines 27 to 35 in Figure 10-14.

```
1    Protected void Page_Load(object sender, EventArgs e)
2    {
3        if (!IsPostBack)     // If it is not a postback of the page,
4        {
5            ddlPerformance.DataBind();    // bind the DropDownList
6            ddlPerformance.Items.Insert(0, new ListItem("", "")); // add an empty item
7            // if transferring from another page, get the name of that page
8            string previousPageName = "";
9            if (PreviousPage != null)
10               previousPageName = Request.UrlReferrer.AbsolutePath;
11               ViewState ["PreviousPageName"] = previousPageName;
12           // If transferring from Confirmation page
13           if (previousPageName == "/Confirmation-4.aspx")
14           {   // Get reservation data from the list
15               List <Reservation> myReservations =
                 (List<Reservation>)Session["MyReservations"];
16               int selectedResIndex = (int)Session["SelectedResIndex"];
17              Reservation res = myReservations[selectedResIndex];
18               // set control properties to previously selected values
19               ddlPerformance.SelectedIndex = (int)Session["PerformanceIndex"];
20               ddlDate.SelectedValue = res.PerformDate.ToString();
21               rblZones.SelectedIndex = (int)Session["ZoneIndex"];
22               chkMember.Checked = (bool)Session["IsMember"];
23               txtNoOfTickets.Text = Session["NoOfTickets"].ToString();
24
25               // Display prices and discount
26               lblBasePrice.Text = Session["BasePrice"].ToString();
27               lblZonePrice.Text = Session["ZonePrice"].ToString();
28               lblDiscount.Text = Session["Discount"].ToString();
29               lblUnitPrice.Text = Session["UnitPrice"].ToString();
30           }
31           else if ((Session["SourcePage"] != null) && (Session["SourcePage"].ToString()
              == "Events") && (Request.QueryString["performanceId"] != null))
32           {
33               string PerformanceId = Request.QueryString["performanceId"];
34               ddlPerformance.Items.FindByValue(PerformanceId).Selected = true;
35               Session["BasePrice"] = (decimal)0;
36               Session["ZonePrice"] = (decimal)0;
37               Session["Discount"] = (decimal)0;
38               ComputeBasePrice();
39           }
40           else   // If not transferring from Confirmation/Events page,
41           {      // create session variables
42               Session["BasePrice"] = (decimal)0;
43               Session["ZonePrice"] = (decimal)0;
44               Session["Discount"] = (decimal)0;
45           }
46       }
47       UnobtrusiveValidationModeUnobtrusiveValidationMode=UnobtrusiveValidationMode.None;
48   }
```

Figure 10-14: The Page-Load event handler of the reservation page

Display the Events page and click the Reserve button for any performance.

The Reservation-5 page should open.

The performance should be selected in the drop-down list and the base price and total price displayed in the labels so the user can complete the rest of the reservation.

It's time to practice! Do Exercise 10-3 (see the end of the chapter).

Review Question

4. Compare the Command event of the Button control to the Click event.

10.6 Displaying Selected Item(s) of a Data-Bound List

We will look at two ways to display information on selected items from a data-bound list: (1) Use dynamically generated hyperlinks to scroll to an item, and (2) filter items by changing the SQL Select statement associated with the DataSource.

10.6.1 Use Hyperlinks to Scroll to an Item of a List

A Grand Night
Charlie Berens
Hamlet
Harmonious Wail
Justin Hayward
Ladies of Laughter
Loudon Wainwright III
Oshkosh Orchestra
Steve Earl & The Dukes
TedxOshkosh
The Hunts
The Tin Woman
Young Irelanders

Figure 10-15: Hyperlinks to items in the list

When information presented in a list is too long to be visible in the browser window, you may want to provide the option to scroll to a specific item of the list by clicking a hyperlink. For example, users may want to scroll to the description of a performance by clicking a hyperlink, like those shown in Figure 10-15.

This can be done by wrapping each item in an element like <div> or <h3> and linking the hyperlink to the id of the element. When you have a list that is generated from a database table, it is convenient to use the identifier of each item (primary key of each record) as the id of the element that wraps the item.

Step 22: Wrap each item in an identifiable element.

Insert the <div> element in the aspx code for the Events page as shown below.

The <div> element wraps the elements that display the performance name (line 3) and description (line 4) of each performance. Line 2 sets the id of the <div> element to be the value of the PerformanceId of the current performance, so that the hyperlink can link to the div element using its id. The div element serves as a "bookmark" for each performance.

```
1    <ItemTemplate>
2        <div id="<%# Eval("PerformanceId")  %>" >
3            <h1 style="color: darkblue"> <%# Eval("PerformanceName") %>   </h1>
4            <%# Eval( "Description") %> <br /><br />
5        </div>
```

Next, you will generate hyperlinks that link to the div elements.

Generating Hyperlinks

In general, you can scroll to any element in a page by setting the value of href to the Id of the element preceded by the "#" sign.

Step 23: Generate hyperlinks.

The hyperlinks shown in Figure 10-15 will be initially displayed at the top of the page. Later you will use the CSS to change the layout.

Because each div element is identified by the value of a PerformanceId, the anchor element (<a>) that creates the hyperlink also should have its href attribute set to the same value of PerformanceId. For example, if you were to create a static hyperlink for "The Hunts," you would create it using the anchor element with the href set to "18," which is the Id of the "The Hunts."

```
<a href="#18"> The Hunts </a>
```

The bookmark "18" is the value of the id attribute of the <div> element that contains information on "The Hunts."

To generate hyperlinks for the items in the list, use a Repeater that sets the value of each `href` to the value of a `PerformanceId`.

In the Source view of the Events page, add the code from Figure 10-16 above the code for rptEvents (`<asp:Repeater ID="rptEvents" runat="server">`).

```
1   <br /><br />
2   <asp:Repeater ID="rptLinks" runat="server" DataSourceID="SqlDataSource_Performance">
3   <ItemTemplate>
4      <a href="#<%# Eval("PerformanceId") %>" > <%# Eval("PerformanceName") %></a>
5      <br/>
6   </ItemTemplate>
7   </asp:Repeater> <br /><br />
```

Figure 10-16: Repeater to generate hyperlinks

In line 4, the value of `href` consists of two parts: the # sign and the code block `<%# Eval ("PerformanceId") %>` that gets the value of `PerformanceId`.

Display the page and click the hyperlinks to test the code. The page should scroll, if necessary, to make the performance visible on the browser window.

It's time to practice! Do Exercise 10-4 (see the end of the chapter).

10.6.2 Filtering a List of Items

When a large list of items is displayed on a page, you may want to let the user filter the list using different criteria. For example, a list of products could be filtered based on size, price, or category. This can be done by changing the WHERE clause in the SelectCommand (SQL) of the data source.

Let's add this feature to display only the performances for the month displayed on a calendar on the Events page by clicking a button.

Step 24: Set up filtering controls.

In the Design view, add the Calendar control to a new row below the hyperlinks but above the Repeater rptEvents, as shown in Figure 10-17.

Set the id property of the Calendar to "calFilterByMonth."

Set the VisibleDate property to "2/1/2021."

Add a button to a new row immediately above the Calendar, set the id property to "btnFilterByMonth," and set the Text property to "Filter By Month."

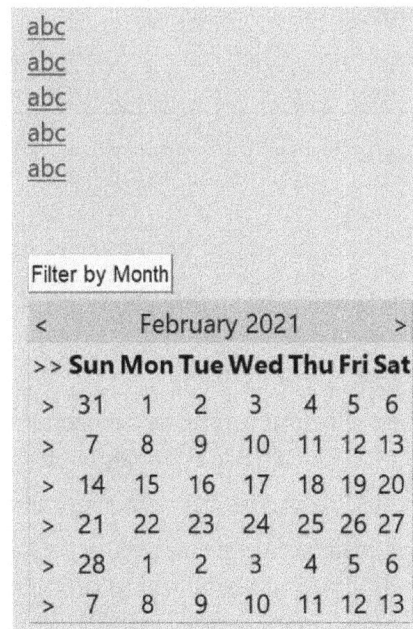

Figure 10-17: The Calendar control

Using Nested SQL to Select Performances for a Month

The performances that are scheduled for the specific month displayed on the Calendar can be found in two steps, as described here, and also in Exercise 14 in Appendix C*.

1. Select from the PerformDate table all PerformanceId's that have a corresponding date in the month displayed on the Calendar. The following SQL would select the PerformanceId's for the month displayed on the calendar:

---

* Appendix can be freely downloaded from https://www.prospectpressvt.com/textbooks/philip-web-development at "Student Resources."

```
Select PerformanceId From PerformDate WHERE Month(PerformDate) = " +
calFilterByMonth.VisibleDate.Month
```

The second line uses the Month() function to get the number of the month (1–12) displayed on the calendar, which is compared to the number of the month in the date from the PerformDate field.

2. Select the records from the Performance table whose PerformanceId matches the set selected in step 1.

To use the set of PerformanceId's from step 1 in the WHERE clause, you would nest the SQL from step 1 in an outer SQL as follows:

```
1  SqlDataSource_Performance.SelectCommand =
2      "SELECT * FROM [Performance] Where PerformanceId IN
3          (Select PerformanceId From PerformDate WHERE Month(PerformDate) = " +
4              calFilterByMonth.VisibleDate.Month + ")";
```

To sort performances by Performance name, you would add the ORDER BY clause to the outer SQL, as shown in Figure 10-18.

Step 25: Add SQL to select performances for a month.

Create a click event handler for the button by double-clicking it, and add the code shown in Figure 10-18 into the event handler.

```
1  SqlDataSource_Performance.SelectCommand =
2      "SELECT * FROM [Performance] Where PerformanceId IN " +
3      "(Select PerformanceId From PerformDate WHERE Month(PerformDate) = " +
4      calFilterByMonth.VisibleDate.Month + ")" + "ORDER BY PerformanceName";
```

Figure 10-18: The ORDER BY clause to sort performances

Display the page and move to the month of February using the arrows at the top of the calendar. Click the *Filter by Month* button. You should see only the performances for February.

Step 26: Reset the list.

To reset the list by displaying all Performance records, use a second button that removes the WHERE clause from the SQL.

Add an <h1> heading, "View Performances," above the hyperlinks.

Add a button between the heading and the hyperlinks, set the id property to "btnAll," and set the Text property to "All Performances." See Figure 10-19.

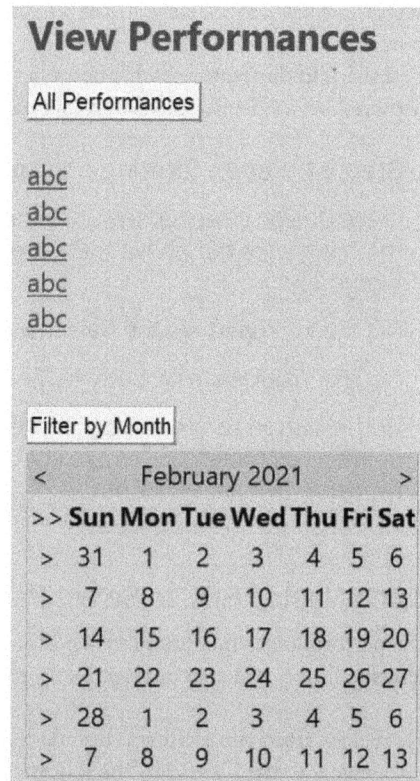

Figure 10-19: The button to select all performances

Create the click event handler of the button and insert the following code to reset the SQL:

```
1    protected void btnAll_Click(object sender, EventArgs e)
2    {
3        SqlDataSource_Performance.SelectCommand =
4            "SELECT * FROM [Performance] ORDER BY PerformanceName";
5    }
```

Display the page, filter the list by a month, and click the *All Performances* button.

The list should be reset to display all performances.

Step 27: Filter the performances by the presenter.

Provide a drop-down list in the left panel, which lets the user select a presenter and filter the list to display only the performances for the selected presenter. Note that each performance has a presenter, and a presenter may present multiple performances as shown in the Performance table.

It's time to practice! Do Step 27 and/or Exercise 10-5 (see the end of the chapter).

Review Question

5. Consider the following hyperlink:

```
<a href="#20"> Hamlet </a>
```

What is the effect of setting the href attribute to "#20"?

10.7 Displaying an Item of a List from Another Page Using JavaScript

The previous section added the features within the Events page to display subsets of the list of performances using two methods: (1) filter the list by specifying the criterion in the SQL, and (2) scroll to a specific performance. This section uses the same two approaches to let the user select a performance from the **Reservation page** and display the corresponding information on the Events page.

To filter the list on the Events page, you need to pass the id of the performance to be displayed from the Reservation page to the Events page. You will use client-side JavaScript to do this. You can use JavaScript to scroll to an item as well.

While this section demonstrates an application of JavaScript in ASP.NET web forms, for comparison purposes, we also will discuss the use of server-side C# code to do the same tasks.

10.7.1 Using JavaScript to Pass a Query String to Another Page

You will create a JavaScript function, `ViewDetails`, on the Reservation page to let the user view details about the selected performance on the Events page by clicking a button next to the existing DropDownList, as shown here.

The click event handler of the button will call a JavaScript function named `ViewDetails` that will send the ID of the selected performance as a query string to the Events page. The Events page will use the ID to select and display the specified performance.

Figure 10-20 shows the JavaScript function, `ShowDetails`, as defined within a script element.

```
1   <script>
2       function ShowDetails() {
3           var ddl = document.getElementById("ddlPerformance");
4           var performId = ddl.options[ddl.selectedIndex].value;
5           window.location.href = 'Events.aspx?qsPerformId=' + performId;
6       }
7   </script>
```

Figure 10-20: JavaScript function to send the PerformanceId as a query string

The function gets the Id of the selected performance and sends it as a query string to the Events page. As you learned in Chapter 5, you can access a control on a web form from JavaScript using the function **getElementById** with the id of the control as the argument.

Line 3 gets the DropDownList control and stores a reference to it in the variable ddl.

To get an item from a DropDownList, use the **options** collection of the DropDownList. The options collection contains the items in the list: options[0] represents the first item, options[1] represents the second, and so on. Line 4 uses the index of the selected performance (selectedIndex) as the index of the options collection to get the id of the performance. The value property gives the Performance Id, and the text property gives the performance name.

The window.location.href property lets you redirect to a URL by setting the property to the URL. The window is a client-side browser object that can be accessed in JavaScript. Line 5 sets the href to the URL of the Events page and also passes the Id of the selected performance using a query string.

Step 28: Use JavaScript to pass a query string to another page to filter a list.

Create the script element and the ShowDetails() function within the head element of the Reservation-5 page, as shown in Figure 10-20.

Create an HTML button to the right of the DropDownList, ddlPerformance, using code inserted below the code for the DropDownList:

```
<button type="button" id="btnViewDetail" name="btnViewDetail"
onclick="ShowDetails()"> View Details </button>
```

The onclick property of the button specifies the function that will be called when the button is clicked.

Next, you will add the necessary code in the Page_Load event handler of the Events page to filter the list of performances. The code shown in Figure 10-21 uses the value of the query string, qsPerformId, to select only the performance for the specified value of PerformanceId.

```
1   protected void Page_Load(object sender, EventArgs e)
2   {
3       if (Request.QueryString["qsPerformId"] != null) {
4           string performId = Request.QueryString["qsPerformId"];
5           SqlDataSource_Performance.SelectCommand =
6           "SELECT * FROM [Performance] Where PerformanceId =" + performId;
7       }
8   }
```

Figure 10-21: The Page_Load event handler of the Events page

Step 29: Add code to the **Page_Load** event handler of the Events page to filter the list.

Add the code from Figure 10-21 to the Page_Load event handler of the Events page.

Display the Reservation-5 page and select a performance from the DropDownList.

Click the View Details button. You should see the Events page with the selected performance displayed in it.

Scrolling to an Item of a List Using window.location.href

An alternative to passing the PerformanceId as a query string and then filtering the list is to use the `window.location.href` property to redirect to a page and scroll to an item within the page. You will use this method to redirect to the Events page and scroll to the `div` element that contains information on the performance selected in the Drop-DownList. Recall that the id of each `div` element is set to the id of the performance contained within the element.

To scroll to an element on another page, set the `window.location.href` property to the URL of the page followed by the "#" sign and the `id` of the element, as shown in the modified version of the `ShowDetails()` function in Figure 10-22.

```
1   function ShowDetails() {
2       var ddl = document.getElementById("ddlPerformance");
3       var performId = ddl.options[ddl.selectedIndex].value;
4       // window.location.href = 'Events.aspx?qsPerformId=' + performId;
5       window.location.href = 'Events.aspx#' + performId; // scroll to a bookmark
6   }
```

Figure 10-22: JavaScript to scroll to a performance on the Events page

Step 30: Scroll to a performance using window.location.href property.

Change the `ShowDetails()` function by commenting out line 4 and adding line 5 as shown in Figure 10-22.

Display the Reservation-5 page, select a performance, and click the View Details button.

The Events page should be displayed with the browser window scrolled to a position where the selected performance is visible.

Uncomment line 4 and comment out line 5 so that the function will pass the query string.

Using a JavaScript File to Store Functions

Rather than cluttering the aspx code with JavaScript, you may store the functions in one or more JavaScript files in the Scripts folder of the project, and specify a link to the file within the aspx file.

For example, you may store the `ShowDetails()` function in a file named JavaScript.js (or any meaningful name) and provide the following link in the head element of the aspx page:

```
<script type="text/javascript" src="Scripts/JavaScript.js"></script>
```

The `src` attribute specifies the folder and file that contains JavaScript functions.

Step 31: Create a JavaScript file and copy the ShowDetails function into it.

Right-click the project name, Theater, and select Add > JavaScript File.

Click OK to accept the default name JavaScript.

Copy the ShowDetails function into the JavaScript.js file, and save it.

Step 32: Add the following link within the head element:

```
<script type="text/javascript" src="Scripts/JavaScript.js"></script>
```

Comment out the script element that contains the ShowDetails function (using `<%-- --%>`).

Display the Reservation-5 page, select a performance, and click the Show Details button to test the code.

10.8 Using Server-Side Code to Display an Item of a List on Another Page

Instead of using client-side JavaScript, you may use server-side code to display an item of a list on another page. Using server-side code has the advantage that you don't have to know JavaScript, but in general, client-side code tends to be faster to execute.

Let's consider the same task discussed in the previous section: displaying the details of the performance selected in the DropDownList on the Reservation-5 page. Similar to the approaches used with JavaScript, you can filter the items by passing the PerformanceId as a query string or scroll to an item.

A difference is that the server-side code is added to the click event handler of a web server Button control, instead of using an HTML button to invoke a client-side JavaScript function.

For comparison purposes, Figure 10-23 shows the JavaScript you developed in the previous section with the corresponding server-side code shown in Figure 10-24.

```
1   function ShowDetails() {
2       var ddl = document.getElementById("ddlPerformance");
3       var performId = ddl.options[ddl.selectedIndex].value;
4       window.location.href = 'Events.aspx?qsPerformId=' + performId;//pass query string
5       // window.location.href = 'Events.aspx#' + performId; // scroll to a div element
6   }
```

Figure 10-23: JavaScript function to pass a query string or scroll to an item

```
1   protected void btnViewDetail_Click(object sender, EventArgs e)
2   {
3       string performId = ddlPerformance.SelectedValue.ToString();
4       Response.Redirect("Events.aspx?qsPerformId=" + performId); // pass query string
5       // Response.Redirect("Events.aspx#" + performId);  // scroll to a div element
6   }
```

Figure 10-24: Server-side C# code to pass a query string or scroll to an item

Note that the server-side code can access the DropDownList directly (line 3), whereas the client-side code uses the `getElemenetById` function to access it.

In place of the `window.location.href` property in the JavaScript, the server-side code uses the `Response.Redirect` method, discussed earlier, to redirect the browser to the Events page and also to pass the performance id using a query string (line 4) or scroll to a bookmark (line 5).

Step 33: Add a web server Button control to the Reservation-5 page.

Comment out the aspx code for the HTML button, `btnViewDetail`, as shown here:

```
1   <%--
2   <button type="button" id="btnViewDetail" name="btnViewDetail"  onclick="ShowDetails()">
    View Details </button>
3   --%>
```

In the Design view, add a Button control to the right of the ddlPerformance.

Set the properties of the Button:

ID:	btnViewDetail
Text:	View Details
CauseValidation:	False

It is important to set the CauseValidation property to False. Otherwise, the validation of the controls will prevent moving to the Events page without entering all the required data.

Step 34: Set up the event handler.

Create the click event handler of btnViewDetail, and add the code from Figure 10-24.

Display the Reservation-5 page, select a performance, and click the View Details button to test the code.

Comment out line 4, uncomment line 5, and test again.

It's time to practice! Do Exercise 10-6 (see the end of the chapter).

10.9 Applying CSS to the Events Page

This section applies CSS, which you learned about in Chapter 3, to specify the layout that displays the hyperlinks and the Calendar in a side panel on the left, along with the list on the right, as shown earlier in Figure 10-1. You will use the flexbox layout with two items using two div elements: one to include the elements on the left panel, and another to include all the elements to display the list on the right, as shown in Figure 10-25. All elements within the form are included in a section element (class=eventsbox) so that you can apply certain styles for the entire page.

```
1   <form id="form1" runat="server">
2   <section class="eventsbox">
3       <br /><br /><br />
4   <div id="search">
5       <h2>View Performances</h2>
6       <asp:Button ID="btnAll" … OnClick="btnAll_Click" Text="All Performances" … />
7       ...
8       ...
9       <asp:Calendar ID="calFilterByMonth" … </asp:Calendar>
10  </div>
11
12  <div id="list">
13      <asp:Repeater ID="rptEvents" … >
14          ...
15          ...
16      </asp:Repeater>
17  </div>
18  <br />
19  <asp:SqlDataSource ID="SqlDataSource_Performance" …> </asp:SqlDataSource>  <br />
20  </section>
21  </form>
```

Figure 10-25: Section and div elements to apply the CSS

The two div elements have id's "search" and "list," respectively. Figure 10-26 shows the CSS to apply the layout.

```
1   .eventsbox {
2       display: flex;
3       flex-direction: row;
4       border: 5px solid gray;
5       width: 1300px;
6       margin: 20px;
7       padding: 20px;
8       background-color: lightgray;
9   }
10          .eventsbox #search {
11          border: 2px solid black;
12          width: 350px;
13          padding: 20px;
14          background-color: lightblue;
15          }
16          .eventsbox  #list {
17          border: 2px solid black;
18          width: 850px;
19          padding: 20px;
20          background-color: lightblue;
21          }
```

Figure 10-26: CSS for the Events page

Step 35: Apply the CSS to the Events page.

Add a section element that includes the entire content of the form element, and add two `div` elements as shown in Figure 10-25.

Open styles.css and add the code from Figure 10-26.

Make sure the head section in the Events page contains a link to the Styles.css file, as shown here:

```
<link rel="stylesheet" type="text/css" href="stylesheets/styles.css" />
```

Display the page; it should have the layout shown in Figure 10-1.

Exercises

Exercise 10-1: Recipe List

Create a project named Recipes and develop a page named Breakfast.aspx that displays the name, image, and description of each breakfast recipe stored in the database table, Recipe (RecipeId, RecipeName, Description, Category, Image), within the database PI-Traders. The values of Category includes breakfast, dinner, and snack. The PI-Traders database can be found in the folder Project_starts/DataFiles. Make sure you create an ASP.NET folder named App_Data in the project directory and copy the database into the folder. It would help to view the Recipe data from the table before you develop the code to display it.

Currently, none of the recipes have an image in the Image field. Upload the image to the Image field of the oatmeal recipe.

Make sure you format the data to make it easy and attractive for users to view.

Exercise 10-2: Recipe List with Ingredients

This exercise expands the page developed in Exercise 10-1. Do Exercise 10-1 if you haven't already.

In addition to the name, image, and description displayed in Exercise 10-1, display the list of ingredients for each recipe (each ingredient on a separate line), immediately following the description of the recipe. The ingredients for each recipe are provided in the table RecipeIngredient (RecipeIngId, RecipeId, Ingredient).

Exercise 10-3: Favorite Recipes

You may do this exercise by adding code to the Breakfast page developed in Exercise 10-1 or 10-2. Do Exercise 10-1 if you haven't already.

Dynamically generate a button for each recipe when the page is displayed. Display the button below the description of each recipe. The user should be able to click the button to add the name of the recipe to a drop-down list of favorites in the panel on the left side of the page. Use a Command event handler to add the recipe name to the list. The button may have a caption like "Add to favorites."

Exercise 10-4: Hyperlinks

You may do this exercise by adding code to the page developed in Exercise 10-1 or any later version of it. Do Exercise 10-1 if you haven't already.

Generate a list of hyperlinks that lets the user scroll to the description of a specific recipe. The text for the hyperlinks should be the names of the recipes. Display the hyperlinks with one hyperlink in each row.

Exercise 10-5: Filtering Recipes

You may do this exercise by adding code to the page developed in Exercise 10-1 or any later version of it. Do Exercise 10-1 if you haven't already.

Provide a RadioButtonList that gives the user the option to select one of the following ingredients: oats, rice, wheat, chicken, beef, pork, egg. When the user selects an ingredient, filter the list of recipes so the list displays only information on recipes that have the selected item as an ingredient.

Exercise 10-6: Default Page

You may do this exercise by adding code to the page developed in Exercise 10-1 or any later version of it. Do Exercise 10-1 if you haven't already.

Add a new form named Default.aspx to the Recipe project. Provide a DropDownList on the Default page to let the user select a recipe from a list of all recipe names. Provide a button next to the DropDownList that lets the user filter the Recipes on the Recipe page to display information on only the recipe selected in the DropDownList. Do it using client-side JavsScript and also using server-side code.

Exercise 10-7: Converting an .html Page to an .aspx Page

The TheGrand.html page was created in Exercises 1-1 and 1-2 in Chapter 1 as a static HTML page to help you learn the basics of HTML. A drawback of the static page is that some content, like the list of performances and scheduled dates, is difficult to maintain. To address this problem, you will create a dynamic version of this page, named **Default. aspx**, within the Theater project. Instead of displaying static data on performances, the Default.aspx page will have hyperlinks to the performance information displayed on the Events.aspx page, as shown here:

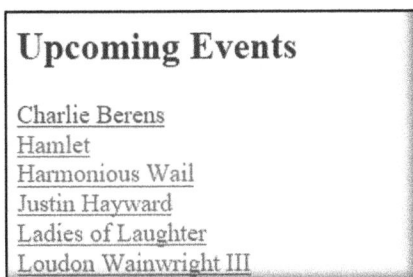

Upcoming Events

Charlie Berens
Hamlet
Harmonious Wail
Justin Hayward
Ladies of Laughter
Loudon Wainwright III

To do this, you will work with the DataSource and Repeater controls.

You may do the following steps:

Create a new web form named Default.aspx.

Copy the header, main, and footer elements, including their contents, from TheGrand.html, and paste them inside the form element in Default.aspx.

Delete the "Current Performances" section from the Default.aspx page, including the table element with its contents (performances and dates/times).

Create the DataSource for performance by copying the following code from the Events page and pasting it into the Default.aspx page between the code for the seat map and Box Office information:

```
1  <asp:SqlDataSource ID="SqlDataSource_Performance" runat="server"
2      ConnectionString="<%$ ConnectionStrings:ConnectionString %>"
3      SelectCommand="SELECT * FROM [Performance] ORDER BY [PerformanceName]">
4  </asp:SqlDataSource>
```

To create the Repeater control to generate the hyperlinks to performances, you may copy the code for the Repeater control, rptLinks, from the Events page and paste it into Default.aspx page above the code for the DataSource control.

Assignments

E-Commerce Assignment: Part E (ASP.NET)—Lists

This assignment expands E-Commerce Assignment: Part C (ASP.NET)—Using Databases (see Chapter 8) to add a new page named **laptops**, which lets the user view the list of laptop models and filter them. Make sure first you develop the web page **laptop-selection** as specified in Part C, if you haven't already done so.

Develop a new web form named *laptops*. When the page is opened, it should display the complete list of laptops. For each model, display the model name, description, and available configurations of the model. You may use a layout similar to the one shown in the following figure:

Laptop Models

Model: B780

A light-weight 2-in-one 15-inch touch-screen performance laptop featuring 8th Gen Intel® Core™ processors; ideal for any office.

Configurations

RAM	SSD	Hard Drive	Price
8 GB	0	1 TB	$995
8 GB	512 GB	0	$1095
16GB	512 GB	500 GB	$1195

Select this model

Model: G780

14 inch touch-screen gaming laptop featuring NVIDIA® GeForce® GTX 1060 graphics and 8th Gen Intel® Quad-and-Hex Core™ CPU

Configurations

RAM	SSD	Hard Drive	Price
8 GB	256 GB	0	$950
16 GB	256 GB	500 GB	$1150

Select this model

The *Select this model* link button should display the laptop-selection page, with the DropDownList displaying the name of the selected model.

The web page should provide the following additional features to filter the list:

- A set of hyperlinks should let the user scroll to a specific model.

- Let the user select a CPU from a DropDownList to display only laptop models with that CPU.

- Let the user select the number of GBs of RAM (4 GB, 8 GB, 12 GB, or 16 GB) from a RadioButtonList or DropDownList and display only models with the selected amount of RAM.

Auto Rental Assignment: Part E (ASP.NET)—Lists

This assignment expands E-Commerce: Part C (ASP.NET)—Using Databases (see Chapter 8) to add a new page named **fleet-vehicles**. Make sure you develop the **Auto-selection** page as specified in Part C, with the exception that the page shouldn't have a button to compute the total cost.

Develop a new web form named *fleet-vehicles*. The page should display the complete list of auto types available for rental. For each auto type in the fleet, display the auto type, passenger capacity, rental rate per day, and rental rate per week. In addition, for each auto-type, display the models available (model name and cargo capacity). You may use a layout similar to the one shown in the following figure:

Auto types and models

Standard SUV

Seating Capacity: 5
Daily rate: 45
Weekly rate: 250
Models

Model name	Cargo space
Honda CRV	39
Ford Escape	34

Select this type

Full-size SUV

Seating Capacity: 7
Daily rate: 65
Weekly rate: 400
Models

Model name	Cargo space
Chevy Suburban	39
Ford Expedition MAX	38

Select this type

The *Select this type* link button should display the auto-selection page, with the DropDownList displaying the selected auto type.

Provide the following features to filter the list:

- A set of hyperlinks should let the user scroll to a specific model.

- Let the user select a passenger capacity (2, 5, 7, 12, or 15) DropDownList or RadioButtonList and filter the list to display information for only auto types with the selected passenger capacity.

- Let the user select a ModelName from a DropDownList and filter the list to display information for only the selected model.

## Chapter 11

Application State and Cookies

In previous chapters, you learned how to use the **view state** to share data within the same page and **session state** to share data between pages within the same session. This chapter describes how to share global data across all sessions and users using **application state** and **caching** in combination with databases. You will use the application state to provide additional features in the Theater web project. You will also learn how to use **cookies** to store user-specific information on the client.

Learning Objectives

After studying this chapter, you should be able to

- Describe application state, cache, and cookies.
- Create and use application state items to share data across sessions and users.
- Use Application and Session events to work with an application state and cookies.
- Read database records using the Command object and DataReader.
- Use Lists to store multiple database records in an application state item.
- Use GridView control to display and update a database table.
- Create cookies and retrieve their values.
- Use cookies to keep track of user visits to a website and user activities.

11.1 Application State

Web applications may require sharing data across all sessions of an application and all users. For example, you may want to count the number of current users of a website or share a common sales tax rate among all users. You will use the application state to share such global data by storing it in the server's memory from the start of an application to its end.

Because the application state stores data in the server memory, it provides faster access compared to data stored in databases and files. However, storing large amounts of data in the application state may negatively impact the performance of the server.

When an application starts, ASP.NET creates an application state object (from the `HttpApplicationState` class) that you can use to store data and access from all web pages of the application by all users until the application ends. An application ends when the web server is shut down, and also when it is restarted after rebuilding the application due to changes in source code or after changing certain folders/files like the web.config file and the bin folder.

It should be noted that the application state is not shared between multiple servers when an application is processed on multiple servers.

11.1.1 How to Use the Application State

Using the application state is very similar to using the session state. You can add items to the collection contained in the application state using the `Application` property of the `Page` class as follows:

```
Application.Add("name", value);
```

or

```
Application["name"] = value
```

Here is an example that shows how to create an item, also called an **application variable**, named "Discount" and store the value 10 in it, using each syntax:

```
Application.Add("Discount", 10);
```

or

```
Application["Discount"] = 10;
```

Just like the `Session` variable, you don't have to declare the `Application` variable. However, when you retrieve a value, you have to cast it to the appropriate type, determined by the type of the value assigned to it. To retrieve the value of the application variable `Discount`, you may use its name or its index in the collection:

```
decimal discount = (decimal)Application["Discount"];
```

or

```
decimal discount = (decimal)Application [index];
```

You may access the application state object from a class other than the `Page` class (or a class that inherits the `Page` class) using the `Application` property of the `HttpContext` object:

```
HttpContext.Current.Application["Discount"] = 10;
```

In addition to the properties and methods of the session state, the application state has two methods for locking and unlocking the application state collection, as shown in Table 11-1. You will learn how to work with these later in the chapter.

Table 11-1: Properties/methods of the application state

Properties/methods	Description
Count	Number of items in the collection e.g., `int count = Application.Count;`
Add (name, value)	Adds an item to the collection, or replaces it if it exists e.g., `Application.Add("Discount", 10);`
Clear()	Removes all items from the collection e.g., `Application.Clear();`
Remove (name)	Removes the specified item from the collection e.g., `Application.Remove("Discount");`
Lock()	Locks the application state object e.g., `Application.Lock();`
Unlock()	Unlocks the application state object e.g., `Application.Unlock();`

11.2 Using a Cache Object to Share Global Data

An alternative to using an application state is using *Caching*, which allows you to keep data in the server memory for a specified period of time and allows the server to clear the data when the memory is low.

You can use the Cache object in the same way as you use the application state. In other words, you can use the Cache property of the page instead of the Application property. For example, the following statements show how to add an item named Discount to the collection of items of the Cache object:

```
Cache.Add("Discount", 10);
```

or

```
Cache["Discount"] = 10;
```

Cache, however, lets you specify an expiration time for the item you add. You can use this option to specify that an item like a car rental rate is valid for only a week. Typically, you would use a process to read the rate from a database or compute the rate, and update the cache.

To specify the expiration date, you need to use the Insert method instead of the Add method, using the syntax

```
Insert(name, value, dependency, absolute expiration, sliding expiration)
```

The *name* and *value* parameters are the same as in an application state. The third parameter, *dependency*, specifies an object associated with the item, which determines the removal of the item. For example, you can make the removal of a cache item dependent on changes in a specified file.

Absolute expiration specifies the absolute expiration time, and *sliding expiration* specifies the maximum duration of time the item will be in memory since the last time it was accessed. To share a discount that expires in two days, you would store it in Cache as follows:

```
Cache.Insert("Discount", 10, null, DateTime.Now.AddDays(2),
        System.Web.Caching.Cache.NoSlidingExpiration);
```

The examples in this chapter use the application state to share data. However, you can replace the application state with Cache in these examples.

11.3 Working with Application Events

Two important events raised by the application state object are the **Application_Start** and **Application_End** events. The Application_Start event is raised the first time a page of the application is requested by any user. An Application_End is raised when the application ends. As stated earlier, an application ends when the web server is shut down, and also when it is restarted after rebuilding the application or after changing certain folders/files.

Let's look at some examples of how these events are used. A common use of the Application_Start event is to initialize application state items. For example, to count the number of current users using an application state variable, you would set its value to zero when the application starts and then increment its value by one when a user starts a session. Whenever a user ends a session, you would decrement the value by one. Conversely, to count the total number of users who visited a site, you would increment an application state variable for each new user. You can then use the Application_End event to save this value to a data store like a file or database to be accessed when the application starts again. On Application_Start, you would then restore the saved value and continue incrementing for each new user.

To store an item like a tax rate or discount in the application state, you may read the initial value from a data store when the application starts, as in the previous example. However, in such cases, you should provide a way to change the value of the application state by an administrator. Further, such changes should be saved to the data store when the application state is changed or when the application ends, so that the application state has the updated value when the application is restarted.

To work with application and session events, add the Global.asax.cs file. You can do this by right-clicking the Project file, Theater, and selecting *Add>New Item>Global Application Class*.

The Global.asax file lets you write code that responds to application and session events, and contains the declarations for several standard event handlers. Some common events are shown in Figure 11-1.

```
 1  public class Global : System.Web.HttpApplication
 2  {
 3      protected void Application_Start(object sender, EventArgs e)
 4      {
 5      }
 6      protected void Session_Start(object sender, EventArgs e)
 7      {
 8      }
 9      protected void Application_AuthenticateRequest(object sender, EventArgs e)
10      {
11      }
12      protected void Application_Error(object sender, EventArgs e)
13      {
14      }
15      protected void Session_End(object sender, EventArgs e)
16      {
17      }
18      protected void Application_End(object sender, EventArgs e)
19      {
20      }
21  }
```

Figure 11-1: Default code for Global.asax.cs

Application_Error event is raised whenever an unhandled exception occurs in the application. You can use this event to catch and/or handle errors, as well as to log error conditions for later troubleshooting.

Application_AuthenticationRequest event is raised when a request is ready to be authenticated. You can use this event to check the user's roles and rights.

The Session_Start event is raised when a user starts a session, and Session_End is raised when the session ends. These events can be used to perform actions in preparation for the user to start working. For example, you might have data that should be loaded into the Session object.

Tutorial: Application State and Cookies

To help you understand the different ways the application state is used, we will look at three different scenarios using the application state:

1. Display the number of current user sessions of the Theater website on the Reservation page.
2. Share the member discount using the application state rather than hard-coding it or reading from a file/database for each reservation.
3. Share zone prices using the application state rather than reading them from the database for each reservation.

You will continue to work with the Reservation-5 form from Chapter 10. However, if you didn't develop the Reservation-5 page in Chapter 10, you may make a copy of Reservation-4 and rename it as Reservation-5.

11.3.1 Application State Example 1: Displaying the Number of Current User Sessions

To count the number of current users who have an open session, you will use an application state variable (Current-Users) and set its value to zero when the application starts, increment its value by one when a user starts a session, and decrement its value by one when the user ends a session, as shown in Figure 11-2.

Note that before you update an application state variable, you need to lock the application state object to prevent simultaneous update of the variable by multiple users, which will negate all updates except the last one. In a multiuser system, locking a shared data item (or a group of data items) by a user before updating makes sure that a second user cannot update the same data before the first user unlocks the data after completing the update process. This is not an issue when updating the session state, because each session state item can be accessed by only one user.

```
1   public class Global : System.Web.HttpApplication
2   {
3       protected void Application_Start(object sender, EventArgs e)
4       {
5           Application["CurrentUsers"] = 0;
6       }
7       protected void Session_Start(object sender, EventArgs e)
8       {
9           Application.Lock();
10              Application["CurrentUsers"] = (int)(Application["CurrentUsers"]) + 1;
11          Application.UnLock();
12      }
13      protected void Session_End(object sender, EventArgs e)
14      {
15          Application.Lock();
16              Application["CurrentUsers"] = (int)(Application["CurrentUsers"]) - 1;
17          Application.UnLock();
18      }
19  }
```

Figure 11-2: Using the application state to track the number of current sessions

After you create the application state variable `CurrentUsers`, you can access it from any page and display it. You can access it from the Reservation-5 form and then display it on the form.

Step 1: Add the Global.asax.cs file to the Theater project.

Right-click the project, Theater, and select *Add>New Item>Visual C#>Web>Global Application Class*.

Step 2: Add code to the Global.asax.cs file.

Insert the code from Figure 11-2 into the `Application_Start`, `Session_Start`, and `Session_End` event handlers to create, initialize, and update `CurrentUsers`.

Step 3: Display the number of current users in a Label on the Reservation-5 form.

Open Reservation-5.aspx. Below the Price/Ticket label, add the code in lines 5–7 in Figure 11-3.

```
1   <p>
2       Price/ticket <br />
3       <asp:Label ID="lblUnitPrice" runat="server"></asp:Label>
4   </p>
5   <p>
6       Current Users: <asp:Label ID="lblCurrentUsers" runat="server"></asp:Label>
7   </p>
```

Figure 11-3: CurrentUsers label

To display the value of `CurrentUsers`, add the following line of code to the end of the `Page_Load` event handler of Reservation-5.aspx.cs:

```
lblCurrentUsers.Text = ((int)Application["CurrentUsers"]).ToString();
```

Display the Reservation-5 form in the browser.

You should see a value of 1 for current users.

Copy the URL of the web page from the browser.

Open another browser, paste the URL and click the *Enter* key.

You should see the Reservation page with the number of users increased to 2.

You can start new sessions by opening the same site in different browsers or in an incognito/private window of the same browser. You should see the number of users increase each time you start a new session. If you go back to one of the sessions you previously started and refresh the page, you should see the number of Current Users change to reflect the number of sessions you've started.

11.3.2 Application State Example 2: Sharing Member Discount

Currently, the discount offered to members ($10) is hard-coded in the event handler for the check box `chkMember`. To make it easy to change the discount, you could store the amount in a data store like a file or database and read it for each user session. A more efficient alternative that avoids repeated reading from a data source is to read the discount once from a data store and store it in the server memory using an application state item created in the `Application_Start` event handler, as shown in Figure 11-4.

The method of reading the values from a data store and storing it in an application state is illustrated in Figure 11-4. In this example, we store the value 10 in the application state (line 6) so that it can be accessed by all users.

```
1  public class Global : System.Web.HttpApplication
2  {
3      protected void Application_Start(object sender, EventArgs e)
4      {
5          Application["CurrentUsers"] = 0
6          Application["Discount"] = 10;
7          // you may read the value from a file or database
8      }
```

Figure 11-4: Storing the discount amount to the application state

Now you can use the discount amount on any page. You will use it in the Reservation-5 form to compute the discount.

Step 4: Store the discount in the application state.

Add the code to the `Application_Start` event handler to store the discount amount in the application state variable *Discount*, as shown in line 6, Figure 11-4.

Step 5: Read the Discount amount and display on page.

Modify the code in the `Checked_Changed` event handler of `chkMember` in Reservation-5.aspx.cs to get the discount amount from the application state variable, as shown in Figure 11-5.

322 • Chapter 11 / Application State and Cookies

```
1   protected void chkMember_CheckedChanged(object sender, EventArgs e)
2   {
3       // Compute and display discount for members
4       if (chkMember.Checked)
5           // discount = 10;
6           discount = decimal.Parse(Application["Discount"].ToString());
```

Figure 11-5: Checked_Changed event handler of chkMember

Display the Reservation-5 form and verify that the discount is computed correctly.

Changing the Value of an Application State Variable while the Application Is Running

The Application_Start event is raised only when the application starts. But how can you change the member discount specified in the Application_Start without restarting the application? The application state can be easily changed in any code segment. You will create a form named **Admin.aspx**, as shown in Figure 11-6, which lets the administrator enter the discount amount into the textbox and change the application state by executing code in the event handler of the button.

It should be noted that if an application runs on multiple processors or multiple servers (e.g., a server farm), the application state is not shared across servers. Updating the application state on one processor/server doesn't change it on others.

Update member discount

Member Discount: [   ]   [Update Discount]

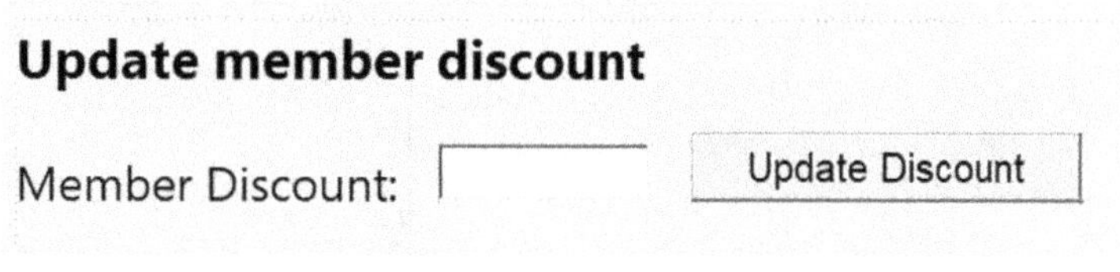

Figure 11-6: Form to change the member discount

Figure 11-7 shows the code to set Discount to the value entered into the TextBox. The administrator can execute this code and update the application state by clicking the button, without having to restart the application.

```
1   protected void btnUpdateDiscount_Click(object sender, EventArgs e)
2       {
3           Application["Discount"] = Convert.ToDecimal(txtDiscount.Text);
4           // If the discont is stored in a database, it also should be updated
5           // in this event handler or in the Application_End event handler
6       }
```

Figure 11-7: Updating the application state on the Admin page

Step 6: Set up the Admin page.

Create a new form named Admin.aspx with a TextBox and a Button, as shown in Figure 11-6.

Set the ID of the TextBox to txtDiscount and the ID of the Button to btnDiscount.

In addition, copy the link to the stylesheet and the <header> element from Reservation-5 to the Admin page so that the Admin page will have the links to display other pages. Update the link to the Reservation page from Reservation-4 to Reservation-5.

Step 7: Update the application state.

Create the Click event handler of btnUpdateDiscount, as shown in Figure 11-7.

Note that if the discount is stored in a database, the database also should be updated. Otherwise, if the application is restarted, it will read the original value from the database. Updating the database is illustrated in the third example.

To test the page, set Admin as the start page and display it.

Enter a value for discount, different from the current value, and click the **Update Discount** button to change the value. From now on, the updated discount will be available for all users.

Click the link for the Reservation page. Check the CheckBox to indicate member status.

Verify that the updated discount is displayed in the Label on the right pane.

Later, you will learn how to authenticate the user so that only the administrator will have access to the Admin page.

11.3.3 Application State Example 3: Share Zone Prices

To find the zone price for the selected zone, the reservation form reads the zone prices from the Zone table, once in each user session. You can avoid accessing the database multiple times by reading the Zone table once in the Application_Start event handler and storing it in an application state item.

In order to store the entire Zone table in a single application state variable, you can store each record in an object and store the objects in a List. Then, you can store the entire List in a single application variable, similar to how you stored multiple reservations in a List.

Step 8: Add a Zone class.

To store a Zone record, create the Zone class shown in Figure 11-8 in the App_Code folder. Once the class is created, check its properties and make sure the Build Action is set to Compile.

```
1    namespace Theater
2    {
3        public class Zone
4        {
5            public string Id { get; set; }
6            public string Name { get; set; }
7            public decimal Price { get; set; }
8            public Zone(string id, string name, decimal price)
9            {
10               this.Id = id;
11               this.Name = name;
12               this.Price = price;
13           }
14       }
15   }
```

Figure 11-8: The Zone class

Reading Records Using the Command Object and DataReader

In previous versions of the reservation form, you used the SqlDataSource control along with the DataView and DataRowView objects to retrieve data from the Performance and Zone tables. However, the Global.asax file doesn't have a designer window where you can add the DataSource control. One option is to create the DataSource and specify its properties using code. A simpler and more commonly used method to sequentially read records from a table is to use an SqlCommand object along with an SqlDataReader object. This method is used also in Chapter 16. You will use this method to read the Zone records and store them in a List.

The SqlCommand object lets you store an SQL statement in its CommandText property and execute the SQL to retrieve the records from the database.

A SqlDataReader provides an efficient way to sequentially access data retrieved by the Command object. As the SQL executes, each row retrieved by the SQL is stored temporarily in the client's buffer, accessed sequentially by the Read method of the DataReader, and then removed from the buffer. Thus DataReader can help reduce memory usage and improve performance. But you cannot move back to previous records; it has forward-only access, and the

DataReader does not have methods to update, delete, or add records. The connection is kept open for exclusive use by the DataReader until the DataReader is closed.

Figure 11-9 shows how to set this up. First, lines 1 and 2 create the SqlCommand and specify the SQL to retrieve the Zone records. Second, lines 3–6 create the Connection object, specify the connection string, and open the connection:

```
1   SqlCommand cmdZone = new SqlCommand();
2   cmdZone.CommandText = "Select * From Zone";
3   cmdZone.Connection = new SqlConnection();
4   cmdZone.Connection.ConnectionString = @"Data Source=(LocalDB)\MSSQLLocalDB;
        AttachDbFilename=|DataDirectory|\Theater.mdf";
5   cmdZone.Connection.Open();
```

Figure 11-9: Connecting to the database

Line 1 creates the object, cmdZone, from the SqlCommand class.

Line 2 sets the CommandText property of the object to the SQL that selects all Zone records.

Line 3 creates a Connection object from the SqlConnection class and sets a reference to it in the Connection property of the SqlCommand object.

Line 4 stores the connection string in the ConnectionString property of the SqlCommand object. In the current project, an easy way to specify the connection string is to copy it from SqlDataSource_Zone on the reservation form.

Line 5 opens the connection.

Third, you declare a reference variable for the Zone object that stores the Zone record, and create the List to store the objects, as follows:

```
1   Zone zone;
2   List<Zone> listOfZones = new List<Zone>();
```

Fourth, you use the SqlCommand object to retrieve Zone records and use the DataReader to access each record. Each Zone record is stored in a Zone object and added to the List.

```
1   SqlDataReader drZone = cmdZone.ExecuteReader();
2   while (drZone.Read() == true)
3   {   // store each row in a Zone object
4       string id = drZone["ZoneId"].ToString();
5       string name = drZone["ZoneName"].ToString();
6       decimal price = Convert.ToDecimal(drZone["ZonePrice"]);
7       zone = new Zone(id, name, price);
8       listOfZones.Add(zone);  // Add the Zone object to the List
9   }
10  Application["ListOfZones"] = listOfZones; //Save the List in application state
11  cmdZone.Connection.Close()
```

Line 1 uses the ExecuteReader method of the SqlCommand object to execute the SQL stored in the CommandText property and return the result set as a DataReader object. The left side of the statement creates the reference variable, drZone, that is used to read each row of the data set retrieved by the SQL.

Line 2 uses a loop to read each row using the Read() method of the DataReader. The Read() method returns the value *false* when there are no more rows to read.

Lines 4–6 get the values of ZoneId, ZoneName, and ZonePrice from the current row and store them in variables.

Line 7 creates the zone object, and line 8 adds the object to the List.

Line 10 saves the List in the application state variable, ListOfZones.

Figure 11-10 shows the complete code in the Application_Start event handler, which creates the application state items CurrentUsers, Discount, and ListOfZones.

```
1   protected void Application_Start(object sender, EventArgs e)
2   {
3       Application["CurrentUsers"] = 0;
4           // you may read the initial value from a file or database
5       Application["Discount"] = 10;
6       // Create an SqlCommand object
7       SqlCommand cmdZone = new SqlCommand();
8       cmdZone.CommandText = "Select * From Zone"; // specify the SQL
9       // Create the Connection object and open the connection
10      cmdZone.Connection = new SqlConnection();
11      cmdZone.Connection.ConnectionString = @"Data Source=(LocalDB)\MSSQLLocalDB;
            AttachDbFilename=|DataDirectory|\Theater.mdf";
12      cmdZone.Connection.Open();
13      // Create the List of Zones
14      Zone zone;
15      List<Zone> listOfZones = new List<Zone>();
16      // Use the ExecuteReader method of the Command object to execute the SQL, and
17      // Create the SqlDataReader to read each row of the data set
18      SqlDataReader drZone = cmdZone.ExecuteReader();
19      while (drZone.Read() == true)
20      {   // store each row in a Zone object
21          string id = drZone["ZoneId"].ToString();
22          string name = drZone["ZoneName"].ToString();
23          decimal price = Convert.ToDecimal(drZone["ZonePrice"]);
24          zone = new Zone(id, name, price);
25          listOfZones.Add(zone);  // Add the Zone object to the List
26      }
27      Application["ListOfZones"] = listOfZones; //Save the List in application state
28      cmdZone.Connection.Close();
29  }
```

Figure 11-10: Complete code for the Application_Start event handler.

Step 9: Store the list of Zone objects in the application state.

Add the following `using` statements:

```
using System.Data.SqlClient;
using Theater.App_Code;
```

Add the code from Figure 11-10 to the `Application_Start` event handler to create the `List` of `Zone` objects and store the `List` in the application state variable, `ListOfObjects`.

Now that you have stored the zone records in an application variable, you can use it to find the price for the zone selected by the user on the reservation form, without having to read from the database each time. You will replace the current code, which uses the `DataSource` object, with new code that gets the zone price from the List, as shown in lines 8–16 in Figure 11-11.

```
1   protected void rblZones_SelectedIndexChanged(object sender, EventArgs e)
2   {
3       //DataView dvZoneTable =
            (DataView)SqlDataSource_Zone.Select(DataSourceSelectArguments.Empty);
4       //dvZoneTable.RowFilter = "ZoneId = " + rblZones.SelectedValue.ToString();
5       //DataRowView drvZoneRow = (DataRowView)dvZoneTable[0];
6       //zonePrice = (decimal)drvZoneRow["ZonePrice"];
7
8       List<Zone> listOfZones = (List<Zone>)Application["ListOfZones"];
9       foreach (Zone zone in listOfZones)
10      {
11          if (zone.Id == rblZones.SelectedValue.ToString())
12          {
13              zonePrice = zone.Price;
14              break;
15          }
16      }
```

Figure 11-11: Radio button event handler using application state data to set the zone price

Line 8 gets the List of Zones from the application state, casts it to List<Zone>, and stores it in the reference variable listOfZones.

Line 9 uses a foreach loop to access each object in the List.

Line 11 checks whether the Id property of the object matches the ZoneId of the selected zone (given by the SelectedValue of the RadioButtonList). If they match, line 13 gets the price and line 14 breaks out of the loop.

Step 10: Update the SelectedIndexChanged event handler of rblZones to use the List of Zones.

Open Reservation-5.aspx.cs and comment out or delete the code shown in lines 3–6 in Figure 11-11.

Add the code from lines 8–16.

Leave the rest of the code as is.

To test the code, display the reservation form and select a zone. Verify that the correct zone price is displayed in the Label.

Updating the List of Zones and the Zone Database Table

Next, we look at how to let the administrator change the zone prices in the application state without having to restart the application. The process is similar to the one you used for updating the discount. The difference is that the application state item for zone prices is a list of objects, whereas the discount is a simple data item.

In addition, when you change the zone prices in the application state, you also need to change them in the database so that if the application is restarted, the updated prices are loaded into the application state. You will update the application state and the database table together so that the two always will be in sync.

Updating Database Tables Using GridView Control

A relatively easy way to edit and update small database tables is to use a GridView control that is bound to a Data-Source. In Chapter 6, you used the GridView control with a Select button to select rows. You also can update a database table using Edit and Delete buttons that can be added to the GridView.

You will set up the GridView control to display data from the Zone table so that the administrator can change the data, update the table, and update the application state.

Step 11: Add a data source to the Admin page.

Add a SqlDataSource named SqlDataSource_Zone to the Admin form and configure it to select all data from the Zone table, as shown here:

```
    <asp:SqlDataSource ID="SqlDataSource_Zone" runat="server"
        ConnectionString="<%$ ConnectionStrings:ConnectionString %>"
    SelectCommand="SELECT * FROM Zone"></asp:SqlDataSource>
```

Then, click the *Smart Tag* button for the control and select *Configure Data Source*. Click *Next* and then the *Advanced* button. Select the option to generate *INSERT*, *UPDATE*, and *DELETE* statements. Click *OK*. For more detail on this process, refer to Chapter 6.

Next, you will add text and controls to the Admin page, as shown in Figure 11-12.

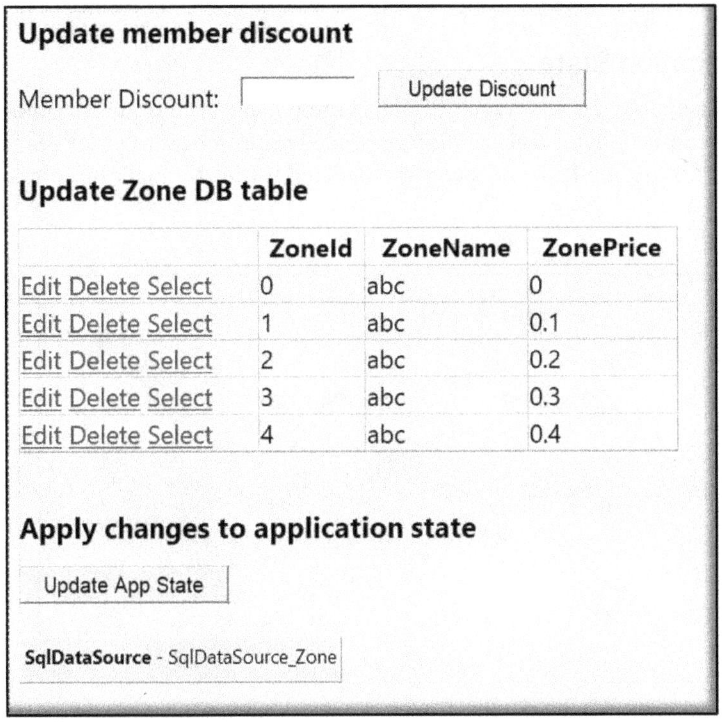

Update member discount

Member Discount: [   ]   [Update Discount]

Update Zone DB table

	ZoneId	ZoneName	ZonePrice
Edit Delete Select	0	abc	0
Edit Delete Select	1	abc	0.1
Edit Delete Select	2	abc	0.2
Edit Delete Select	3	abc	0.3
Edit Delete Select	4	abc	0.4

Apply changes to application state

[Update App State]

SqlDataSource - SqlDataSource_Zone

Figure 11-12: Admin page with a GridView to edit Zone data.

Step 12: Set up GridView on the Admin page

Add a GridView to the Admin form and set it up as follows:

ID: gvZone

DataSourceId: SqlDataSource_Zone

Click the smart tag (which resembles a > symbol) at the top right, and select *Edit Columns…*. Then expand the CommandField and add the three fields for Edit, Select, and Delete. The GridView should look like Figure 11-12 when shown in Design View.

To specify a back color for the selected row, select BackColor from the SelectedRowStyle property in the list of properties for the entire GridView and set it to a light color. Here's what the GridView should look like in code when you're done:

```
<asp:GridView ID="gvZone" DataSourceID="SqlDataSource_Zone" runat="server"
    SelectedRowStyle-BackColor="#ffffcc">
    <Columns>
        <asp:CommandField ShowEditButton="True"></asp:CommandField>
        <asp:CommandField ShowSelectButton="True"></asp:CommandField>
        <asp:CommandField ShowDeleteButton="True"></asp:CommandField>
    </Columns>
</asp:GridView>
```

Step 13: Test the GridView control.

Display the Admin form and click the *Edit* button for a row, which changes to two buttons: *Update* and *Cancel*. Change the price for the selected zone. Click *Update* to update the database.

Clicking the *Update* button updates the Zone table with the new price. However, you haven't changed the application state yet. So, if you display the Reservation page and select the zone you changed, it will display the old price.

To test, display the Reservation page. Select the zone whose price you changed. Notice that the zone price displayed in the Label is the old price.

Updating the List Stored in the Application State

To update the List stored in the application state ListOfZones, you will add a button named btnUpdateZones to the Admin page and add the following code to its Click event handler. The code creates a new List of Zone objects, adds the updated data from the GridView to the List, and stores the updated list in the application state (Figure 11-13).

```
1   protected void btnUpdateZones_Click(object sender, EventArgs e)
2   {
3       List<Zone> listOfZones = new List<Zone>();
4       foreach (GridViewRow row in gvZone.Rows)
5       {
6           string id = row.Cells[3].Text;  // Get ZoneId from column 4 of current row
7           string name = row.Cells[4].Text;
8           decimal price = Convert.ToDecimal(row.Cells[5].Text);
9
10          Zone zone = new Zone(id, name, price); // Create Zone object
11          listOfZones.Add(zone); // Add the zone object to the List
12      }
13      Application["ListOfZones"] = listOfZones; // Store the List in application state
14  }
```

Figure 11-13: Updating the application state

Line 3 creates the List object.

Line 4 uses a foreach loop to loop through each row of the Rows collection of the GridView. Note that each row in the Rows collection is of type GridViewRow.

Lines 6–8 get the data from the columns of the row using the Cells[].Text property. Because the first three columns (indices 0, 1, and 2) are buttons, the ZoneId, ZoneName, and ZonePrice have indices 3, 4, and 5, respectively.

Step 14: Update the Application State.

Add the *Update App State* button, as shown in Figure 11-12, to apply the changes in the GridView to the application state.

Change its ID to *btnUpdateAppState*.

Type in the headings above the button and the GridView, as shown in Figure 11-12.

Add a using statement to Admin.aspx.cs for Theater.App_Code.

Add the code from Figure 11-13 to the Click event handler of btnUpdateAppState.

Step 15: Test the code.

Display the Admin page. Click the Edit button and change the price for a zone.

Click the Update link in the grid view to update the database.

Click the Update App State button to update the application state item, ListOfZones.

Click the link for the Reservation form to display the page.

Select the zone whose price you changed. You should see the updated price in the Label, indicating that the application state was updated.

It's time to practice! Do Exercise 10.2 (see the end of the chapter).

Review Questions

1. What are two differences between the application state and the session state?
2. State two characteristics of cache that distinguish it from application state.
3. Explain why it is necessary to lock the Application object when you update an application state variable.
4. The following statement is used to create an application state item:

```
Application.Add("Discount", 10.5);
```

Provide the statement to get the value of the application state variable and store it in a variable of type decimal, named *discount*.

11.4 Cookies

Cookies are commonly used by websites to store information about users and their visits, such as a user's Id, date/time visited, or the product looked at. Such information is commonly used to customize the content for a user. For example, you may welcome back a user who has previously visited the site or suggest products based on previous purchases.

Typically, each cookie is a small bit of text stored by a website on the user's hard disk. A cookie created in a browser is specific to that browser; it is not generally accessible by other browsers.

A website creates cookie(s) when a user takes a specific action, like visiting a site for the first time, making a reservation, ordering a product, submitting a survey, or just visiting a page. To create the cookie, you may use such information as the login information, data entered by the user into a TextBox, or the id of a product ordered by the user.

Every time a user requests a page from a website (e.g., clicking a link to a page on the site), the browser looks on the hard disk for any cookie(s) associated with the site. If applicable, the browser sends the cookies, along with the request, to the server so that the application can access the cookies.

Because browsers typically limit the combined size of all cookies stored by a website/domain (about 4000 bytes) and the number of cookies allowed from each website, cookies are not suitable for storing large amounts of data. If a website's cookies exceed the limit set by the browser, there is the risk that the older cookies may get discarded. Browsers also may limit the combined total number of cookies from all sites. So, you should keep information stored in cookies limited—usually just to ids that can be used to look up the rest of the information from a file/database. You also should keep in mind that users can delete or disable cookies. Thus they are not suitable for storing critical information.

Just like session and application states, a cookie stores data as a key-value pair. As described earlier, you also can specify an expiration date. If you don't specify an expiration date, the cookie is stored in the server memory and removed when the user session ends. Such a cookie is called a **non-persistent (transient) cookie**, whereas a cookie that has an expiration date is called a **persistent cookie**. As you learned in Chapter 9, the session state is stored in a non-persistent cookie. Thus, non-persistent cookies are also called **session cookies**.

11.4.1 Creating Cookies

You create cookies from the HttpCookie class using server-side code and store them in the cookies collection (HttpCookieCollection). When the server sends the response to the client, it also sends the cookies collection in the HttpResponse object, to be stored on the client's computer. So, when you create a cookie, you would add it to the Cookies collection in the HttpResponse object. You can access the HttpResponse object using the Response property of the page.

Figure 11-14 shows how to create a cookie named UserName and store the user's name (from the login information) in the cookie.

```
1   HttpCookie userName;
2   userName = new HttpCookie("UserName");
3   userName.Value = HttpContext.Current.Request.LogonUserIdentity.Name.ToString();
4   userName.Expires = DateTime.Now.AddYears(2);
5   Response.Cookies.Add(userName);
```

Figure 11-14: Creating a cookie

Line 1 creates a reference variable of type HttpCookie.

Line 2 creates a cookie named UserName and stores a reference to it in userName.

Line 3 sets the Value property of the cookie to a login user name.

Line 4 sets the expiration date to two years from now.

Line 5 adds the cookie to the cookies collection of the HttpResponse object.

Table 11-2 shows the properties of the HttpCookie class.

Table 11-2: Properties of the HttpCookie class

Property	Description
Name	Name of the cookie
Value	The value (string) assigned to the cookie
Expires	The date/time when the cookie expires
Secure	A Boolean value that indicates whether a secure connection is required to send the cookie
Domain	The domain for which the cookie is valid. The domain may include multiple servers.

Table 11-3 shows the properties/methods of the Cookies collection.

Table 11-3: Properties/methods of HttpCookie collection

Property/method	Description	Example using the cookie, UserName
Count	The number of cookies in the collection	Cookies.Count
Add(cookie)	Adds a cookie to the collection	Response.Cookies.Add(userName)
Remove(cookie name)	Removes the cookie from the HttpRequest/ HttpResponse collection, but doesn't delete the cookie from the client	Response.Cookies. Remove("UserName")
Clear()	Removes all cookies from the HttpRequest/ HttpResponse collection, but doesn't delete the cookies from the client	Request.Cookies.Clear() Response.Cookies.Clear()

To delete a cookie from the client, set the Expires property to zero or a negative value.

11.4.2 Retrieving the Value of a Cookie

You can retrieve the value of a cookie, which is sent with the request by the browser to the server, using the cookies collection in the HttpRequest object. You can refer to this object using the Request property of the page. You specify the name of the cookie to access it from the collection as follows:

```
Request.Cookies["UserName"].Value
```

We will look at two different examples to help you understand how cookies are used.

11.4.3 Example 1: Keep Track of User Visits

The first example uses a cookie to display a message that includes the user's name to welcome back users who have previously visited the website, as shown here:

Welcome back, John!

For now, we will assume that the message is to be displayed only on a specific page, like the Reservation page. The message will be displayed only the first time the page is displayed in a session, and not when the page is revisited during the same session. Thus you need to check whether the user has visited the website in a previous session and whether the user is visiting the page for the first time in the current session.

To check whether a user has visited the site previously, you would store the name of each user in a cookie when the user starts a session, if the cookie doesn't exist. Creating the cookie in the `Session_start` event ensures that the cookie exists when any page on the website is loaded. So, you can use the `Page_Load` event of any page to display the welcome back message if it is the first visit to the page in the current session.

The next two steps show the `Session_Start` and `Page_Load` event handlers.

> ### Step 16: Create a cookie to store the user's name.

Add the code to the `Session_start` event handler to create a cookie named `UserName` and store the user's name in it, as shown in lines 6–21 of Figure 11-15.

```
1  protected void Session_Start(object sender, EventArgs e)
2  {
3      Application.Lock();
4          Application["CurrentUsers"] = (int)(Application["CurrentUsers"]) + 1;
5      Application.UnLock();
6      // Initialize variable to help identify returning visitors
7      Session["IsFirstVisit"] = false;
8      // Create UserName cookie, if it doesn't exist, or it is a different user
9      string currentUser =
10         HttpContext.Current.Request.LogonUserIdentity.Name.ToString();
10     HttpCookie userName;
11     if ((Request.Cookies["UserName"] == null) ||
       (Request.Cookies["UserName"].Value != currentUser))
12     {
13         userName = new HttpCookie("UserName");
14         userName.Value =
15             HttpContext.Current.Request.LogonUserIdentity.Name.ToString();
16         userName.Expires = DateTime.Now.AddYears(50);
17         Response.Cookies.Add(userName);
18         Session["IsFirstVisit"] = true;
19     }
20     // Initialize variable to identify first visit to a page in a session
21     Session["IsSessionOpener"] = true;
22 }
```

Figure 11-15: Storing a user name in a cookie

The code in lines 10–17 that creates the cookie is the same as that presented earlier in Figure 11-14, except that line 11 uses an `if` statement to ensure that the cookie is created only if it doesn't exist or the current user's name is different from that in the cookie. To simplify the code, it is assumed that if two different users access the same website from the same client and browser, only the last user's name will be stored in the cookie. So, any later visit by the first user is treated as his first visit, and hence he won't get a welcome back message for that visit.

Line 7 sets the session variable `IsFirstVisit` to `false` to help display a welcome back message only to returning visitors to the website, and line 18 sets it to `true` if it is the first visit, as indicated by a null value for the cookie.

Line 21 sets the session variable `IsSessionOpener` to *true* to help identify the first visit to a page in a session. The welcome back message is to be displayed only the first time a page is displayed in a session.

Next, you will add the code in the Page_Load event handler of the Reservation page to display the welcome back message the first time the user visits the Reservation page during a session.

Step 17: Display the welcome back message on the Reservation page if it is visited for the first time in a session.

Add a label with the ID lblWelcome immediately below the h1 element at the top of the Reservation-5 page.

Add the code from lines 5–9 in Figure 11-16 to the Page_Load event handler of Reservation5.

```
1    protected void Page_Load(object sender, EventArgs e)
2    {
3        if (!IsPostBack)       // If it is not a postback of the page,
4        {
5            if (((bool)(Session["IsFirstVisit"]) == false) &&
                 (bool)Session["IsSessionOpener"] == true)
6            {
7                lblWelcome.Text = "Welcome back, " +
                                    Request.Cookies["UserName"].Value + "!";
8                Session["IsSessionOpener"] = false;
9            }
```

Figure 11-16: Using a cookie to post a welcome back message

The if statement in line 5 ensures that the message is displayed only if it is not the first visit by the user and that it is the first time the page is displayed in the current session.

Line 8 sets IsSessionOpener to false so the message is not displayed in subsequent visits to the page.

Step 18: Test the code.

If the web project is running, click the **Stop** button.

Click the **Run** button to display the Reservation-5 form. The form should not display the message because IsFirstVisit would have the value true.

Click the **Stop** button and display the Reservation page again. It should display the welcome back message because IsFirstVisit would be false and IsSessionOpener would be true.

Click the **Home** link in the navigation bar to display it, and go back to the Reservation page by clicking the Reservation link. It should not display the message because IsSessionOpener was set to false the first time the page was loaded.

Displaying the Message on Any Page

To display the message on any one of multiple pages, you would copy the code from lines 5–9 in Figure 11-16 to the Page_Load event handler of those pages. In this case, the message will be displayed only on the first page displayed from the group in a session. You may test it by copying the code to the Default.aspx code. In the next chapter, you will see how you can add common code and content to a Master Page without having to copy it between pages.

11.4.4 Example 2: Keeping Track of User Activities

Another typical application of cookies is keeping track of user activities. To help understand such applications, you will use a cookie to display the previous performance reserved by a user, when she comes back to the site.

When the user confirms the reservations on the confirmation page, get the PerformanceId of the last reservation from the List, MyReservations, and store it in a cookie. When the user visits the Reservation page the next time, get the PerformanceId from the cookie and select the corresponding performance in the DropDownList.

Creating the Cookie

Figure 11-17 shows the modified Click event handler for the Confirm button on the Confirmation page, which includes the code to create the cookie (lines 7–20).

```
1   protected void btnConfirm_Click(object sender, EventArgs e)
2   {
3       lblConfirm.Text = "Your reservation is confirmed!";
4       Response.Write("<script> alert('Your reservation is confirmed') </script>");
5       btnRemove.Enabled = false;
6
7       //Store the PerformanceId from the last of the current reservations in a cookie
8       myReservations = (List<Reservation>)Session["MyReservations"];
9       int lastIndex = myReservations.Count-1;
10      string lastPerformId = myReservations[lastIndex].PerformanceId;
11      // Create the cookie
12      HttpCookie performId;
13      if ((Request.Cookies["PerformId"] == null) ||
14             (Request.Cookies["PerformId"].Value != lastPerformId ))
15      {
16          performId = new HttpCookie("PerformId");
17          performId.Value = lastPerformId;
18          performId.Expires = DateTime.Now.AddYears(50);
19          Response.Cookies.Add(performId);
20      }
21  }
```

Figure 11-17: Modified event handler for the Confirm button.

Line 9 gets the index of the last reservation in the List, and line 10 uses the index to get the PerformanceId from the last reservation.

The if statement in line 13 makes sure that the cookie is created only if the cookie doesn't exist or the cookie contains a PerformanceId that is different from the PerformanceId of the last reservation in the current List.

Lines 16–19 create the cookie.

Step 19: Add code to create the cookie.

Add the code from lines 7–20 in Figure 11-17 to the Click event handler of btnConfirm on the Confirmation-4 page.

Using the Cookie to Select a Performance on the Reservation Page

Keep in mind that cookies should not be used to select a performance in the DropDownList if the user is moving from the Confirmation-4 page or the Events page after selecting a specific performance to edit/reserve it. Thus you will add the code in the else clause of the if statement in the Page_Load event handler of the Reservation-5 page, as shown in lines 6–11 of Figure 11-18.

```
1  else    // If not transferring from Confirmation-4 page for editing,
2  {       // or from Evens page to make a reservation, create session variables
3       Session["BasePrice"] = (decimal)0;
4       Session["ZonePrice"] = (decimal)0;
5       Session["Discount"] = (decimal)0;
6       // Select the  last reservation in the dropdown list
7       if (Request.Cookies["PerformId"] != null)
8       {
9           ddlPerformance.SelectedValue = Request.Cookies["PerformId"].Value.ToString();
10          ComputeBasePrice();
11      }
12 }
```

Figure 11-18: Using the cookie value

Note that line 9 sets the `SelectedValue` of the `DropDownList` to the `PerformanceId` from the cookie. What is displayed in the `DropDownList` would be the `SelectedText`, which would be the name of the performance.

Step 20: Add code to use the cookie.

Add the code from lines 6–11 in Figure 11-18 to the Page_Load event handler of Reservation5.

Step 21: Test the code.

Display Reservation-5 and add two reservations and click the **Confirm** button on the Confirmation-4 page. Note the second performance you reserved. Click the **Stop** button in the browser.

Display Reservation-5 again. The `DropDownList` should show the last (second) performance you reserved.

Review Questions

5. What are two differences between a persistent cookie and a non-persistent cookie?

6. Which of the following two statements would you use to display the user name stored in a cookie named UserName?

```
lblUserName.Text = Request.Cookies["UserName"].Value
```

Or,

```
lblUserName.Text = Response.Cookies["UserName"].Value
```

Exercises

Exercise 11.1

Modify the Reservation-5 page to display the number of times users have accessed the Theater website—that is, the number of user sessions (the number of hits). You may display the count on the Reservation-5 page below the number of current users.

Exercise 11.2

Currently, every time a user selects a performance on the Reservation-5 page, the ComputeBasePrice() method reads the Performance table and finds the corresponding base price. Update the Theater project as follows:

Read the PerformanceId, PerformanceName, and BasePrice from the Performance table when the application starts and store the records in the application state. Use the application state in the ComputeBasePrice() method to find the base price. Use a GridView control to let the administrator update the base price(s), save the changes to the Performance table, and also update the application state.

Exercise 11.3

Use the application state to keep track of the number of times users have added a performance to their shopping cart, as well as how many performances have been confirmed. Display these values on the Admin page.

Exercise 11.4

Add a page where customers can add their name, address, and telephone number. Use one or more cookies to store the information and bring it back the next time the user visits the page.

Exercise 11.5

Create a new website where you use the application state to track and display each new user session on a page of the site. Include the time that the session started and ended, along with a calculation of the duration in minutes of each session.

Exercise 11.6

Use the cache to track the number of user sessions on a site in the past hour.

Exercise 11.7

Create a unique ID for each user session and store it in a cookie. Then use this information to display the length of time for each user's session on the site.

## Chapter 12

Common Site Content and Bootstrap Form Layout

Modern websites have content that is repeated on many pages, such as navigation controls, headers, footers, and more. Having all this content repeated on every page on the site can prove quite difficult to manage. Fortunately, there are several controls and elements available in the ASP.NET framework that make it easier to maintain common site elements.

In this chapter, you will see several of these elements, along with some that make it easier to control the look and feel across an entire site.

Master pages allow you to separate common content and controls into one page, and then attach the content pages of the site to this common page. The master page is flexible enough that you can customize its look and feel for each content page (for example, by having a navigation control placed on the master page showing which content page is currently being displayed).

By using Bootstrap form controls, you gain a common look and feel that is modern and adds additional features, especially with regard to data validation.

Learning Objectives

After studying this chapter, you should be able to:

- Describe how master pages allow for separating common content.
- Create master pages.
- Create content pages.
- Convert existing pages to content pages based on a master page.
- Describe how master pages and content pages interact to create a dynamic experience.
- Create forms with Bootstrap controls.
- Lay out forms using the Bootstrap grid as well as Bootstrap spacing utilities.
- Apply Bootstrap classes to ASP.NET controls.
- Describe the differences between ASP.NET and HTML data validation.

12.1 Creating and Using Master Pages

A master page is a special ASP.NET page that in many ways functions just like the regular .aspx pages you have used thus far. It has the same structure, with an html-based .aspx page that includes the content of the page, a designer page, and a code-behind page. However, the master page only contains those elements that are shared across many pages in your website, with placeholders for page-specific content. A master page can contain any number of placeholders for content, but a typical setup has placeholders for headers, menus, and footers. Your site can have several different master pages used in different parts of your site. For example, you could have one master page for pages for customers and another master page for pages used by internal users. You can also create dynamic content on your master pages to customize the content. For example, you can dynamically indicate in the master page menu what the current page is.

Tutorial: Set Up Master Pages

In this tutorial, you will expand the Theater system to use a master page and several content pages using that master.

Step 1: Open the project for this chapter's tutorial.

Open the Chapter 12 tutorial project. This project is specially prepared for this chapter, so you cannot continue from your own project from the previous chapter. Run the website. The site will launch on the Events page. You can click on the *Reservation* tab and see that the Reservation page is also included in the project. However, the Home and About links in the Navigation bar do not work. Clicking the *Reservation* link while on the Reservation page also will not work. This is because it points to the old version of the Reservation page (Reservation-4.aspx).

When creating this version of the project, we copied pages from the old project, but failed to fix all the links in the navigation bar. We would have had to fix the navigation bar on every single page in the project, since the HTML code for the navigation bar is repeated on every page. Using master pages, we can put the navigation bar in a single place and avoid the inconsistencies that exist now.

Stop the site from running and return to Visual Studio.

Step 2: Add a master page.

Right-click the project name in Solution Explorer and select *Add > New Item...*. Select *Web > Web Forms > Web Forms Master Page*. Name the form *Main.Master*. Since a project can contain multiple master pages, you should make sure you give your master pages appropriate names. Figure 12-1 shows the default code that is generated when you create a master page.

```
1   <%@ Master Language="C#" AutoEventWireup="true" CodeBehind="Main.master.cs"
    Inherits="Ch11_MasterPages.Main" %>
2
3   <!DOCTYPE html>
4
5   <html>
6   <head runat="server">
7       <title></title>
8       <asp:ContentPlaceHolder ID="head" runat="server">
9       </asp:ContentPlaceHolder>
10  </head>
11  <body>
12      <form id="form1" runat="server">
13          <div>
14              <asp:ContentPlaceHolder ID="ContentPlaceHolder1" runat="server">
15              </asp:ContentPlaceHolder>
16          </div>
17      </form>
18  </body>
19  </html>
```

Figure 12-1: Default code for a master page

You will notice that this is very similar to a regular web form, but notice a few important details. In line 1, the keyword `Master` is used to indicate that this is a master page. Regular pages have the keyword `Page` instead. In lines 8 and 9, as well as in lines 14 and 15, there are two content placeholders. These are the places where content pages will insert specific content. By default, there is a content placeholder in the head of the HTML document and one in the `form` portion of the page body. However, you can add additional placeholders as needed. For example, if you need every page to have a two-column layout, you can set up the layout in the master page with a placeholder in each column for the page-specific content.

Next, you will add the content that is common across all pages to the master page. You can refer to Figure 12-2 for the code to add. The following tutorial steps will present the various parts of the code you'll be adding.

Step 3: Reference Bootstrap and the stylesheet.

First, since the page title is going to be unique to each page, you can comment out line 7.

Since we will be using Bootstrap and Font Awesome, add the standard Bootstrap references in lines 8–11 and JavaScript plugins in lines 49–51. To get the latest version of the Bootstrap framework, you can go to www.bootstrapcdn.com and copy the link elements. If you choose the JavaScript Bundle, you can use that in place of lines 50 and 51.

You will get warning messages that `integrity` and `crossorigin` aren't valid HTML elements. This is a Visual Studio problem and can be ignored. The two elements allow browsers to check if the externally loaded frameworks are valid and the code hasn't been modified by a hacker.

The jQuery link in line 49 can be retrieved at code.jquery.com. Choose the slim minified version.

The links are also available in the bootstrap-links.txt file in the Chapter 12 starter folder.

In addition, the project has a stylesheet that you will want to reference, as shown in line 12.

Step 4: Add a common header and footer.

To ensure the header looks the same across all pages, add lines 19–23 just inside the `body` element. The `w-100` Bootstrap class gives the image a 100% width. You can also specify widths of 25%, 50%, and 75% using this class.

Next, add the footer shown in lines 44–47.

Step 5: Add a common navigation bar.

To make sure navigation is consistent across all pages, add the navigation bar shown in lines 24–37. This is a standard collapsible navigation bar that sticks to the top of the page when scrolling (see Chapter 3 for more details).

Step 6: Rename content placeholders.

Change the IDs for the two content placeholders to `HeadContentPlaceHolder` and `MainContentPlaceHolder`, as shown in lines 15 and 40. This will make it easier to distinguish between them on the content pages.

Step 7: Add a content page.

To illustrate how the master page works, add a content page to the project. Right-click the project in the Solution Explorer and select *New Item > Web > Web Forms > Web Form with Master Page*, as shown in Figure 12-3. Name the page *Default*.

```
1   <%@ Master Language="C#" AutoEventWireup="true" CodeBehind="Main.master.cs"
    Inherits="Ch09_Theater_MasterPages.TheGrand" %>
2
3   <!DOCTYPE html>
4
5   <html>
6   <head runat="server">
7       <!-- <title></title> -->
8       <!-- Bootstrap stylesheet-->
9       <link rel="stylesheet"
    href="https://stackpath.bootstrapcdn.com/bootstrap/4.1.3/css/bootstrap.min.css"
    integrity="sha384-MCw98/SFnGE8fJT3GXwEOngsV7Zt27NXFoaoApmYm81iuXoPkFOJwJ8ERdknLPMO"
    crossorigin="anonymous">
10      <!-- Font Awesome icons -->
11      <link rel="stylesheet"
    href="https://use.fontawesome.com/releases/v5.3.1/css/all.css"
    integrity="sha384-mzrmE5qonljUremFsqc01SB46JvROS7bZs3IO2EmfFsd15uHvIt+Y8vEf7N7fWAU"
    crossorigin="anonymous">
12      <link rel="stylesheet" type="text/css" href="stylesheets/styles.css" />
13      <meta charset="utf-8" />
14
15      <asp:ContentPlaceHolder ID="HeadContentPlaceHolder" runat="server">
16      </asp:ContentPlaceHolder>
17  </head>
18  <body>
19      <header>
20          <div class="w-100" style="background-color: darkmagenta;">
21              <img src="img/Grand_Logo.png" alt="The Grand Oshkosh Logo" />
22          </div>
23      </header>
24      <nav class="navbar navbar-expand-md bg-light sticky-top">
25          <!-- Toggler/collapsible Button -->
26          <button class="navbar-toggler" type="button"
27              data-toggle="collapse" data-target="#collapsibleNavbar">
28              <span class="fas fa-bars"></span>
29          </button>
30
31          <div class="navbar-nav nav-tabs collapse navbar-collapse"
                         id="collapsibleNavbar">
32              <a class="nav-item nav-link" href="Default.aspx">Home</a>
33              <a class="nav-item nav-link" href="About.aspx">About</a>
34              <a class="nav-item nav-link" href="Reservation-5.aspx">Reservations</a>
35              <a class="nav-item nav-link" href="Events.aspx">Events</a>
36          </div>
37      </nav>
38      <form id="form1" runat="server">
39          <div>
40              <asp:ContentPlaceHolder ID="MainContentPlaceHolder" runat="server">
41              </asp:ContentPlaceHolder>
42          </div>
43      </form>
44      <footer>
45          Copyright © 2018 The Grand Oshkosh. All Rights Reserved.
46          Site development and hosting by interGen web solutions.
47      </footer>
48      <!-- JavaScripts needed for some Bootstrap functionality -->
49      <script
    src="https://code.jquery.com/jquery-3.3.1.slim.min.js"
    integrity="sha384-q8i/X+965DzO0rT7abK41JStQIAqVgRVzpbzo5smXKp4YfRvH+8abtTE1Pi6jizo"
    crossorigin="anonymous"></script>
50      <script
    src="https://cdnjs.cloudflare.com/ajax/libs/popper.js/1.14.3/umd/popper.min.js"
    integrity="sha384-ZMP7rVo3mIykV+2+9J3UJ46jBk0WLaUAdn689aCwoqbBJiSnjAK/l8WvCWPIPm49"
    crossorigin="anonymous"></script>
51      <script
    src="https://stackpath.bootstrapcdn.com/bootstrap/4.1.3/js/bootstrap.min.js"
    integrity="sha384-ChfqqxuZUCnJSK3+MXmPNIyE6ZbWh2IMqE241rYiqJxyMiZ6OW/JmZQ5stwEULTy"
    crossorigin="anonymous"></script>
52  </body>
53  </html>
```

Figure 12-2: Master page with a header, navigation, and a footer.

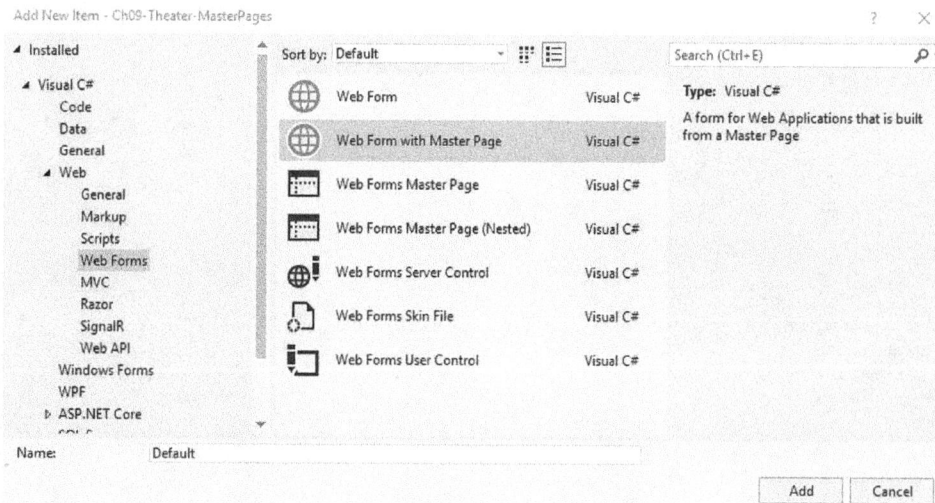

Figure 12-3: Adding a content page to the project

On the next page, you choose which master page you want the content page to use. Since there is only one in this project, the choice is simple: select **Main.Master** and click **OK**.

This will generate the default code shown in Figure 12-4. Notice the reference to `MasterPageFile` in line 1, which points to `Main.Master`. The two `asp:Content` elements allow you to place content that is specific to this page.

```
1  <%@ Page Title="" Language="C#" MasterPageFile="~/Main.Master" AutoEventWireup=
   "true" CodeBehind="Default.aspx.cs" Inherits="Ch09_Theater_MasterPages.Default" %>
2  <asp:Content ID="Content1" ContentPlaceHolderID="HeadContentPlaceHolder"
   runat="server">
3  </asp:Content>
4  <asp:Content ID="Content2" ContentPlaceHolderID="MainContentPlaceHolder"
   runat="server">
5  </asp:Content>
```

Figure 12-4: Default code for the content page

Set the title for the home page by adding this line to the `HeadContentPlaceHolder` between lines 2 and 3:

```
<title>The Grand Oshkosh Home</title>
```

You can accomplish the same thing by adding the title to the `Title` attribute in the first line of the file.

Next, add the following content to `HeadContentPlaceHolder` between lines 4 and 5:

```
<h1>Where Arts and Entertainment Come Together</h1>
<p>Join us in supporting the arts in Oshkosh today.</p>
```

Step 8: Update the CSS styles.

To make it easier to distinguish between the master and content pages, you will need to update the CSS in styles.css. Figure 12-5 shows the code you should add.

```
1   footer {
2       border: 2px solid black;
3       padding: 10px;
4       margin: 3px;
5   }
6
7   #MainContentStyle {
8       border: 2px solid red;
9       padding: 10px;
10      margin: 3px;
11  }
```

Figure 12-5: Changes to the CSS

The footer is styled with a black border (lines 1-5), and the last rule in lines 7–11 specifies a red border to be used around the main content placeholder.

Step 9: Add a style around the main content placeholder.

The MainContentStyle rule can't be applied directly to the content placeholder element on the master page. Instead, you will need to target the div surrounding it. Add an ID of MainContentStyle to the div in line 13 of Figure 12-1 (this is in Main.Master).

You can now set Default as the start page and run the site. It should look like Figure 12-6. Notice the red and black boxes surrounding the main content and the footer. For more insight into what's going on, you can take a look at Figure 12-7, which shows the main content placeholder "popped out" from the master page content.

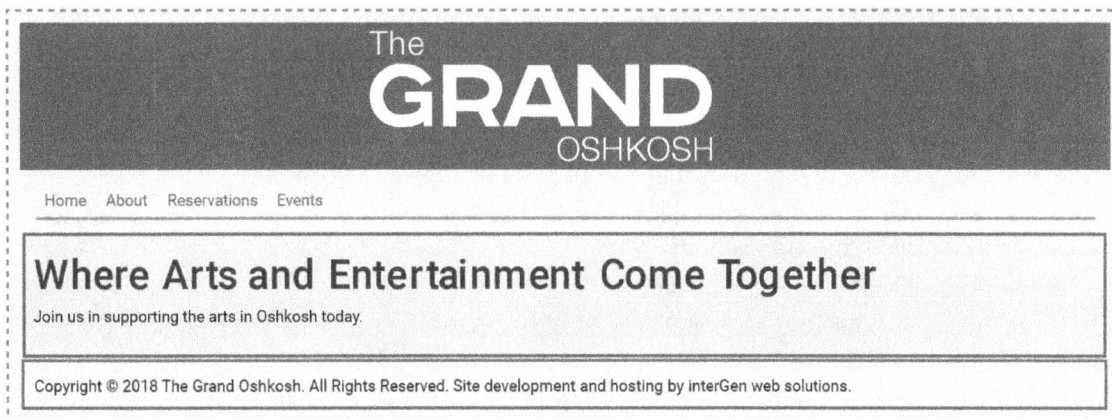

Figure 12-6: The web page with the master and content page.

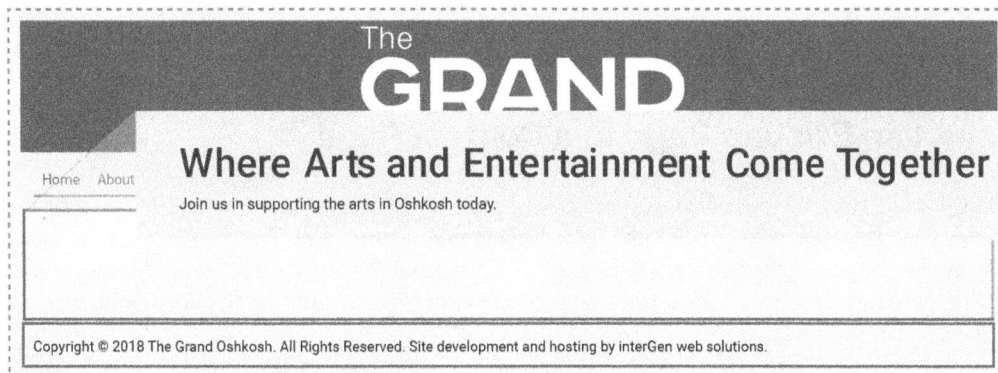

Figure 12-7: The web page with the content section "popped out"

It's time to practice! Add the About page as a content page and add content to make it look like Figure 12-8. Be sure to also adjust the navigation bar to have the About menu point to your new page. The image is in the *img* folder and is named *InsideGrand.jpg*.

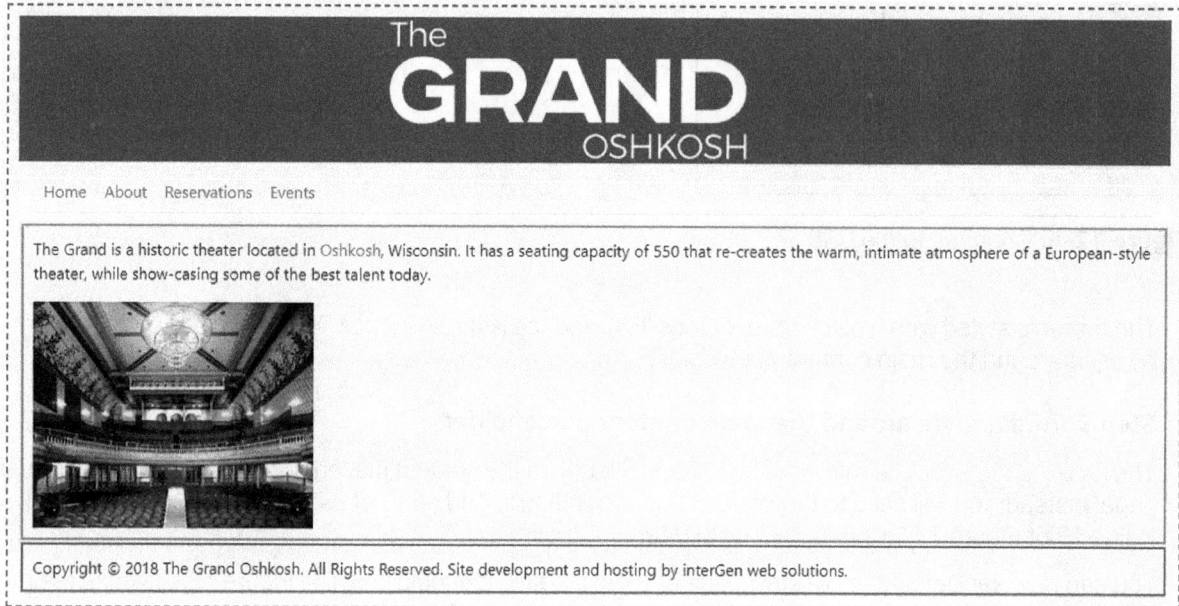

Figure 12-8: The About page

12.2 Converting Existing Pages to Content Pages

In many projects, you will have a number of existing pages that may need to be converted to a master page/content page project. This is relatively simple to do. The basic steps are as follows:

1. Add a reference to the master page in the first line of the content page.
2. Remove the content from line 2 through the end of the head element. You will also need to remove the closing `body` and `html` elements at the end of the file.
3. Add a content placeholder for the head element.
4. Remove any content inside the body element that is already on the master page. This will typically include a header and navigation elements.
5. Add content placeholder elements to match those that are on the master page. Be sure to also remove any `form` elements, as a page can only have one form and the master already has one.
6. Move existing content into the content placeholder elements.

In the following tutorial section, you will get a chance to practice converting an existing page. You should notice that the code on the content page is shorter and simpler because common elements are removed.

Tutorial: Convert an Existing Page to a Content Page

In the steps that follow, you will see how you can convert the Events page to a content page based on the Main master page. Figure 12-9 shows what the page looks like before it is converted to a content page. You'll notice that the menu looks very different from the one you created on the master page. While scanning the site, you would also notice that the links aren't working quite right. Figure 12-10 shows the page after conversion, with the updated menu and red border around the content section.

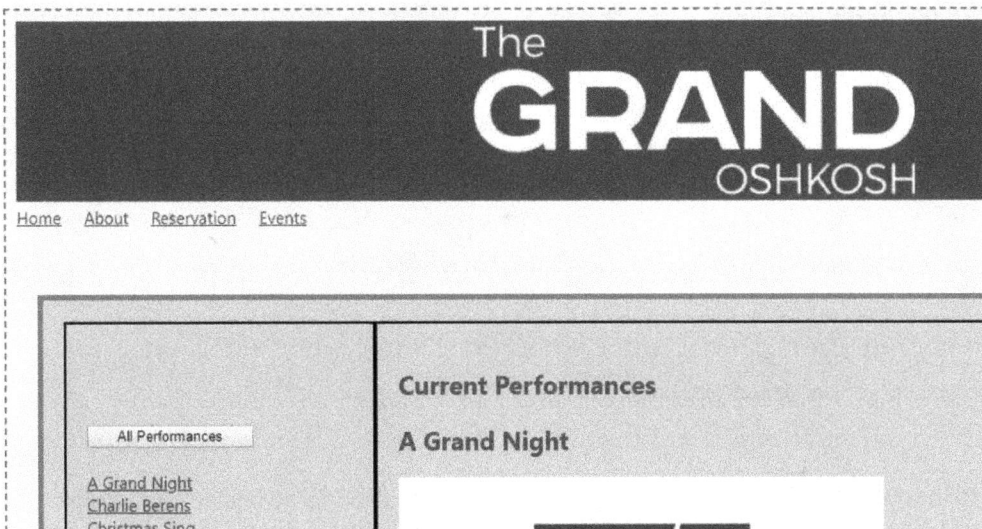

Figure 12-9: The existing Events page before conversion to a content page

Figure 12-10: The Events page after conversion to a content page

Step 10: Add a reference to the master page.

In the very first line in the Events.aspx file, insert the `MasterPageFile` attribute as shown here:

```
<%@ Page Language="C#" MasterPageFile="~/Main.Master" AutoEventWireup="true"
CodeBehind="Events.aspx.cs" Inherits="Ch11_MasterPages.Events" %>
```

Step 11: Change the head element.

Comment out everything between the first line and the end of the head section of the file, and replace it with this code:

```
1    <%--<!DOCTYPE html>
2    <html xmlns="http://www.w3.org/1999/xhtml">
3    <head runat="server">
4        <title></title>
5        <link rel="stylesheet" type="text/css" href="stylesheets/styles.css" />
6    </head>--%>
7
8    <asp:Content ID="Content1" ContentPlaceHolderID="HeadContentPlaceHolder"
     runat="server">
9        <title>Events at The Grand Oshkosh</title>
10   </asp:Content>
```

This adds a title to the page (line 9) but also removes the reference to the CSS stylesheet. Regardless, the stylesheet will still be applied to the page, since it is referenced on the main page. If you have a special stylesheet you need to apply to only some of the pages on the site, you can still reference it inside the content placeholder.

Step 12: Convert the page header.

Next, you need to replace the page-specific menu with the one on the master page. Comment out lines 1–12 as shown in the following code snippet, and add line 13.

```
1    <%--<body>
2        <header>
3            <img src="img/Grand_Logo.png"
             style="background-color: darkmagenta; width: 100%"
             alt="The Grand Oshkosh" />
4            <nav>
5                <a href="default.aspx">Home</a>
6                <a href="about.html">About</a>
7                <a href="Reservation-5.aspx">Reservation</a>
8                <a href="events.aspx">Events</a>
9            </nav>
10       </header>
11       <br /><br />
12       <form id="form1" runat="server">--%>
13   <asp:Content ID="Content2" ContentPlaceHolderID="MainContentPlaceHolder"
     runat="server">
```

Since the header image and the menu are both set up on the master page, you can remove those elements from the content page, as shown above. Make sure you also comment out the form element. A page can only contain one form element, and the master page already has one, so you must remove the one in the content page. Line 13 is the beginning of the content placeholder that will surround all the content on the page. Be sure to use the correct ContentPlaceHolderID.

Scroll to the very bottom of the page and comment out the ending form, body, and html elements, and add the closing asp:Content element:

```
1    <%--</form>
2    </body>
3    </html>--%>
4    </asp:Content>
```

You can now run the page and see that it has changed to conform to the master page, as shown in Figure 12-10.

It's time to practice! Use the techniques shown above to convert the Confirmation-4.aspx and Reservation-5.aspx pages to use the master page.

12.3 Interactions between Master and Content Pages

In situations where most but not all of the pages in a site have the same content, you can add default content to the master page and then override that for specific pages. There are a few ways you can add default content. First, you can add content to a placeholder on the master page. This content will exist on all the pages where that placeholder has not been added. To override the content, simply add the placeholder to the page and add page-specific content. Second, if you add a `title` element to the master page, you can override this by using the `Title` attribute in the `Page` directive. Third, you can add a public property to the master page and use that to modify content dynamically from a content page. For example, it is a recommended search engine optimization (SEO) practice that every page has an `h1` element at the top of the page. By adding the element to the master page with some default content and a public property, you can set the value of the element from each content page that needs to override it.

12.3.1 Master Page Public Properties

A master page is at its core just another user control, so we can create public properties that can be used by content pages. This can be useful for overriding default content. For example, imagine you need to insert unique text in the footer of some content pages but not others. You can achieve this by adding an empty label to the footer and then creating a public property in the master page that can be accessed from content pages.

Tutorial: Overriding Default Content

This part of the tutorial shows a few different ways in which content and master pages can interact by creating a footer element that can be customized on different pages.

Step 13: Setting the page title.

To ensure every page has a title, uncomment the `title` element in the `head` section of Main.Master:

```
1    <head runat="server">
2        <title>The Grand Oshkosh</title>
```

If you run the site now, you'll notice that every page has the title that is set in the master page, even though several content pages also include their own `title` element. To override the master title, you will need to move these titles into the `Title` attribute in the `Page` directive in the first line of each content page.

Go ahead and do this for every content page with a `title` element.

Step 14: Add a footer label.

Open Main.Master and add the following line of code as the last line inside the `footer` element:

```
<asp:Label ID="lblContactInfo" runat="server" Text=""></asp:Label>
```

This sets up a regular Label control with empty text.

Step 15: Add a public property.

Open Main.Master.cs and add the following code to create a public property that sets and reads the value of the Label control:

```
1    public string FooterText
2    {
3        get { return lblContactInfo.Text; }
4        set { lblContactInfo.Text = value; }
5    }
```

Step 16: Modify content pages via Virtual Path.

Here you'll modify the label on the About page. Add the following line as the second line in About.aspx:

```
<%@ MasterType VirtualPath="Main.Master" %>
```

Adding this line creates a strongly typed `Master` property in the content page, allowing you to access any of the public properties on the master page. However, it doesn't allow you to access the controls, which is why you needed to add the public Label property above to expose the control.

Step 17: Set the Footer Label text.

Switch to About.aspx.cs and add the following to the Page_Load method:

```
1   protected void Page_Load(object sender, EventArgs e)
2   {
3       Master.FooterText = "Please call us at 920-555-4455";
4   }
```

You can now use the `Master` property to access the public `FooterText` property on the master page and thus control the content of the `lblContactInfo` control on the master page.

Start the site and navigate between the About page and other pages on the site. Notice how the phone number message only appears in the footer of the About page.

12.3.2 Indicate the Active Page in the Navigation Bar

At this point, you have learned how to add a menu to the master page, but it would be nice to be able to indicate the currently active page that the user is viewing, despite the master page and the menu being the same for all content pages. There are several approaches you can take to dynamically indicate the active page. All these approaches rely on applying the `active` Bootstrap class to the currently activated menu item, like this:

```
<a class="nav-item nav-link active" href="About.aspx">About</a>
```

Of course, you can't just include this class in the master page HTML code, as it wouldn't change based on which page is actually being viewed. Instead, the active class will have to be applied dynamically as each page is loaded. This can be done client-side using JavaScript or server-side using C#. In either case, you can set up a loop to go through each of the items in the menu and compare the file name of the page being loaded to the target URL of the menu item, then modify the HTML to apply the `active` class to the element that matches it.

Regardless of which approach you choose, the results will match Figure 12-11, which shows the About page with the tab for that page highlighted in the menu bar. The following two tutorial sections will illustrate how to implement both solutions.

There is no functional difference between the two approaches, so the choice of which one to use is likely a matter of how much you want to depend on client-side JavaScript versus server-side C# in the site as a whole, and which code-base is easier to maintain.

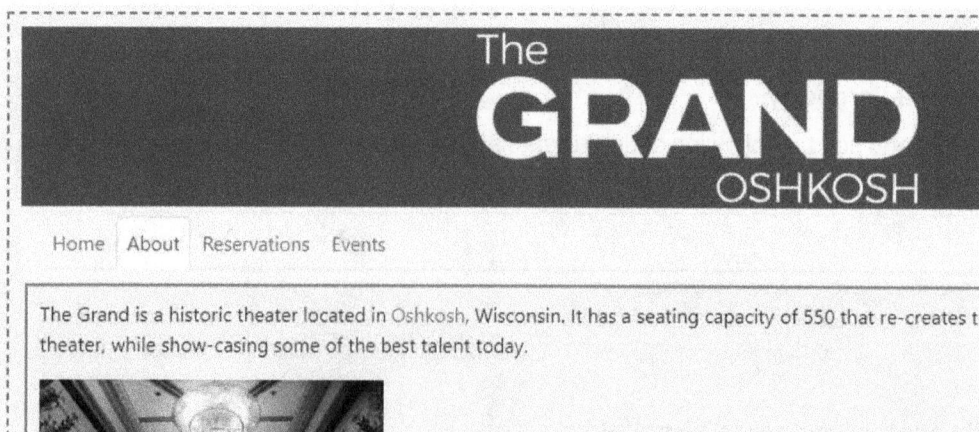

Figure 12-11: Example of the active page being highlighted

Tutorial: Indicate the Active Menu Page Using JavaScript

Say you want to add JavaScript that needs to access the elements of a page and run automatically upon the page's loading. Here the script will need to be included at the bottom of the HTML code, so that all the elements are loaded and available when it runs.

Step 18: Ensure the menu is correct.

This tutorial relies on the HTML for the navigation menu being on a particular form, so you should make sure that the menu in the master page is as follows:

```
1    <div class="navbar-nav nav-tabs collapse navbar-collapse"
2                 id="collapsibleNavbar">
3      <a class="nav-item nav-link" href="Default.aspx">Home</a>
4      <a class="nav-item nav-link" href="About.aspx">About</a>
5      <a class="nav-item nav-link" href="Reservation-5.aspx">Reservations</a>
6      <a class="nav-item nav-link" href="Events.aspx">Events</a>
7    </div>
```

As you can see, the entire menu has an `id` of `collapsibleNavbar`, and each item in the menu is an anchor element with the classes `nav-item` and `nav-link` applied and the `href` attribute pointing to the page that the user will navigate to when clicking the link.

Step 19: Add the JavaScript to the master page

Open the master page and add the following code just before the closing `</body>` element of the master page:

```
1    <script>
2        var currentPage = location.pathname;
3        var menu = document.getElementById("collapsibleNavbar");
4        var items = menu.getElementsByClassName("nav-item");
5        for (var i = 0; i < items.length; i++) {
6            if (items[i].href.endsWith(currentPage)) {
7                items[i].classList.add("active");
8            }
9            else {
10               items[i].className = items[i].className.replace(" active", "");
11           }
12       }
13   </script>
```

This code starts in line 2 by getting the file name of the current page. (We will need this later for comparison to the file names used in the navigation menu HTML code.) Line 3 then sets up a variable that contains the navigation menu, which has the ID `collapsibleNavbar`. This is used in line 4 to extract all the elements inside the menu with the class `nav-item` into a collection. These elements contain links to the actual pages in the navigation menu. With all the relevant elements extracted, the rest of the code is a loop that iterates over those elements. Line 6 checks whether the `href` attribute points to the same place as the current page. If this is the case, then the item being iterated over is the active page and we can add the active class to the item, which is what happens in line 7. If there is no match, then line 10 makes sure that the item doesn't have the `active` class applied by replacing it with an empty string.

When you run the site, the current page being viewed should be highlighted, as shown in Figure 12-11. Notice how the JavaScript manipulates the HTML sent from the server before it is rendered by the browser.

Tutorial: Indicate the Active Menu Page Using C#

This part of the tutorial shows how to achieve the same result using C# on the server. As mentioned earlier, the JavaScript and C# approaches are functionally the same, so choosing between them will likely be a matter of which approach is easier to maintain on your particular site.

Step 20: Comment out the JavaScript.

Comment out the JavaScript you added above to indicate the active page, so it won't interfere with the next part of the tutorial.

Step 21: Modify the HTML to be accessible to code-behind code.

In order for an HTML element to be able to be targeted by the C# code, it needs to have the `runat="server"` attribute applied. Start by modifying the menu to add this to both the div for the entire menu as well as each anchor element, as shown here:

```
1    <div class="navbar-nav nav-tabs collapse navbar-collapse"
2                        id="collapsibleNavbar" runat="server">
3      <a class="nav-item nav-link" href="Default.aspx" runat="server">Home</a>
4      <a class="nav-item nav-link" href="About.aspx" runat="server">About</a>
5      <a class="nav-item nav-link" href="Reservation-5.aspx"
                            runat="server">Reservations</a>
6      <a class="nav-item nav-link" href="Events.aspx" runat="server">Events</a>
7      <a class="nav-item nav-link" href="ContactUs.aspx" runat="server">Contact Us</a>
8    </div>
```

Step 22: Add the C# code.

Open the code-behind page for the master page and add the following two `using` statements:

```
1    using System.IO;
2    using System.Web.UI.HtmlControls;
```

Then add the following code to the `Page_Load` method:

```
1    protected void Page_Load(object sender, EventArgs e)
2    {
3        string pageName = Path.GetFileNameWithoutExtension(Page.AppRelativeVirtualPath);
4        HtmlGenericControl menu = collapsibleNavbar;
5        foreach (HtmlAnchor item in menu.Controls.OfType<HtmlAnchor>())
6        {
7            string cssClass = item.Attributes["class"].ToString();
8            if (item.HRef.Contains(pageName))
9                item.Attributes.Add("class", cssClass + " active");
10           else
11               item.Attributes["class"].Replace(" active", "");
12       }
13   }
```

This code follows the same structure as the JavaScript code you saw above. Line 3 gets the page name of the currently active page. Then we get the file name without its extension. The menu is extracted in line 4 by referencing the id directly as a variable name. (This works because you added `runat="server"`). The loop starting in line 5 casts the items in the menu to `HtmlAnchor`, allowing line 8 to access the `HRef` attribute and check to see if the `HRef` attribute contains the file name without its extension. Lines 9 and 11 manipulate the attributes for the navigation item by adding or removing the `active` class.

Run the page again and make sure the menu bar indicates the active page.

12.4 Bootstrap Forms

In Chapter 3, you explored how Bootstrap can be used to format pages to provide a common look and feel and be responsive to screen size. In Chapter 4, you saw how to create form fields using HTML controls, and in Chapter 6 you learned how to use ASP.NET form controls. In this section, you will see how to combine these three approaches to set up forms and apply Bootstrap classes to ASP.NET form controls for a responsive experience.

12.5 Bootstrap Form Controls

Bootstrap formatting can be applied to many ASP.NET controls, but how to go about setting them up to render as expected will vary. Sometimes you will need to use HTML controls as you saw in Chapter 4, and at other times you can use the regular ASP.NET controls with Bootstrap classes applied. In this chapter, you will see examples of both. However, this chapter will not contain a comprehensive listing, so you may have to do some research on your own for specific situations.

In HTML5, the input element is used to set up many controls. In most cases, you can use an `asp:TextBox` control with the Bootstrap class `form-control` applied, and in other cases, you will need a particular `TextMode` to create a control with a particular input type. However, there are some cases where neither of these approaches work.

Table 12-1 shows how many common controls can be created in both HTML5 and ASP.NET. The table shows you how to create HTML5 input types in ASP.NET. In most cases, this is simply a matter of setting the `Text-Mode` on an `asp:TextBox` to match the input type and then applying the `form-control` Bootstrap class. However, a few instance are different in that they require different controls (`asp:Button`, `asp:FileUpload`, and `asp:ImageButton`).

Table 12-1: Overview of how HTML5 input types can be created using ASP.NET controls

HTML5 input type	ASP.NET control	TextMode	Bootstrap class
button	asp:Button	N/A	btn, btn-default etc.
color	asp:TextBox	Color	form-control
date	asp:TextBox	Date	form-control
datetime-local	asp:TextBox	DateTimeLocal	form-control
email	asp:TextBox	Email	form-control
file	asp:FileUpload	N/A	form-control-file
image	asp:ImageButton	N/A	N/A
month	asp:TextBox	Month	form-control
number	asp:TextBox	Number	form-control
password	asp:TextBox	Password	form-control
range	asp:TextBox	Range	form-control
search	asp:TextBox	Search	form-control
tel	asp:TextBox	Phone	form-control
text	asp:TextBox	SingleLine (or omit)	form-control
time	asp:TextBox	Time	form-control
url	asp:TextBox	Url	form-control
week	asp:TextBox	Week	form-control

Here are a few simple examples of how to set this up. First, a simple textbox:

```
<asp:TextBox ID="txtName" TextMode="SingleLine" runat="server"
    CssClass="form-control"></asp:TextBox>
```

As noted in Table 12-1, the TextMode can be omitted here. Next, here is a date picker:

```
<asp:TextBox ID="startDate" TextMode="Date" runat="server"
        CssClass="form-control"></asp:TextBox>
```

Finally, here is a file upload control:

```
<asp:FileUpload ID="upload" runat="server"
                CssClass="form-control-file"></asp:FileUpload>
```

A few other common controls are approached differently. To set up a dropdown list in HTML, you would use the select element. This corresponds to the asp:DropDownList control. The only thing you need to do is to apply the form-control class.

Radio buttons and checklist controls are relatively tricky within an ASP.NET environment. There are two approaches for this. You can either use the HTML input element with type radio or checkbox, or use the corresponding ASP.NET controls. Using the HTML controls makes applying Bootstrap classes simple, but reading and setting the HTML control values in C# isn't as easy and straightforward as working with the ASP.NET controls. However, the ASP.NET controls don't allow you to properly apply the Bootstrap classes. Fortunately, in practice, the differences between how the HTML and ASP.NET radio button and checkbox controls are rendered are minimal, so you may choose to use the approach that makes the most sense in a given situation. If you need to do your processing on the server in C#, it is likely more convenient to use the ASP.NET controls, but if you can do your processing client-side, the HTML controls may be a better choice.

Tutorial: Contact Us Page

To illustrate how to work with Bootstrap controls in an ASP.NET environment, you will be adding a Contact Us page to the Grand Oshkosh website. This page will illustrate many different controls, as well as how to capture and process data in C#.

Step 23: Add a new page.

Add a new web form with a master page based on the Main.master. Name the new form ContactUs.aspx.

Step 24: Add controls.

Add the controls shown in Figure 12-12 inside the Content2 placeholder. This code is also available in ContactUs.txt in the Chapter 12 starter folder.

```
1   <h1>Contact Us</h1>
2   <p>Please enter your request below and we will be back in touch shortly.</p>
3   <div class="form-group">
4       <label for="txtFirstName">First Name</label>
5       <asp:TextBox ID="txtFirstName" runat="server"
                             CssClass="form-control"></asp:TextBox>
6   </div>
7   <div class="form-group">
8       <label for="txtFirstName">Last Name</label>
9       <asp:TextBox ID="txtLastName" runat="server"
                             CssClass="form-control"></asp:TextBox>
10  </div>
11  <div class="form-group">
12      <label for="txtCity">City</label>
13      <asp:TextBox ID="txtCity" runat="server" CssClass="form-control"></asp:TextBox>
14  </div>
15  <div class="form-group">
16      <label for="txtState">State</label>
17      <asp:TextBox ID="txtState" runat="server"
                             CssClass="form-control"></asp:TextBox>
18  </div>
19  <div class="form-group">
20      <label for="txtEmail">Email Address</label>
21      <asp:TextBox ID="txtEmail" runat="server" CssClass="form-
    control" TextMode="Email"></asp:TextBox>
22  </div>
```

Figure 12-12. HTML code for Bootstrap controls (*continues*)

```
23 <div class="form-group">
24     <label for="txtPhone">Phone Number</label>
25     <asp:TextBox ID="txtPhone" runat="server" CssClass="form-
   control" TextMode="Phone"></asp:TextBox>
26 </div>
27 <div class="form-group">
28     <label for="ddlContact">How would you like to be contacted?</label>
29     <asp:DropDownList CssClass="form-control" ID="ddlContact" runat="server">
30         <asp:ListItem>Email</asp:ListItem>
31         <asp:ListItem>Phone</asp:ListItem>
32         <asp:ListItem>Text</asp:ListItem>
33     </asp:DropDownList>
34 </div>
35 <fieldset>
36     Are you a member?
37     <div class="form-check-inline">
38         <input class="form-check-input" type="radio"
                    name="rblMember" id="rbYes" value="Yes" required>
39         <label class="form-check-label" for="rbYes">Yes</label>
40     </div>
41     <div class="form-check-inline">
42         <input class="form-check-input" type="radio"
                    name="rblMember" id="rbNo" value="No">
43         <label class="form-check-label" for="rbNo">No</label>
44     </div>
45 </fieldset>
46 <div class="form-group">
47     <label for="txtDeadline">When do you need a response by?</label>
48     <asp:TextBox ID="txtDeadline" runat="server" CssClass="form-control"
                     TextMode="Date"></asp:TextBox>
49 </div>
50 <div class="form-group">
51     <label for="txtMessage">Enter your message here</label>
52     <asp:TextBox ID="txtMessage" runat="server" CssClass="form-control"
                     TextMode="MultiLine"></asp:TextBox>
53
54 </div>
55 <br />
56 <asp:Button ID="btnSubmit" runat="server" CssClass="btn-primary" Text="Submit"  />
57 <br />
58 <asp:Label ID="lblConfirmation" runat="server"></asp:Label>
```

Figure 12-12. HTML code for Bootstrap controls (*continued*)

Figure 12-13 shows the output of this form. Notice that all the controls are stacked at 100% width and textboxes have rounded corners and other slight design cues that are different from standard ASP.NET controls.

Most of the controls follow a similar pattern with a `label` followed by a control, with both of those inside a `div` element. For example, lines 3–5 show the code for the First Name field. They are in a `div` with the class `form-group`, which is a typical way to group a label and its corresponding control. The textbox where the user can enter their first name is an `asp:TextBox`, with the class `form-control` applied. The `form-control` class is all you need to have the textbox rendered as a Bootstrap control. The controls for first name, last name, city, and state are all simple text input, so nothing further is needed.

The email field (line 21) is an example of using the `TextMode` property to provide validation. Before the form can be submitted, it will check if the field contains a proper email address.

The deadline field (line 48) uses the `TextMode` of `Date`, which renders it for date entry and adds validation for whether all elements of the date are included and valid.

Line 50 shows a multi-line text field by giving `TextMode` a value of `MultiLine`, which allows the user to drag one corner to make the field smaller or larger.

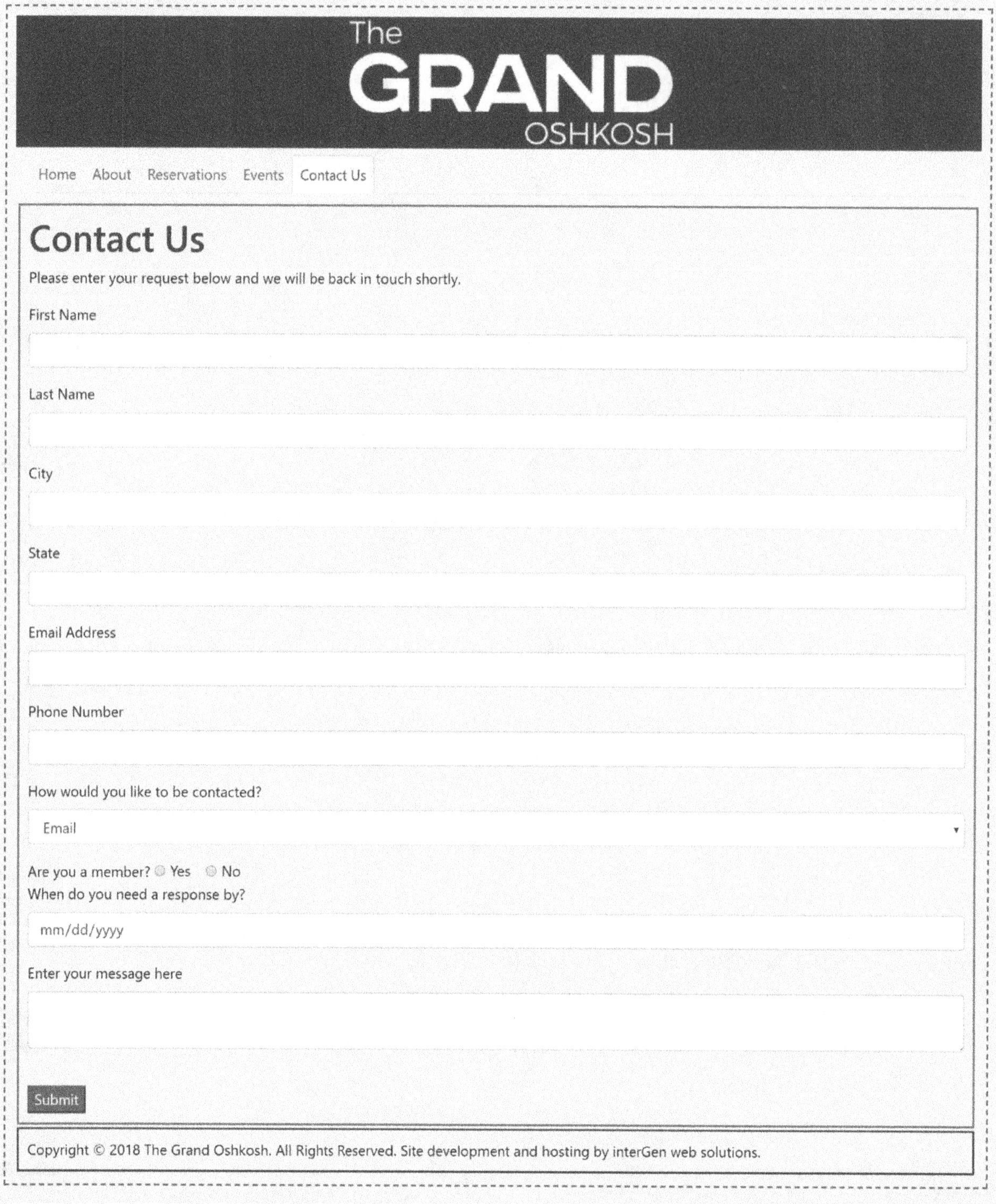

Figure 12-13. Bootstrap form

Lines 29–33 illustrate how you can set up a dropdown list. This is easy to do: simply apply the `form-control` class to an `asp:DropDownList` control.

Two of the more complex controls to set up are checkboxes and radio buttons, as `asp:RadioButtonList` and `asp:CheckboxList` don't render properly by simply applying `form-control`. You will thus have to use HTML controls, which makes the C# code for capturing and setting form values quite different. In this form, you have an example of a set of radio buttons (lines 37–44). Line 37 sets up the list by applying `form-check-inline` to a `div`. This places the radio buttons horizontally. If you wanted vertical placement, you would use `form-check` instead. Each radio button is created with an `input` element of type `radio` (to create checkboxes, you would use type `check` instead) using the class `form-check-input`. Group the radio boxes together by giving each radio button the same `name`, so only one of them can be activated at a time. The radio button also shows the use of the `required` attribute, which sets up validation to ensure the user checks one of the options.

Buttons are created using `asp:Button` with one of the Bootstrap button classes (`btn`, `btn-primary`, `btn-danger`, etc.). Line 56 shows one with `btn-primary` applied.

It's time to practice! Add a Contact Us page to the navigation menu.
In Main.Master, add a new entry in the navigation menu for the new page you just created.
Run the site to display the Contact Us page.

Step 25: Add C# code.

Open ContactUs.aspx in code view and find the `asp:Button` control toward the bottom. Add the `Onclick` attribute and select *Create New Event* in the pop-up menu to generate the event method.

Switch to ContactUs.aspx.cs and enter the code shown in Figure 12-14 in the `btnSubmit_Click` method.

```
1  protected void btnSubmit_Click(object sender, EventArgs e)
2  {
3      string fName = txtFirstName.Text;
4      string lName = txtLastName.Text;
5      string city = txtCity.Text;
6      string state = txtState.Text;
7      string phone = txtPhone.Text;
8      string email = txtEmail.Text;
9      string contactMethod = ddlContact.SelectedValue;
10     string strMember = Request.Form["rblMember"].ToString();
11     Boolean member = (strMember == "Yes");
12     DateTime deadline = DateTime.Parse(txtDeadline.Text);
13     string message = txtMessage.Text;
14     string msg = "Hello " + fName + " " + lName +
                 ",<br> Thank you for your request. We will be sure to have ";
15     msg += "someone contact you before " + deadline.ToShortDateString() +
                  " by " + contactMethod + " at ";
16         switch (contactMethod)
17     {
18         case "Email": msg += email;
19             break;
20         case "Phone": msg += phone;
21             break;
22         case "Text": msg += phone + " (SMS rates may apply)";
23             break;
24     }
25     msg += ".<br>";
26     if (member)
27         msg += "Thank you for your support of our mission.";
28     else
29         msg += "Please consider joining our membership program.";
30     msg += "<br>Here's the message you wrote:<br>" + message;
31     lblConfirmation.Text = msg;
32 }
```

Figure 12-14. Code to capture values from the Bootstrap form

Most of this code is straightforward, as it just grabs values from the textboxes on the form and then constructs a message from the values that is output on the `lblConfirmation` label. A few elements require some additional explanation, however. The selected radio button value is found by selecting the `rblMember` control value from the form (line 10). Remember that this is the name given to each of the radio buttons, only one of which can be selected. The string `strMember` contains the contents of the `value` attribute (either Yes or No in this case). Line 11 then just compares the value to "Yes," which will give us a Boolean true if Yes was selected and false if any other value was selected. Of course, if you had more options in your radio button list, you would have to add more sophisticated logic.

Line 12 shows that you will need to parse the value from the deadline field, as it just contains a string with the date.

The last part of the code, from lines 14 through 30, creates a message to the user from the captured data and merely displays it back on the screen. In a real application, the data would be saved to a database and other processes would have already begun.

12.6 Responsive Layout of Forms

The form you created above has all the controls stacked at 100% width, which may be a good layout on a narrow phone screen, but for wider screens, it makes for an odd look. There are a few ways to manage the placement of the various controls in the form. Due to its responsive nature, you can set up Bootstrap forms to have different layouts on different screen widths (e.g., controls stack on phone screens but some fields appear side by side on larger screens).

The simplest approach to add structure to forms is to use the form-group class. This is primarily used to pair a label with its corresponding control. Since it is a class, you can apply it to both <fieldset> and <div> elements, as well as to other elements that can be used to group elements. Here's a simple example of what this looks like:

```
1  <div class="form-group">
2      <label for="txtFirstName">First Name</label>
3      <asp:TextBox ID="txtFirstName" runat="server"
                   CssClass="form-control"></asp:TextBox>
4  </div>
```

Notice how the for attribute of the label element points to the ID of the TextBox element.

12.6.1 Single Line Forms and Spacing

You can surround form elements by a div with the form-inline class applied to place all those controls in a single row. Figure 12-15 shows an example of how this would work for first and last name controls.

```
1  <div class="form-inline">
2      <div class="form-group">
3          <label for="txtFirstName">First Name</label>
4          <asp:TextBox ID="txtFirstName" runat="server"
                       CssClass="form-control"></asp:TextBox>
5      </div>
6      <div class="form-group">
7          <label for="txtLastName">Last Name</label>
8          <asp:TextBox ID="txtLastName" runat="server"
                       CssClass="form-control"></asp:TextBox>
9      </div>
10 </div>
```

Figure 12-15: Placing form elements in a row

This produces the following output:

First Name [   ]   Last Name [   ]

Step 26: Create controls in line on the form.

Add a `div` element for `form-inline`, as shown in line 1 in Figure 12-15, and the closing `div` in line 10.

This places the controls on a single line but doesn't put any spacing between them. This can be fixed by using the Bootstrap utility classes for adding spacing by controlling either the margin or padding. They are written as follows on the form:

{property}{sides}-{size}

The first of these, *property*, can have two different values: **m** to set margin, and **p** to set padding. The second, *sides*, can have a number of different values:

- **t** — top
- **b** — bottom
- **l** — left
- **r** — right
- **x** — both left and right (horizontal)
- **y** — both top and bottom (vertical)
- blank — all four sides of the element

Finally, *size* determines how much space to add. This is measured as a multiplier of the variable $spacer, which by default is set to 1 rem—the default font size of the html element. While there's a lot of detail to discuss around rem units and spacing, essentially what this means is that 1 rem is the equivalent of the size of one character. You can specify the size as follows:

- **0** — eliminates the spacing by setting it to 0
- **1** — 0.25 rem
- **2** — 0.5 rem
- **3** — 1 rem
- **4** — 1.5 rem
- **5** — 3 rem
- **auto** — sets the margin to auto

For example, `mr-2` sets the margin on the right side of the element to 0.5 rem, whereas `py-4` sets the padding on the top and bottom to 1.5 rem.

We can thus fix the spacing in the first and last name controls above by adding classes to the labels as shown in lines 3 and 7 below. In this case, we add 0.5 rem to the right of each of the labels and 3 rem between the first name textbox and the last name label.

```
1  <div class="form-inline">
2     <div class="form-group">
3        <label class="mr-2" for="txtFirstName">First Name</label>
4        <asp:TextBox ID="txtFirstName" runat="server"
                    CssClass="form-control"></asp:TextBox>
5     </div>
6     <div class="form-group">
7        <label class="mr-2 ml-5" for="txtLastName">Last Name</label>
8        <asp:TextBox ID="txtLastName" runat="server"
                    CssClass="form-control"></asp:TextBox>
9     </div>
10 </div>
```

Figure 12-16: Fixing spacing between form elements

First Name [ ]   Last Name [ ]

Step 27: Fix the spacing in the form.

Add the spacing adjustments shown in lines 3 and 7 in Figure 12-16. Run the page to see the effect of making the adjustments.

12.6.2 Using the Bootstrap Grid to Place Form Elements

Using the Bootstrap grid gives you very fine-grained control over the layout of a form on different screen sizes. With this approach, you can surround each row of controls with a `div` with the class `form-row` applied. Then each control can be given a number of columns to occupy. Figure 12-17 shows an example of how you can use the Bootstrap 12-column grid to set up a form. This code is rendered as shown in Figure 12-18. Notice that this has four rows, as designated by `form-row` in lines 1, 8, 16, and 21. By default, the rows will not have any spacing between them. If you would like to add a little spacing, you can use the spacing classes as shown in lines 1 and 8. This has added some spacing between the first three rows, but not between the last two rows. Here you can see the Email and Phone textboxes are touching each other.

Line 11 shows an example of inserting blank space by adding an empty div with a column width applied to it. This inserts a bit of space between the City textbox and the State label. This is an alternative to adding margins to either of the two controls for separation.

In this example, the columns are specified with breakpoints at small screen sizes, so for extra small screens, the controls will stack at 100% width. In the first row of controls, the two labels occupy two columns and the textboxes occupy four columns.

The class `col-form-label` applied to the labels ensures that the labels are vertically centered with their corresponding textboxes.

```
1   <div class="form-row my-1">
2       <label class="col-form-label col-sm-2" for="txtFirstName">First Name</label>
3       <asp:TextBox ID="txtFirstName" runat="server"
                    CssClass="form-control col-sm-4"></asp:TextBox>
4       <label class="col-form-label col-sm-2" for="txtLastName">Last Name</label>
5       <asp:TextBox ID="txtLastName" runat="server"
                        CssClass="form-control col-sm-4"></asp:TextBox>
6   </div>
7
8   <div class="form-row my-1">
9       <label class="col-sm-2" for="txtCity">City</label>
10      <asp:TextBox ID="txtCity" runat="server"
                        CssClass="form-control col-sm-6 "></asp:TextBox>
11      <div class="col-sm-1"></div>
12      <label class="col-form-label col-sm-1" for="txtState">State</label>
13      <asp:TextBox ID="txtState" runat="server"
                        CssClass="form-control col-sm-2"></asp:TextBox>
14  </div>
15
16  <div class="form-row">
17      <label class="col-form-label col-sm-3" for="txtEmail">Email Address</label>
18      <asp:TextBox ID="txtEmail" runat="server"
                        CssClass="form-control col-sm-9" TextMode="Email"></asp:TextBox>
19  </div>
20
21  <div class="form-row">
22      <label class="col-form-label col-sm-3" for="txtPhone">Phone Number</label>
23      <asp:TextBox ID="txtPhone" runat="server"
                        CssClass="form-control col-sm-9" TextMode="Phone"></asp:TextBox>
24  </div>
```

Figure 12-17: Code for the form using the Bootstrap grid

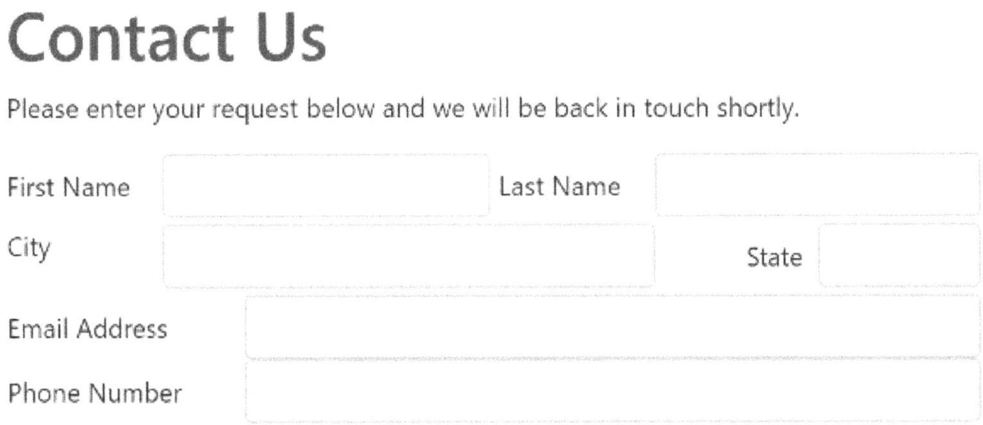

Contact Us

Please enter your request below and we will be back in touch shortly.

First Name   Last Name

City   State

Email Address

Phone Number

Figure 12-18: Example of the form using the Bootstrap grid

Step 28: Adjust the form to use rows.

Make the changes to the controls shown in Figure 12-17. Run the page to make sure it looks as shown in Figure 12-18. You can now try making the window narrower until the controls move back to being placed in a stack. This shows how the page would look when loaded on a mobile phone screen.

12.7 Data Validation: HTML versus ASP.NET

One of the new features added in HTML5 is built-in data validation of many of the input controls. For example, if you specify that a textbox accepts input of type `email`, the user will get an error message and the form can't be submitted if the email address isn't properly entered. Similarly, you can include the attribute `required` to ensure user entry in a particular field.

However, ASP.NET also includes several data validation controls you can add to a page, and offers both server-side and client-side validation. It may not be obvious which to use (server-side or client-side). This section will outline some of the differences between the two approaches to help you arrive at your decision.

12.7.1 Server-Side versus Client-Side Validation

Before we get to the comparison of the two approaches, it will be helpful to understand that validation can be done on both the client and the server. Validation done on the client-side is more responsive to the user, as it doesn't require a trip to the server. This allows for giving user feedback upon leaving a field and avoiding network traffic. However, client-side validation often requires JavaScript, which may be disabled by the user. It is also possible for a malicious actor to bypass the validation and send data to the server that doesn't adhere to the validation. Server-side validation can take two forms:

1. Checking that validation was properly performed on the client
2. Checking specific data values to ensure they adhere to the rules set up

Some validation may not be possible on the client. For example, if validation of an entered value requires comparison to a database value, the database would have to be queried in order for the validation to take place.

12.7.2 HTML5 Validation

HTML5 introduced the validation of form data. This happens entirely on the browser and is fairly simple to set up. Some of the validation is done by simply applying a type to an input element (and TextMode to an ASP.NET textbox control). For example, by using type date, only valid dates can be entered (e.g., February 30th will not be allowed). The following validation options are available:

- *Require a field by specifying the required attribute.* Be sure to also show the user which fields are required—and only require fields that you truly need to collect data from.
- *Data types.* For example, only numbers can be entered with the type `number`.
- *Ranges.* Number fields can be constrained to maximum and minimum values.
- *Length of entry.* Text fields can have the number of characters constrained.
- *Pattern matching.* Regular expressions can be used for more advanced patterns. For example, if a product code always has two letters followed by five digits, you can set up a regular expression to match this pattern

HTML5 validation can prevent a form from being submitted but relies entirely on the browser. Server validation requires separate code to be written on the server, independently of the client validation.

Since the `asp:TextBox` control gets rendered as an HTML input element, you can specify most of the HTML validation attributes in the same way as you would for HTML elements. For example, the following code will create a textbox that allows exactly two characters with a custom error message:

```
1   <asp:TextBox ID="txtState" runat="server" MinLength="2" MaxLength="2"
2       CssClass="form-control"
3       onvalid="this.setCustomValidity('')"
4       oninvalid="this.setCustomValidity('State field must be exactly 2 characters')">
5   </asp:TextBox>
```

12.7.3 ASP.NET Validation

Whereas HTML5 validation is built-in to the input controls, ASP.NET validation (as discussed in Chapter 4) relies on separate validation controls that can provide all the same types of validation as for HTML validation. In addition, ASP.NET includes a `ValidationSummary` control that allows for providing all validation errors on one part of the page.

ASP.NET controls incorporate both client-side and server-side validation. When the page is rendered, it includes JavaScript that will do client-side validation. When a page is submitted, you can check on the server-side if it passed validation by checking the `IsValid` attribute of the `Page` object. It is common to see code on this form in the server-side code:

```
1   if (IsValid)
2   {
3       ...
4   }
```

By checking the `IsValid` property, you can be confident that all the validations were carried out on the client-side and will not need to repeat those checks.

12.7.4 Choosing a Validation Approach

It is vitally important to thoroughly validate data entered by the user to avoid bad user experiences and potential security issues. Since the HTML control validation is so easy to add, you should always include the proper `TextMode` on any `TextBox` you add. However, there are some limitations to the HTML control validation, such as not being able to guarantee that the page has been validated when processing the controls on the server. Table 12-2 compares the pros and cons of each of the two approaches.

Table 12-2: Comparing the HTML and ASP.NET validation approaches

	HTML validation	**ASP.NET validation**
Pro	• Very easy to add to controls	• A full-fledged validation control system • The validation summary collects all error messages in one area • Can check on server-side that the page was validated
Con	• Difficult to control the look and feel of error messages • Cannot handle errors involving multiple controls (e.g., start date having to be before the end date) • Cannot guarantee that the page was validated	• Additional complexity with additional controls on the page

Exercises

Exercise 12.1: Contact Us Page

On the Contact Us page, specify that the phone number field should only allow numbers in the form (999) 999-9999 and that the state field should be exactly two characters.

Change the design of the page using the Bootstrap grid to create a more appropriate form layout on larger screens.

Exercise 12.2: Events Page Controls

Convert the controls on the Events page to use Bootstrap.

Exercise 12.3: Events Page Responsive Design

Convert the design of the content on the Events page to be responsive to smaller screens. On small size screens, the controls in the left column should be at the top of the page, with the events listed below.

Exercise 12.4: Reservation Page Controls

Convert the controls on the Reservation page to use Bootstrap.

Exercise 12.5: Reservation Page Design

Convert the design of the content on the reservation page to be responsive to smaller screens. On small size screens, the controls in the left column should be at the top of the page, with the price summary and seat map listed below.

Exercise 12.6: Confirmation Page

Convert the controls on the Confirmation page to use Bootstrap.

Exercise 12.7: New Master Page

Create a new master page with a different layout; then add a new page using the new master page.

Chapter 13

User Authentication

As users, we increasingly live and work on the web. We expect websites to provide a customized experience and keep the information we create/provide secure from other users. For example, e-commerce sites need to allow users to save a shopping cart and check on the status of previously placed orders. Such a site would also need to allow site administrators to adjust product availability, pricing, and so forth.

In this chapter, you will learn how to set up a system where users can create an account to allow the site to remember what they were working on during their previous visit. You'll also see how you can restrict certain parts of the site to some of the site's users.

Learning Objectives

After studying this chapter, you should be able to

- Describe the need for security on a website.
- Describe the difference between authorization and authentication.
- Describe different methods to implement authorization.
- Implement the ASP.NET Identity system on a website.
- Add login and account management controls to a site.
- Describe the benefits of role-based authentication schemes.
- Control access to parts of a website based on user roles.

13.1 Authentication and Authorization

If your website needs to restrict access and remember its users, you can use *authentication* to have users prove that they are who they claim to be. This is typically done by creating an account for them and then having them log in when they visit your site. Once a user has been authenticated, *authorization* allows you to specify rules for which resources the user is allowed to access. For example, you can use authorization to restrict access to a support page on your website to your existing customers or to the administrator interface to employees.

When authenticating a user, several approaches are available. The most common one involves a username and password. In this scheme, we assume that the valid user is the only one able to present the correct combination of the username and password. However, this means the password has to be complex enough that it is difficult to guess and long enough that a computer is unable to crack it by attempting every possible combination in what is called a *brute-force attack*. If a website forces users to have very long and complex passwords, chances are that the users will write their passwords down, which reduces security because the piece of paper with the password might be found. Users might also use the same username and password combination on multiple sites, which is also problematic because if one site is compromised, all of a user's passwords are at risk. The best advice for users then is to use a password manager, which can securely store all passwords and fill them in for the user as needed.

Some websites, like banking and many workplace sites, need far more security than can be provided with usernames and passwords. They instead turn to multi-factor authentication (MFA). In this scheme, a user is asked to present more than one set of credentials to prove their identity. In MFA systems, the factors are chosen from three separate categories:

1. What you know—typically usernames and passwords
2. What you are—typically some form of biometrics
3. What you have—typically some kind of token or a message sent to a smartphone

With a combination of at least two of these factors, it becomes much more difficult for a would-be attacker to impersonate a user. Almost all web systems rely on passwords but may combine them with one of the other categories for additional security. It is beyond the scope of this book to cover more than username/password authentication, but it is important to be aware of how this fits into the larger picture of authentication.

ASP.NET applications can use three different types of authentication mechanisms, as described in Table 13-1. The rest of this chapter will focus on Individual User Accounts. With this approach, as a developer, you are responsible for adding controls to the site that allow users to register for an account, log in, log out, and manage their account. You have considerable flexibility in how to do this. With the Individual User Accounts approach, your application will store usernames and encrypted passwords, which poses an added complexity and potential security risk compared to the other two approaches. With Windows-based authentication, credentials are typically stored in an active directory, and with third-party authentication, the credentials and account management are controlled by the provider.

Table 13-1: Authentication mechanisms supported by ASP.NET

Windows-Based Authentication	Individual User Accounts	Third-Party Authentication
• Based on Windows user accounts/active directory • Appropriate for internal applications/intranets	• Developer creates code and pages to allow users to create accounts • Application determines how to authenticate users • Username and encrypted passwords stored by application • Appropriate for public and internal applications	• Accounts are managed through third-party providers like OpenID, Google, Facebook, and others • User credentials are not stored by application • Appropriate for public and internal applications

To authorize users to access certain pages, you will add XML code to Web.config files that contain rules for which users are allowed to access certain files and directories. These rules are flexible and fairly easy to write. You will see them covered in more detail later in the chapter. For now, we will describe how the browser and server interact when the user requests a protected page that requires user authentication.

Figure 13-1 shows what happens when a browser requests a protected page from a web server. The server will determine if the request includes an authentication cookie, and if not, it will send the browser a redirect message to go to the login page. Once the browser has requested and the server has sent the login page, the browser will display the page, allowing the user to enter a username and password. Those credentials are then sent to the server for evaluation. If they are deemed to be valid, the server will generate an authentication cookie, which will be sent to the browser along with a redirect to the originally requested page. On every subsequent request to the server, the browser will include the authentication cookie, so the server knows it can send protected pages on future requests.

As a developer, you don't have to worry about most of this process, as it happens automatically within the ASP.NET framework. However, there are several configuration settings you can modify. You will have to specify the file name and location of the login page. You can also control the timeout parameters of the authentication cookie—that is, how long a user can stay logged in before having to re-authenticate.

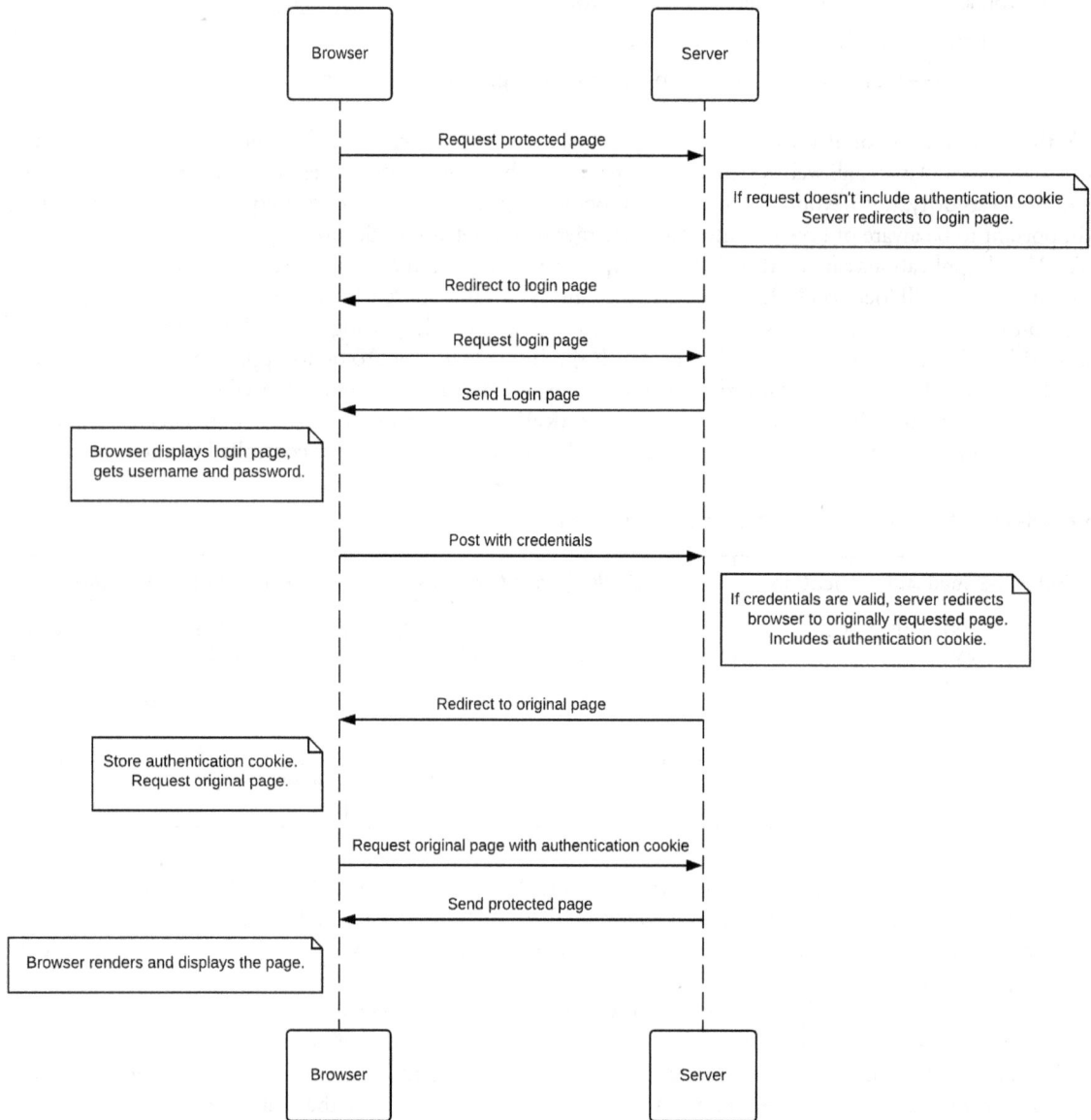

Figure 13-1: HTTP requests and responses when requesting a protected page using Individual User Accounts

Review Questions

1. Describe the difference between *authentication* and *authorization*.
2. With Windows-based authentication, how are accounts managed?
3. Which type(s) of authentication is appropriate for a public-facing e-commerce website?
4. Which file contains authorization rules?
5. If a user isn't logged in, what happens if they try to request a page that requires them to be logged in?

13.2 ASP.NET Identity System

Microsoft introduced the Identity system in 2013 as a replacement for the previous Membership system. The Identity system can be used with all the ASP.NET frameworks, including Webforms and MVC, and is quite flexible. It is based on the open-source *Open Web Interface for .Net (OWIN)* middleware and is distributed as a NuGet package, making it easy to update. Table 13-2 shows some of the key objects used to manage the Identity system.

Table 13-2: Key objects in the Identity system

Object	Description
OwinContext	Returns a context that allows for interacting with the OWIN middleware layer. A context is an object that allows for interacting with the Identity data stored in the database.
IdentityDbContext	The base class for the Entity Framework database context used for Identity. Allows for getting access to data stored about users and roles.
IdentityUser	Represents a user in the Identity system
UserStore	Provides access to the users stored in the IdentityDBContext
UserManager	Allows for managing a single user
IdentityRole	Represents a role in the Identity system
RoleManager	Allows for managing roles
IdentityUserRole	Represents the link between a user and a role, allowing users to be assigned to particular roles
SignInManager	Manages logins for users
IdentityResult	Represents the result of an identity operation

Tutorial: Creating a Website That Authenticates Users

In this tutorial, you will see how to create a website that uses Individual User Accounts to implement user authentication. You will also get an opportunity to explore the various controls and objects used to implement the solution.

Step 1: Create a project.

In Visual Studio 2019, create a new project. Select the *ASP.NET Web Application (.Net Framework)* template. Name the project *Theater*; then choose *Web Forms* and click the *Change* link to choose Individual User Accounts (see Figure 13-2).

Step 2: Run the project.

Launch the project by clicking the *green triangle* at the top of Visual Studio.

Once the site launches in the browser, you'll notice that the navigation bar at the top of the screen includes links to Register and Login, as shown in the top part of Figure 13-3. Click each of these in turn to see the pages with forms for each function.

Step 3: Register a user.

Go to the Register page and register a user. It may take a few minutes for the registration process to finish, since the database to store the user information is created when the first user is registered. If you get an error message that the operation timed out, use the browser back button and try again.

Once the user is created, you will see that you are logged into the site, as the email address you used for registration is shown in the upper-right-hand corner of the page (where the login link was), as shown in the bottom part of Figure 13-3.

Create a new ASP.NET Web Application

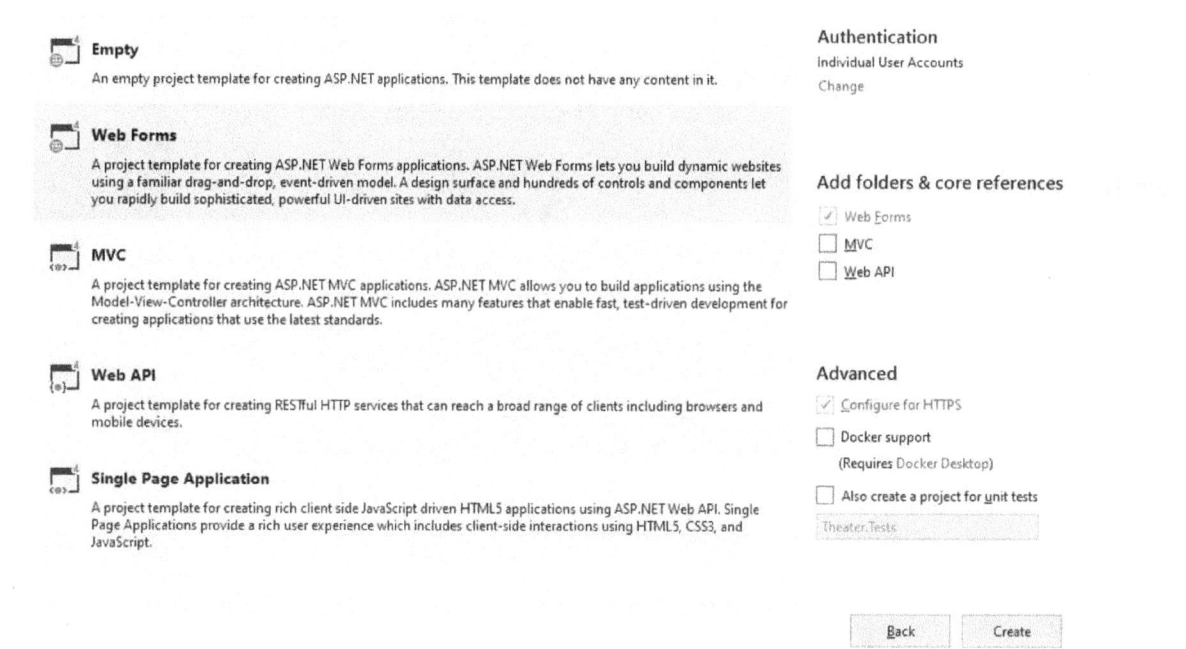

Empty
An empty project template for creating ASP.NET applications. This template does not have any content in it.

Web Forms
A project template for creating ASP.NET Web Forms applications. ASP.NET Web Forms lets you build dynamic websites using a familiar drag-and-drop, event-driven model. A design surface and hundreds of controls and components let you rapidly build sophisticated, powerful UI-driven sites with data access.

MVC
A project template for creating ASP.NET MVC applications. ASP.NET MVC allows you to build applications using the Model-View-Controller architecture. ASP.NET MVC includes many features that enable fast, test-driven development for creating applications that use the latest standards.

Web API
A project template for creating RESTful HTTP services that can reach a broad range of clients including browsers and mobile devices.

Single Page Application
A project template for creating rich client side JavaScript driven HTML5 applications using ASP.NET Web API. Single Page Applications provide a rich user experience which includes client-side interactions using HTML5, CSS3, and JavaScript.

Authentication
Individual User Accounts
Change

Add folders & core references
☑ Web Forms
☐ MVC
☐ Web API

Advanced
☑ Configure for HTTPS
☐ Docker support
    (Requires Docker Desktop)
☐ Also create a project for unit tests
    Theater.Tests

Back   Create

Figure 13-2: Creating a default project that uses Individual User Accounts for authentication

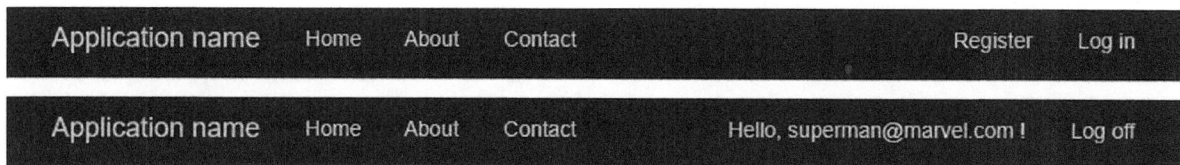

Figure 13-3: Navigation bar shown with an anonymous and logged-in user.

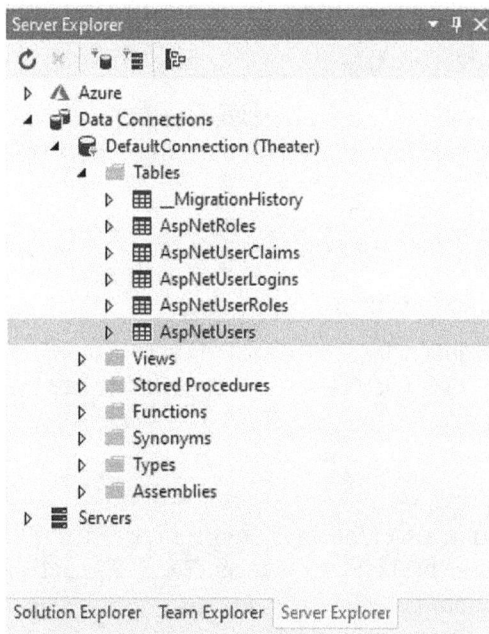

Figure 13-4: Server Explorer with generated ASP.NET Identity database

Log off and log back in to ensure that the user registration system is working.

Step 4: Examine the user database.

Return to Visual Studio and stop the application if you were debugging. Then open *Server Explorer* (should be a tab next to Solution Explorer). Expand *DefaultConnection (Theater)* and then *Tables*. You should now see six tables, as shown in Figure 13-4. If you don't see the data connection, you can try clicking the refresh button on the upper left. You can also look for the database in the SQL Server Object Explorer.

Right-click the *AspNetUsers* table and select *Show Table Data*. This will show all the users that have registered on the site (probably just one in your case). The two most important fields in the table are UserName and PasswordHash. The UserName field contains the username (email address by default) of the registered user, and the PasswordHash is the password after it has been run through a one-way hash function.

By hashing the password, the Identity system avoids the need to store the password in readable plain text. The way the login system works is that when the user enters their password, it will be run through the same hash function as when the user registered. The result will then be compared to the stored hash value, and if the two match, then the original password is deemed to be the same as what the user used to register. If a hacker were to gain access to the AspNetUser database, the passwords would contain relatively little value, as they cannot be entered into a password field to gain access to the system.

Step 5: Configure the password settings.

You can modify the behavior of the Identity system by modifying IdentityConfig.cs, which is in the App_Start folder in Solution Explorer. The `Create` method in this file allows you to change settings like the length of the password and whether numbers and special characters are required. Figure 13-5 shows the relevant code. Change some of the parameters and experiment with creating new users with the revised settings.

```
1   // Configure validation logic for passwords
2   manager.PasswordValidator = new PasswordValidator
3   {
4       RequiredLength = 6,
5       RequireNonLetterOrDigit = true,
6       RequireDigit = true,
7       RequireLowercase = true,
8       RequireUppercase = true,
9   };
```

Figure 13-5: Code to set password parameters

Step 6: Retrieve the username.

It is often necessary to find the username of the currently logged-in user. The approach you take depends on whether the code you are writing is in a code-behind page or non-page code.

For a code-behind page, you will have access to the `User` class directly and can access the username. To illustrate, add an `h2` element just below the `h1` element on the Default page. Then add an `asp:Label` with an ID of `lblUser`. Switch to the code-behind and modify the `Page_Load` method with the following code:

```
1   protected void Page_Load(object sender, EventArgs e)
2   {
3       if (User.Identity.IsAuthenticated)
4           lblUser.Text = User.Identity.Name;
5       else
6           lblUser.Visible = false;
7   }
```

This shows two different ways of using the `User` class. First, in line 3, check if the current user is authenticated. If this is false, the current session is not associated with a logged-in user, and in this case, you hide the label (line 6). Otherwise, in line 4, the username is added as the `Text` property of the label.

To retrieve the username in non-page code, use `HttpContext.Current.User`.

Run the page and log in as a user you have registered on the site. Then navigate to the Default page. You should see the username of the logged-in user.

13.3 Modifying the User Profile

The default user profile stores very limited data about the user—only an email address. In most applications, it is necessary to store more data about the user, such as name and address. There are two ways to expand the data that can be stored about the user:

1. Create a separate table for application-specific user data and use the UserName as a foreign key to link the two tables together.
2. Expand the AspNetUser table with additional fields.

The first approach may be desirable if you already have an extensive database created and the website is just being added to a bigger system. The latter approach should be favored if there is no existing database or if the website will be the primary way that the system will function. We will address the latter approach in the following steps.

The Identity system is based on Code First Entity Framework, so that is the approach we will use to expand the functionality of the database. It is beyond the scope of this book to fully cover Entity Framework, so we will only show enough so that you will be able to expand the database.

Step 7: Enable code-first migration.

Go to *Tools > Nuget Package Manager > Package Manager Console*.

Type the `Enable-Migrations` command on the console command line and press *Enter*.

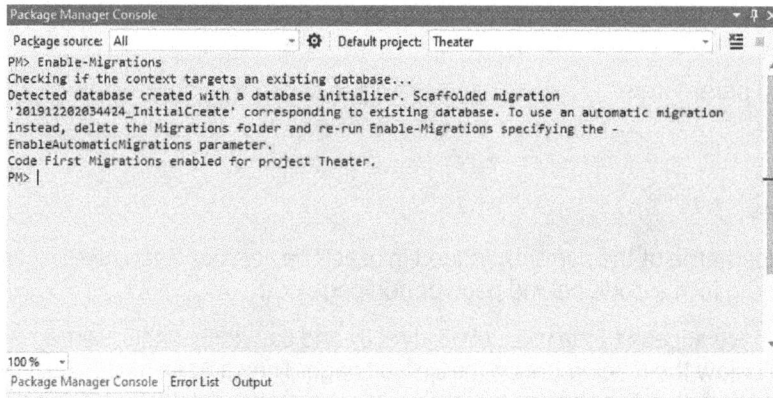

This will create a folder in the project called *Migrations*. Inside that folder, you will find two files: Configuration.cs and a file with a lot of numbers followed by _InitialCreate.cs.

Configuration.cs allows for seeding the database with initial data when launching the application. This can be useful for testing an application without having to type in test data. The second file contains code that creates all the tables and fields in the database. Entity Framework allows for updating the database with new fields without losing data, as well as for migrating the database back and forth between different versions. We are only going to use the feature of updating a database with new fields.

Step 8: Add new fields.

The User profile is set up in the file IdentiyModels.cs in the Models folder. Inside this file is a class called ApplicationUser, which is inherited from IdentityUser. You can modify ApplicationUser with additional fields to store in the database. Add the following three fields to ApplicationUser:

```
1   public string FirstName { get; set; }
2   public string LastName { get; set; }
3   public string City { get; set; }
```

In the Package Manager Console, execute the following command:

```
Add-Migration NameCity
```

This will create a new file in the Migrations folder that contains instructions for how to modify the database with the three new fields. To update the database, run the following command:

```
Update-Database
```

Right-click the *AspNetUsers* table in Server Explorer and click *Refresh*. You should now see that the new fields have been added to the database.

Step 9: Modify the Register page.

Next, you'll need to modify the Register page to include the new fields. The Register page is in the Account folder. Figure 13-6 shows the finished page, and Figure 13-7 shows the code for the new fields. This code is inserted just before the form-group that contains the button. You will likely wind up inserting the new code immediately after line 38 in the file.

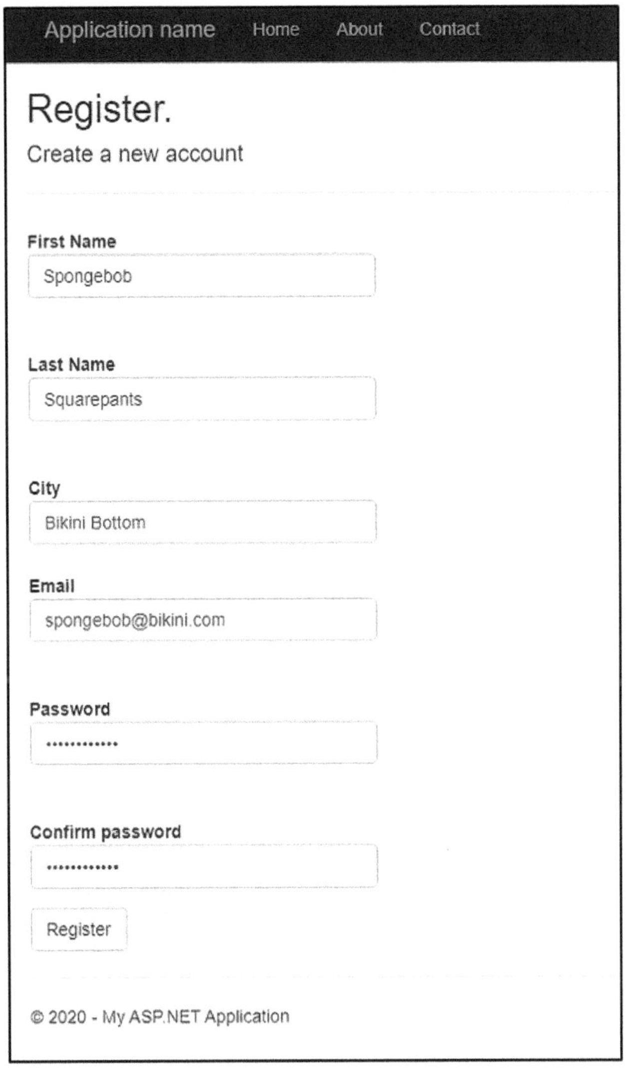

Figure 13-6: Registration Page with added fields.

```
1  <div class="form-group">
2      <asp:Label runat="server" AssociatedControlID="txtFirstName"
                CssClass="col-md-2 control-label">First Name</asp:Label>
3      <div class="col-md-10">
4          <asp:TextBox runat="server" ID="txtFirstName" CssClass="form-control" />
5          <asp:RequiredFieldValidator runat="server" ControlToValidate="txtFirstName"
6              CssClass="text-danger"
                ErrorMessage="The first name field is required." />
7      </div>
8  </div>
9  <div class="form-group">
10     <asp:Label runat="server" AssociatedControlID="txtLastName"
                       CssClass="col-md-2 control-label">Last Name</asp:Label>
11     <div class="col-md-10">
12         <asp:TextBox runat="server" ID="txtLastName" CssClass="form-control" />
13         <asp:RequiredFieldValidator runat="server" ControlToValidate="txtLastName"
14             CssClass="text-danger"
                ErrorMessage="The last name field is required." />
15     </div>
16 </div>
17 <div class="form-group">
18     <asp:Label runat="server" AssociatedControlID="txtCity"
                       CssClass="col-md-2 control-label">City</asp:Label>
19     <div class="col-md-10">
20         <asp:TextBox runat="server" ID="txtCity" CssClass="form-control" />
21     </div>
22 </div>
23 <div class="form-group">
24     <asp:Label runat="server" AssociatedControlID="Email"
                CssClass="col-md-2 control-label">Email</asp:Label>
25     <div class="col-md-10">
26         <asp:TextBox runat="server" ID="Email" CssClass="form-control"
                      TextMode="Email" />
27         <asp:RequiredFieldValidator runat="server" ControlToValidate="Email"
28             CssClass="text-danger" ErrorMessage="The email field is required." />
29 </div>
```

Figure 13-7: Code for additional fields on the Register page

The new fields follow the pattern from the other fields that were already on the page. Each label/textbox pair is surrounded by a form-group. For example, lines 9–16 have the controls for the last name. Line 9 sets up a div with class form-group applied. Then the label is set up in line 10 to point to the associated textbox through the AssociatedControlID property. Lines 11–15 set up the textbox with a RequiredFieldValidator.

Step 10: Modify the code-behind page.

In order for the data from the new fields to be added to the database, you will need to modify the code-behind page for the Register page. Figure 13-8 shows the code that creates the new user. You will need to insert lines 9–11, which take the values from the three new fields in the user interface and assign them to the corresponding values in the user object. You may find that the existing code to create a user (lines 5–8) is all in one line in the existing code. You can just add a comma after Email.Text and paste lines 9–10 after it.

```
1   protected void CreateUser_Click(object sender, EventArgs e)
2   {
3       var manager = Context.GetOwinContext().GetUserManager<ApplicationUserManager>();
4       var signInManager = Context.GetOwinContext().Get<ApplicationSignInManager>();
5       var user = new ApplicationUser()
6       {
7           UserName = Email.Text,
8           Email = Email.Text,
9           FirstName = txtFirstName.Text,
10          LastName = txtLastName.Text,
11          City = txtCity.Text
12      };
13      IdentityResult result = manager.Create(user, Password.Text);
14      if (result.Succeeded)
15      {
16          signInManager.SignIn( user, isPersistent: false, rememberBrowser: false);
17          IdentityHelper.RedirectToReturnUrl(Request.QueryString["ReturnUrl"], Response);
18      }
19      else
20      {
21          ErrorMessage.Text = result.Errors.FirstOrDefault();
22      }
23  }
```

Figure 13-8: Code to create a new user record

Step 11: Add a page to show the user information.

Before you can test the new functionality, you will need to add functionality to display the added user information. To do this, you'll add a simple message to the already-existing About page.

Open About.aspx and add an `asp:Label` with the ID `lblUserInfo` as the last line in the `asp:Content` element.

Open **About.aspx.cs** and add the following three using statements:

```
1   using Microsoft.AspNet.Identity;
2   using Microsoft.AspNet.Identity.Owin;
3   using Theater.Models;
```

Modify the `PageLoad` method as follows:

```
1   protected void Page_Load(object sender, EventArgs e)
2   {
3       if (User.Identity.IsAuthenticated)
4       {
5           var context = System.Web.HttpContext.Current.GetOwinContext();
6           var userManager = context.GetUserManager<ApplicationUserManager>();
7           string userID = User.Identity.GetUserId();
8           ApplicationUser user = userManager.FindById(userID);
9           lblUserInfo.Text = string.Format("Name: {0} {1}.<br>City: {2}",
                                    user.FirstName, user.LastName, user.City);
10      }
11  }
```

The method first checks to see if the user is authenticated in line 3. Then line 5 gets the OwinContext that can be used to interact with the authentication middleware. This context is used in line 6 to get the application user manager, which in turn can be used in line 8 to find the currently logged-in user's application object. Line 9 simply employs the user object to pull out the new fields and add them to a string to display in the label.

Review Questions

6. Describe the two approaches to expanding the Identity system with additional fields.

13.4 Adding Identity to an Existing Project

Now that you have seen how the Identity system works in the default template, we will take a look at how to add the Identity functionality to an existing project. This will allow us to expand the Grand Opera House system to allow staff to log in and make changes to the upcoming events. Expanding an existing system to allow for users to log in requires you to add several NuGet packages to the project, as shown in Table 13-3.

Table 13-3: NuGet packages needed to implement the Identity Framework

NuGet Package	Description
`Microsoft.AspNet.Identity.EntityFramework`	The Identity system is based on Entity Framework. Installing this package will also install Entity Framework and ASP.NET Identity Core.
`Microsoft.AspNet.Identity.Owin`	Contains a set of OWIN extension classes to manage and configure OWIN authentication middleware to be consumed by ASP.NET Identity Core packages
`Microsoft.Owin.Host.SystemWeb`	Contains an OWIN server that enables OWIN-based applications to run on IIS using the ASP.NET request pipeline

The most straightforward way to enable user registration and login for an existing project is to copy the needed elements from the project you created using the Visual Studio template that includes authentication, and then customize it for your needs. The following tutorial will show you which parts to import and how to modify them to fit the needs of the Grand Opera House website.

Tutorial: Adding Login Controls to the Grand Opera House Website

To complete this tutorial, you can continue working on the project you completed in the previous chapter, or you can download the Chapter 13 starter project.

In this part of the tutorial, you will complete the following tasks:

1. Set up and configure the Identity system (steps 12–17).
2. Add a page to register new users (steps 18–21).
3. Add a page to manage login and logout functionality (steps 22–26).

Step 12: Install NuGet packages.

To prepare your project, install the three packages listed in Table 13-3. You do this by right-clicking on the project in Solution Explorer and selecting *Manage NuGet Packages...*.

Click the *Browse* tab at the top and use the search to find each of the three packages. Be sure to install the right one. Some packages have very similar names.

Select the package and click *Install*. See Figure 13-9 for details. You may get a dialog showing additional required packages that will be installed. Click *OK* on this dialog.

Click *I Accept* on any license agreements.

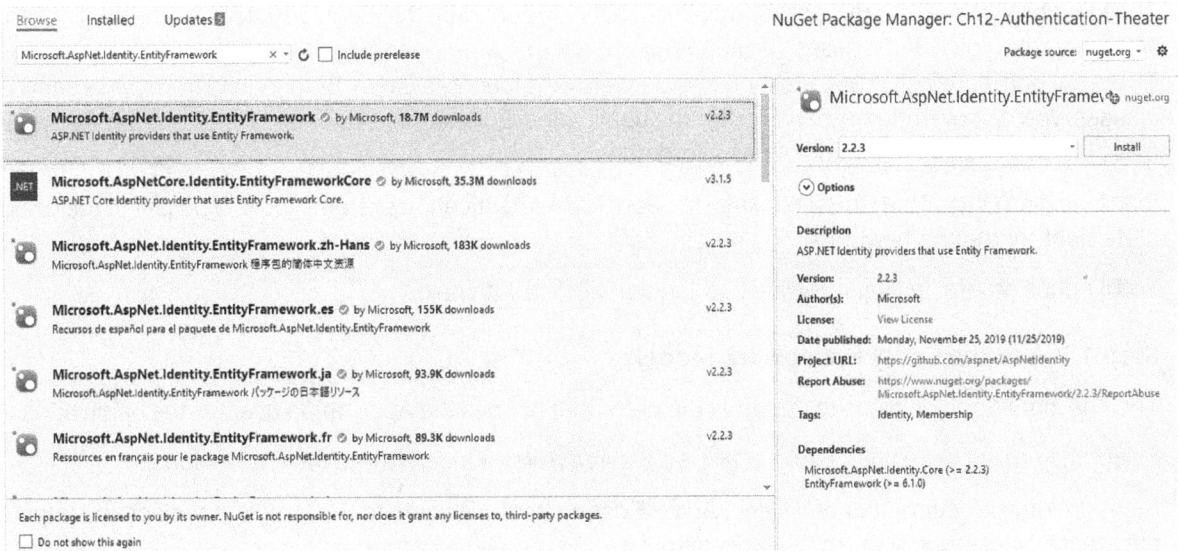

Figure 13-9: Installing NuGet packages

Step 13: Copy the OWIN Startup file from the template project.

The Identity system is configured in a few files in the template project that you created previously in this chapter. Rather than create these from scratch, you will import them from that project. The simplest way to import is to right-click the project in Solution Explorer and select **Add > Existing Item…**. Then navigate to the Theater template project and find the Startup.cs file; click **Add**.

Open the file—the code should match Figure 13-10. Line 4 tells the OWIN system that this is the starting point for the Identity system and will automatically call the Configuration method in line 8. Currently, the ConfigureAuth method call in line 9 will have a red squiggly line. This will be fixed in the next step when you import Startup.Auth.cs.

```
1    using Microsoft.Owin;
2    using Owin;
3
4    [assembly: OwinStartupAttribute(typeof(Theater.Startup))]
5    namespace Theater
6    {
7        public partial class Startup {
8            public void Configuration(IAppBuilder app) {
9                ConfigureAuth(app);
10           }
11       }
12   }
```

Figure 13-10: Startup.cs file

Step 14: Copy Startup.Auth.cs.

Right-click the **App_Start** folder and select **Add > Existing Item…**. Select **IdentityConfig.cs** and **Startup.Auth. cs** by holding down the Control key.

Once the files are imported, the red squiggly lines in Startup.cs will disappear. If you open Start.Auth.cs, you will see that the first method in this class is ConfigureAuth, so this method will also automatically be executed. This method sets up authentication to use cookies to store information about the signed-in user. It also sets up an object that connects to the database as well as the user manager and sign-in manager (see Table 13-2).

The IdentityConfig file contains several classes that help configure the Identity system for the application. Most notably, you'll find classes for `ApplicationUserManager` and `ApplicationSignInManager`. These are both subclasses of corresponding classes in the Identity system. By inheriting from these generic classes, you can override the behavior for the application. For example, the `ApplicationUserManager` class includes the configuration of password rules.

Since we didn't install the NuGet package for Google's authentication system, you should delete the using statement for that framework.

You'll notice several red squiggly lines, which we will deal with next.

Step 15: Add the Entity Framework Models.

The final configuration piece to import is the class that defines the Application user for the application.

Select the project in Solution Explorer and select *Add > New Folder*. Name the folder *Models*.

Right-click the *Models* folder and select *Add > Existing Item*. Navigate to the template project and import the IdentityModels.cs file from the Models folder.

You already worked with this file previously when configuring user fields. You will modify this slightly in this project as well. For the Grand Opera House, the `City` field isn't relevant, so you can delete line 18 in the `ApplicationUser` class.

The project should now compile without any syntax errors.

Step 16: Add a connection string.

Since the username and password information is stored in a database, you will need to add a connection string entry to the Web.Config file. The connection string by default is named `DefaultConnection` and is shown in lines 7–13 of Figure 13-11. You can also copy this from the template project, but you should rename the database file (line 9) and initial catalog (line 10) to avoid conflicts.

```
1    <connectionStrings>
2      <add name="ConnectionString"
3            connectionString="Data Source=(LocalDB)\MSSQLLocalDB;
4            AttachDbFilename=|DataDirectory|\Theater.mdf;
5            Integrated Security=True;Connect Timeout=30"
6            providerName="System.Data.SqlClient" />
7      <add name="DefaultConnection"
8            connectionString="Data Source=(LocalDb)\MSSQLLocalDB;
9            AttachDbFilename=|DataDirectory|\LoginDB.mdf;
10           Initial Catalog= LoginDB;
11           Integrated Security=True;
12           MultipleActiveResultSets=True"
13        providerName="System.Data.SqlClient" />
14   </connectionStrings>
```

Figure 13-11: Connection strings

With these elements in place, you are now ready to build the user registration and login functionality to take advantage of the Identity system. In this system, these pages will be created in a folder called *Account*.

Step 17: Add the Account folder.

Right-click the project and select *Add > New Folder*.

Name the folder *Account*. You will be adding all the login-related pages to this folder.

Step 18: Add a registration page.

Next, you will add a page that allows new users to register.

Right-click the *Account* folder and select *Add > Web form with master page*.... Name the form Register. aspx.

Add the code as shown in Figure 13-12. This is a very simple registration form with fields for a user's first name, last name, and email (which will be used as the username); password and confirm password textboxes; and a Register button. Figure 13-13 shows the final form.

```
1   <%@ Page Title="Register New User" Language="C#" MasterPageFile=
    "~/Main.Master" AutoEventWireup="true" CodeBehind="Register.aspx.cs"
    Inherits="Theater.Account.Register" %>
2
3   <asp:Content ID="Content1" ContentPlaceHolderID="HeadContentPlaceHolder"
    runat="server">
4   </asp:Content>
5   <asp:Content ID="Content2" ContentPlaceHolderID="MainContentPlaceHolder"
6       runat="server">
7       <h2>Register a new user</h2>
8       <asp:PlaceHolder runat="server" ID="StatusMessage" Visible="false">
9           <p>
10              <asp:Label runat="server" ID="StatusText" />
11          </p>
12      </asp:PlaceHolder>
13
14      <div class="form-group">
15          <label for="txtFirstName" class="col-md-2">First name</label>
16          <asp:TextBox runat="server" ID="txtFirstName" />
17      </div>
18      <div class="form-group">
19          <label for="txtLastName" class="col-md-2">Last name</label>
20          <asp:TextBox runat="server" ID="txtLastName" />
21      </div>
22      <div class="form-group">
23          <label for="txtEmail" class="col-md-2">Email address</label>
24          <asp:TextBox runat="server" ID="txtEmail" />
25      </div>
26      <div class="form-group">
27          <label for="txtPassword" class="col-md-2">Password</label>
28          <asp:TextBox runat="server" ID="txtPassword" TextMode="Password" />
29      </div>
30      <div class="form-group">
31          <label for="ConfirmPassword"
32              class="col-md-2">
33              Confirm password</label>
34          <asp:TextBox runat="server" ID="ConfirmPassword"
35              TextMode="Password" />
36      </div>
37      <br />
38      <div>
39          <asp:Button CssClass="btn" runat="server" OnClick="CreateUser_Click"
40              Text="Register" />
41      </div>
42  </asp:Content>
```

Figure 13-12: HTML code for the registration page

The code in this form is standard HTML. We use the Placeholder control to be able to show and hide different parts of the form as needed. You may notice the use of TextMode="Password" for the two password fields (lines 24 and 29). This will hide the text as it is typed to increase the security of the form.

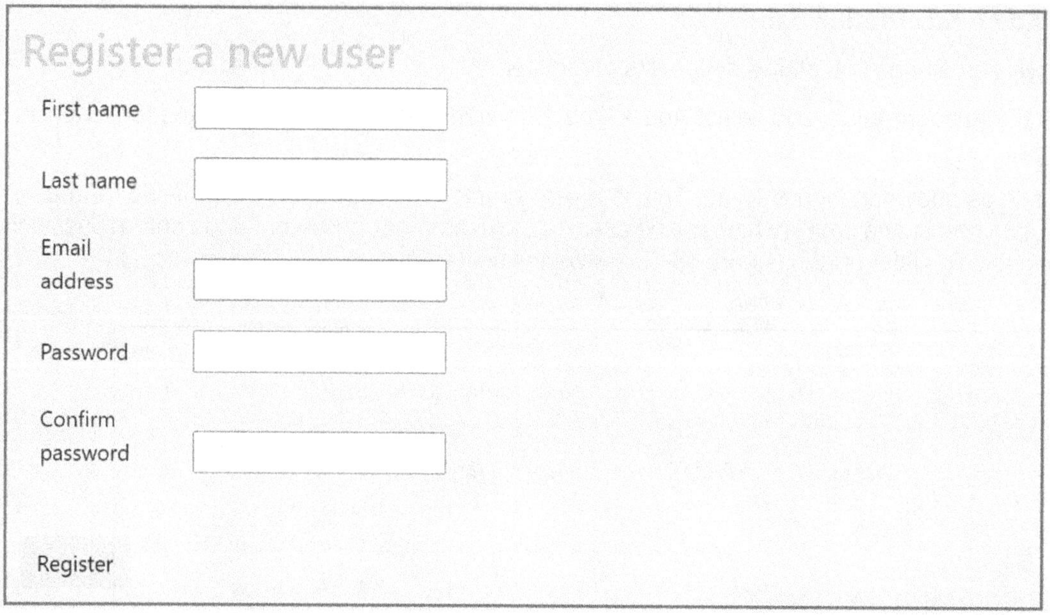

Figure 13-13: Completed registration form

Step 19: Add the code-behind page for the registration form.

Open *Register.aspx.cs* and add the following using statements:

```
1   using Microsoft.AspNet.Identity;
2   using Microsoft.AspNet.Identity.Owin;
3   using Theater.Models;
```

Next, add the CreateUser_Click method:

```
1   protected void CreateUser_Click(object sender, EventArgs e)
2   {
3       var userManager = Context.GetOwinContext().GetUserManager<ApplicationUserManager>();
4       var signInManager = Context.GetOwinContext().Get<ApplicationSignInManager>();
5       var user = new ApplicationUser()
6       {
7           UserName = txtEmail.Text,
8           Email = txtEmail.Text,
9           FirstName = txtFirstName.Text,
10          LastName = txtLastName.Text,
11      };
12      IdentityResult result = userManager.Create(user, txtPassword.Text);
13      StatusMessage.Visible = true;
14      if (result.Succeeded)
15      {
16          signInManager.SignIn(user, isPersistent: false, rememberBrowser: false);
17          IdentityHelper.RedirectToReturnUrl(Request.QueryString["ReturnUrl"], Response);
18      }
19      else
20      {
21          StatusText.CssClass = "text-danger";
22          StatusText.Text = result.Errors.FirstOrDefault();
23      }
24  }
```

Line 3 gets a reference to the user manager object that is utilized to create the user object in line 12, and line 4 gets a reference to the sign-in manager object that will sign in the user in line 16.

Lines 5–11 create a new ApplicationUser object and initialize the fields with values from the form. Then line 12 will attempt to create the user object in the Identity system. This will return an IdentityResult object that can be used to determine if a user was actually created. If the creation of the user was unsuccessful, lines 23–24 display an error message (for example, if a user with the same username already exists or the password rules were not followed). If the creation is successful, line 16 uses the sign-in manager to log the user in, and line 17 redirects the user to a `ReturnUrl` if the `QueryString` contains one—or the home page if it doesn't. This corresponds to the "Redirect to original page" step in Figure 13-1, where the user is redirected to the originally requested page. If the user could not be created, lines 21–22 display a status message.

Step 20: Test the user registration page.

You can now test that the user registration page works.

Right-click the **Register.aspx** file and select **Set as Start Page**. Press Ctrl+F5 to run the website.

You should see the registration page as shown in Figure 13-13. Try creating a new user. If the user is created successfully, you will be redirected back to the home page. Otherwise, you should see an error message above the form.

You can also test the different errors that can occur, such as trying to create a user with a name that already exists.

Step 21: Check the database content.

As the first user is created in the system, Entity Framework will create a database and any necessary tables. You can check the content of the database, which should be similar to what you saw in Figure 13-4.

Now that you have enabled user registration, you will need to add a page where users can actually log in.

13.5 Login Controls

Now that you have added the ability to register a user in the database, you will need to add a form for already-registered users to log into the application. This is done on a separate page, Login.aspx, created with standard ASP.NET controls. To log in a user, use the OWIN middleware to find and authenticate them.

It is also helpful to be able to display status information about the user in the header or footer of the page. This allows for a consistent place where users who aren't logged in can do so and users who are logged in can manage their accounts. You will use the `LoginView` and `LoginName` controls added to the footer of the master page to allow this.

Step 22: Add a login page.

Right-click the **Account** folder and add a new Web Form with a master page. Name the page **Login.aspx**.

Add the code shown in Figure 13-14. This is also available as Login.aspx.txt.

```
1   <%@ Page Title="Login" Language="C#" MasterPageFile="~/Main.Master"
2       AutoEventWireup="true" CodeBehind="Login.aspx.cs"
3       Inherits="Theater.Account.Login" %>
4   <asp:Content ID="Content1" ContentPlaceHolderID="HeadContentPlaceHolder"
5       runat="server">
6   </asp:Content>
7   <asp:Content ID="Content2" ContentPlaceHolderID="MainContentPlaceHolder"
8       runat="server">
9       <h2>Log In</h2>
10      <asp:PlaceHolder runat="server" ID="StatusMessage" Visible="false">
11          <p>
12              <asp:Label runat="server" ID="StatusText" /><br />
13          </p>
14      </asp:PlaceHolder>
15      <asp:PlaceHolder runat="server" ID="LoginForm">
16          <div class="form-group">
17              <label for="txtEmail" class="col-md-2">Email address</label>
18              <asp:TextBox runat="server" ID="txtEmail" />
19          </div>
20          <div class="form-group">
21              <label for="Password" class="col-md-2">Password</label>
22              <asp:TextBox runat="server" ID="Password" TextMode="Password" />
23          </div>
24          <br />
25          <div>
26              <asp:Button runat="server" OnClick="SignIn" Text="Log in"
27                  CssClass="btn btn-primary" />
28          </div>
29          <p>
30              <a href="Register.aspx">Register new user.</a>
31          </p>
32      </asp:PlaceHolder>
33      <asp:Button runat="server" ID="btnSignout"
34          OnClick="SignOut" Text="Log out" CssClass="btn btn-primary" />
35  </asp:Content>
36
```

Figure 13-14: Login form

The completed form can be seen in Figure 13-18, with the placeholders shown and hidden, based on their status.

Step 23: Add a SignIn method.

Open Login.aspx.cs and add the following using statements:

```
1   using Microsoft.AspNet.Identity;
2   using Microsoft.AspNet.Identity.Owin;
3   using Theater.Models;
```

Then add the SignIn method as shown in Figure 13-15.

```
1   protected void SignIn(object sender, EventArgs e)
2   {
3       var owinContext = Context.GetOwinContext();
4       var signInManager = owinContext.GetUserManager<ApplicationSignInManager>();
5       SignInStatus result = signInManager.PasswordSignIn(txtEmail.Text,
6           Password.Text, isPersistent: false, shouldLockout: true);
7       switch (result)
8       {
9           case SignInStatus.Success:
10              string queryString = Request.QueryString["ReturnUrl"];
11              IdentityHelper.RedirectToReturnUrl(queryString, Response);
12              break;
13          case SignInStatus.LockedOut:
14              StatusText.Text = "User account is locked.";
15              StatusText.CssClass = "text-warning";
16              StatusMessage.Visible = true;
17              break;
18          case SignInStatus.Failure:
19          default:
20              StatusText.Text = "Invalid login attempt";
21              StatusText.CssClass = "text-danger";
22              StatusMessage.Visible = true;
23              break;
24      }
25  }
```

Figure 13-15. SignIn method

To log in a user, you need to get the OWIN context (line 3), which is used in line 4 to get a reference to the sign-in manager. In line 5, you call the `PasswordSignIn` method passing in the username and password, along with a couple of settings for whether the login attempt is persistent and whether failed login attempts should count toward locking the user account. In this case, we set the login to not be persistent. That means if the user closes the browser (session), they will have to log in again. Depending on the nature of the application, you can set `IsPersistent` to *true* to reduce the number of times the user is asked to log in.

The login attempt returns a `SignInStatus` object. This is used in the switch statement in line 7 to determine how to proceed. If the attempt succeeded, lines 10–11 redirect the user to the `ReturnUrl` query string. Otherwise, two different error messages are displayed, depending on whether the user account is locked (lines 13–17) or otherwise failed (lines 18–23).

Step 24: Implement the PageLoad method.

If the user is already logged in, we want to hide the login form and instead show a message to the user, along with a logout button. Add the code in Figure 13-16 to the `Page_Load` method to accomplish this.

```
1   protected void Page_Load(object sender, EventArgs e)
2   {
3       if (!IsPostBack)
4       {
5           if (User.Identity.IsAuthenticated)
6           {
7               var context = Context.GetOwinContext();
8               var userManager = context.GetUserManager<ApplicationUserManager>();
9               string userID = Context.User.Identity.GetUserId();
10              ApplicationUser user = userManager.FindById(userID);
11              StatusText.Text = string.Format("Hello {0} {1}!",
12                  user.FirstName, user.LastName);
13              StatusText.CssClass = "text-info";
14              StatusMessage.Visible = true;
15              LoginForm.Visible = false;
16              btnSignout.Visible = true;
17          }
18          else
19          {
20              LoginForm.Visible = true;
21              btnSignout.Visible = false;
22          }
23      }
24  }
25
```

Figure 13-16. PageLoad method on the Login page

Line 5 shows how you can check if the request is coming from an authenticated user. If that's the case, lines 7–10 show how you can get information about the logged-in user from the ApplicationUser object. This allows us to show the first and last name of the user in the status message. Lines 14–16 make the status message and logout button visible and hides the login form. If the user isn't authenticated, lines 20–21 show the login form and hide the logout button. The status message placeholder is hidden by default, so it doesn't need to be hidden in this case.

Step 25: Add a Signout method.

The last portion of the Login page allows the user to log out. Add the SignOut method, as shown in Figure 13-17.

```
1   protected void SignOut(object sender, EventArgs e)
2   {
3       var authenticationManager = Context.GetOwinContext().Authentication;
4       authenticationManager.SignOut(DefaultAuthenticationTypes.ApplicationCookie);
5       Response.Redirect("~/Account/Login.aspx");
6   }
```

Figure 13-17: SignOut method

This method starts by getting the OWIN authentication manager in line 3. Then in line 4, it calls the SignOut method, which does the actual logout. Line 5 redirects back to the Login page to display the Login form again.

Step 26: Test the Login functionality.

Run the application and access the Login page. Try logging in and out with a registered user account. Also, try a failed login attempt to make sure you see the various parts of the Login form, as shown in Figure 13-18.

Figure 13-18. Completed Login form

Since you haven't built any functionality for unlocking a user account, you will have to access the user table directly to unlock the user. The AspNetUsers table has a column named LockoutEndDateUtc, which will have a timestamp indicating the end of the lockout period. To unlock the user account, you can either change the timestamp to be in the past or replace the cell value with NULL.

The number of failed login attempts leading to lockout, as well as the duration of the lockout period, is set in IdentityConfig.cs. If you wanted to unlock the account programmatically, the easiest way is to have the user reset their password, which automatically removes the lock. However, you can also use something like the following:

```
1   var owinContext = Context.GetOwinContext();
2   var userManager = owinContext.GetUserManager<ApplicationUserManager>();
3   string userID = txtEmail.Text;
4   userManager.ResetAccessFailedCount(userID);
5   userManager.SetLockoutEndDate(userID, DateTime.Now);
```

Line 4 resets the counter for the number of failed attempts, and line 5 sets the lockout end time to now, which effectively unlocks the account. It also gives you a record of when the account was unlocked.

It is also worth noting that a successful login doesn't reset the count of failed logins, so you may want to include line 4 above when the user successfully logs in.

Review Questions

7. What is the difference between the `LoginView` and `LoginName` controls?

8. How can you unlock a locked account?

13.6 Showing Login Status

ASP.NET includes a few standard controls that you can use to manage the login experience for users. In this section, we will discuss two of these, LoginView and LoginName. You will be adding these to the master page footer to provide easy access to the login controls on every page. In this application, the end customers will not be creating accounts and logging in. This is for staff at the Grand Opera House to be able to manage the Performances, so the login status information will be shown in the footer to make them less prominent for the majority of users. An alternative design would be to create a separate set of pages with a different master page just for staff. This setup will be left as an exercise.

The LoginName control is quite simple in that it simply displays the username for a logged-in user. It is easy to use but limited in that it can only display the username and not the user's actual name, as you saw earlier in the chapter.

The LoginView control allows you to display different information based on the login status of the user. For example, if the user isn't logged in, you can show a link to the login page or even an embedded login form, and if they are logged in, you can show their username and a link to manage their account. This functionality is implemented through the use of templates. The AnonymousTemplate contains the information shown to users who aren't logged in, whereas the LoggedInTemplate content is shown to users who have been authenticated.

Step 27: Add LoginView to the master page.

Open the Main.Master page and move the `</form>` tag after the `</footer>` tag, so the `<footer>` element is included in the `<form>` element.

Add lines 5–19 in Figure 13-19 to the `footer` element.

```
1   <footer>
2       Copyright © 2018 The Grand Oshkosh. All Rights Reserved.
3       Site development and hosting by interGen web solutions.
4       <asp:Label ID="lblContactInfo" runat="server" Text=""></asp:Label>
5       <br />
6       <asp:LoginView ID="loginView" runat="server">
7           <AnonymousTemplate>
8               <a href="/Account/Login.aspx">Admin Login</a>
9           </AnonymousTemplate>
10          <LoggedInTemplate>
11              Hello <asp:LoginName ID="ctlUserName" runat="server" />.
12              <br />
13              <asp:Button runat="server" ID="btnLogout"
14                  OnClick="btnLogout_Click" CssClass="btn btn-primary"
15                  Text="Logout" />
16              <a href="/Admin/AddEvent.aspx" class="btn btn-primary">
17                  Add Event</a>
18          </LoggedInTemplate>
19      </asp:LoginView>
20  </footer>
21  </form>
```

Figure 13-19: LoginView and LoginName in the footer of the master page

Line 6 starts the LoginView, which has two templates, AnonymousTemplate and LoggedInTemplate, that are displayed depending on the user's login status. Line 8 contains a link to the Login page that's displayed to non-authenticated (anonymous) users. Lines 11–17 in the LoggedInTemplate are shown to users who have logged in. Line 11 shows the LoginName control that shows the username. In addition, this also includes a button to logout and a link styled as a button that takes the user to a page for adding new events. You will build this page in the next section of the tutorial.

It's time to practice! Do Exercise 13.8 (see the end of the chapter).

Test the site and check that you can log in and see the LoginView change based on your login status.

Review Questions

9. What is the difference between `LoggedInTemplate` and `AnonymousTemplate` in the `LoginView` control? How is each used?

10. What are the limitations of the `LoginName` control?

13.7 Controlling Access

Currently, there is no difference between what a logged-in user and an anonymous user are able to do on the Grand Opera House site. What we have done is implement a way to do *authentication*—that is, asking the user to present credentials that prove who they are. Of course, one main reason for asking a user to authenticate is to provide access to certain resources based on knowing who they are. This process is called *authorization* and is common in many systems. The ASP.NET Identity system supports two modes for authorizing users: individually, based on their username, or as part of some group they belong to.

The group-based approach is called *role-based authorization* in the Identity system and allows you to specify access based on the roles you assign to the user. This allows for a much more flexible and easier to maintain system of authorization. For example, if a user leaves the organization, you won't have to find all the places where they were given the authorization to access a document or a folder. Instead, you can simply remove them from the roles they had been added to. Similarly, for adding new users, you can just add them to the necessary roles, and they will be able to access all the relevant resources. The assumption underlying this system is that users change more frequently than roles.

Controlling access to pages and directories can be done quite simply in an ASP.NET website by using an `authorization` element in a Web.config file that can be added in each directory you wish to protect. As an example, the following Web.Config file provides access to any files in the directory it is placed for the users `donald` and `andy`, as well as any user who is given the role `engineering`. Any other users are denied access via the asterisk, as shown in line 7.

```
1   <?xml version="1.0"?>
2   <configuration>
3     <system.web>
4       <authorization>
5         <allow users="donald, andy" />
6         <allow roles="engineering"/>
7         <deny users="*" />
8       </authorization>
9     </system.web>
10  </configuration>
```

The `allow` and `deny` rules are processed from top to bottom, and as soon as a rule that matches the current user is detected, the user is either denied or granted access and any further rules are not evaluated. If no rule matches the user, the user is granted access. It is best practice to ensure that all users have a rule that explicitly allows or denies access.

In addition to the asterisk, you can also use a question mark to indicate all anonymous users (see Table 13-4). The following rule will deny access to anonymous users and grant access to any authenticated user:

```
1   <authorization>
2     <deny users="?" />
3     <allow users="*" />
4   </authorization>
```

The `authorization` rules can be defined in the `system.web` element in the Web.config file for the site or in the Web.config file in a subdirectory. If they are added in

Table 13-4: Wildcards for the user attribute

Wildcard	Description
?	Matches any unauthenticated (anonymous) users
*	Matches any user, whether authenticated or not

the top-level Web.config, the rules apply to the entire site but can be overridden for a folder by adding a Web.config with an `authorization` element in that folder. You can also specify a location that governs the entire `system.web` element. For example, Figure 13-20 shows three examples of controlling access from the root level Web.config file. Lines 1–7 controls access to the entire Admin directory by denying anonymous users but allowing everyone else (that is, only authenticated users have access to the directory). Lines 9–16 show how you can provide access based on roles. In this case, users with the manager role are granted access to the ManageUsers.aspx file, and everyone else is denied access. Lines 19–26 show how you can control access to a specific file. In this case, the Register.aspx file in the Account directory is accessible to anonymous users but denied for everyone else (so you can only register if you aren't already logged in).

There is no functional difference between writing the authorization rules in the root-level Web.config file or in each directory. Choosing which method is best all depends on the specific website. Keeping the rules in one place makes it easier to obtain an overview of all of them but may make the root-level Web.config file unwieldy.

```
1   <location path="Admin">
2     <system.web>
3       <authorization>
4         <deny users="?"/>
5         <allow users="*"/>
6       </authorization>
7     </system.web>
8   </location>
9
10  <location path="Admin/ManageUsers.aspx">
11    <system.web>
12      <authorization>
13        <allow roles="manager"/>
14        <deny users="*"/>
15      </authorization>
16    </system.web>
17  </location>
18
19  <location path="Account/Register.aspx">
20    <system.web>
21      <authorization>
22        <allow users="?"/>
23        <deny users="*"/>
24      </authorization>
25    </system.web>
26  </location>
```

Figure 13-20: Examples of controlling access to different directories

Tutorial: Restricting Access

In this tutorial, you will see how you can restrict access to some of the pages and directories in the system. We will be creating an Admin directory with functionality that is only available to registered users. Here we will restrict access to the Admin folder, based on both specific usernames and all authenticated users. Similarly, you will see how to restrict access to the Register page to only allow users who aren't logged in.

Step 28: Create the Admin directory.

Right-click the project in the *Solution Explorer* and select *Add > New Folder*.

Name the folder *Admin*.

Step 29: Create the AddEvent page.

Right-click the *Admin* folder and add a new Web Form with a master page.

Name the new page *AddEvent*.

Add the .aspx code shown in Figure 13-21. This is also available as AddEvent.aspx.txt.

```
1    <%@ Page Title="Add New Event" Language="C#" MasterPageFile="~/Main.Master"
2        AutoEventWireup="true" CodeBehind="AddEvent.aspx.cs"
3        Inherits="Theater.Admin.AddEvent" %>
4
5    <asp:Content ID="Content1" ContentPlaceHolderID="HeadContentPlaceHolder"
6        runat="server">
7    </asp:Content>
8    <asp:Content ID="Content2" ContentPlaceHolderID="MainContentPlaceHolder"
9        runat="server">
10       <div class="form-group">
11           <label class="col-sm-3" for="txtPerformanceName">Performance Name</label>
12           <asp:TextBox runat="server" ID="txtPerformanceName" />
13       </div>
14       <div class="form-group">
15           <label class="col-sm-3" for="txtBasePrice">Base Price</label>
16           <asp:TextBox runat="server" ID="txtBasePrice" />
17       </div>
18       <div class="form-group">
19           <label class="col-sm-3" for="txtDescription">Description</label>
20           <asp:TextBox TextMode="MultiLine" runat="server" ID="txtDescription" />
21       </div>
22       <div class="form-group">
23           <label class="col-sm-3" for="txtPresenter">Performer</label>
24           <asp:TextBox runat="server" ID="txtPresenter" />
25       </div>
26       <div class="form-group">
27           <label class="col-sm-3" for="imgUpload">Image</label>
28           <asp:FileUpload ID="imgUpload" runat="server" />
29       </div>
30       <br />
31       <asp:Button Text="Save" runat="server" CssClass="btn"
32           ID="btnSaveEvent" OnClick="btnSaveEvent_Click" />
33
34       <asp:PlaceHolder runat="server" ID="StatusMessage" Visible="false">
35           <p>
36               <asp:Label runat="server" ID="StatusText" /><br />
37           </p>
38       </asp:PlaceHolder>
39   </asp:Content>
```

Figure 13-21: Code for the AddEvent.aspx page

This is a fairly standard form, the output of which is shown in Figure 13-22. One notable aspect of the form is the inclusion of a file upload control in line 28. You will learn how to process the uploaded file shortly.

Step 30: Add the code to save an event.

Add the following using statements to AddEvent.aspx.cs:

```
1    using System.Configuration;
2    using System.Data;
3    using System.Data.SqlClient;
```

Add the code for the btnSaveEvent_Click method in the code-behind file, as shown in Figure 13-23. This is available as btnSaveEvent.txt.

Figure 13-22: Form to add a new event

```
1   protected void btnSaveEvent_Click(object sender, EventArgs e)
2   {
3       string connectionString =
                ConfigurationManager.ConnectionStrings["ConnectionString"].ConnectionString;
4       var connection = new SqlConnection(connectionString);
5       connection.Open();
6       byte[] imageArray = null;
7       if (imgUpload.HasFile)
8       {
9           int fileLength = imgUpload.PostedFile.ContentLength;
10          imageArray = new byte[fileLength];
11          HttpPostedFile image = imgUpload.PostedFile;
12          image.InputStream.Read(imageArray, 0, fileLength);
13      }
14      //Find highest ID
15      string sqlMax = "SELECT Max(PerformanceId) FROM Performance";
16      var commandMax = new SqlCommand(sqlMax, connection);
17      string sql = "insert into Performance " +
18          "(PerformanceId, PerformanceName, BasePrice, Description, Presenter, image) " +
19          "Values(@Id, @Name, @BasePrice, @Description, @Presenter, @Image)";
20      var command = new SqlCommand(sql, connection);
21
22      try
23      {
24          SqlDataReader reader = commandMax.ExecuteReader();
25          reader.Read();
26          int maxID = int.Parse(reader[0].ToString()); //Highest ID
27          maxID++;//Increment
28          reader.Close();
29          //Insert parameter values
30          command.Parameters.AddWithValue("@Id", SqlDbType.Int).Value = maxID;
31          command.Parameters.AddWithValue("@Name", SqlDbType.VarChar).Value =
                    txtPerformanceName.Text;
32          command.Parameters.AddWithValue("@BasePrice", SqlDbType.Decimal).Value =
                    txtBasePrice.Text;
33          command.Parameters.AddWithValue("@Description", SqlDbType.VarChar).Value =
                    txtDescription.Text;
```

Figure 13-23: Code to save a performance (*continues*)

```
34          command.Parameters.AddWithValue("@Presenter", SqlDbType.VarChar).Value =
                 txtPresenter.Text;
35          command.Parameters.AddWithValue("@Image", SqlDbType.Image).Value = imageArray;
36
37          //Insert record
38          command.ExecuteNonQuery();
39          //Display message
40          StatusMessage.Visible = true;
41          StatusText.CssClass = "text-success";
42          StatusText.Text = "Record saved";
43          //Clear form
44          txtBasePrice.Text = "";
45          txtDescription.Text = "";
46          txtPerformanceName.Text = "";
47          txtPresenter.Text = "";
48       }
49       catch (Exception ex)
50       {
51          StatusMessage.Visible = true;
52          StatusText.CssClass = "text-danger";
53          StatusText.Text = "Record not saved: " + ex.Message;
54       }
55       finally
56       {
57          connection.Close();
58       }
59  }
```

Figure 13-23: Code to save a performance (*continued*)

This method is fairly lengthy and carries out the following functions:

Line 3 gets the connection string from Web.Config.

Lines 4–5 open a connection to the database.

Lines 6–12 process the uploaded file. Line 7 checks if a file was selected, to avoid null pointer errors. Lines 9 and 10 use the file length to create a byte array to hold the file data. Line 10 gets a reference to the uploaded file through the PostedFile property of the image control. The Read method in line 12 reads the file and stores it in the image byte array, which is then stored in the SQL parameter in line 35.

The Performance table includes a PerformanceId as a primary key. To ensure we store a unique value, lines 15 and 16 set up a query and SQL command to find the highest value already stored in the field. This query is executed in line 24, and the resulting value is read in lines 25–26. Line 27 increments the value, and line 30 inserts it as a parameter.

Line 17 sets up the primary SQL query to store all the values in the database. We use parameters in this query, and those parameters are populated with values in lines 30–35, based on the values from the controls.

Line 38 executes the query to insert the new record.

If the insertion is successful, lines 40–42 show a success status message and clear the form. However, if an exception is thrown anywhere along the way, the catch statement in lines 49–53 will display the error message.

The finally clause in lines 55–57 ensures that the database connection is closed, regardless of whether the operation was successful.

Step 31: Test the Add Event page.

Run the application and navigate to the Add Event Page. If you are logged in, you can use the button in the footer.

Try adding a new event to the database. If you are successful, you can navigate to the Events page to see the new event show up.

To have the error message show up, you can leave the image upload blank, as this field is required in the database.

Step 32: Restrict the Admin directory to authenticated users.

Right-click the *Admin* directory and select *Add > New Item...*. Use the search to find the Web Configuration File item. Click *Add*.

Add an `authorization` element and an element that denies anonymous users (see lines 3–6 in Figure 13-20 for the code to add).

Step 33: Change the path for the Login page.

OWIN needs to know where your login page is located, so you'll need to make sure this is specified correctly in Startup.Auth.cs, which is in the App_Start folder.

Look for the `LoginPath` line and make sure it is set up to point to your Login page, including the following .aspx extension:

```
LoginPath = new PathString("/Account/Login.aspx"),
```

Step 34: Test the authorization.

Launch the website (Ctrl+F5).

Check that you are not logged in; then navigate to the Admin/EditEvents.aspx page. You should be redirected to the login page.

After logging in, you should be redirected back to the AddEvent page.

Review Questions

11. How does the Identity system implement authorization?
12. What are the advantages of a role-based authorization system?
13. How are anonymous users indicated in authorization rules?
14. What happens if a user doesn't match any authorization rules?

13.8 Working with Roles

As described previously, the Identity system allows for role-based authorization, where users can be added to roles and the authorization of resources that are accessible can be done by roles instead of usernames. This makes it easier to maintain who has access to what in the system. However, it also requires a facility to manage the roles and which users are added to the roles. Working with roles is similar to working with users. The OWIN context contains a role manager that you can use to create and edit the roles. You use the user manager to add and remove roles for a particular user.

Before you can use the role manager, you need to add the `ApplicationRoleManager` as a subclass of `RoleManager` to the IdentityConfig.cs file and add a statement to the `ConfigureAuth` method of the Startup class to ensure that the role manager is included in the OWIN context. Once all the setup is done, you can get a reference to the role manager using the `Get` method, as follows:

```
ApplicationRoleManager roleManager =
                    Context.GetOwinContext().Get<ApplicationRoleManager>();
```

As you manage users and roles, you will work extensively with the `UserManager` and `RoleManager` classes. Table 13-5 shows some of the most commonly used methods and properties from the two classes as you work with roles and users.

Table 13-5: Methods and properties of the UserManager and RoleManager classes

Class	Method/Property	Description
UserManager	Create(user)	Creates a user object Returns an `IdentityResult` object with status of the operation
	Update(user)	Makes changes to a user object Returns an `IdentityResult` object
	Delete(user)	Deletes a user from the system Returns an `IdentityResult` object
	Users	Property that provides access to an `IQueryable` object of all the users in the system
	AddToRole(userID, role)	Adds the role to the user with the given user ID Returns an `IdentityResult` object
	RemoveFromRole(userID, role)	Removes the role from the user Returns an `IdentityResult` object
	IsInRole(userID, role)	Returns a Boolean indicating whether the role is associated with the user
RoleManager	Create(role)	Creates a new role with the given name Returns an `IdentityResult` object
	Update(role)	Makes changes to a role Returns an `IdentityResult` object
	Delete(role)	Deletes a role from the system Returns an `IdentityResult` object
	Roles	Property that provides access to an `IQueryable` object of all the roles in the system

Tutorial: A Page for Managing Roles and Users

In this tutorial, you will build a page that will allow you to manage the roles in the system and add and remove users from roles. Figure 13-24 shows the finished page. At the top is a `GridView` showing all the users who are in the system. Selecting a user will show the user's name in the Assign Roles section. In the middle part of the screen, all roles are listed in a GridView on the left side. Selecting a role will show the role name in the Update Roles textbox, as well as in the Selected Role label under Assigning Roles. The Update Roles textbox can be used to change the name of the role. New roles can be added by typing them in the Add New Role textbox and clicking the corresponding button. After selecting a user and a role, the role can be assigned to the user. Implementation of the Remove Role functionality is left as an exercise.

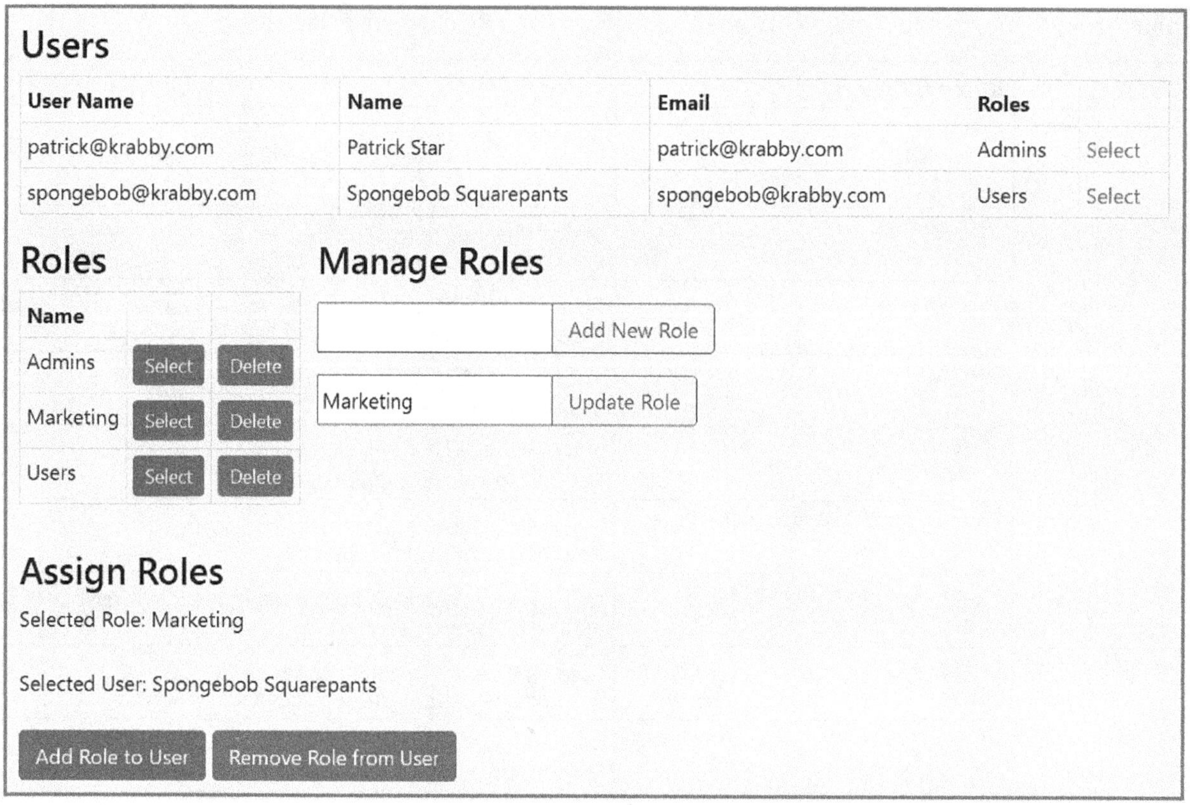

Users

User Name	Name	Email	Roles	
patrick@krabby.com	Patrick Star	patrick@krabby.com	Admins	Select
spongebob@krabby.com	Spongebob Squarepants	spongebob@krabby.com	Users	Select

Roles

Name		
Admins	Select	Delete
Marketing	Select	Delete
Users	Select	Delete

Manage Roles

	Add New Role

Marketing	Update Role

Assign Roles

Selected Role: Marketing

Selected User: Spongebob Squarepants

Add Role to User	Remove Role from User

Figure 13-24: Completed screen for managing roles and assigning them to users

Step 35: Create the ApplicationRoleManager class.

Open IdentityConfig.cs in the *App_Start* folder and add the code in Figure 13-25 below the ApplicationSignInManger class.

```
1  public class ApplicationRoleManager : RoleManager<IdentityRole>
2  {
3      public ApplicationRoleManager(IRoleStore<IdentityRole, string> store) : base(store)
4      {
5
6      }
7
8      public static ApplicationRoleManager
                     Create(IdentityFactoryOptions<ApplicationRoleManager> options,
                         IOwinContext context)
9      {
10         var roleStore = new RoleStore<IdentityRole>(context.Get<ApplicationDbContext>());
11         return new ApplicationRoleManager(roleStore);
12     }
13 }
```

Figure 13-25: ApplicationRoleManager code

The ApplicationRoleManager class is a subclass of RoleManager. The constructor in line 3 accepts a role store and passes it to the constructor of the RoleManager class. The class also contains a Create method that takes an IdentityFactoryOption and an OWIN context. It uses the OWIN context to create a role store that in turn is used to create and return an ApplicationRoleManager object.

Step 36: Modify the Startup class.

Open the Startup.Auth.cs file and add line 9 from Figure 13-26.

```
1   public void ConfigureAuth(IAppBuilder app)
2   {
3       // Configure the db context, user manager and signin manager to use a single
        instance per request
4       app.CreatePerOwinContext(ApplicationDbContext.Create);
5       app.CreatePerOwinContext<ApplicationUserManager>(ApplicationUserManager.Create);
6       app.CreatePerOwinContext<ApplicationSignInManager>(ApplicationSignInManager.Create);
7
8       //Configure the role m anager to use a single instance per request
9       app.CreatePerOwinContext<ApplicationRoleManager>(ApplicationRoleManager.Create);
10
11      [ ... ]
12
13  }
```

Figure 13-26: Adding an ApplicationRoleManager to Startup.Auth.cs

Step 37: Add the UserRoles page.

Add a new web form with a master page to the Account directory called *UserRoles.aspx*. Add the code shown in lines 8–32 in Figure 13-27 to the `MainContentPlaceholder`. This will show all the registered users in a Gridview.

```
1   <%@ Page Title="Users and Roles" Language="C#" MasterPageFile="~/Main.Master"
2       AutoEventWireup="true" CodeBehind="UserRoles.aspx.cs"
3       Inherits="Theater.Account.UserRoles" %>
4
5   <asp:Content ID="Content1" ContentPlaceHolderID="HeadContentPlaceHolder" runat="server">
6   </asp:Content>
7   <asp:Content ID="Content2" ContentPlaceHolderID="MainContentPlaceHolder" runat="server">
8       <h3>Users</h3>
9
10      <asp:GridView ID="grdUsers" runat="server" DataKeyNames="Id"
11          SelectMethod="grdUsers_GetData" AutoGenerateColumns="false"
12          ItemType="Theater.Models.ApplicationUser"
13          OnSelectedIndexChanged="grdUsers_SelectedIndexChanged"
14          cssClass="table table-sm table-bordered">
15          <Columns>
16              <asp:BoundField HeaderText="Username" DataField="UserName" />
17              <asp:TemplateField HeaderText="Name">
18                  <ItemTemplate>
19                      <asp:Label runat="server"
20                          Text='<%#Eval("FirstName")+ " " + Eval("LastName")%>'></asp:Label>
21                  </ItemTemplate>
22              </asp:TemplateField>
23              <asp:BoundField HeaderText="Email" DataField="Email" />
24              <asp:TemplateField HeaderText="Roles">
25                  <ItemTemplate>
26                      <asp:Label runat="server"
27                          Text='<%# ListRoles(Item.Roles) %>'></asp:Label>
28                  </ItemTemplate>
29              </asp:TemplateField>
30              <asp:CommandField ShowSelectButton="true" />
31          </Columns>
32      </asp:GridView>
33  </asp:Content>
```

Figure 13-27: GridView for showing the users

Line 11 has the name of the method that will provide the data for the `GridView`. Line 12 shows the data type that items in the `GridView` will display. This allows us to use the properties of the `ApplicationUser` class directly in the `GridView`. Because roles are a list of items for the user, you will create a special method (`ListRoles`) that will turn them into a single string. The roles are displayed in the `GridView` using a `TemplateField` in lines 24–29. Line 30 ensures a Select button is shown.

Step 38: Add code-behind for the users GridView.

Open the code-behind page and add the following using statements:

```
1   using Microsoft.AspNet.Identity.Owin;
2   using Microsoft.AspNet.Identity;
3   using Theater.Models;
4   using Microsoft.AspNet.Identity.EntityFramework;
```

Figure 13-28 shows the code you will need to enter into the code-behind page.

```
1   public partial class UserRoles : System.Web.UI.Page
2   {
3       ApplicationUserManager userManager;
4       ApplicationRoleManager roleManager;
5
6       protected void Page_Load(object sender, EventArgs e)
7       {
8           userManager = Context.GetOwinContext().GetUserManager<ApplicationUserManager>();
9           roleManager = Context.GetOwinContext().Get<ApplicationRoleManager>();
10      }
11
12      public IQueryable<ApplicationUser> grdUsers_GetData()
13      {
14          return userManager.Users;
15      }
16
17      public string ListRoles(ICollection<IdentityUserRole> userRoles)
18      {
19          var names = new List<string>();
20          foreach (var userRole in userRoles)
21          {
22              foreach (var role in roleManager.Roles)
23              {
24                  if (role.Id == userRole.RoleId)
25                      names.Add(role.Name);
26              }
27          }
28          return string.Join(", ", names);
29      }
30
31      protected void grdUsers_SelectedIndexChanged(object sender, EventArgs e)
32      {
33      }
34  }
```

Figure 13-28: Methods for the GridView

Lines 3 and 4 set up references to the user manager and role manager that will be used in various methods in the class. The `Page_Load` method populates the variables with the two objects from the OWIN context.

Lines 12–15 contain the `grdUsers_GetData` method that employs the `Users` property from the user manager to provide the data necessary to populate the `GridView`.

Lines 17–29 show the `ListRoles` method that converts the list of roles for the user into a string to be displayed. The method receives the `Roles` property associated with the `ApplicationUser` object as a parameter. Then, two nested `foreach` loops go through the list of roles for the user and match them up with all the roles in the system exposed through the role manager's `Roles` property. Line 28 joins the strings in the `names` string array with a comma.

Lines 31–33 contain the method that will be used later to determine which user was selected.

Set the UserRoles page as the start page and run the application. Now you should see all the users who are registered in the system, as shown in Figure 13-29. You will add functionality later to populate the Roles column.

Users

User Name	Name	Email	Roles	
patrick@krabby.com	Patrick Star	patrick@krabby.com		Select
spongebob@krabby.com	Spongebob Squarepants	spongebob@krabby.com		Select

Figure 13-29: GridView of users in the system with roles

Step 39: Add a data source for the role table.

The next `GridView` will simply show the contents of the AspNetRoles table and allow for selecting a record for further processing. For this, you will add a `SqlDataSource`.

With UserRoles.aspx open, place the cursor immediately following the closing tag of the `GridView`. In the Toolbox, expand the Data section and double-click `SqlDataSource`.

Change the ID to `sqlRoles`.

Then click the context menu and select *Configure Data Source*, as shown in Figure 13-30.

```
<asp:SqlDataSource ID="sqlRoles" runat="server"></asp:SqlDataSource>
```
⌐ **SqlDataSource Tasks**
  Format Element
  Configure Data Source...

Figure 13-30: Configuring SqlDataSource

In the dialog, use the dropdown to select DefaultConnection for the data connection. Click *Next*.

Choose the AspNetRoles table from the dropdown. Keep the checkmark by the asterisk.

Click *Advanced...* and select both checkboxes to generate INSERT, UPDATE, and DELETE statements, as well as to use optimistic concurrency.

Click *Next*, and then click *Finish*. Figure 13-31 shows the code that is generated for the completed `SqlDataSource` control.

```
1   <asp:SqlDataSource ID="sqlRoles" runat="server" ConflictDetection="CompareAllValues"
2         ConnectionString='<%$ ConnectionStrings:DefaultConnection %>'
3         DeleteCommand="DELETE FROM [AspNetRoles] WHERE [Id] = @original_Id AND [Name] =
                     @original_Name"
4         InsertCommand="INSERT INTO [AspNetRoles] ([Id], [Name]) VALUES (@Id, @Name)"
5         OldValuesParameterFormatString="original_{0}"
6         SelectCommand="SELECT * FROM [AspNetRoles]"
7         UpdateCommand="UPDATE [AspNetRoles] SET [Name] = @Name WHERE [Id] =
                     @original_Id AND [Name] = @original_Name">
8         <DeleteParameters>
9             <asp:Parameter Name="original_Id" Type="String"></asp:Parameter>
10            <asp:Parameter Name="original_Name" Type="String"></asp:Parameter>
11        </DeleteParameters>
12        <InsertParameters>
13            <asp:Parameter Name="Id" Type="String"></asp:Parameter>
14            <asp:Parameter Name="Name" Type="String"></asp:Parameter>
15        </InsertParameters>
16        <UpdateParameters>
17            <asp:Parameter Name="Name" Type="String"></asp:Parameter>
18            <asp:Parameter Name="original_Id" Type="String"></asp:Parameter>
19            <asp:Parameter Name="original_Name" Type="String"></asp:Parameter>
20        </UpdateParameters>
21    </asp:SqlDataSource>
```

Figure 13-31: Completed SqlDataSource for roles

Step 40: Add a GridView for roles.

Next, you'll add a `GridView` to display the roles in the table, along with controls to add, update, and delete roles.

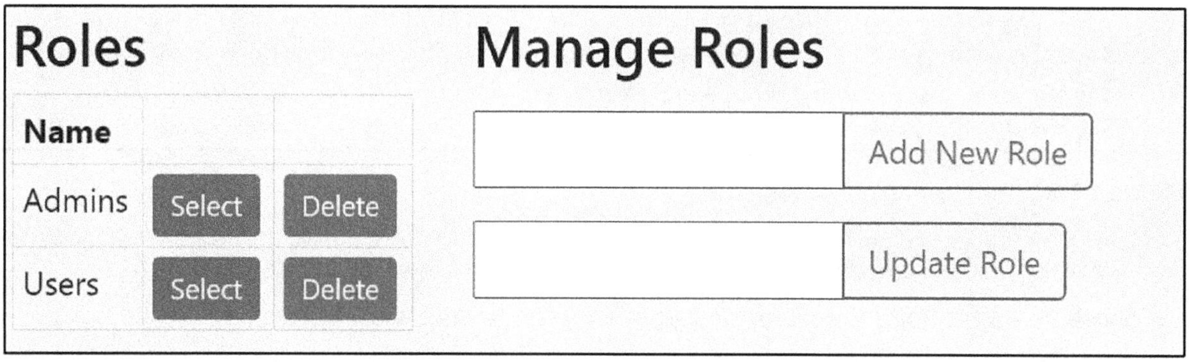

Figure 13-32: Interface for managing roles

Add the code shown in Figure 13-33 below the GridView that displays users.

```
1   <div class="row mb-3">
2       <div class="col-sm-3">
3           <h3>Roles</h3>
4           <asp:GridView ID="grdRoles" runat="server" AutoGenerateColumns="False"
5               DataKeyNames="Id" DataSourceID="sqlRoles"
6               OnSelectedIndexChanged="grdRoles_SelectedIndexChanged"
7               OnRowDeleted="grdRoles_RowDeleted"
8               CssClass="table table-sm table-bordered">
9               <Columns>
10                  <asp:BoundField DataField="Name" HeaderText="Name"
11                      SortExpression="Name" />
12                  <asp:CommandField ShowSelectButton="True"
13                      ControlStyle-CssClass="btn btn-sm btn-primary" />
14                  <asp:CommandField ShowDeleteButton="true"
15                      ControlStyle-CssClass="btn btn-sm btn-danger" />
16              </Columns>
17          </asp:GridView>
18      </div>
19      <div class="col-sm-9">
20          <h3>Manage Roles</h3>
21          <div class="input-group mt-3">
22              <asp:TextBox ID="txtNewRole" runat="server"></asp:TextBox>
23              <div class="input-group-append">
24                  <asp:Button runat="server" Text="Add New Role"
25                      CssClass="btn btn-outline-secondary"
26                      OnClick="btnNewRole_Click" />
27              </div>
28          </div>
29
30          <div class="input-group mt-3">
31              <asp:TextBox ID="txtRoleToUpdate" runat="server"></asp:TextBox>
32              <div class="input-group-append">
33                  <asp:Button runat="server" ID="btnUpdateRole" Text="Update Role"
34                      OnClick="btnUpdateRole_Click"
35                      CssClass="btn btn-outline-secondary" />
36              </div>
37              <asp:Label ID="lblSelectedRole" runat="server"
38                  Visible="false"></asp:Label>
39          </div>
40      </div>
41  </div>
```

Figure 13-33: Code for displaying and managing roles

This contains two blocks of controls in one row. The first column contains a GridView of roles in lines 4–17. Note the DataSourceId in line 5 points to the SqlDataSource you created previously. The table contains two fields, Id and Name. We use the Id field as the DataKeyName for the GridView (line 5) and display the Name field (lines 10–11). In addition, lines 12–13 add a Select button, and lines 14–15 add a Delete button. The action methods for the two buttons are defined in lines 6 and 7.

The second column (starting in line 19) contains two input groups that combine a textbox and a button for adding a new role (lines 21–28) and updating a selected role (lines 30–36). The label in lines 37–38 is needed to store the name of the selected role to find the original role to update.

Step 41: Add code-behind to manage roles.

Switch to UserRoles.aspx.cs and add the code for the methods referenced in Figure 13-33. You will be adding the following methods (see Figure 13-34):

- btnNewRole_Click. Lines 10–17. Creates a new role with the name entered in the txtNewRole textbox. It starts by checking with the role manager to see if the role already exists and, if not, creates it by passing the name of the role to the Create method.

- `grdRoles_SelectedIndexChanged`. Lines 19–25. Adds the name of the selected role to the textbox for updating, as well as to the hidden label. The method gets the selected Id from the GridView (line 21). This is used in line 22 to get the role that corresponds to the Id. Lines 23 and 24 then add the text of the name to the textbox and label.

- `btnUpdateRole_Click`. Lines 27–36. Updates a selected role with a new name. Line 29 finds the role based on the name stored in the hidden label. If a role was found, line 32 updates its name based on the textbox and line 33 updates it with the role manager.

- `grdRoles_RowDeleted`. Lines 38–47. Deletes the corresponding role. Line 40 gets the value from the deleted row. The value contains the role ID, which is then used in line 41 to find the corresponding role. If the role exists, the role manager deletes it in line 44.

- `refresh`. Lines 1–8. Resets and refreshes the user interface. This method is called from each of the action methods. It starts by data binding the two GridView controls, which will ensure they're showing updated data. Then it simply sets the two textboxes to empty values.

You should start by adding the refresh method, since it will be referenced in each of the other methods.

```
1   private void refresh()
2   {
3       grdRoles.DataBind();
4       grdUsers.DataBind();
5       lblSelectedRole.Text = "";
6       txtRoleToUpdate.Text = "";
7       txtNewRole.Text = "";
8   }
9
10  protected void btnNewRole_Click(object sender, EventArgs e)
11  {
12      if (!roleManager.RoleExists(txtNewRole.Text))
13      {
14          roleManager.Create(new IdentityRole(txtNewRole.Text));
15          refresh();
16      }
17  }
18
19  protected void grdRoles_SelectedIndexChanged(object sender, EventArgs e)
20  {
21      string roleId = grdRoles.SelectedValue.ToString();
22      var role = roleManager.FindById(roleId);
23      lblSelectedRole.Text = role.Name;
24      txtRoleToUpdate.Text = role.Name;
25  }
26
27  protected void btnUpdateRole_Click(object sender, EventArgs e)
28  {
29      var role = roleManager.FindByName(lblSelectedRole.Text);
30      if (role != null)
31      {
32          role.Name = txtRoleToUpdate.Text;
33          roleManager.Update(role);
34          refresh();
35      }
36  }
37
38  protected void grdRoles_RowDeleted(object sender, GridViewDeletedEventArgs e)
39  {
40      string roleId = e.Values[0].ToString();
41      var role = roleManager.FindById(roleId);
42      if (role != null)
43      {
44          roleManager.Delete(role);
45          refresh();
46      }
47  }
```

Figure 13-34: Methods for managing roles

Run the website to test that you can add, delete, and update roles.

Step 42: Assign roles to users on the .aspx page.

The last part of the user interface is to be able to assign roles to users. To do this, you will add a new section to the user interface. In UserRoles.aspx, add the code in lines 1–13 of Figure 13-35 just before the last </asp:Content> tag (line 14).

```
1   <div>
2       <h3>Assign Roles</h3>
3       <div class="form-group col-sm-3">
4           <label for="lblSelectedRole">Selected Role:</label>
5           <asp:Label ID="lblSelectedRole" runat="server"></asp:Label>
6       </div>
7       <div class="form-group col-sm-3">
8           <label for="lblSelectedUser">Selected User:</label>
9           <asp:Label ID="lblSelectedUser" runat="server"></asp:Label>
10      </div>
11      <asp:Button ID="btnAddRoleToUser" runat="server" Text="Add Role to User"
12      OnClick="btnAddRoleToUser_Click" CssClass="btn btn-primary" />
13  </div>
14  </asp:Content>
```

Figure 13-35: ASPX code to assign roles to users

This contains two labels and a button. Note that the first label in line 5 has the same name as the hidden label you used earlier for updating the roles. You will need to delete or comment out that previous label. But the code-behind will still work in the same way as before—now you will just get to see the selected role.

Step 43: Add code-behind for assigning users to roles.

To assign a user to a particular role, you'll need to make several changes to the code-behind file:

1. Determine which user was selected in the GridView.

2. Implement the action method for the button to actually assign the role selected in the role GridView to the user.

3. Update the refresh method to clear the selected user.

First, open UserRoles.aspx.cs and implement the grdUsers_SelectedIndexChanged method, which was created earlier when you created the Users GridView, by adding lines 3–4 in Figure 13-36. This method is called whenever the user clicks the *Select* link in the GridView containing the user information, as shown in line 13 in Figure 13-27.

```
1   protected void grdUsers_SelectedIndexChanged(object sender, EventArgs e)
2   {
3       string userId = grdUsers.SelectedValue.ToString();
4       var user = userManager.FindById(userId);
5       lblSelectedUser.Text = user.FirstName + " " + user.LastName;
6   }
```

Figure 13-36: Code to show the selected user

Because the GridView's DataKeyNames attribute is set to Id, the value recorded when a row is selected is that of the Id property of the ApplicationUser class. Line 3 gets this value and converts it to a string before storing it in the `userId` variable. In line 4, the User Manager finds the user object corresponding to the user id, and in line 5 the user's first and last name is added to the label.

Next, implement the `btnAddRoleToUser_Click` method as shown in Figure 13-37. This is called whenever the button is clicked to assign the role selected in the role `GridView` to the user selected in the user `GridView`.

```
1   protected void btnAddRoleToUser_Click(object sender, EventArgs e)
2   {
3       if (grdUsers.SelectedValue != null & lblSelectedRole.Text != null)
4       {
5           string userId = grdUsers.SelectedValue.ToString();
6           string role = lblSelectedRole.Text;
7           userManager.AddToRole(userId, role);
8           refresh();
9       }
10  }
```

Figure 13-37: Adding a role to a user

The method first checks if both a user and a role are selected in line 3. If so, it gets the `userId` and `roleId` in lines 5 and 6. The key line is line 7, which uses the user manager to add the role to the user by calling the `AddToRole` method. This method just needs the user ID and role name, and not the actual objects, so you don't need to first find the objects; instead, you can just use the simple string values. Finally, the refresh method is called to update the user interface after the role has been added. This will update the user `GridView` with the new role and clear out all the controls.

The final piece is to add a line to the refresh method to clear the selected user's name from the label. To do this, add the following line to the refresh method:

```
lblSelectedUser.Text = "";
```

Test the page to make sure you can assign a role to a user.

Review Questions

15. Which object is used to manage roles in the system?
16. Which object is used to manage users in the system?
17. Which object is used to assign and remove roles from a user?

Exercises

Exercise 13.1: New Master Page

Add a master page for staff and have the admin pages use that page instead of Main.Master.

Exercise 13.2: Manage Performance Details

Create a page for managing the details of a single performance (description, title, image etc.). Ensure that only users with admin roles are able to access the page and update the new page.

Exercise 13.3: Protect Account Pages

Protect the account pages. (Don't allow logged-in users to register.)

Exercise 13.4: Add Role

Create a new role and distinguish between the two roles (manager and worker) on certain pages.

Exercise 13.5: Security Issues

Explain what security issue arises from having the user registration page publicly available in the Grand Opera House system. How can this issue be fixed?

Exercise 13.6: User Data Upload

Create a page where an admin user can upload a CSV file with data about users and then create user accounts based on the information in the file. Include the default password in the file.

Exercise 13.7: Generate Passwords

Modify the page in the previous exercise to generate initial passwords for the users and display them to the admin user.

Exercise 13.8: Logout from the Master Page

Modify the master page to allow the user to log out directly by clicking a logout button.

Exercise 13.9: Password Reset

Create a password reset page. Then require users to reset their password on their first login.

Exercise 13.10: Remove Role

Add a button to remove a role from a user on the UserRoles page. This can be done by calling the RemoveFromRole method on the user manager object.

Exercise 13.11: Customer Account

Expand the Grand Opera House system to allow customers to create an account and store their previous purchases. This will require changing the database, as well as adding and modifying several pages.

## Chapter 14

Interactive Web Pages with Ajax

As the web has matured over the past several years, one trend has been toward creating pages that are more interactive and mimic what was traditionally only possible using desktop applications. This allows for a much smoother and richer user experience. Web pages with deep user interactivity are sometimes called *rich internet applications* (*RIA*). Such applications allow the user to interact with the page without causing a postback of the entire page for every action taken. One of the most common technologies for implementing an RIA is *Asynchronous JavaScript and XML* (Ajax). This technology is used in some form on most modern websites and allows you to do things as simple as showing a spinner while an operation is in progress to allowing for the drag-and-drop of elements on the page.

Learning Objectives

After studying this chapter, you should be able to

- Describe the architecture used to implement Ajax.
- Describe the difference between full and partial postbacks.
- Use the five Ajax server controls to implement partial postbacks on a web page.
- Implement custom error handling pages.
- Handle errors that occur during partial postbacks.
- Use the controls in the Ajax Control Toolkit to implement rich website functionality.

14.1 Introduction to Ajax

Ajax is built on top of the traditional hypertext transfer protocol (HTTP) that you have seen in previous chapters, which is the foundation of the web. When a user takes an action on a web page, such as clicking on a button, this causes a postback to the same page. This leads the browser to send an HTTP request to the server that includes the viewstate of the page. After processing, the server returns an HTTP response that contains an updated version of the page with a new viewstate, which the browser then renders. In the traditional model, every time the user takes an action the entire page is rendered. For the user, this looks like the page blinks off and back on again with the updated content.

With Ajax, you set up regions of the page that can be updated independently of the overall page. Thus the user can take an action that updates only a small part of the page, whereas the rest of the page isn't impacted.

This is implemented through an object that is a part of all modern web browsers, called XmlHttpRequest. This object manages the interaction with the server and can send an *asynchronous postback*, which will result in an update of only the viewstate of the controls—and not the entire page. Figure 14-1 shows the difference between regular HTTP requests/responses and Ajax requests/responses. With Ajax, the browser's XMLHttpRequest object sends an XMLHttpRequest message to the server, which in turn responds with data instead of a fully formed HTML page.

In ASP.NET, Ajax is available to use out of the box through five server controls that are available in the ToolBox: ScriptManager, ScriptManagerProxy, UpdatePanel, Timer, and UpdateProgress. In addition to these basic objects, you can also download the Ajax Control Toolkit, which contains rich controls and extensions to existing controls, allowing them to be manipulated in new and richer ways.

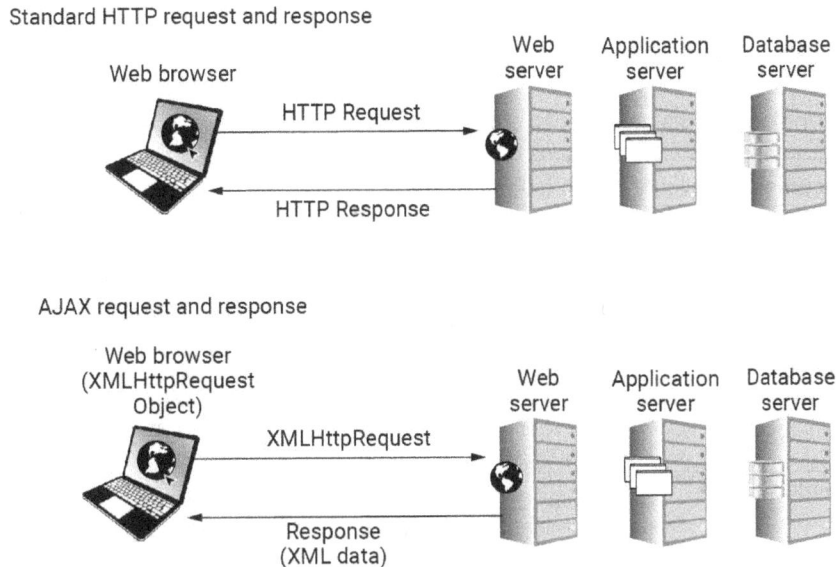

Figure 14-1: Standard HTTP versus Ajax requests and responses

14.2 ASP.NET Ajax Server Controls

As mentioned previously, ASP.NET comes with five controls in the Toolbox that allows you to implement Ajax in your website. In this section, we will describe each of these objects and how to use them.

14.2.1 ScriptManager

The `ScriptManager` must be present on each page that needs to use Ajax. It is often added to the master page so that it's available on all pages on the site. In addition to enabling Ajax, it also allows you to load JavaScript files as well as to register web services for use on the page. You can only have one `ScriptManager` control for a page, so if you have added it to the master page, you cannot add it to a content page. The `ScriptManager` control for enabling Ajax is quite simple and can be added like this:

```
<asp:ScriptManager ID="ScriptManager1" runat="server"></asp:ScriptManager>
```

Table 14-1 shows some commonly used properties and events of the `ScriptManager` control. You will see examples of how to use some of these properties later in the chapter.

To register JavaScript scripts, add a `<Scripts>` element inside the `ScriptManager` and then add an `asp:ScriptReference` element for each script to register, as shown here:

```
1  <asp:ScriptManager ID="ScriptManager1" runat="server">
2      <Scripts>
3          <asp:ScriptReference Path="MyScript.js" />
4          <asp:ScriptReference Assembly="SampleControl"
                                 Name="SampleControl.UpdatePanelAnimation.js" />
5      </Scripts>
6  </asp:ScriptManager>
```

Line 3 registers a JavaScript file called MyScript.js. Line 4 registers a JavaScript file that is embedded in the control assembly for a custom control.

The code for registering a web service is similar to this:

```
1  <asp:ScriptManager ID="ScriptManager1" runat="server">
2      <Services>
3          <asp:ServiceReference Path="~/Services/WebService.asmx" />
4          <asp:ServiceReference Path="http://myserver.com/Services/WebService.svc" />
5      </Services>
6  </asp:ScriptManager>
```

In line 3, a WCF web service is defined in WebServce.asmx. Line 4 shows how to register based on a web service located on a remote server in a .svc file.

Table 14-1: Common properties and events of the ScriptManager control

ScriptManager properties and events	Description
AsyncPostBackErrorMessage	Gets or sets the error message that is sent to the client when an unhandled server exception occurs during an asynchronous postback
AsyncPostBackTimeout	Gets or sets a value that indicates the time, in seconds, before asynchronous postbacks time out if no response is received
OnAsyncPostBackError	Event that raises the AsyncPostBack event
IsInAsyncPostBack	Gets a value that indicates whether the current postback is being executed in partial-rendering mode
EnablePartialRendering	Gets or sets a value that enables the partial rendering of a page, which in turn enables you to update regions of the page individually by using UpdatePanel controls
SupportsPartialRendering	Gets a value that indicates whether the client supports partial-page rendering

14.2.2 ScriptManagerProxy

The ScriptManagerProxy control is used to extend the functionality of the ScriptManager control. It is primarily used when the ScriptManager is added to the master page, and you need to load additional JavaScript or register web services on a content page. You can have more than one ScriptManagerProxy on a page, but you cannot change the settings already set for the ScriptManager. The ScriptManagerProxy can only be used to register additional JavaScript or web services.

This shows a ScriptManagerProxy that registers both a script and a web service:

```
1  <asp:ScriptManagerProxy ID="Proxy1" runat="server">
2     <Scripts>
3        <asp:ScriptReference Path="~/Scripts/MyScript.js" />
4     </Scripts>
5     <Services>
6        <asp:ServiceReference Path="~/Services/MyService.asmx" />
7     </Services>
8  </asp:ScriptManagerProxy>
```

14.2.3 UpdatePanel

The UpdatePanel is used to enable partial page updates on asynchronous postbacks. Controls placed inside the same UpdatePanel are all updated together, without affecting the rest of the page. Each page can contain multiple UpdatePanel controls that can be updated independently of each other.

The controls that will be updated are placed inside a ContentTemplate element. Asynchronous postbacks that update the controls can be triggered by any control inside the update panel or by adding a Triggers element that will contain references to controls outside the update panel that can trigger an asynchronous postback.

The UpdatePanel has two important properties that control how the panel is updated:

- **ChildrenAsTriggers:** This Boolean property determines if controls placed inside the panel will trigger an asynchronous postback for the panel. The default is True.

- **UpdateMode:** You can use this property to control how the panel should be updated. If you set this to Always, which is the default, the content of the panel will be updated when any control on the page causes a postback. On the other hand, if you set it to Conditional, the panel will only be updated when one of its own triggers causes a postback. Note that if an UpdatePanel is nested inside another UpdatePanel, it is automatically updated when the parent panel is updated.

Here's an example of a simple UpdatePanel:

```
1  <asp:UpdatePanel ID="UpdatePanel2" runat="server">
2      <ContentTemplate>
3          <asp:Button ID="Button1" runat="server" Text="Save" />
4          Result: <asp:Label ID="lblResult" runat="server" Text=""></asp:Label>
5      </ContentTemplate>
6      <Triggers>
7          <asp:AsyncPostBackTrigger ControlID="ddlEvent"
8              EventName="SelectedIndexChanged" />
9          <asp:PostBackTrigger ControlID="Button1" />
10     </Triggers>
11 </asp:UpdatePanel>
```

In this example, there are two controls in the ContentTemplate: a button and a label. You can use the Triggers element to indicate controls that will cause the panel to be updated. The AsyncPostbackTrigger element in line 7 shows how a control outside the panel can cause an asynchronous postback that updates the panel. The PostBackTrigger in line 9 shows how a control inside the panel can cause a full postback of the page.

14.2.4 Timer

The Timer control allows you to update a panel on a periodic basis, without the user having to trigger the update. You will typically place the Timer control inside an UpdatePanel that you want to update at a regular interval. If you place it outside an UpdatePanel, it will cause a full postback.

Setting up a Timer is fairly simple, as shown here:

```
1  <asp:UpdatePanel ID="UpdatePanel2" runat="server" UpdateMode="Conditional">
2      <ContentTemplate>
3          <asp:Timer ID="Timer1" runat="server" Interval="5000" Enabled="true"></asp:Timer>
4      </ContentTemplate>
5  </asp:UpdatePanel>
```

The Interval property determines the time in milliseconds between postbacks. The 5000 in this example indicates "every five seconds." If you set the interval too short, you may cause too much load on the server, with a high number of postbacks. You do not have control over the accuracy of the timer, as this depends on the JavaScript implementation in the user's browser.

14.2.5 UpdateProgress

The UpdateProgress control is used to display a status message to the user if an asynchronous postback takes a while to finish. This could happen if it involves a call to a web service that takes time to complete or if a large file is being uploaded. Here is an example of an UpdateProgress control placed inside an UpdatePanel:

```
1  <asp:UpdatePanel ID="UpdatePanel2" runat="server" UpdateMode="Conditional">
2      <ContentTemplate>
3          <asp:UpdateProgress ID="UpdateProgress1" runat="server">
4              <ProgressTemplate>
5                  Update in progress. Please be patient.
6              </ProgressTemplate>
7          </asp:UpdateProgress>
8      </ContentTemplate>
9  </asp:UpdatePanel>
```

The content to be displayed is included in a ProgressTemplate element (line 4). By default, the content will display after 500 ms (0.5 seconds). You can adjust this time using the DisplayAfter property.

You can also place the UpdateProgress control outside an UpdatePanel, as shown in this example:

```
1  <asp:UpdateProgress ID="UpdateProgress2" runat="server"
2      AssociatedUpdatePanelID="UpdatePanel3"
3      DynamicLayout="false">
4      <ProgressTemplate>
5          <div class="spinner-border text-primary" role="status">
6              <span class="sr-only">Loading...</span>
7          </div>
8      </ProgressTemplate>
9  </asp:UpdateProgress>
```

The `AssociatedUpdatePanelID` property allows you to indicate which panel the progress update will be displayed for. This example also shows the `DynamicLayout` property. If set to false, space will always be allocated to the progress update. By default, it is set to true and space is only allocated when the content is displayed. This example also shows how you can use Bootstrap to display a spinner (line 5). Line 6 is a text message that is displayed for screen readers for people with visual impairments.

Review Questions

1. Describe the difference between asynchronous and regular postbacks.
2. How do you control which part of the page is updated with an asynchronous postback?
3. Describe three functionalities of the ScriptManager control.
4. Where do you place a Timer control to update a portion of a page?
5. How do you display a message to a user if an operation is taking a long time to complete?

Tutorial: A Simple Example of Ajax

In this tutorial, you will set up a new page that uses the Ajax controls you have just read about. In this tutorial, you will create a page that the Grand Oshkosh Theater managers can use to calculate how many tickets to sell for an event to be profitable. Figure 14-2 shows the final page layout.

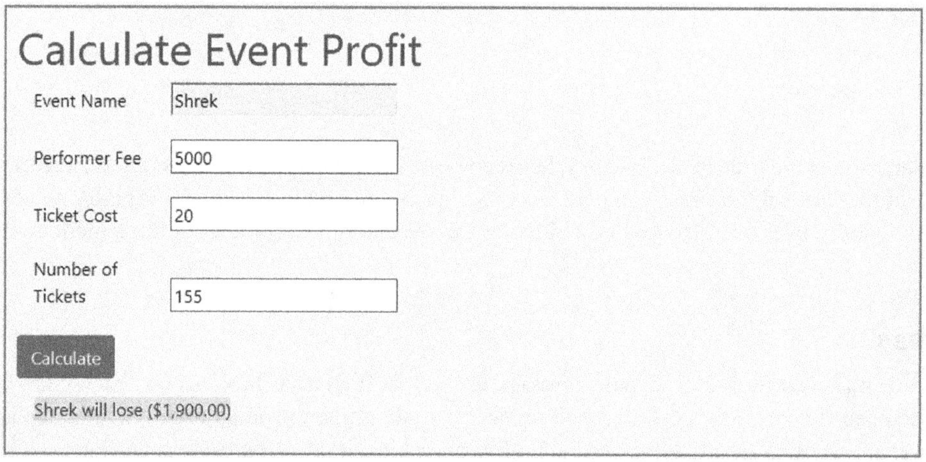

Figure 14-2: Completed form to calculate event profit

Step 1: Open the project.

You can continue from the project you worked on in the previous chapter or download the Chapter 14 starter project.

Step 2: Create the page.

Create a new page named *ProfitCalculator* based on the master page in the *Admin* folder of the website.

Step 3: Create the form.

Set the Title attribute in the `@Page` directive to *Calculate Event Profit*.

Implement the form fields by adding the code shown in lines 3–28 of Figure 14-3 inside the second content placeholder. This code is also available as ProfitCalculator.txt.

This is a pretty straightforward form. Clicking the button in lines 22–24 will show the result of the calculation in the label in line 27, based on the values from the textboxes.

```
1  <asp:Content ID="Content2" ContentPlaceHolderID="MainContentPlaceHolder"
2      runat="server">
3      <h1>Calculate Event Profit</h1>
4      <div class="form-group">
5          <label for="txtEventName" class="col-md-2">Event Name</label>
6          <asp:TextBox runat="server" ID="txtEventName"></asp:TextBox>
7      </div>
8      <div class="form-group">
9          <label for="txtPerformerFee" class="col-md-2">Performer Fee</label>
10         <asp:TextBox runat="server" ID="txtPerformerFee"></asp:TextBox>
11     </div>
12     <div class="form-group">
13         <label for="txtTicketCost" class="col-md-2">Ticket Cost</label>
14         <asp:TextBox runat="server" ID="txtTicketCost"></asp:TextBox>
15     </div>
16     <div class="form-group">
17         <label for="txtNumTickets"
18             class="col-md-2">Number of Tickets</label>
19         <asp:TextBox runat="server" ID="txtNumTickets"></asp:TextBox>
20     </div>
21     <div>
22         <asp:Button runat="server" ID="btnCalculate"
23             CssClass="btn btn-primary" Text="Calculate"
24             OnClick="btnCalculate_Click" />
25     </div>
26     <div class="m-3">
27         <asp:Label runat="server" ID="lblResult"></asp:Label>
28     </div>
29 </asp:Content>
```

Figure 14-3: Form to calculate profit

Step 4: Implement the code-behind page.

Open ProfitCalculator.aspx.cs and add the code in the btnCalculate_Click method shown in Figure 14-4. If you typed the code in Figure 14-3, you would have been prompted in line 24 to create a new method. This would generate a method stub in the code-behind file, and you would just have to enter lines 3–24.

```
1  protected void btnCalculate_Click(object sender, EventArgs e)
2  {
3      double performerCost = Double.Parse(txtPerformerFee.Text);
4      int numTickets = int.Parse(txtNumTickets.Text);
5      double ticketCost = Double.Parse(txtTicketCost.Text);
6      double profit = numTickets * ticketCost - performerCost;
7      if (profit > 0)
8      {
9          lblResult.Text = string.Format("{0} will bring a profit of {1:c}",
10             txtEventName.Text, profit);
11         lblResult.CssClass = "alert-success";
12     }
13     else if (profit <0)
14     {
15         lblResult.Text = string.Format("{0} will lose {1:c}",
16             txtEventName.Text, profit);
17         lblResult.CssClass = "alert-danger";
18     }
19     else
20     {
21         lblResult.Text = string.Format("{0} will break even",
22             txtEventName.Text);
23         lblResult.CssClass = "alert-info";
24     }
25 }
```

Figure 14-4: Method to calculate profit

This method reads the values from the textboxes and uses them to calculate the profit for the show in line 6. Lines 7–24 show a message to the user, based on whether the profit is positive.

Step 5: Test the page.

Set the page as the startup page.

Run the page (Ctrl+F5) and enter data in the form. Fill in the fields and click the button to show a message at the bottom of the page.

Resize the browser and scroll down so the top of the page is no longer visible. When you click the button now, the page will refresh and will scroll to the top. This is because the button causes a full postback. Next, we will explore how a partial page refresh with an asynchronous postback only refreshes a part of the page and doesn't scroll you to the top.

Step 6: Add a ScriptManager to the master page.

To allow for Ajax functionality on every page of the site, you will add a `ScriptManager` control to the master page.

Open Main.master and add a `ScriptManager` control at the top of the `form` element before the `div` element surrounding the content placeholder. You can add it from the Toolbox by placing the cursor at the right spot and double-clicking the control in the dropbox. Or you can type the code:

```
<asp:ScriptManager ID="ScriptManager1" runat="server"></asp:ScriptManager>
```

Step 7: Add an UpdatePanel.

Switch back to ProfitCalculator.aspx and add an `UpdatePanel` with a `ContentTemplate` after the last form-group (lines 16–20 in Figure 14-3). Then move the code in lines 21–28 in Figure 14-3 into the `ContentTemplate`, as shown in Figure 14-5.

```
1   <asp:UpdatePanel ID="UpdatePanel1" runat="server">
2       <ContentTemplate>
3           <div>
4               <asp:Button runat="server" ID="btnCalculate"
5                   CssClass="btn btn-primary" Text="Calculate"
6                   OnClick="btnCalculate_Click" />
7           </div>
8           <div class="m-3">
9               <asp:Label runat="server" ID="lblResult"></asp:Label>
10          </div>
11      </ContentTemplate>
12  </asp:UpdatePanel>
```

Figure 14-5: UpdatePanel with controls

You can now test the page again. It should function just like before, but you should notice that when you click the button, only the label changes and you are not scrolled anywhere else on the page. That's because you are now using an asynchronous postback and only updating the controls in the UpdatePanel.

Step 8: Calling the event method from the textboxes.

The managers at the Grand Oshkosh are busy people, so they would like a more streamlined form that updates the result every time one of the numbers is changed in a textbox, without having to click the Calculate button.

To do this, you will need to enable the textboxes to cause a postback and call the event method to update the label. Modify each of the three textboxes that take a number with the two properties `AutoPostBack` and `OnTextChanged`, as shown here:

```
1    <asp:TextBox runat="server" ID="txtPerformerFee"
2       AutoPostBack="true" OnTextChanged="btnCalculate_Click"></asp:TextBox>
```

Setting `AutoPostBack` to True will cause a postback from the textbox. The `OnTextChanged` property indicates that the button's event method (`btnCalculate_Click`) will be called on that postback.

To avoid an error where the method tries to calculate the values before they're all typed in, add the `if` statement in lines 3–11 in Figure 14-6 to the beginning of the `btnCalculate_Click` method. You'll also need to add an extra curly brace at the end of the method. Note that line 12 is the current first line in the method and should not be entered again.

```
1    protected void btnCalculate_Click(object sender, EventArgs e)
2    {
3        if (txtEventName.Text == "" ||
4            txtNumTickets.Text == "" ||
5            txtPerformerFee.Text == "" ||
6            txtTicketCost.Text == "")
7        {
8            lblResult.Text = "";
9        }
10       else
11       {
12           double performerCost = Double.Parse(txtPerformerFee.Text);
13           [ Rest of method ... ]
14       }
15   }
```

Figure 14-6: If statement to ensure the calculation is only performed on complete data

If you run the page now, the result should be updated whenever a number is changed in one of the textboxes. However, this will be done as a full postback, so the page will scroll to the top every time the user changes a number and the tab placement will also be lost. You can use asynchronous postbacks to improve this user experience.

Step 9: Update the panel by external control.

Open ProfitCalculator.aspx and add a `Triggers` element to the `UpdatePanel`, as shown in lines 5–12 in Figure 14-7.

```
1    <asp:UpdatePanel ID="UpdatePanel1" runat="server">
2        <ContentTemplate>
3            [ ... ]
4        </ContentTemplate>
5        <Triggers>
6            <asp:AsyncPostBackTrigger
7                ControlID="txtPerformerFee" EventName="TextChanged" />
8            <asp:AsyncPostBackTrigger
9                ControlID="txtTicketCost" EventName="TextChanged" />
10           <asp:AsyncPostBackTrigger
11               ControlID="txtNumTickets" EventName="TextChanged" />
12       </Triggers>
13   </asp:UpdatePanel>
```

Figure 14-7: External triggers in UpdatePanel

Add an `AsyncPostBackTrigger` for each of the texboxes' `TextChanged` event. Whenever those events fire, it will cause an asynchronous postback and just update the content in the panel—not the entire page. This means the scroll position and tab placement will be preserved. The user can either press Enter to stay in a textbox after making a change or press Tab to move to the next textbox. This provides a much more fluid and engaging user experience.

Step 10: Adding an UpdateProgress control in an existing UpdatePanel.

The page is very responsive in its current form, as the functionality is very simple. However, for more complex pages, or requests that require sending a request to an external web service, the response may take some time to come back. In those situations, you can use an `UpdateProgress` control to provide feedback to the user that data is being transferred. Here's how to add it to your page as part of an existing `UpdatePanel`.

Inside the `ContentTemplate` element of the `UpdatePanel`, add the `UpdateProgress` element, as shown in lines 4–10 in Figure 14-8.

```
1   <asp:UpdatePanel ID="UpdatePanel1" runat="server">
2       <ContentTemplate>
3           [ ... ]
4           <asp:UpdateProgress runat="server">
5               <ProgressTemplate>
6                   <div class="spinner-border text-info" role="status"
>
7                       <span class="sr-only">Loading...</span>
8                   </div>
9               </ProgressTemplate>
10          </asp:UpdateProgress>
11      </ContentTemplate>
12      <Triggers>
13          [ ... ]
14      </Triggers>
15  </asp:UpdatePanel>
```

Figure 14-8: UpdateProgress inside an UpdatePanel

In this case, the progress template will display a Bootstrap spinner (line 6). Line 7 will show a message for screen readers.

In order for the spinner to show up, you will need to add a delay to simulate some complex calculation or lookup function. Switch to the code-behind file and add the code in line 13 of Figure 14-9 to the `btnCalculate_Click` method. This will add 3,000 milliseconds (3 seconds) of delay to the process before completing the calculation.

```
1   protected void btnCalculate_Click(object sender, EventArgs e)
2   {
3       if (txtEventName.Text == "" ||
4           txtNumTickets.Text == "" ||
5           txtPerformerFee.Text == "" ||
6           txtTicketCost.Text == "")
7       {
8           lblResult.Text = "";
9       }
10      else
11      {
12          //Add delay to allow for demonstrating progress control
13          System.Threading.Thread.Sleep(3000);
14          double performerCost = Double.Parse(txtPerformerFee.Text);
15          int numTickets = int.Parse(txtNumTickets.Text);
16          [ ... ]
17      }
18  }
```

Figure 14-9: Add a delay to simulate a slow process

Now when you run the page, you should see a spinner show up below the Calculate button, whenever the button is clicked or the numbers are changed in the textboxes.

If the spinner doesn't show up, it may be that you need to update the version of Bootstrap you are using. The spinner was added in version 4.2.

Step 11: Add an UpateProgress control outside the UpdatePanel.

Sometimes you may want to show the progress in a different part of the screen from the `UpdatePanel`, where the delay is caused. To do this, you can add a stand-alone `UpdateProgress` element anywhere on the page. Figure 14-10 shows it added immediately following the page-top `h1` element and before the form begins. The only difference between this control and the one in Figure 14-8 is the color of the spinner (`text-danger` versus `text-info`) in line 5, allowing you to easily distinguish between the two. Figure 14-11 shows both spinners on the page.

```
1   <h1>Calculate Event Profit</h1>
2
3   <asp:UpdateProgress runat="server">
4       <ProgressTemplate>
5           <div class="spinner-border text-danger" role="status">
6               <span class="sr-only">Loading...</span>
7           </div>
8       </ProgressTemplate>
9   </asp:UpdateProgress>
10
11  <div class="form-group">
12  [ ... ]
```

Figure 14-10: Stand-alone UpdateProgress control

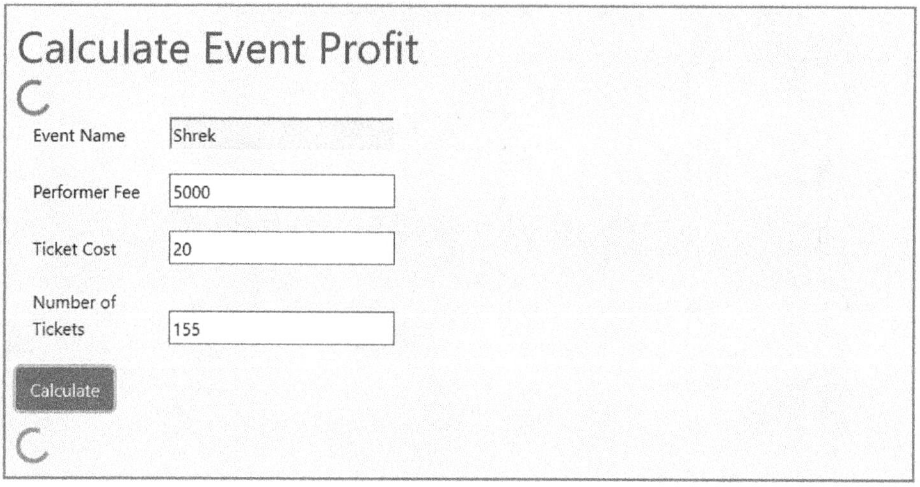

Figure 14-11: Form with two UpdateProgress spinners

14.3 Error Handling

When errors occur on a web page, the user is typically taken to a separate page with error messages. You have probably seen this in cases where you test a website and it crashes when an error is thrown. However, taking the user to a separate page can produce a bad user experience in terms of the kinds of pages you would build with partial postbacks. Most users' first instinct when seeing an error page would be to use the browser back button to get back to what they were working on. However, this can cause problems with the state of the controls, which means the user could lose their progress. Instead, ASP.NET by default will not take action on errors that occur within a page with partial page rendering. This means that you will have to add your own functionality to handle error messages. This section will discuss how you can customize error handling to show appropriate user error messages, as well as how to handle errors when you use a master page.

14.3.1 Custom Error Handling in ASP.NET

You have no doubt seen the "Yellow Screen of Death (YSOD)" that is shown when your application throws an uncaught exception. This page is very useful to developers, as it contains valuable information to help you fix the problem. However, to the average user, it is not useful and quite off-putting. And what's worse, the page can contain sensitive information that could be used by malicious actors to compromise your website. So, once your site is in production, you should not show this page to users.

ASP.NET allows for showing three different types of error pages:

1. **The Exception Details Yellow Screen of Death error page.** This page is shown to local users and contains details of the error that occurred, along with a stack trace.

2. **The Runtime Error Yellow Screen of Death error page.** This page is by default shown to remote users and hides the technical details of the error, but it doesn't look very professional and contains instructions on how to modify the web.config file to show detailed errors.

3. **A custom error page.** This is a page that is created by the developer to show a custom error message to the user. This can be customized to mirror the site layout and could contain links to move on or report a problem to the site administrator. You can use a master page for custom error pages. Many real-world sites spend quite a bit of energy customizing their error pages.

You can manage custom error pages in the Web.config file with a `customErrors` element. This element has two attributes that control which error page is displayed:

- **DefaultRedirect.** This optional attribute specifies the location of the custom error page to display.
- **Mode.** This required attribute accepts three different values—`On`, `Off`, and `RemoteOnly`—and determines in which situations the custom page is displayed.

- On: All users, regardless of whether they are local or remote, will get the custom error page.

- Off: The YSOD page is displayed to all users.

- RemoteOnly: Local users will get the YSOD page, while remote users will get the custom error page. This is the default setting.

Here's an example of a snippet from Web.Config that configures the website to show a custom error page named Ope.aspx to all users:

```
1   <configuration>
2       [...]
3       <system.web>
4           <customErrors mode="On"
5                           defaultRedirect="~/Errors/Ope.aspx" />
6           </customErrors>
7           [...]
8       </system.web>
9   </configuration>
```

The custom error pages are web pages on your site, so you simply create a directory and add pages to that location.

You should never display the YSOD to a remote user for a production site. However, without the error pages, you will have no way to find out what errors occurred on the site. So, in addition to custom error pages, you should implement a solution that logs any error that occurs. There are several ways to do this, but one approach is to install the Error Logging Modules and Handlers (ELMAH) package using NuGet. This will log all unhandled exceptions and provide a separate web page to view the details of the errors.

As discussed previously, HTTP provides various status messages with the response that is sent back to the client. The two most common codes are 200, which indicates success, and 400, which indicates a file wasn't found on the server. You can use these error codes to provide more meaningful error messages to the user. To specify a different page, depending on the HTTP status code, add an error element to the customErrors element, as shown here:

```
1   <customErrors mode="On"
2                   defaultRedirect="~/Errors/Ope.aspx" />
3       <error statusCode="404" redirect="~/Errors/404.aspx" />
4   </customErrors>
```

14.3.2 Error Handling in Ajax

To override the custom error pages on Ajax pages, you need to add an endRequest event handler to the PageRequestManager object, which manages partial-page rendering in the browser. To set this up, you would add the JavaScript to the head section of the page, as shown in Figure 14-12.

```
1   <script>
2       function pageLoad() {
3           let manager = Sys.WebForms.PageRequestManager.getInstance();
4           manager.add_endRequest(endRequestHandler);
5       }
6
7       function endRequestHandler(sender, args) {
8           if (args.get_error() != undefined) {
9               let Error = args.get_error();
10              //document.getElementById("divError").innerHTML = Error.message;
11              alert(Error.message);
12              args.set_errorHandled(true);
13          }
14      }
15  </script>
```

Figure 14-12: JavaScript to enable Ajax error handling

The `pageload` method is run whenever the page is loaded and gets a reference to the `PageRequestManager` object in line 3. Line 4 then adds the `endRequest` event handler. The function name in line 7 has to match the name of the event handler in line 4. Lines 10 and 11 show two different ways to display the message—either in a `MessageBox` (line 11) or in an HTML element on the page (line 10). Line 12 sets the `ErrorHandled` property to true to indicate the error has been handled.

In order to override the custom error page redirect, you will also need to modify the `ScriptManager` as shown in Figure 14-13.

```
1   <asp:ScriptManager ID="ScriptManager1" runat="server"
2       AllowCustomErrorsRedirect="false"
3       OnAsyncPostBackError="ScriptManager1_AsyncPostBackError">
4   </asp:ScriptManager>
```

Figure 14-13: ScriptManager modified to handle errors

By setting `AllowCustomRedirect` to false, you block the custom error page from loading. The `OnAsyncPostbackError` property specifies an event method to execute when an error occurs. The event method can be used to set an appropriate error message, as shown in Figure 14-14.

```
1   protected void ScriptManager1_AsyncPostBackError(object sender, AsyncPostBackErrorEventArgs e)
2   {
3       ScriptManager1.AsyncPostBackErrorMessage = "Error occurred<br>" + e.Exception.Message;
4   }
```

Figure 14-14: Setting the error message

This will display a fairly technical error message, which may not be appropriate. Instead, you could also log the technical error and just let the user know that something problematic occurred.

14.3.3 Ajax Error Handling on Master Pages

If the script manager is placed on a master page, the error handling also will have to be done on the master page. This can make it a little more difficult to provide appropriate messages and error handling. However, the approach is the same. You will need to add the JavaScript above to the master page and modify the script manager on the main page.

Tutorial: Managing Errors in the Profit Calculation

In this part of the tutorial, you will add custom error pages to the website. Then you will experiment with Ajax error handling—both with and without master pages involved.

Step 12: Create a custom error Page.

The first step in setting up custom error pages is to create the page that will be displayed when an error occurs. This is just like creating any other page. In this case, create a folder called *Errors,* and inside this folder, create a page called *Error.aspx* using the regular master page.

Add a generic error message like this in the main content placeholder (`Content2`):

```
1   <h1>Problem Occurred</h1>
2       <p>Unfortunately, something bad happened. Please try again.</p>
```

Step 13: Modify Web.config.

Add the following code to the `system.web` element in the Web.config file:

```
1    <customErrors defaultRedirect="~/Errors/Error.aspx" mode="On">
2    </customErrors>
```

Note that the `mode` in this case is set to `On`, so you can see the effect of the custom error pages even while working locally.

Run the site and generate an error page. This could be as simple as trying to access a page that doesn't exist. You can also enter a string into a field that expects a number on the profit calculation page. You should see your error page displayed.

Step 14: Enable Ajax error handling without a master page.

If you tried causing an error on the profit calculation page, you would have experienced a problem with the custom error approach. In the middle of the interactive experience, you would have been taken to a completely different page, and going back would have lost all the data already entered on the profit calculation form. There are two approaches to fixing this: without and with a master page.

Start by creating a new page named *ProfitCalcNoMaster.aspx* in the *Admin* directory. Be sure to not specify a master page.

Copy all content inside the `MainContentPlaceHolder` of the original profit calculator page to the `form` element of the new page.

Add the `ScriptManager` shown in Figure 14-13 at the top of the form element.

Next, add the code shown in Figure 14-12 to the `head` section of the page, but uncomment line 10 and comment out line 11, to show the error message in the `divError` HTML element on the page instead of as a `MessageBox`.

Step 15: Add functionality to the code-behind page.

Copy the `btnCalculate_Click` method from the code-behind page of the original profit calculation page to the new one. Comment out the three-second delay in the `btnCalculate_Click` method.

Add the method shown in Figure 14-14. This will set the error message to be displayed. In this case, the exception message is still shown, but you might choose to show even less information and instead log the detailed exception information. If you wanted to allow users to report on or get more information about a specific error message, you could generate a unique code to show to the user and log the detailed exception information. This would allow you to examine the log and provide more detail to the user, as needed.

Add the following line to the end of the `btnCalculate_Click` method to make sure an error message gets cleared on subsequent postbacks:

```
ScriptManager1.AsyncPostBackErrorMessage = "";
```

Step 16: Add an error message element.

On the ASPX page, add a `div` element to display error messages below the label used to display results, as shown in line 3:

```
1    <div class="m-3">
2        <asp:Label runat="server" ID="lblResult"></asp:Label>
3        <div class="text-danger" id="divError"></div>
4    </div>
```

Step 17: Control the error messages.

If you run the program now, you will see that the error messages shown are not particularly helpful. In order to control the error messages shown, you can catch and throw new error messages with more helpful error messages.

To do this, add a `try/catch` block around the code to further control and customize the error message displayed, as shown in Figure 14-15.

```
1   try
2   {
3       double performerCost = Double.Parse(txtPerformerFee.Text);
4       int numTickets = int.Parse(txtNumTickets.Text);
5       double ticketCost = Double.Parse(txtTicketCost.Text);
6       double profit = numTickets * ticketCost - performerCost;
7       if (profit > 0)
8       {
9           lblResult.Text = string.Format("{0} will bring a profit of {1:c}.",
10              txtEventName.Text, profit, multiplier);
11          lblResult.CssClass = "alert-success";
12      }
13      else if (profit < 0)
14      {
15          lblLoss.Text = string.Format("{0} will lose {1:c}.",
16              txtEventName.Text, profit);
17          lblLoss.CssClass = "alert-danger";
18      }
19      else
20      {
21          lblResult.Text = string.Format("{0} will break even.",
22              txtEventName.Text);
23          lblResult.CssClass = "alert-info";
24      }
25      ScriptManager1.AsyncPostBackErrorMessage = "";
26
27  }
28  catch (FormatException ex)
29  {
30      throw new Exception("Error Code 101: Enter a number.");
31  }
```

Figure 14-15: try/catch block to handle errors

The `try` section covers the code that converts the data entered in the textboxes so any exception that occurs if the user enters a non-number will be caught in line 30. The new exception thrown in line 32 is very generic and just informs the user to enter a number. It also shows an error code as an example of how you might develop your own set of error codes to refer to specific error conditions in your code.

This code moves the line of code that clears out the error message from the `Page_Load` method to line 25 at the very end of the `try` block. This way, the error message will be cleared when no exceptions occur.

When you run the code, you will see that the error message in line 30 is added to the error message you created in the `ScriptManager1_AsyncPostBackError` method.

14.3.4 Catching Error Messages on Master Pages

When you use master pages, you will need to move the code for handling and displaying errors to the master page. In this last part of the tutorial, you will modify the original profit calculation page to handle errors—and, by extension, handle asynchronous errors that happen on any page that uses the same master page. The steps are similar to what you went through for the page with no master, but there are some specific differences.

Step 18: Modify the script manager.

First, since the script manager is located on the master page, you will have to implement the error handling on the master page. You can modify the script manager on the Main.Master page by adding lines 2 and 3 from Figure 14-13.

Step 19: Add error handling code.

Add the error handling method in Figure 14-14 in Main.Master.cs.

Step 20: Add error display code.

Add the JavaScript in Figure 14-12 in the `head` element of the master page (uncomment line 10 and comment out line 11) and the `divError` element immediately following the end `nav` element and before the `form` element. When you handle error messages on the master page, the placement of the error message will be the same across all pages, so you are limited to placing it either above or below the content of the regular pages.

You can copy all this code from the ProfitCalcNoMaster page.

At this point, any exceptions on any page will be caught and displayed just below the navigation header on the main page. You can then manage what gets displayed on each content page, like you did above by catching errors and rethrowing with the specific error messages you wish to display to the user.

Run the site and go to the ProfitCalculator.aspx page to check that any errors generated on this page are handled with an error message below the navigation bar. Errors that aren't handled on a page, such as "page not found" (Error 404), will still go to the default error page.

Review Questions

6. Describe the three types of error pages that can be displayed by ASP.NET.

7. Why should you avoid displaying detailed error messages to external end users?

8. Why should you block custom error pages from loading when using asynchronous postbacks?

14.4 Using the Ajax Control Toolkit to Create Rich Sites

In addition to the built-in Ajax controls that allow for managing asynchronous postbacks, Ajax also allows for creating controls that make websites much more responsive and interactive. This can be done using the Ajax Control Toolkit, which contains a large number of rich and interactive controls that can be added to your site. The toolkit contains both specific Ajax-ified controls as well as extenders that modify the behavior of existing controls. The extenders can turn a textbox into a calendar picker or allow you to implement drag-and-drop on the page.

The Ajax Control Toolkit is an open-source toolkit maintained by DevExpress. It can be installed in two different ways:

1. **Download from the DevExpress website** (https://www.devexpress.com/products/ajax-control-toolkit/). This will install the toolkit into the Toolbox to allow for drag-and-drop of the controls to the page and some support in configuring the controls. You will need to have permission to install software on the machine you're working on.

2. **NuGet.** This will install the components into the project. This approach is simple and straightforward. However, it does not provide Toolbox integration, so you will need to configure the controls using code. No software installation permissions are required, as the installation is done in the project within Visual Studio.

Support for the toolkit is available on GitHub (https://github.com/DevExpress/AjaxControlToolkit/wiki), where you can find a list of each of the controls and extenders with full documentation of properties, as well as a demo page, where you can see how each control would look when implemented.

The controls in the toolkit come in two varieties: full controls and extenders. The full controls are stand-alone controls that you add to the page and that then will be displayed on their own. Extenders are added to existing ASP.NET controls to augment the functionality of the controls. Table 14-2 shows examples of both kinds of controls.

Table 14-2: Selected Ajax Control Toolkit controls and extenders

Ajax toolkit control	Type of control	Description
Accordion	Full Control	A series of panes that can be viewed one at a time so content in the other panes collapse out of sight
AjaxFileUpload	Full Control	Allows for asynchronously uploading files to the server
TabContainer	Full Control	Created a set of tabs for organizing page content
TabPanel	Full Control	The content for each tab in a TabContainer
Twitter	Full Control	Shows tweets from Twitter, either from a particular user account or based on a search
Calendar	Extender	Attaches to a TextBox to provide a UI for picking a date
ColorPicker	Extender	Attaches to a TextBox to provide a UI for picking a color
ConfirmButton	Extender	Attaches to a button and displays a message to the user for confirming the action on the original button
DropDown	Extender	Attaches to most ASP.NET controls to provide a SharePoint–style hover dropdown menu
ModalPopup	Extender	Shows a dialog that shows up on top of the page content and doesn't allow for interacting with the underlying page until the dialog is dismissed
MultiHandleSlider	Extender	Attaches to a TextBox to display a slider with one or more handles and corresponding multiple input values
NumericUpDown	Extender	Attaches to a TextBox to provide up/down arrows that can be used to increment/decrement a numerical value. Can also be used to cycle through custom lists, like the names of the months.
TextBoxWatermark	Extender	Attaches to a textbox to display a message to the user when the textbox is blank

Tutorial: Working with the Ajax Control Toolkit

In this tutorial, you will see how to install the Ajax Control Toolkit and work with both a full control and an extender.

Step 21: Install the Ajax Control Toolkit from the DevExpress website.

Go to https://www.devexpress.com/products/ajax-control-toolkit/ and select the download link. After the download completes, shut down all instances of Visual Studio and run the installer. Follow the instructions to install the toolkit.

Once the installation is complete, restart Visual Studio. You will now see a new Toolbox section with all the controls and extenders in the toolkit, as shown in Figure 14-16.

Step 22: Install the Ajax Control Toolkit with NuGet (alternative).

If you had problems installing the toolkit from the DevExpress website, you can also install it using NuGet. If you were able to complete the previous step, you should *not* complete this step.

Right-click the project in the Solution Explorer and select *Manage NuGet packages…*.

In the dialog that opens, select *Browse* on the left and type *Ajax Control* in the search bar. This should bring up the Ajax Control Toolkit. Select it and click *Install* on the right.

This will add the toolkit as a reference in the project (see Figure 14-17).

To use the toolkit, you will need to register it on the page where you use it by adding this code line:

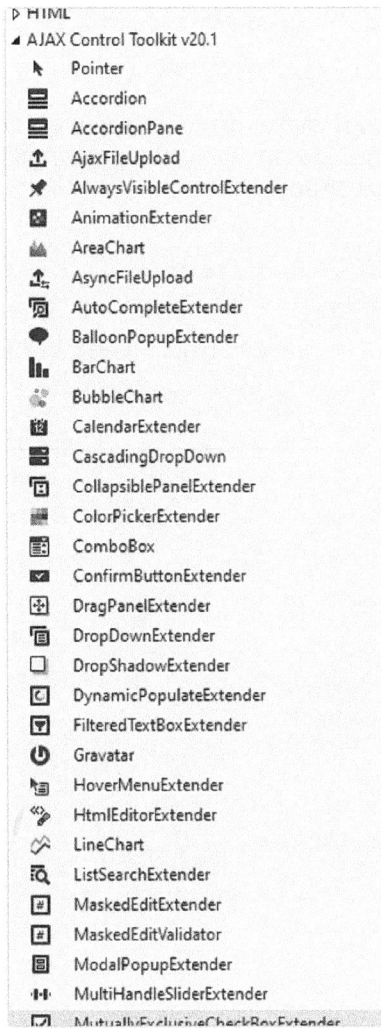

Figure 14-16: Selection of controls from the Ajax Control Toolkit shown in the Toolbox.

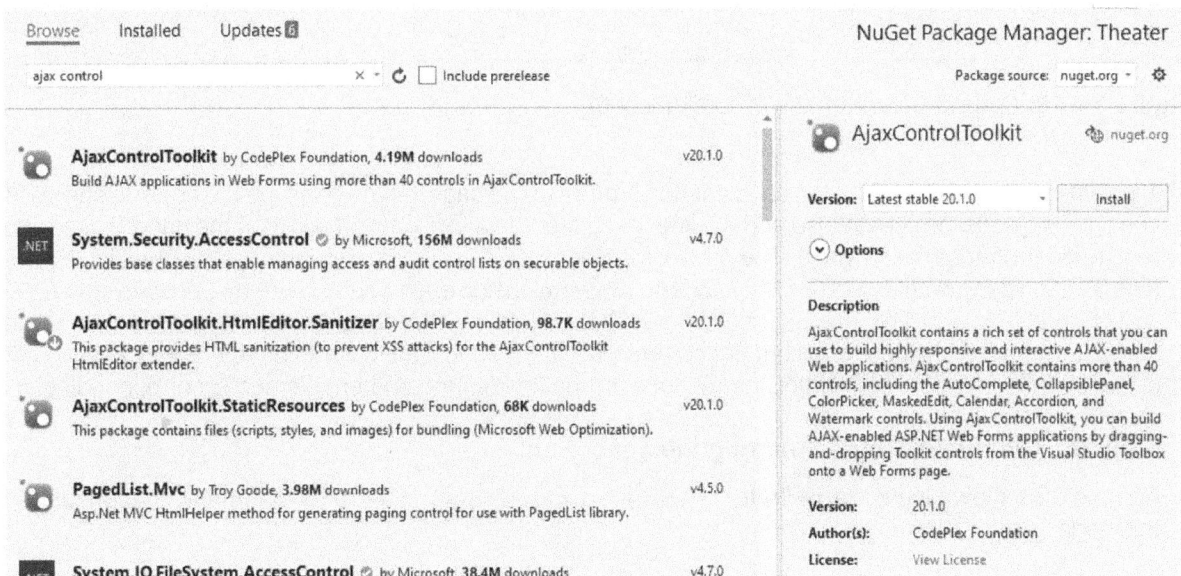

Figure 14-17: Installing the Ajax Control Toolkit as a NuGet package

```
<%@ Register TagPrefix="ajaxToolkit" Namespace="AjaxControlToolkit" Assembly-
="AjaxControlToolkit" %>
```

If you installed this via the downloaded installer, dragging a control from the toolbox will automatically add the line to the page. You can place this line on the master page, making the toolkit available on all pages that use the master. The lines should be added before the opening `html` element.

Step 23: Use the accordion control.

You will now see how to use the accordion control to organize text on the About page.

Open the About.aspx page and add the code in Figure 14-18 inside the `MainContentPlaceHolder`, leaving the existing content in place.

```
1  <ajaxToolkit:Accordion ID="Accordion1" runat="server"
2      HeaderCssClass="accordionHeader" ContentCssClass="accordionContent">
3      <Panes>
4          <ajaxToolkit:AccordionPane runat="server">
5              <Header>Overview</Header>
6              <Content>
7
8              </Content>
9          </ajaxToolkit:AccordionPane>
10         <ajaxToolkit:AccordionPane runat="server">
11             <Header>Early Years</Header>
12             <Content>
13
14             </Content>
15         </ajaxToolkit:AccordionPane>
16         <ajaxToolkit:AccordionPane runat="server">
17             <Header>Renovations and Reopening</Header>
18             <Content>
19
20             </Content>
21         </ajaxToolkit:AccordionPane>
22         <ajaxToolkit:AccordionPane runat="server">
23             <Header>The Historic Grand Opera House Today</Header>
24             <Content>
25
26              </Content>
27         </ajaxToolkit:AccordionPane>
28     </Panes>
29 </ajaxToolkit:Accordion>
```

Figure 14-18: Code structure for the accordion control

To create this code, you can drag an accordion control to the page, or you can type the code as shown. If you type the code, be sure to also add the line to register the Ajax Control Toolkit on the page, unless you registered it on the master page. The parts of the accordion are contained in an element called Panes (line 3). Each pane is an `ajaxToolkit:AccordionPane`. Inside each `AccordionPane` are two elements: `Header` and `Content`. The `Header` is the content that is always visible for all panes. `Content` contains the content that will be hidden for all but the currently selected pane. Line 2 contains references to two CSS classes that contain the formatting instructions for the header and content elements, correspondingly.

Step 24: Add the CSS for formatting the accordion.

Add the CSS shown in Figure 14-19 to the `HeadContentPlaceHolder` to format the accordion just on this page.

```
1  <style>
2      .accordionHeader {
3          width: 50%;
4          background-color: darkmagenta;
5          border-color: white;
6          border-width: 1px;
7          border-style: solid;
8          font-size: large;
9          font-weight: bold;
10         padding: 5px;
11         color: white;
12     }
13
14     .accordionContent {
15         width: 50%;
16         background-color: burlywood;
17         padding: 15px;
18         color: brown;
19     }
20 </style>
```

Figure 14-19: CSS for styling the accordion.

The two classes, `accordionHeader` and `accordionContent`, correspond to what was referenced in the accordion.

Step 25: Add content to the accordion.

The `Content` elements in Figure 14-18 were left blank to provide a better overview of the code structure. To fill in the content, first copy the existing content on the About page into the first `AccordionPane`. For the following three panes, copy the text from the History page on the Grand Oshkosh actual website (https://thegrandoshkosh.org/history/). You may omit the images and just add the text—or take on the added challenge of also including the images.

Run the website and check that the output resembles Figure 14-20 and that each pane opens when you click on the header.

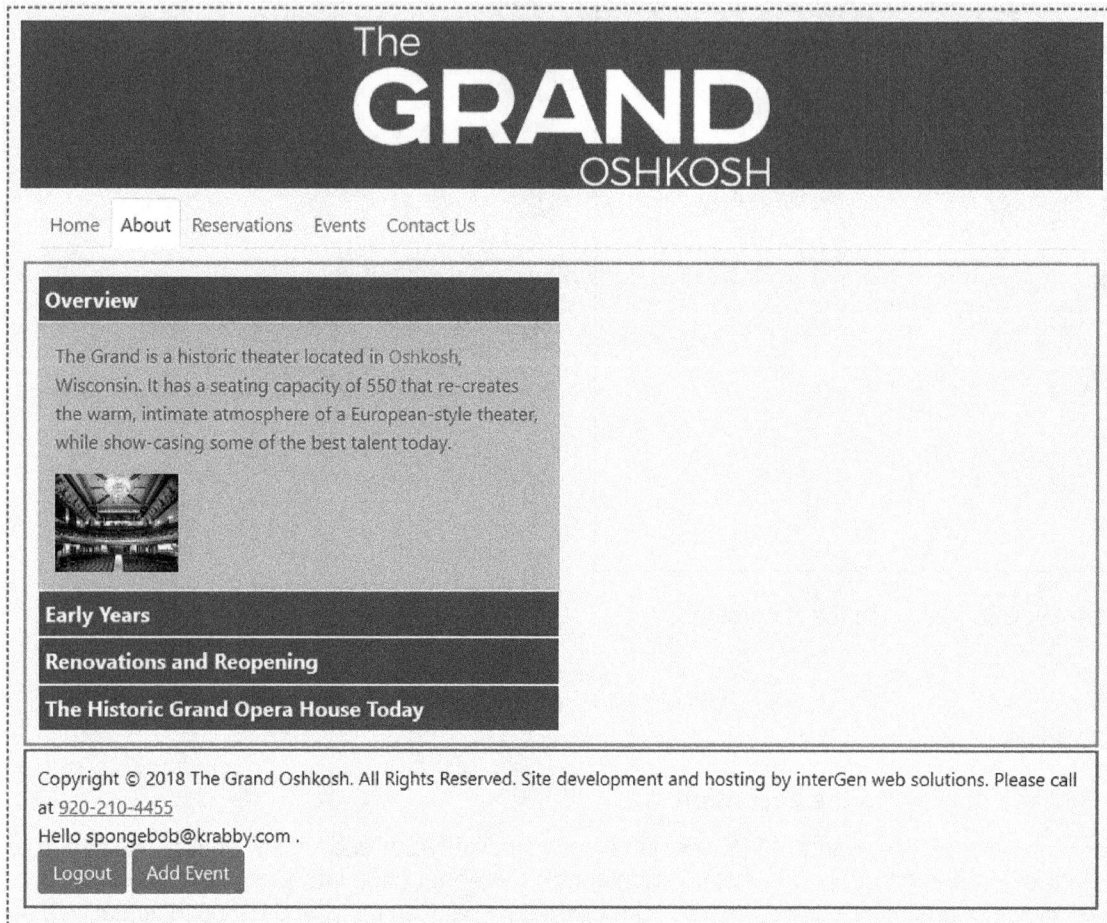

Figure 14-20: The finalized accordion

Step 26: Using the slider extender.

To illustrate how to use an extender, you will convert the Base Price input on the AddEvent page to use a slider instead of a textbox. This would have the benefit that the user can only input numerical values, reducing the risk of errors when parsing data.

Switch to the AddEvent page in the Admin folder and find the `txtBasePrice` textbox. When you hover over the element, you will get a little arrow in a box at the beginning of the line. Click this box and then select *Add Extender...* (see Figure 14-21).

In the dialog that comes up, scroll to the right until you find the SliderExtender. Select the extender and click OK, as shown in Figure 14-22.

```
<label class="col-sm-3" for="txtBasePrice">Base Price</label>
<asp:TextBox runat="server" ID="txtBasePrice" />
```

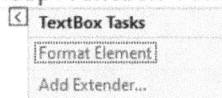

Figure 14-21: Adding an extender to a control

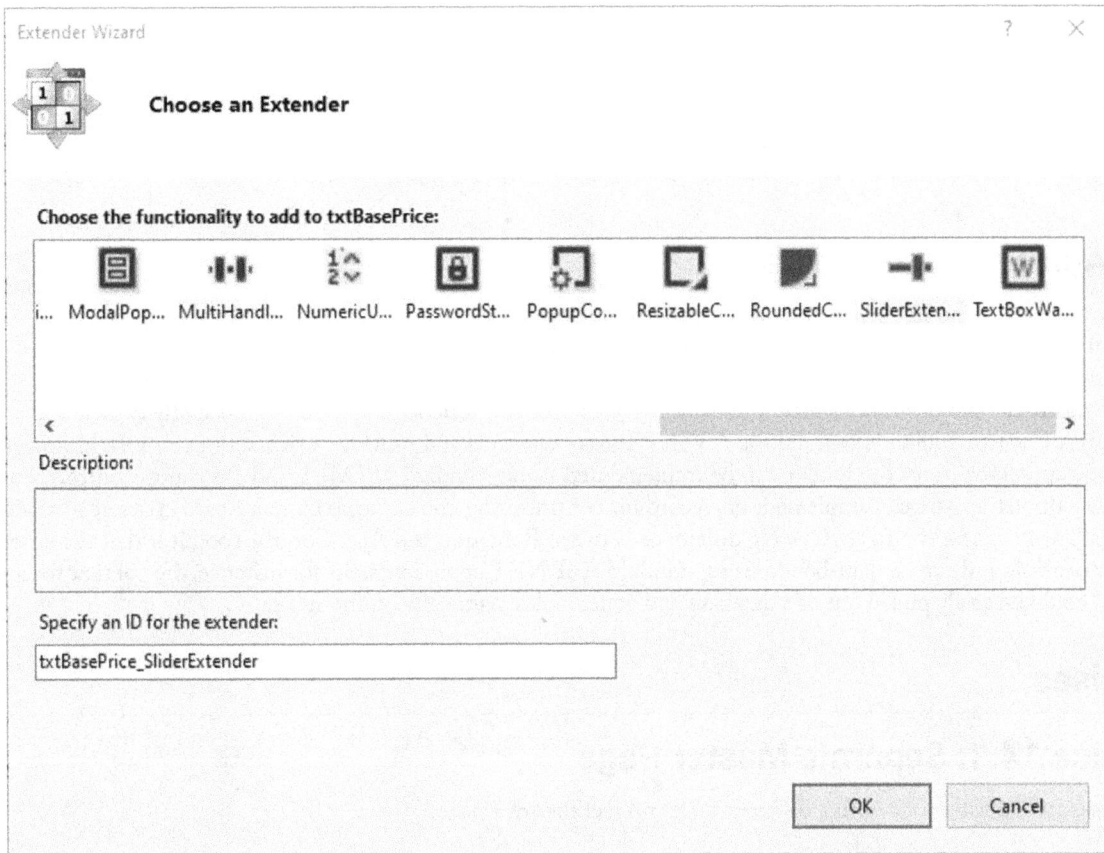

Figure 14-22: Choosing an extender

This will generate code for the extender that will replace the textbox with a slider when the page is rendered.

Step 27: Configure the slider extender.

By default, the slider will allow values between 0 and 100, but it will not display the numerical value selected. You will change both of these next.

Start by changing the value for Maximum to 200, as shown in line 4 in Figure 14-23. This will allow for values between 0 and 200 for the base price. Next, add the code in lines 7–9 inside the opening tag for the slider extender (note the closing angle bracket at the end of line 9). Add line 10 after the extender.

```
1   <div class="form-group">
2       <label class="col-sm-3" for="txtBasePrice">Base Price</label>
3       <asp:TextBox runat="server" ID="txtBasePrice" />
4       <ajaxToolkit:SliderExtender runat="server" Minimum="0" Maximum="200"
5           BehaviorID="txtBasePrice_SliderExtender"
6           TargetControlID="txtBasePrice" ID="txtBasePrice_SliderExtender"
7           Length="300"
8           BoundControlID="lblPrice"
9           Decimals="2">
10      </ajaxToolkit:SliderExtender>
11      <label id="lblPrice"></label>
12  </div>
```

Figure 14-23: Code for the slider extender.

Line 7 changes the length of the slider to 300 pixels. Line 8 specifies the name of a label control that will display the value currently selected in the slider. In this example, the label is specified in line 11. Line 9 indicates how many decimals the selected value will have. Here, we specify two decimals.

If you run the site, you should see the slider appear as shown in Figure 14-24.

14.5 Ajax versus Bootstrap

The Ajax controls are an important tool for creating responsive websites that update in real time and provide feedback to the user without reloading or taking the user to a separate page. Asynchronous postbacks can be implemented in most websites, to provide a richer and more user-friendly experience.

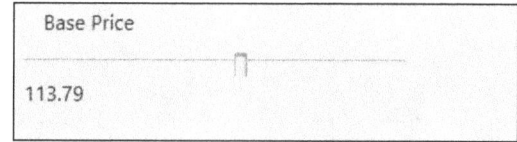

Figure 14-24: Completed slider

The Ajax Control Toolkit is built on the ASP.NET Ajax Controls and provides a rich set of controls. However, most of the functionality in the toolkit can be implemented using standard HTML5 and Bootstrap controls. For example, in Bootstrap you can implement an accordion control using the `accordion` and card `classes`, along with a `collapse data-toggle`. One big difference between Bootstrap and Ajax Control Toolkit is that the latter are server controls and can be data-bound using standard ASP.NET approaches. So, for instance, the content for an accordion could be easily pulled out of a database and generated dynamically on the server.

Exercises

Exercise 14.1: Separate Master Page

Create a separate master page with a different look and feel for error pages.

Exercise 14.2: Custom Error URL

The URL for custom error pages indicates which page the user requested that caused an error. Incorporate this information into the information displayed to the user.

Exercise 14.3: Restrict Custom Error Page

Add a custom error page that is only displayed when the user requests a page that doesn't exist on the server (HTTP error code 404).

Exercise 14.4: Partial Postbacks

Use partial postbacks to provide a better user experience on the UserRoles page.

Exercise 14.5: TabContainer and TabPanel

Use the TabContainer and TabPanel controls to organize the content on the About page.

Exercise 14.6: Expand Accordion

Add the images from the Oshkosh Grand history page to the accordion panes on the About page and add any formatting you see fit to improve the display of the page.

Exercise 14.7: Formatting Slider

Use CSS to format the handle and rail of the slider.

Chapter 15

Web Services

So far in this book, you have created websites where users request a page from your web server and interact with functionality on that server. For many websites, this works perfectly well. However, there are also situations where the data or functionality you need doesn't reside on the web server itself. You might also find that you have data or functionality that you would like to make available to other developers to create new websites from. In addition, some websites are so complex that functionality gets split between multiple different web servers. In all these cases, you will need a way for the servers to communicate with each other. Since you can't just call a C# function on a different server to return some data, you have to take extra steps to set up the communication between the servers. This chapter shows you several different methods to set this up using ASP.NET.

Learning Objectives

After studying this chapter, you should be able to

- Describe the difference between regular web architecture and the architecture used for web services.
- Discuss the differences between the two dominant methods for web services used with ASP.NET: WCF and Web API (REST).
- Create a web service using both WCF and Web API.
- Consume the web services in a client application.
- Discuss ways that web services can be used in a website.

15.1 Introduction to Web Services

As the web has become increasingly commonplace, many websites have found the need to use data or services that are created in other places. For instance, an e-commerce website might need to verify that an address is entered correctly before shipping a package to a street address that doesn't exist within the specified city. It would be very time-consuming to create such a service for all addresses. Fortunately, the US Postal Service has created such a service that can be used by any website.[1] Similarly, the National Weather Service provides weather data for any location within the United States that can be integrated into websites and other services.[2] By using such services, web developers can reduce the time they have to spend creating new websites.

At the same time, many companies also have services that are spread out across different systems that can be referenced using web services. For example, an e-commerce site would need access to inventory data to show users whether a particular product is in stock. The inventory system and the e-commerce system could be created independently and connected to each other using web services. This could allow for creating and optimizing each system independently of the other.

A web service is a web-based system created with a public interface that defines which operations the server makes available to other systems. These operations can then be called from any server across the Internet, similar to how methods can be called within a local system. Web services communicate using standard data formats—typically eXtensible Markup Language (XML) or JavaScript Object Notation (JSON).

---

1. https://www.usps.com/business/web-tools-apis/
2. https://www.weather.gov/documentation/services-web-api#

As more and more business systems are created as web services, it becomes increasingly valuable to connect these systems to each other and create integrated services. Many legacy systems have also been retrofitted with web service interfaces that allow websites to interact with data that resides in COBOL mainframe systems that were created long before the web was even created.

15.2 Web Services Architecture

ASP.NET offers two different approaches to create a web service: Windows Communication Foundation (WCF) and Web Application Programming Interface (API) services. These two approaches are similar in that the server provides an object that you can make a request to and get data back. Figure 15-1 shows at a high level the architecture used for web services where a web server can request data from another web server with the communication between the two taking place using XML or JSON data.

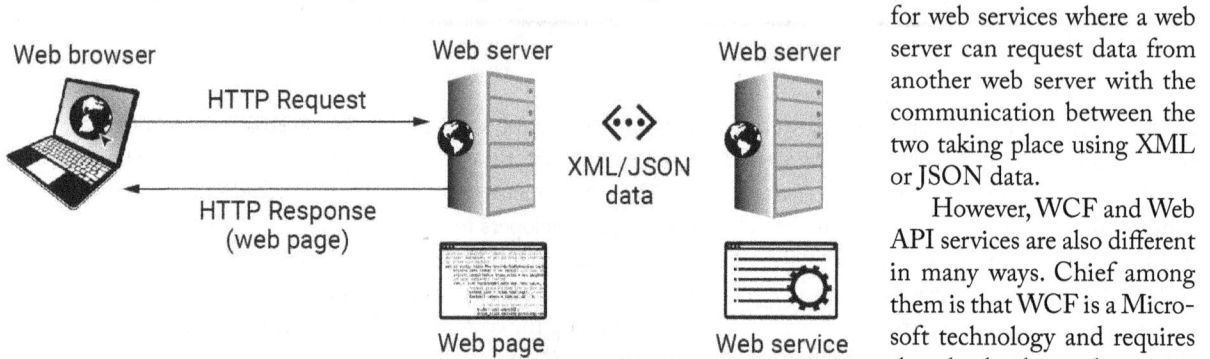

Figure 15-1: Web service architecture

However, WCF and Web API services are also different in many ways. Chief among them is that WCF is a Microsoft technology and requires that both the web service server and the computer making the requests have a version of the .Net Framework installed that supports WCF, whereas Web API services use standard HTTP requests to URLs on the server, so no specific software is needed. Table 15-1 shows some of the key differences between the two technologies.

Table 15-1: Comparison of WCF and Web API Services

	WCF	**Web API services**
Communication	Simple Object Access Protocol (SOAP)	REpresentational State Transfer (REST)
Requests	Via WCF Proxy	HTTP to URL on server
Responses	XML	XML or JSON
Frameworks	Both clients and servers must have .Net Framework installed.	None needed
Architecture	Client code interfaces with the WCF Proxy, which makes a request to the server.	Client code sends an HTTP request to a specific URL directly.

When you connect to a WCF web service in Visual Studio, a middle layer is installed that includes an object representing the available methods on the web service. This means that with WCF, you interact with the web service in your code like any other object by calling methods and passing parameters. Figure 15-2 shows an overview of this architecture. The WCF Proxy is the middle layer in the client that makes the web service transparent to the rest of the client code. The WCF Proxy sends messages to a specific public method (endpoint) on the web service. The communication between the server and the client is done using Simple Object Access Protocol, and data is returned from the server in XML format.

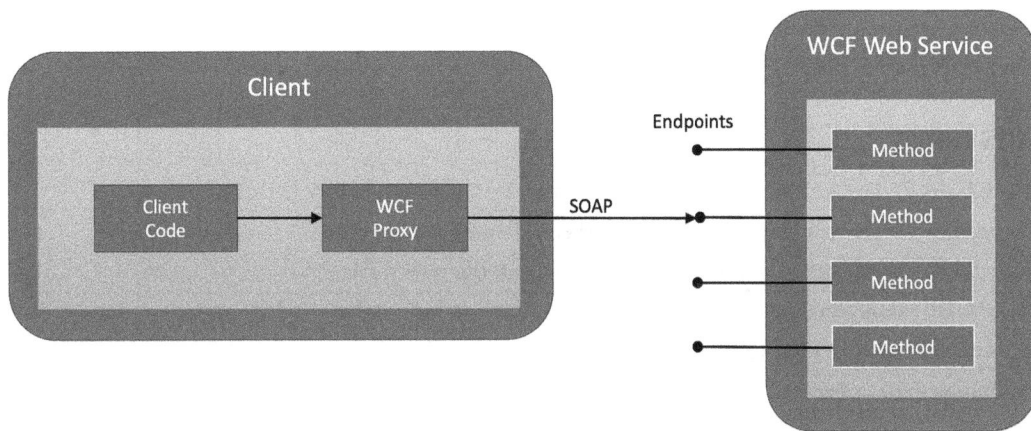

Figure 15-2: WCF web service architecture

With Web API (REST), you will be interacting more closely with the actual URL and setting up HTTP requests when interacting with the service. Figure 15-3 shows how the client code needs to know the URL of the service to directly interact with an endpoint method on the server. In this approach, you will also need to know which HTTP action method was used to implement each method and construct an HTTP request that specifies which action method to use, as well as the necessary data to be passed as parameters. You will frequently use four different HTTP action methods:

- **GET method**: Request data from the server. The structure of the URL is used to specify which method is called on the server. For example, you will see later in the chapter that you can request all performances using a URL like this: http://localhost:44567/api/performances. You also can request a specific performance with this URL: http://localhost:44567/api/performances/14. In addition, you can pass other parameters using a query string like this: http://localhost:44567/api/performances?name1=value1&name2=value2.
- **POST method**: This method is used to send data to the server to create or update resources on the server.
- **PUT method**: This method is used to send data to the server to create or update resources on the server.
- **DELETE method**: This method deletes a resource from the server.

The difference between POST and PUT is that PUT is *idempotent*—that is, calling PUT repeatedly with the same data will not cause the server data to change. Calling POST repeatedly will create a new resource with every call. For that reason, POST is typically used to create new resources, and PUT is used to update existing resources.

One limitation of the GET method is that all data is sent in the URL, which is limited to 2048 characters. Because data is sent as part of the URL, it is less secure than the other methods because the data is sent as part of the URL, so you should not use a GET request with sensitive data.

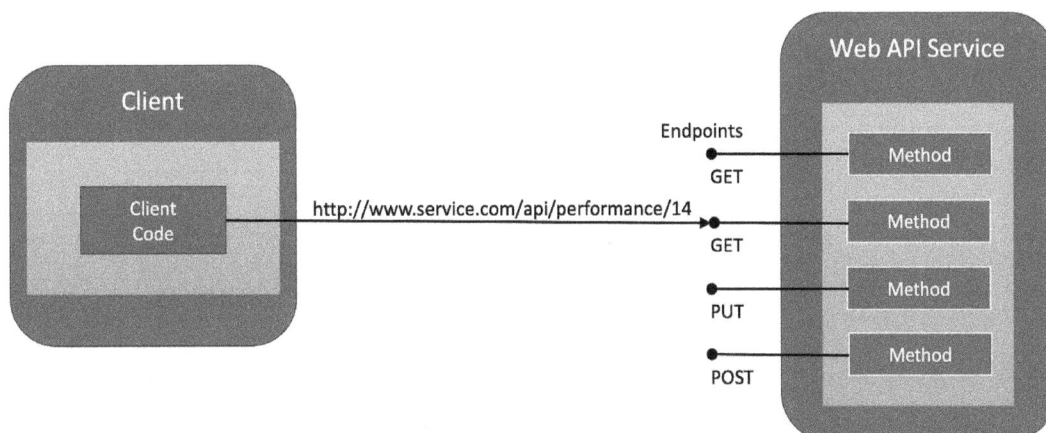

Figure 15-3: Architecture for a Web API service using REST

Web API services are becoming the most common way to use web services, but the WCF approach is still relevant, so in this chapter you will create web services using both.

Review Questions

1. Why would a web developer use a web service rather than create the functionality herself?
2. What are the primary differences between WCF and Web API web services?
3. In WCF, what allows a developer to write C# code that calls on the web service?
4. In Web API, how are parameters passed to the web service?
5. What is the difference between the POST and PUT methods?
6. What are two things to be concerned about when calling a GET method?

Tutorial: Building a WCF Web Service

In this tutorial, you will build a web service to manage performances for the Oshkosh Grand website. This will require first building a web service solution, and then referencing this from the Oshkosh Grand website you have been building in previous chapters. During development, both projects will exist on your development computer, but for deployment, they can be published to different web servers. This allows any website to access your web service project.

Step 1: Install the WCF Project Template (needed for Visual Studio 2019).

In Visual Studio 2019, the template to create a WCF service application isn't installed by default. To check if it is installed for you, go to *File > New Project...*. Then type *WCF* in the search box. If you see *WCF Service Application* as an option, select it. Otherwise, click *Install more tools and features*. Click *Yes* if you are asked if Visual Studio Installer can make changes.

When the installer comes up, select *Individual components* at the top. Use the search box to find *Windows Communication Foundation* (see Figure 15-4) and then select it. Click the *Modify* button on the bottom right. If you are asked for permission to close apps, click *Continue*. When the installation is complete, close the Installer window and open *Create a new project window* in Visual Studio.

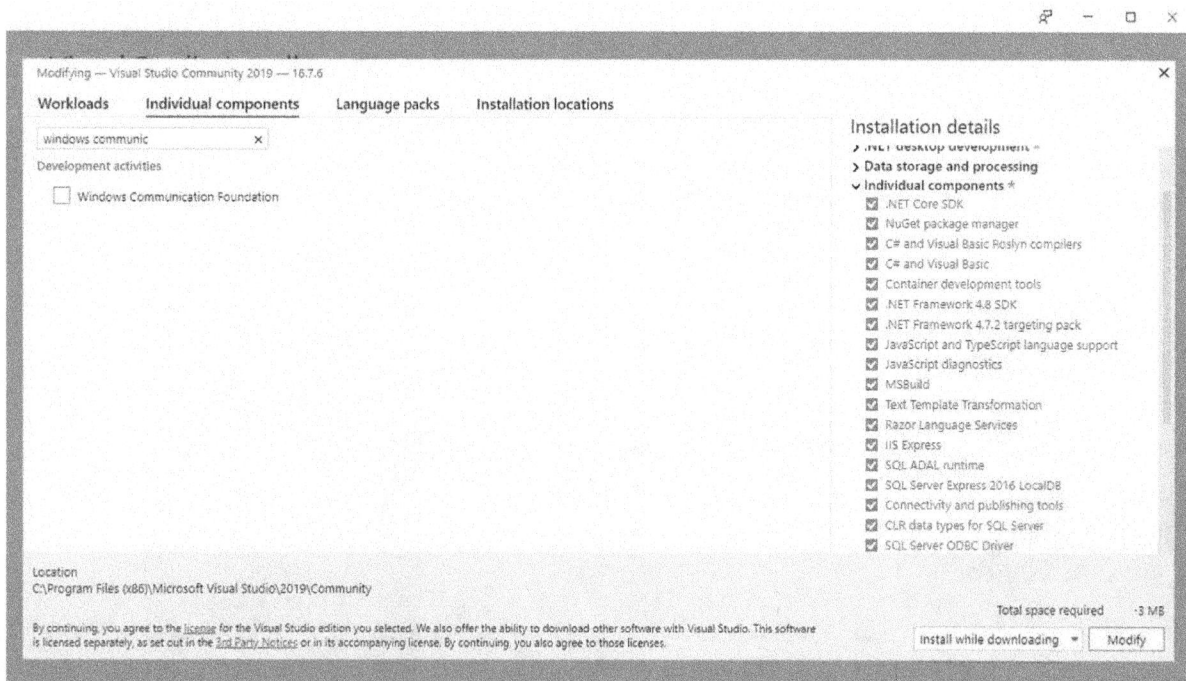

Figure 15-4: Installing the WCF Project Template in Visual Studio 2019

Step 2: Create a web service project.

In the *Create new project* dialog, select the *WCF Service Application* template and click *Next*. Name the service *PerformanceService*, choose an appropriate location, and click *Create*.

Step 3: Set up the project.

Delete the files *IService1.cs* and *Service1.svc*.

Right-click the project and select *Add > New Item*.

Select the *WCF Service* option. Name the new service *PerformanceService*. This will create two files: *PerformanceService.svc* and *IPerformanceService.cs*.

Step 4: Add the database.

Right-click the *AppData* folder and select *Add > Existing Item*. Navigate to the *Tutorials Start* folder and select the *Theater.mdf* file. Click *Add* to copy the file into the project.

Step 5: Set up a service contract.

The web service operations you want to make available are defined in a service contract. This is coded in the IPerformanceService interface as a series of methods decorated with the `OperationContract` attribute. Enter the code shown in Figure 15-5 in IPerformanceService.cs to set up the service contract.

```
1    [ServiceContract]
2       public interface IPerformanceService
3       {
4            [OperationContract]
5            List<Performance> GetPerformances();
6
7            [OperationContract]
8            Performance GetPerformanceById(string id);
9
10           [OperationContract]
11           int InsertPerformance(Performance p);
12
13           [OperationContract]
14           int UpdatePerformance(Performance p);
15
16           [OperationContract]
17           int DeletePerformance(string id);
18
19       }
```

Figure 15-5: Service contract for the PerformanceService

The interface is decorated with the `ServiceContract` attribute in line 1 to indicate that this defines a web service. Each of the methods is then decorated with the `OperationContract` attribute to indicate that they are a part of the service interface. This interface specifies five operations to be made publicly available on the web service. Two of these will read information to be returned to the client (`GetPerformance` in line 5 and `GetPerformanceById` in line 8). The last three will insert, update, and delete a performance on the server, respectively.

Most of the methods reference a `Performance` type, which is a custom type you will also create in the IPerformanceService.cs file, as shown in Figure 15-6. You may find it easier to create the `Performance` class first and then the `IPerformanceService` interface second, so IntelliSense is aware of the `Performance` class.

```
1   [DataContract]
2      public class Performance
3      {
4          [DataMember]
5          public int PerformanceId { get; set; }
6
7          [DataMember]
8          public string PerformanceName { get; set; }
9
10         [DataMember]
11         public double BasePrice { get; set; }
12
13         [DataMember]
14         public string Description { get; set; }
15
16         [DataMember]
17         public string Presenter { get; set; }
18
19         [DataMember]
20         public byte[] Image { get; set; }
21      }
```

Figure 15-6: Performance data contract

The `DataContract` attribute in line 1 indicates that objects of this class contain data to be exchanged over the web service. Then, each property in the class is decorated with the `DataMember` attribute. In this case, the `Performance` class reflects the data fields in the performance table in the Theater database. If you use Entity Framework to access the database, you can also reference those generated classes in the service contract, without having to create your own custom types.

Step 6: Implement data access.

Access to the database is implemented in a dedicated class with methods that correspond to each of the methods in the service contract in Figure 15-5.

Right-click the project and add a new folder called **Models**. Inside this folder, add a class called **PerformanceDB.cs**. This class will contain the code that will access the database. Then add the code shown in Figure 15-7. This is also available as PerformanceDB.txt.

For this code to work, you also need a few using statements:

```
1   using System.Collections.Generic;
2   using System.Configuration;
3   using System.Data;
4   using System.Data.SqlClient;
```

```
1   public class PerformanceDB
2   {
3      public int GetMaxId()
4      {
5          var connection = GetConnection();
6          //Find highest ID
7          string sqlMax = "SELECT Max(PerformanceId) FROM Performance";
8          var cmd = new SqlCommand(sqlMax, connection);
9          SqlDataReader reader = cmd.ExecuteReader();
10         reader.Read();
11         int result = int.Parse(reader[0].ToString()); //Highest ID
12         reader.Close();
13         connection.Close();
14         return result;
```

Figure 15-7: Data access code (*continues*)

```
15      }
16
17      public Performance GetPerformanceById(string id)
18      {
19          var connection = GetConnection();
20          string sql = "SELECT * FROM Performance WHERE PerformanceId = @PerformanceId";
21          var cmd = new SqlCommand(sql, connection);
22          //Insert parameter
23          cmd.Parameters.AddWithValue("PerformanceId", id);
24          SqlDataReader reader = cmd.ExecuteReader();
25          Performance result = null;
26          if (reader.Read())
27          {
28              result = new Performance
29              {
30                  PerformanceId = int.Parse(reader["PerformanceId"].ToString()),
31                  PerformanceName = reader["PerformanceName"].ToString(),
32                  BasePrice = double.Parse(reader["BasePrice"].ToString()),
33                  Description = reader["Description"].ToString(),
34                  Presenter = reader["Presenter"].ToString()
35              };
36              reader.Close();
37          }
38          connection.Close();
39          return result;
40      }
41
42
43      public List<Performance> GetPerformances()
44      {
45          var connection = GetConnection();
46          string sql = "SELECT * FROM Performance ORDER BY PerformanceName";
47          var cmd = new SqlCommand(sql, connection);
48          SqlDataReader reader = cmd.ExecuteReader();
49          List<Performance> result = new List<Performance>();
50          while (reader.Read())
51              result.Add(new Performance()
52              {
53                  PerformanceId = int.Parse(reader["PerformanceId"].ToString()),
54                  PerformanceName = reader["PerformanceName"].ToString(),
55                  BasePrice = double.Parse(reader["BasePrice"].ToString()),
56                  Description = reader["Description"].ToString(),
57                  Presenter = reader[4].ToString()
58
59              });
60          reader.Close();
61          connection.Close();
62          return result;
63      }
64
65      public int InsertPerformance(Performance p)
66      {
67          var connection = GetConnection();
68
69          string sql = "insert into Performance " +
70            "(PerformanceId, PerformanceName, BasePrice, Description, Presenter) " +
71            "Values(@PerformanceId, @PerformanceName, @BasePrice, @Description, @Presenter)";
72          var cmd = new SqlCommand(sql, connection);
73          //Insert parameter values
74          cmd.Parameters.AddWithValue("@PerformanceId", SqlDbType.Int).Value =
75              GetMaxId() + 1;  //Make sure ID is unique
76          cmd.Parameters.AddWithValue("@PerformanceName", SqlDbType.VarChar).Value =
77              p.PerformanceName;
78          cmd.Parameters.AddWithValue("@BasePrice", SqlDbType.Decimal).Value =
79              p.BasePrice;
80          cmd.Parameters.AddWithValue("@Description", SqlDbType.VarChar).Value =
81              p.Description;
```

Figure 15-7: Data access code (*continued*)

```
82          cmd.Parameters.AddWithValue("@Presenter", SqlDbType.VarChar).Value =
83              p.Presenter;
84          //Insert record
85          int result = cmd.ExecuteNonQuery();
86          connection.Close();
87          return result;
88      }
89
90      public int UpdatePerformance(Performance p)
91      {
92          var connection = GetConnection();
93          string sql = "UPDATE Performance "
94          + "SET PerformanceName = @PerformanceName, "
95          + "BasePrice = @BasePrice, "
96          + "Description = @Description, "
97          + "Presenter = @Presenter "
98          + "WHERE PerformanceId = @PerformanceId";
99              var cmd = new SqlCommand(sql, connection);
100         //Insert parameter values
101         cmd.Parameters.AddWithValue("@PerformanceId", SqlDbType.Int).Value =
102             p.PerformanceId;
103         cmd.Parameters.AddWithValue("@PerformanceName", SqlDbType.VarChar).Value =
104             p.PerformanceName;
105         cmd.Parameters.AddWithValue("@BasePrice", SqlDbType.Decimal).Value =
106             p.BasePrice;
107         cmd.Parameters.AddWithValue("@Description", SqlDbType.VarChar).Value =
108             p.Description;
109         cmd.Parameters.AddWithValue("@Presenter", SqlDbType.VarChar).Value =
110             p.Presenter;
111         //Update record
112         int result = cmd.ExecuteNonQuery();
113         connection.Close();
114         return result;
115     }
116
117     internal int DeletePerformance(String id)
118     {
119         var connection = GetConnection();
120         string sql = "DELETE FROM Performance WHERE PerformanceId = @PerformanceId";
121         var command = new SqlCommand(sql, connection);
122         //Insert parameter value
123         command.Parameters.AddWithValue("@PerformanceId", SqlDbType.Int).Value = id;
124         //Delete record
125         int result = command.ExecuteNonQuery();
126         connection.Close();
127         return result;
128     }
129
130     private SqlConnection GetConnection()
131     {
132       string connectionString = ConfigurationManager.ConnectionStrings
133               ["DefaultConnection"].ConnectionString;
133         var connection = new SqlConnection(connectionString);
134         connection.Open();
135         return connection;
136     }
137 }
```

Figure 15-7: Data access code (*continued*)

The first method, GetMaxId (lines 3–15), is included to ensure we always get a unique ID for each new record added. By retrieving the largest ID already in the database, we can simply increment the ID for new performances to ensure the ID is unique. The method is used in line 75 when creating a new performance. The remaining methods use a standard data access code with parameterized SQL queries executed against the database referenced in the connection string. The final method in the class, `GetConnection` (lines 130–136), retrieves the connection string from the Web.Config file. This method is called from each of the data access methods, which are also responsible for ensuring that the connection is closed. This method expects the connection string to be named `DefaultConnection`, so you will need to be sure that matches with the name you use for the connection string in the next step.

Step 7: Add a connection string.

Add a connection string to the Web.Config file, pointing to the database you will be using. The connection string should be added to the configuration section of the Web.Config file. Lines 2–6 in Figure 15-8 show the code you'll need to add.

```
1   <configuration>
2     <connectionStrings>
3       <add name="DefaultConnection"
4            connectionString="Data Source=(LocalDB)\MSSQLLocalDB;
                               AttachDbFilename=|DataDirectory|\Theater.mdf"
5            providerName="System.Data.SqlClient"/>
6     </connectionStrings>
7   </configuration>
```

Figure 15-8: Connection string in the Web.Config file

The |DataDirectory| instruction in line 4 points to the designated folder for databases, which by default is the AppData directory. By using this, your project can be copied to a different folder or machine and will not need any changes before being run.

Step 8: Implement the service interface.

The last step in the process of setting up the web service is to implement the service interface that you defined in Figure 15-5. This is relatively straightforward and involves creating each of the methods of the interface and using the functionality of the PerformanceDB data access code.

Open PerformanceService.svc and add the following two lines:

```
1   using PerformanceService.Models;
2   using System.Collections.Generic;
```

Then add the code shown in Figure 15-9.

```
1  public class PerformanceService : IPerformanceService
2  {
3      private PerformanceDB data;
4
5      public PerformanceService()
6      {
7          data = new PerformanceDB();
8      }
9
10     public Performance GetPerformanceById(string id)
11     {
12         return data.GetPerformanceById(id);
13     }
14
15     public List<Performance> GetPerformances()
16     {
17         return data.GetPerformances();
18     }
19
20     public int InsertPerformance(Performance p)
21     {
22         try
23         {
24             return data.InsertPerformance(p);
25         }
26         catch
27         {
28             return -1;
29         }
30     }
31
32     public int UpdatePerformance(Performance p)
33     {
34         return data.UpdatePerformance(p);
35     }
36
37     public int DeletePerformance(string performanceId)
38     {
39         return data.DeletePerformance(performanceId);
40     }
41 }
```

Figure 15-9: Code to implement the performance service contract

As line 1 shows, this implements the `IPerformanceService` interface you declared in Figure 15-5. It does this by relying on the `PerformanceDB` class you implemented in Figure 15-7. Line 3 sets up a reference to the `PerformanceDB` class, and the constructor in lines 5–8 instantiates an object of the class. The rest of the class implements the methods of the interface by calling on the corresponding methods in `PerformanceDB`. The only bit of code to discuss is lines 28–35, where we show an example of using a `try-catch` to capture any exceptions that may be thrown by the data access code when inserting a new performance. This ensures that the service won't crash but will instead return an invalid value that the client can check for. This pattern could be extended to the other methods in the class.

Step 9: Launch and test the web service.

Set PerformanceService.svc as the start page and press Ctrl+F5. This will launch the WCF Test Client that you can use to test the methods you have created as part of the interface. You can double-click any of the methods and then click *Invoke* to run the method. Methods that require a parameter will show input fields in the Request section of the dialog. For example, Figure 15-10 shows the `GetPerformnceById` method as having been invoked with a value of 14 for the `id`. The result is shown below in the Response section.

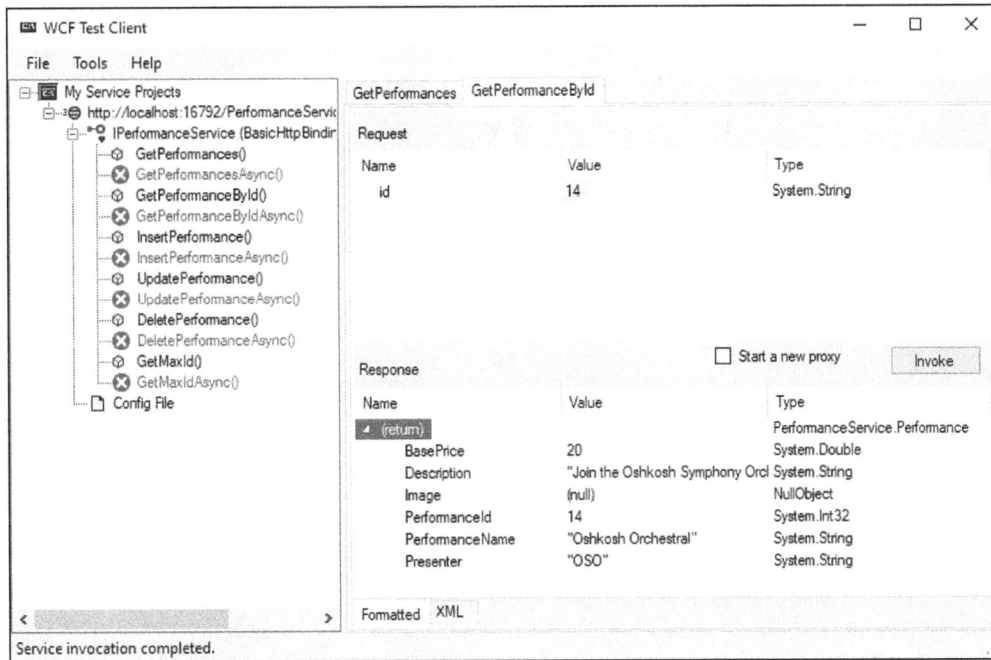

Figure 15-10: WCF test client

Tutorial: Modifying the Client to Use the WCF Web Service

In the next part of the tutorial, you will modify the Oshkosh Grand Theater website to use the web service instead of the local database. To make it simple, you will create a new page that will allow a user to create, display, delete, and modify performances. Figure 15-11 shows the finished page. The form at the top allows for creating a new performance. Clicking the blue *Edit* button under each performance will populate the form, after which clicking *Save* will update the selected performance. The *Delete* button will remove the performance from the database on the server.

Figure 15-11: Manage the performance page using the WCF web service

Step 10: Add a reference to the web service.

Start the PerformanceService web service you created in the previous tutorial. In the WCF Test Client (Figure 15-10), right-click the URL address shown just below My Service Projects in the left part of the dialog and select *Copy Address*.

While keeping the WCF Test Client running, open the Theater project you worked on previously or the one in the Starter folder for this chapter. This project will serve as the client to the PerformanceService web service. To connect the client project to the service project, you will need to add a reference to the service in the client project.

Right-click the *Theater* project and select *Add > Service Reference...*.

Paste the address you copied from the WCF Test Client into the Address textbox and click *Go*. This will connect to the PerformanceService web service and show the available operations. Change the Namespace to PerformanceService, as shown in Figure 15-12. Click *OK*.

The service will show up under Connected Services in the Solution Explorer. If you change anything in the public operations of the service (add/rename a method, change the parameters, etc.), you can right-click the service and select *Update Service Reference...*. You can change the URL of the service by right-clicking and selecting *Configure Service Reference...*. This would be necessary when the service is published to an actual web server.

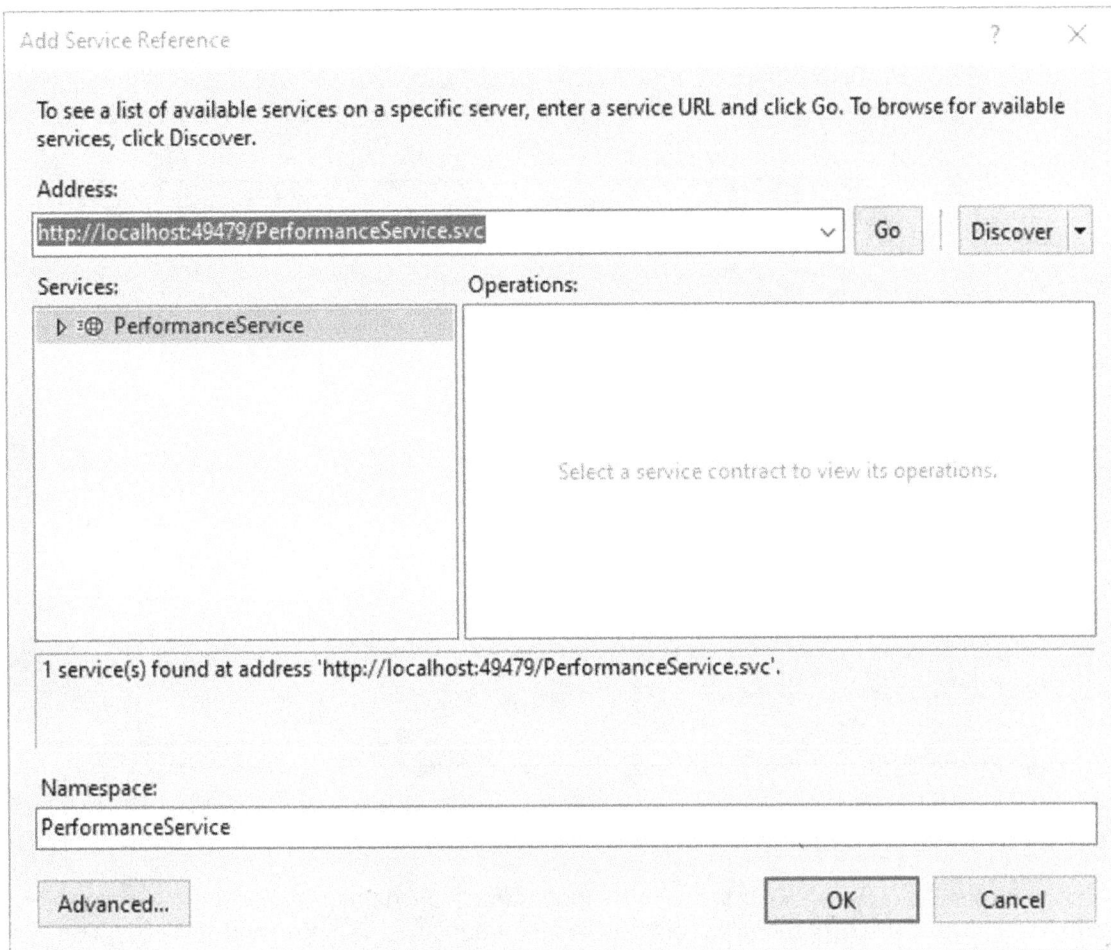

Figure 15-12: Adding a web service reference to the client project

Step 11: Create a web form.

Right-click the project in Solution Explorer and select *Add > Web Form with Master Page*. Name the new page ManagePerformances-WCF.aspx. Set the new page as the start page.

Step 12: Add the data source.

You will use an `ObjectDataSource` to make the data from the web service available on the page. Enter the code in Figure 15-13 inside the `MainContentPlaceHolder`.

```
1   <asp:ObjectDataSource runat="server" ID="ObjDataSource_Performance"
2          TypeName="Theater.PerformanceService.PerformanceServiceClient"
3          DataObjectTypeName="Theater.PerformanceService.Performance"
4          SelectMethod="GetPerformances"></asp:ObjectDataSource>
```

Figure 15-13: ObjectDataSource to connect to PerformanceService.

Line 2 shows the name of the type that the object source interacts with. This is a client of the `PerformanceService`. Line 3 specifies the data type for the objects that the `ObjectDataService` handles. This shows that the `Performance` class you specified in the `PerformanceService` is now available in the client project. Line 4 specifies the `SelectMethod` for the data source to be the method in the service that returns all performances.

Step 13: Display performances.

You can use a `Repeater` to display the performances using the code in Figure 15-14. Enter this immediately below the object data source. As you enter the code in line 3, be sure to select *Create a new event*.

```
1   <asp:Repeater ID="rptEvents" runat="server"
2       DataSourceID="ObjDataSource_Performance"
3       OnItemCommand="rptEvents_ItemCommand"
4       DataMember="DefaultView">
5       <HeaderTemplate>
6           <h1>Current Performances</h1>
7       </HeaderTemplate>
8       <ItemTemplate>
9           <h2> <%# Eval("PerformanceName") %></h2>
10          <%# Eval("Description") %>
11          <br />
12          <asp:Button ID="btnDelete" runat="server" class="btn-danger"
13              CommandName="Delete"
14              CommandArgument='<%#Eval("PerformanceId") %>'
15              Text="Delete"></asp:Button>
16          <asp:Button ID="btnEdit" runat="server" class="btn-primary mb-3"
17              CommandName="Edit"
18              CommandArgument='<%#Eval("PerformanceId") %>'
19              Text="Edit"></asp:Button>
20      </ItemTemplate>
21  </asp:Repeater>
```

Figure 15-14: Displaying performances

Inside the `ItemTemplate` (lines 8–20), the `Eval` method extracts the data from the data source based on the provided field name. In this case, we just display the performance name (line 9) and the description (line 10). Below the description are the two buttons for deleting and editing performances. In the next step, you will see how the `CommandName` is used to identify which button was pressed and the `CommandArgument` is used to identify the ID of the performance to manipulate.

You can run the program now to display all the events in the database. You should note that the page you just created doesn't have any reference to the local database in the Theater project and thus relies instead on the web service. However, the remaining pages on the site still use the local database. In order to avoid duplicate data that is inconsistent, you would need to convert the remaining pages to use the web service as well. This is left as an end-of-chapter exercise.

Step 14: Add a status message to the page.

Next, add the code in Figure 15-15 above the `Repeater`. This is a placeholder that you will use to display messages about the operations carried out.

```
1   <asp:PlaceHolder runat="server" ID="StatusMessage" Visible="false">
2       <p>
3           <asp:Label runat="server" ID="StatusText" /><br />
4       </p>
5   </asp:PlaceHolder>
```

Figure 15-15: Status message

Step 15: Code the method to delete the performances.

Switch to the code-behind file and add the following using statement:

```
using Theater.PerformanceService;
```

Then code the `rptEvents_ItemCommand` method as shown in Figure 15-16 to handle the button presses from the `Repeater` and delete a performance.

```
1   protected void rptEvents_ItemCommand(object source, RepeaterCommandEventArgs e)
2   {
3       PerformanceServiceClient svc = new PerformanceServiceClient();
4       string ID = e.CommandArgument.ToString();
5       switch (e.CommandName)
6       {
7           case "Delete":
8               {
9                   int result = svc.DeletePerformance(ID);
10                  if (result > 0)
11                  {
12                      StatusMessage.Visible = true;
13                      StatusText.CssClass = "text-success";
14                      StatusText.Text = "Record deleted";
15                      rptEvents.DataBind();
16                  }
17                  else
18                  {
19                      StatusMessage.Visible = true;
20                      StatusText.CssClass = "text-danger";
21                      StatusText.Text = "Record not deleted";
22                  }
23              }
24              break;
25          case "Edit":
26              {
27                  //Code to edit the performance will go here
28              }
29              break;
30      }
31  }
32
```

Figure 15-16: Handling button presses in the Repeater

Line 3 creates a `PerformanceServiceClient` object that you can use to call the methods in the web service. This makes the use of the web service completely transparent, allowing you to use it like any other object that you have created locally. All of the methods in the web service have been recreated in your client project, and the calls to the service are now handled by the generated code.

Line 4 gets the ID of the selected performance from the `CommandArgument` parameter from the Repeater. Line 5 sets up a switch statement based on the `CommandName` from the Repeater. See Figure 15-14 for how these were set up in the `Repeater`. If the `CommandName` is "Delete," line 9 will call on the `DeletePerformance` method passing in the ID of the performance. The delete method returns the number of records deleted, so if the operation is successful, lines 12–15 will display a message to that fact and update the data source. Otherwise, lines 19–21 displays a message indicating that the operation didn't succeed.

You will code the Edit button later once the form has been created.

Step 16: Create a form for editing and adding performances.

In order to create new performances and edit existing ones, you will need a form with textboxes for editing the data about the performances. Enter the code in Figure 15-17 between the status message placeholder and the `Repeater`.

```
 1   <h1>Add Performance</h1>
 2   <div class="form-group">
 3       <label class="col-sm-3" for="txtPerformanceName">Performance Name</label>
 4       <asp:TextBox runat="server" ID="txtPerformanceName" />
 5   </div>
 6   <div class="form-group">
 7       <label class="col-sm-3" for="txtBasePrice">Base Price</label>
 8       <asp:TextBox runat="server" ID="txtBasePrice" />
 9       <label id="lblPrice"></label>
10   </div>
11   <div class="form-group">
12       <label class="col-sm-3" for="txtDescription">Description</label>
13       <asp:TextBox TextMode="MultiLine" runat="server" ID="txtDescription" />
14   </div>
15   <div class="form-group">
16       <label class="col-sm-3" for="txtPresenter">Presenter</label>
17       <asp:TextBox runat="server" ID="txtPresenter" />
18   </div>
19   <asp:Button Text="Save" runat="server" CssClass="btn btn-primary mb-3"
20       ID="btnSavePerformance" OnClick="btnSavePerformance_Click" />
```

Figure 15-17: Form for editing performances

This is a standard form with labels and corresponding textboxes for each of the fields in the `Performance` class, except the ID. In line 20, be sure to generate a new `OnClick` method.

Step 17: Code the method to create a performance.

Switch to the code-behind page and add the code in Figure 15-18 to capture the data from the form and insert a `Performance` object to the web service.

```
1   protected void btnSavePerformance_Click(object sender, EventArgs e)
2   {
3       //read values from UI into new Performance object
4       Performance p = new Performance()
5       {
6           BasePrice = double.Parse(txtBasePrice.Text),
7           PerformanceName = txtPerformanceName.Text,
8           Description = txtDescription.Text,
9           Presenter = txtPresenter.Text
10      };
11
12      PerformanceServiceClient svc = new PerformanceServiceClient();
13      //Insert record
14      int result = svc.InsertPerformance(p);
15
16      if (result == 1)
17      {
18          StatusMessage.Visible = true;
19          StatusText.CssClass = "text-success";
20          StatusText.Text = "Record saved";
21          txtBasePrice.Text = "";
22          txtDescription.Text = "";
23          txtPerformanceName.Text = "";
24          txtPresenter.Text = "";
25          rptEvents.DataBind();
26          Session["SelectedId"] = null;
27      }
28
29      else
30      {
31          StatusMessage.Visible = true;
32          StatusText.CssClass = "text-danger";
33          StatusText.Text = "Record not saved";
34      }
35  }
```

Figure 15-18: Code to save a performance

This code starts in lines 4–10 by using the values from the form to create a `Performance` object. Note that in this simplified example, the parsing of the base price in line 6 is not protected and may throw an exception. Line 12 creates an object for the web service that is used in line 14 to call the `InsertPerformance` method to send the `Performance` object to the server to be saved. If the operation is successful, the method returns 1, and the first part of the `if` statement (lines 16–27 shows the status message as "Record Saved" and clears the textboxes. You can run the site now and see that you can insert performances and have them displayed.

Step 18: Code the method to edit a performance.

Editing a performance requires retrieving the details about the performance and populating the form with those values. You will use a session variable to store the ID for the performance to be edited.

The first part of this is done by adding the code in lines 3–8 in Figure 15-19 after line 27 in Figure 15-16.

```
1   case "Edit":
2       {
3           Performance p = svc.GetPerformanceById(ID);
4           txtPerformanceName.Text = p.PerformanceName;
5           txtBasePrice.Text = p.BasePrice.ToString();
6           txtDescription.Text = p.Description;
7           txtPresenter.Text = p.Presenter;
8           Session["SelectedId"] = p.PerformanceId;
9       }
```

Figure 15-19: Code to populate the form with a selected performance

Line 3 retrieves the correct performance from the server based on the ID. Lines 4–7 populate the textboxes in the form with the values from the retrieved object. Line 8 stores the ID of the performance in a session variable so it can be retrieved later when storing the revised values.

Next, you'll need to distinguish between inserting a new and editing an existing performance. You do this by updating lines 12–14 of Figure 15-18 with the code in Figure 15-20 to either insert a new object or save the edited one.

```
1   PerformanceServiceClient svc = new PerformanceServiceClient();
2   int result = 0;
3   string selectedID = Session["SelectedId"].ToString();
4   if (selectedID == null)
5   {
6       //Insert record
7       result = svc.InsertPerformance(p);
8   }
9   else
10  {
11      //Update existing record
12      p.PerformanceId = int.Parse(selectedID);
13      result = svc.UpdatePerformance(p);
14  }
```

Figure 15-20: Code to save performances

The code now brings up the `selectedId` value from the session variable. Line 4 checks if this value is null, which means no performance has been selected and the user is trying to save a new performance. Lines 6–11 are unchanged to save a new performance. The `else` clause is then used to update an existing performance. This starts with adding the performance ID from the session state to the performance object in line 15, and then in line 16, the object with the updated values is sent to the server using the `UpdatePerformance` method.

You can now run the project and check that you can create new performances and edit existing ones.

Tutorial: Building the REST Web API Web Service

In this tutorial, you will create a new web service based on the REST Web API, but with the same functionality as the WCF service you built above. This will be accompanied by a new version of the ManagePerformance web page to consume and use the REST web service.

Step 19: Create the REST Web API project.

Start a new project in Visual Studio, and select an ASP.NET Web Application (.Net Framework) template. Click *Next*, and name the solution *PerformanceService-REST*. Click *Create* and choose the Web API template (see Figure 15-21). Click *Create* to create the project.

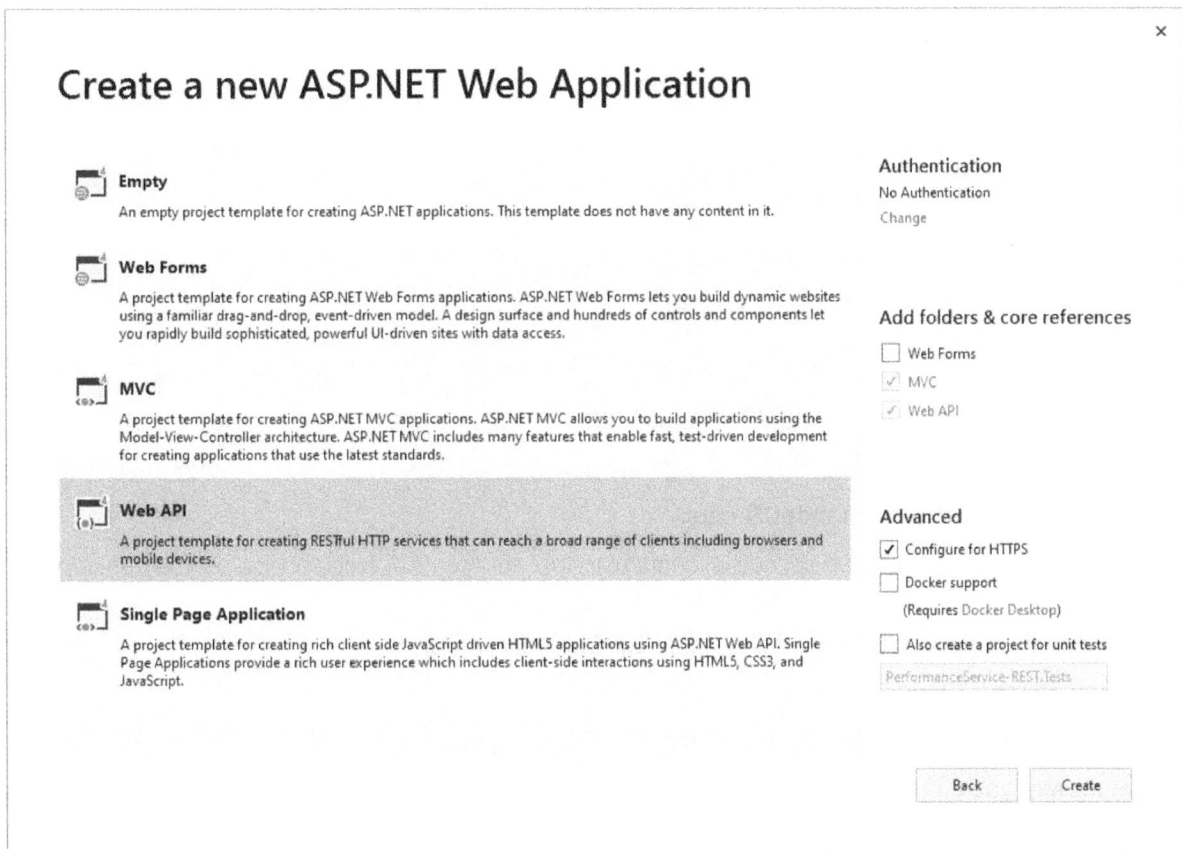

Figure 15-21: Creating the REST Web API project

Step 20: Set up the project.

Once the project has been created, you will need to add a few files.

Right-click the *App_Data* folder and select *Add > Existing Item*. Navigate to the tutorial start folder and select the *Theater.mdf* database.

Add a connection string pointing to the Theater database in the Web.Config file using the code in Figure 15-8.

In the Controllers folder, delete the ValuesController.cs file.

Step 21: Add the Performance class.

Right-click the *Models* folder, and add a new class called *Performance*. Add the code in Figure 15-22 to implement the class. The class contains an automatic property for each of the fields in the table.

```
1   public class Performance
2   {
3       public int PerformanceId { get; set; }
4
5       public string PerformanceName { get; set; }
6
7       public double BasePrice { get; set; }
8
9       public string Description { get; set; }
10
11      public string Presenter { get; set; }
12  }
```

Figure 15-22: Code for the Performance class

Step 22: Add the PerformanceDB class.

Access to the data in this web service is identical to that for the WCF web service, so you can copy the `PerformanceDB` class you created previously into the new project.

Right-click the *Models* folder and select *Add > Existing Item...*.

Navigate to the WCF project and add PerformanceDB.cs.

Change the namespace for the class to `PerformanceService_REST.Models`.

Alternatively, you can create the class in the *Models* folder and then copy the code from the previous project.

Step 23: Add a controller.

Right-click the *Controllers* folder and select *Add > Controller...*. Choose the *Web API 2 Controller – Empty* template. Name the controller *PerformancesController*.

Note that the controller's name must end with Controller. The first part of the controller name (*Performances*) is used to generate the URL pattern that is used when accessing the web service. So, in this case, the service will be accessed at the URL /api/performances.

Step 24: Program the web service controller.

The controller plays the same role as the service contract in the WCF web service. Both of these classes specify the available operations that the web service makes available and how to access those methods. However, where the methods in the service contract for WCF are called using C# as they are written, the Web API controller methods need to follow a specific naming convention to indicate how calls from the clients will be routed to the proper interface. As described in the introduction to this chapter, HTTP defines four basic action methods: GET, PUT, POST, and DELETE. Method names in the controller should start with one of these words to indicate which action method is used. Add this using statement:

```
using PerformanceService_REST.Models;
```

Then, add the code in Figure 15-23 to implement the controller.

```
1   public class PerformancesController : ApiController
2   {
3       PerformanceDB data;
4
5       public PerformancesController()
6       {
7           this.data = new PerformanceDB();
8       }
9       //GET: api/performance
10      public IEnumerable<Performance> GetPerformances()
11      {
12          return data.GetPerformances();
13      }
14      //GET api/performance/12
15      public Performance GetPerformanceByID(string id)
16      {
17          return data.GetPerformanceById(id);
18      }
19
20      //POST(Insert): api/performance
21      public int PostPerformance([FromBody]Performance value)
22      {
23          return data.InsertPerformance(value);
24      }
25
26      //PUT(update): api/performance
27      public int PutPerformance([FromBody] Performance value)
28      {
29          return data.UpdatePerformance(value);
30      }
31
32      //DELETE: api/performance
33      public int DeletePerformance(string id) {
34          return data.DeletePerformance(id);
35      }
```

Figure 15-23: REST web service controller

Each of the methods begins with an HTTP method keyword (GET, POST, PUT, DELETE). For example, the method to insert a new performance is called `PostPerformance` (line 21), and the one to remove a performance is called `DeletePerformance` (line 33).

The POST and PUT methods that create and edit performances respectively (lines 20–34 and 26–30) include a `[FromBody]` attribute indicating that the values for the parameter are to be found in the body of the HTTP request as key-value pairs.

Each of the methods in the controller call on a corresponding method in `PerformanceDB`.

Step 25: Test the web service.

You are now ready to test the web service. Unlike a WCF web service, there is no test client. However, you can use a browser to test the service. When you run the service, it will launch a website with an API link in the toolbar (Figure 15-24).

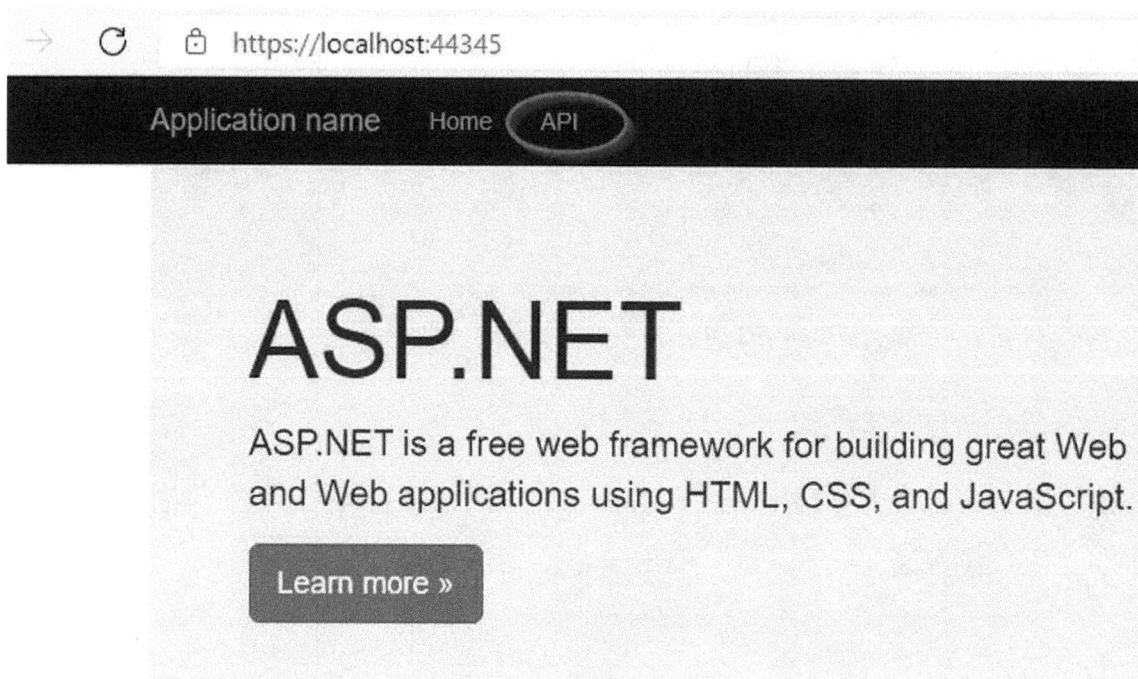

Figure 15-24: Web API website

Clicking the API link will bring you to a help page with documentation for each of the available operations in the API.

To access the operations, you can simply enter the URL in the browser using the pattern shown in the comments in the code in Figure 15-23.

For the server shown here, to get all performances, you would use this URL: https://localhost:44345/api/performances.

Entering this in the browser will return an XML representation of the data, as shown in Figure 15-25. The exact display may differ, depending on the browser you're using. The methods that alter data on the server cannot be called directly from the address bar in the browser, as they rely on being able to change the HTTP method and adding data to the body of the HTTP request. You will see these methods in action in the next tutorial.

Now, you can try adding a number to the end of the URL to retrieve a specific performance by that ID. For example, https://localhost:44345/api/performances/39 will retrieve the performance with ID 39.

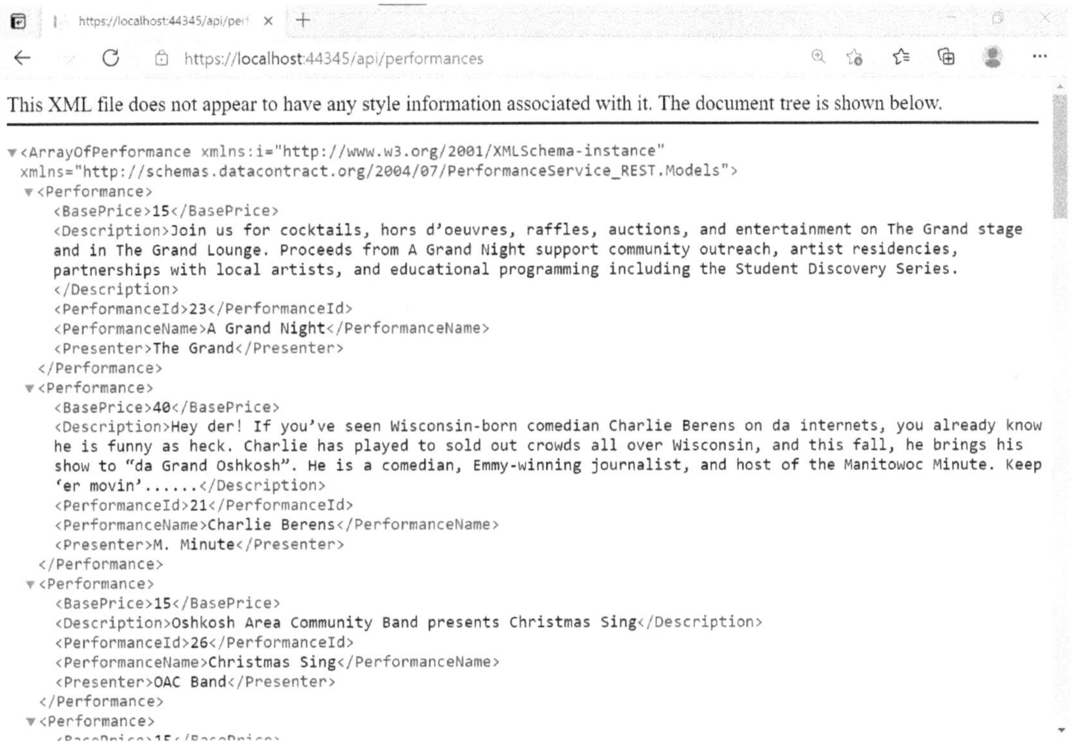

Figure 15-25: XML returned for all performances

Tutorial: Creating the REST Client

In this tutorial, you will build a client that uses the methods in the Web API to replicate the functionality of the ManagePerformance page you built previously for the WCF client. Here you will create two separate pages—one that shows how to use jQuery directly from the browser and one that uses server-side C# to access the Web API service.

Because the Web API returns JSON by default, jQuery is well suited for interacting with a Web API service. Thus most of this tutorial will be devoted to building the page and writing the jQuery code to interact with the service. However, you may also need to interact with a Web API service from C# server-side code, so we will also illustrate this at the end of the tutorial.

15.3 Consuming Web Service Using jQuery

Step 26: Add an ASPX page.

In the Theater project from the previous tutorial, add a new Web Form with Master Page named ManagePerformances-REST.aspx.

Step 27: Add the HTML code.

Add the code shown in Figure 15-26 to the `MainContentPlaceHolder`. This will implement a form that is almost identical to what you created for the WCF client.

```
1   <span id="statusText"> </span>
2   <h1>Add Performance</h1>
3   <div class="form-group">
4       <input type="hidden" id="PerformanceId" name="PerformanceId" />
5   </div>
6   <div class="form-group">
7       <label class="col-sm-3" for="PerformanceName">Performance Name</label>
8       <input type="text" id="PerformanceName" name="PerformanceName" />
9   </div>
10  <div class="form-group">
11      <label class="col-sm-3" for="BasePrice">Base Price</label>
12      <input type="number" id="BasePrice" name="BasePrice" />
13      <label id="lblPrice"></label>
14  </div>
15  <div class="form-group">
16      <label class="col-sm-3" for="Description">Description</label>
17      <textarea id="Description" name="Description" rows="4" cols="50">
18          </textarea>
19  </div>
20  <div class="form-group">
21      <label class="col-sm-3" for="Presenter">Presenter</label>
22      <input type="text" id="Presenter" name="Presenter" />
23  </div>
24  <div class="form-group">
25      <input type="button" value="Add New" class="btn btn-primary"
26          onclick="savePerformance();" />
27      <input type="button" value="Update" class="btn btn-primary"
28          onclick="updatePerformance();" />
29      <input type="button" value="Clear" class="btn btn-outline-danger"
30          onclick="clearAll();" />
31  </div>
32
33  <div id="performances"></div>
```

Figure 15-26: HTML code to create a form for the Web API page, for creating and editing performances

Compared to the page you created previously (Figure 15-17), the primary differences here are that all the controls are standard HTML, as opposed to ASP controls. Instead of using a Repeater, the div in line 32 will be used to programmatically display the performances using a similar format.

The PerformanceId field is hidden in line 4, as the value will need to be assigned to it before sending it to the server for updating, as you will see later.

Step 28: Add the jQuery code.

The jQuery code goes in a JavaScript file, so right-click the Scripts folder and select **Add > JavaScript**. Name the file **performance.js**. Add the code shown in Figure 15-27 to the new file. This code is available in performance.js.txt.

```
1   const api = "https://localhost:44345/api/Performances/";
2
3   $(document).ready(function () {
4       showPerformances();
5   });
6
7   function showPerformances() {
8       $('#statusText').text("Loading data...");
9       $.ajax({
10          headers: {
11              'Content-Type': 'application/x-www-form-urlencoded'
12          },
```

Figure 15-27: jQuery code to manage performances (*continues*)

```
13              dataType: "json",
14              type: "GET",
15              url: api,
16              success: function (data) {
17                  let output = "";
18                  $.each(data, function (key, val) {
19                      output += "<h2>" + val.PerformanceName + "</h2>";
20                      output += "<p>" + val.Description + "</p>";
21                      output += "<input type = 'button' value =";
22                      output += "'Select Performance' class='btn btn-primary mb-3 mr-1'";
23                      output += "onclick='selectPerformance(" + val.PerformanceId + "); ' />";
24                      output += "<input type='button' value='Delete' class='btn btn-danger mb-3'";
25                      output += "onclick='deletePerformance(" + val.PerformanceId + "); ' />";
26                  });
27                  $('#performances').html(output);
28                  clearAll();
29              },
30              error: showError
31          });
32  };
33
34  function updatePerformance() {
35      $('#statusText').text("Updating...");
36      let url = api + $('#PerformanceId').val();
37      $.ajax({
38          type: 'PUT',
39          url: api + $('#PerformanceId').val(),
40          data: $('#form1').serialize(),
41          dataType: "json",
42          success: showPerformances,
43          error: showError
44      });
45  };
46
47  function savePerformance() {
48      $('#statusText').text("Saving...");
49      $.ajax({
50          type: 'POST',
51          url: api,
52          data: $('#form1').serialize(),
53          dataType: "json",
54          success: showPerformances,
55          error: showError
56      });
57  };
58
59  function selectPerformance(id) {
60      let href = api + id;
61      $('#statusText').html("Selecting performance...");
62      $.getJSON(href, function (data) {
63          $('#PerformanceId').val(data.PerformanceId);
64          $('#PerformanceName').val(data.PerformanceName);
65          $('#BasePrice').val(data.BasePrice);
66          $('#Description').val(data.Description);
67          $('#Presenter').val(data.Presenter);
68          $('#statusText').text("");
69      })
70          .fail(showError);
71  };
72
73  function deletePerformance(id) {
74      $('#statusText').text("Deleting...");
75      $.ajax({
76          type: 'DELETE',
77          url: api + id,
78          success: showPerformances,
```

Figure 15-27: jQuery code to manage performances (*continues*)

```
79          error: showError
80      });
81  };
82
83  function showError(request, status, error) {
84      try {
85          let response = JSON.parse(request.responseText);
86          $('#statusText').text(response.ExceptionMessage);
87      } catch (e) {
88          $('#statusText').text("A problem occurred - server unreachable.");
89      }
90  };
91
92  function clearAll() {
93      $('#statusText').text("");
94      $('#PerformanceId').val("");
95      $('#PerformanceName').val("");
96      $('#BasePrice').val("");
97      $('#Description').val("");
98      $('#Presenter').val("");
99  };
```

Figure 15-27: jQuery code to manage performances (*continued*)

Line 1 contains a reference to the base URL for the API. By setting up a constant to hold this, you will only have to change it in one place when the service is moved to a different server. You'll need to update the port number (44345) to the port number shown in the URL in your browser when running your Web API site. The first method is the jQuery `ready` function, which is executed when the page is loaded but before it's displayed. This method calls the `showPerformances` method, which sends an AJAX request to the server. The jQuery `ajax` method specifies the data that is sent to the server and the URL it goes to. If the request is successful, it uses the jQuery `each` function to go through each of the returned records and construct the HTML output in a format similar to the `Repeater` you set up earlier in Figure 15-14. The output string holding the resulting HTML code is added to the `performance's div` in line 29. The code sets up two buttons for each performance—one allows for selecting a performance and the other for deleting it.

When the ***select*** button is clicked, the `selectPerformance` method (lines 62–75) is executed. This uses the performance ID to get the performance from the server using the jQuery `getJSON` method. Line 64 constructs the correct URL to get the data for the correct performance. Lines 67–72 update the form with the values for the performance.

The ***delete*** button executes the `deletePerformance` method (lines 78–89), which sends an AJAX request to the server with the HTTP type DELETE and a URL that includes the selected performance ID. If the request is successful, the `showPerformance` method is called to update the display of performances (without the deleted one).

Saving a new and updating an existing performance are very similar and are implemented in lines 36–48 (updating) and lines 50–60 (saving new). Both of these methods serialize the form (lines 43 and 55). This creates a data record that includes all the data fields and their values. The `name` attribute for each control has to match the field names in the `Performance` class as it is created on the server. This is why we need to include a hidden input field for the update operation, so the ID can be passed along with the rest of the form when updating a performance. There are only two differences between updating and saving a new performance: First, the `updatePerformance` method sends the ID of the performance to be updated in the `URL` (line 42). Second, updating a performance uses the `PUT` method (line 41), whereas saving a new performance uses the `POST` method (line 53).

Recall that `PUT` is used when calling it multiple times will not change the result on the server, whereas `POST` is used when calling the operation multiple times will change the server. This is why we use `PUT` for updates and `POST` for creating new records.

Step 29: Link to a JavaScript file.

In order to load the jQuery code, you will need to add a reference to it in the ASPX code. Add this line to the `HeadContentPlaceHolder`:

```
<script src="/Scripts/performance.js"></script>
```

Step 30: Load the jQuery library.

In order for the jQuery code to work, you need to have jQuery loaded before the code is supposed to run. Currently, the site loads jQuery at the end of `Main.Master`. However, that is too late, so move the jQuery script element to the `head` element. You'll also need to make sure the jQuery library isn't the slim version. If that's the case, go to *code.jquery.com* and get a link to the latest version of the minified version of jQuery.

15.4 Cross-Origin Resource Sharing (CORS)

For security reasons, web browsers restrict web pages from making requests to multiple different servers. But that is exactly what we need to do. The client site hosts the HTML and jQuery code, which then makes a request to the web server, at a different location. In this case, as you are developing, the server and the client are on the same domain (localhost), but they have different port numbers, so the browser will recognize them as different server origins.

To get around this issue, you can loosen the restriction by enabling Cross-Origin Resource Sharing. Note that this will relax security and make your site *less* secure, so you should be careful with how you do this. For example, you can specify which servers are allowed to initiate cross-origin requests to only include those that you control the development of.

Step 31: Install CORS packages.

You can enable CORS on your server by installing a NuGet package in the Web Server project. Switch to the PerformanceService-REST project and go to *Tools > NuGet Package Manager > Package Manager Console*. Then issue the following command:

```
PM> Install-Package Microsoft.AspNet.WebApi.Cors
```

Step 32: Enable CORS.

Open the WebApiConfig.cs file, which is in the App_Start folder. Add the following line of code at the beginning of the `Register` method:

```
config.EnableCors();
```

Next, switch to PerformancesController.cs and add the following using statement:

```
using System.Web.Http.Cors
```

Then add the `[EnableCors]` attribute above the class declaration:

```
[EnableCors(origins: "https://localhost:44379", headers: "*", methods: "*")]
```

You'll need to replace the port number with the one used by your client. You can run the client to get the right port number. You can control which servers are allowed to issue cross-origin requests by listing them in `origins`. In `headers`, you can specify specific headers that must be present in the request to allow cross-origin requests. In `methods`, you can specify which HTTP methods (GET, POST, PUT, DELETE) are allowed. In the example here, we allow all headers and all methods from a single server.

Step 33: Test the service.

Run the web service and the REST client. Once both are fully launched, you should see that the new page will function like the WCF page you created previously.

15.5 Consuming Web API Using C#

For the most part, when using a REST Web API, you will want to consume that using jQuery. However, you can also use C#. This last part of the tutorial will involve retrieving data from the Web API using server-side C# and a Repeater control. You will create a new page to display the performances, similar to how you have done on the previous two pages, but this time the data will be retrieved in XML format from the Web API service using C# and bound to the Repeater control.

Step 34: Create the form.

Start by creating a new web form with a master page named *ManagePerformances-C-Sharp*. Add lines 2–13 of Figure 15-28 to `MainContentPlaceHolder`.

```
1   <asp:Content ID="Content2" ContentPlaceHolderID="MainContentPlaceHolder"
        runat="server">
2       <asp:Repeater ID="rptPerformances" runat="server">
3           <HeaderTemplate>
4               <h1>Current Performances</h1>
5           </HeaderTemplate>
6           <ItemTemplate>
7               <h2><%# Eval("PerformanceName") %></h2>
8               <%# Eval("Description") %>
9               <br />
10              Presented by: <%# Eval("Presenter") %>
11              <br />
12          </ItemTemplate>
13      </asp:Repeater>
14  </asp:Content>
```

Figure 15-28: Repeater to show performances

Step 35: Create the code-behind.

Switch to the code-behind page and enter the code in Figure 15-29 in the `Page_Load` method. This code will require adding three `using` statements for `System.Net`, `System.IO`, and `System.Data`.

```
1   protected void Page_Load(object sender, EventArgs e)
2   {
3       string url = "https://localhost:44345/Api/Performances/";
4       HttpWebRequest request = (HttpWebRequest)WebRequest.Create(url);
5       request.Method = "GET";
6       request.ContentType = "text/xml; encoding='utf-8'";
7
8       //Send request
9       WebResponse response = request.GetResponse();
10      Stream stream = response.GetResponseStream();
11
12      //Read stream
13      DataSet ds = new DataSet();
14      ds.ReadXml(stream);
15
16      //Bind repeater to XML data source
17      rptPerformances.DataSource = ds;
18      rptPerformances.DataBind();
19  }
```

Figure 15-29: C# code to pull data from the Web API

Line 3 sets up a variable for the base URL for the web service, as you've seen before. Be sure to change the port number to match your service. Line 4 creates an object for the HTTP request. Line 5 sets the type to `GET`, indicating you will be retrieving data, and line 6 sets the content type to XML. While it is possible to read JSON in C#, the support for reading XML is stronger. Lines 9 and 10 send the request and connect the response to a stream. The stream is read into a `DataSet` in line 14 using an XML reader. The `DataSet` is then bound to the `Repeater` in lines 17 and 18. If you need to further process your data, you can use regular XML processing techniques.

You can now run the page and observe that the data for the performances are displayed.

Exercises

Exercise 15.1: WCF Web Service

Display all fields of all performances in a table on a new web page using the WCF web service.

Exercise 15.2: Web API Web Service

Display all fields of all performances in a table on a new web page using the Web API web service.

Exercise 15.3: Filter by Presenter—WCF

Add a new method to the WCF web service to retrieve performances by a specific presenter. Then modify the user interface to allow the user to filter by presenter.

Exercise 15.4: Filter by Multiple Presenters—WCF

Expand the previous exercise to allow for filtering by multiple presenters.

Exercise 15.5: Filter by Presenter—Web API

Add a new method to the Web API web service to retrieve performances by a specific presenter. Then modify the user interface to allow the user to filter by presenter.

Exercise 15.6: Filter by Multiple Presenters—Web API

Expand the previous exercise to allow for filtering by multiple presenters.

Exercise 15.7: Convert to Use Web Services

Convert the remaining pages in the Theater solution to use a web service.

Chapter 16

Razor Pages

Chapters 4 and 5 presented HTML forms that used client-side JavaScript to take actions in response to various events. However, client-side scripts cannot and should not do everything. For example, it is neither easy nor safe to use a client-side script to access databases on a server. Similarly, you should not use client-side scripts for applications that require large amounts of memory and/or processing power, which could result in a slow response. Further, client-side scripting generally makes applications less secure. Server-side scripts help address these issues. This chapter presents Razor Pages to run server-side C# code.

Learning Objectives

After studying this chapter, you should be able to

- Create an ASP.NET Core web application using the Razor Pages framework.
- Develop PageModel and content pages.
- Use Razor syntax to embed server-side code into web pages.
- Access databases to dynamically create HTML elements.
- Use model binding to access input data from the server.
- Use session state to store data on the server and share it.
- Share data between PageModel and content pages.
- Restore field values after a page reload.

16.1 Introduction

Razor Pages and Model-View-Controller (MVC) are two related frameworks that are popular for developing web applications involving server-side processing. The ASP.NET web forms discussed in Chapters 6–15 also do server-side processing. Unlike Razor Pages, ASP.NET web forms are designed to work with drag-and-drop server controls, their event handlers, post-backs, and ViewState to maintain the state of controls, making web forms particularly suitable for rapid application development. However, Razor Pages and MVC provide better performance, maintenance, ease of testing, and control over the rendered HTML code.

Razor Pages is designed to create dynamic **HTML** pages that include server-side code and database access. This framework supports cross-platform development and deployment on Windows, Unix, and Mac operating systems.

Compared to ASP.NET web forms, Razor Pages maintains a strict separation between the view (the user interface) and the model (code that specifies the data and business logic). This separation contributes to ease of maintenance and testing. Razor Pages uses a cleaner and easy-to-use markup syntax (Razor syntax) to embed server-side code in HTML. In addition, Razor Pages allows you to use tag helpers to generate HTML elements that may have attributes that are not available in the standard elements.

Both Razor Pages and MVC frameworks are integrated with the **ASP.NET Core**, which is a relatively new open-source cross-platform framework for developing web applications. It should be noted that both Razor Pages and MVC use the Razor view engine to render the HTML for a page. The two frameworks are similar in many respects, but there are differences that make Razor Pages easier to understand and use for many applications. However, the MVC model provides several benefits, including ease of maintenance, especially for larger applications. The MVC model is presented in Chapters 17 and 18.

Razor Pages is the default method of web development in Visual Studio. Because of the simplicity of Razor Pages and its similarity to web forms, Razor Pages is introduced first before we present the MVC approach, to help you learn server-side processing. An understanding of Razor Pages also makes it easier to learn MVC.

You may use Visual Studio Community (for Windows) or Visual Studio for Mac (for MacOS), which provide an easy-to-use environment for developing ASP.NET Core MVC applications. The tutorials in this chapter are developed using ASP.NET Core 3.1.

16.1.1 ASP.NET versus ASP.NET Core

ASP.NET and ASP.NET Core are platforms for developing web applications. ASP.NET Core is relatively new. A major difference between the two platforms is that ASP.NET Core is an open-source cross-platform framework, whereas ASP.NET is Windows-based. ASP.NET Core supports multiple operating systems, including Windows, macOS, and Linux.

ASP.NET supports ASP.NET Web Forms and an earlier implementation of MVC, called ASP.NET MVC, which is significantly different from the newer ASP.NET Core MVC. But ASP.NET does not support Razor Pages and ASP.NET Core MVC. ASP.NET Core does not support ASP.NET Web Forms development. It supports Razor Pages and ASP.NET Core MVC.

Table 16-1 shows the frameworks supported by the two platforms.

Table 16-1: ASP.NET versus ASP.NET Core

Web development platforms	Frameworks supported
ASP.NET	ASP.NET Web Forms ASP.NET MVC
ASP.NET Core	Razor Pages ASP.NET Core MVC

16.2 Working with Razor Pages

The Razor Pages framework stores the view, which represents the user interface, in a file with the extension **.cshtml**. The view typically consists of HTML, JavaScript, CSS, and Tag Helpers. You also may embed C# server-side code within the HTML document or place it outside the document using the Razor syntax.

The C# code for the model component, which consists of the data and business logic, is placed in a separate file (code-behind file) with the extension **.cshtml.cs**.

To help you learn about web development using the Razor Pages framework, you will create a Razor Pages version of the reservation form that you developed in Chapters 4 and 5.

Tutorial: Reservation Page

In Chapters 4 and 5, you developed an HTML form that used client-side JavaScript to compute the prices and to display them on the Reservation and Confirmation pages. In this tutorial, you will create a Razor page to compute the same prices, as in Chapter 5. The Razor version of the reservation form is shown in Figure 16-1, which differs from the HTML form in the following way:

- In the Razor page, the prices and costs are computed on the server, which is more tamper-proof than client-side computing. These computed prices/costs are displayed on the Confirmation page shown in Figure 16-2.

- Instead of using hardcoded data for performance names and base prices, these data items are retrieved from a database and displayed in the dropdown list. Later, you also will retrieve the zone names and prices from the database.

- Some input data (like personal information about the user) is not collected, to keep the code short and to help focus on important Razor features.

You will develop this page in three parts:

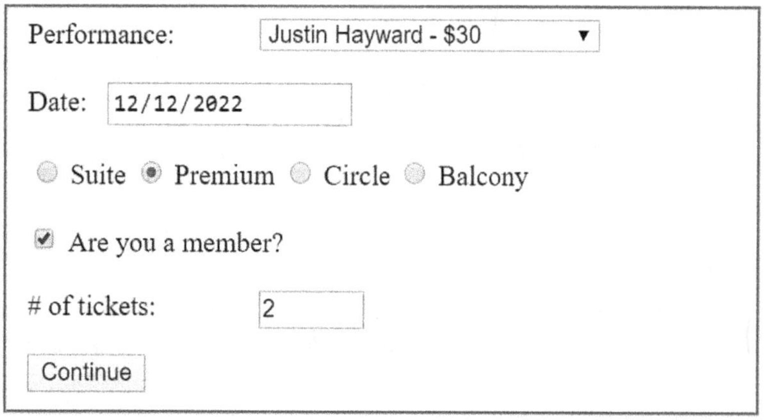

Figure 16-1: Reservation page

- The first part (steps 1–9) builds the basic user interface and develops server-side code to accesses the database and to add items to the dropdown list.
- The second part (steps 10–12) expands the project to let the user proceed to another page named the Confirmation page and displays the reservation data on the Confirmation page.
- The third part (steps 13 to the end) shows how to restore the values of Reservation fields if the user goes back to make changes.

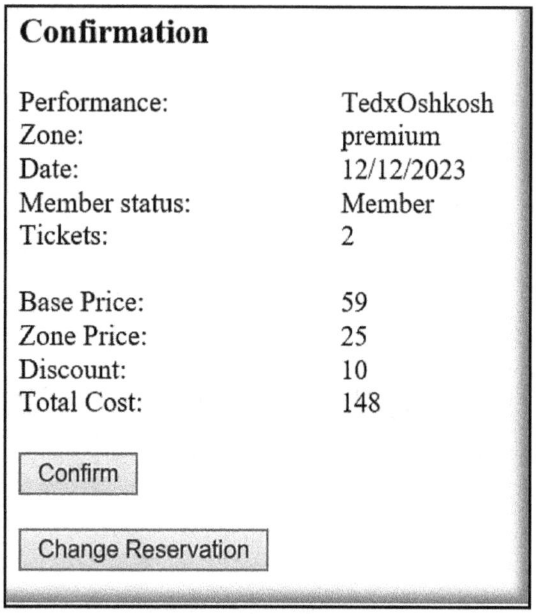

Figure 16-2: The Confirmation page

You will create the Confirmation page that displays the reservation data as shown in Figure 16-2.

16.2.1 Creating an ASP.NET Core Web Application

First, you will create an ASP.NET Core project and add a Razor page. You may use Visual Studio Community (for Windows) or Visual Studio for Mac (for MacOS). An introduction to the Visual Studio (for Windows) environment is provided in Chapter 6.

It should be noted that the initial steps to create a project in Visual Studio for Mac is similar to the steps given here, but there are some differences that make it necessary for you to figure out certain steps yourself. Similarly, for database access, this book uses SQL Server Express LocalDB, which is a limited version of SQL Server. If you are using Visual Studio for Mac, you may use the SQLite database. Again, there are some differences in terms of their use.

Step 1: Create an ASP.NET Core web application.

Create a new folder named *Razor-pages* within the *WebProjects* folder, which is the main folder that you created in Chapter 1 to store all projects.

Create a new ASP.NET Core web application within the *Razor-pages* folder, as follows:

Open Visual Studio. You will see the initial window with different options to get started, including opening a project and creating a project.

Select *Create a new project*. You will see a *Create a new project* window that is similar to the one shown in Figure 16-3, except that the recent project templates may be different.

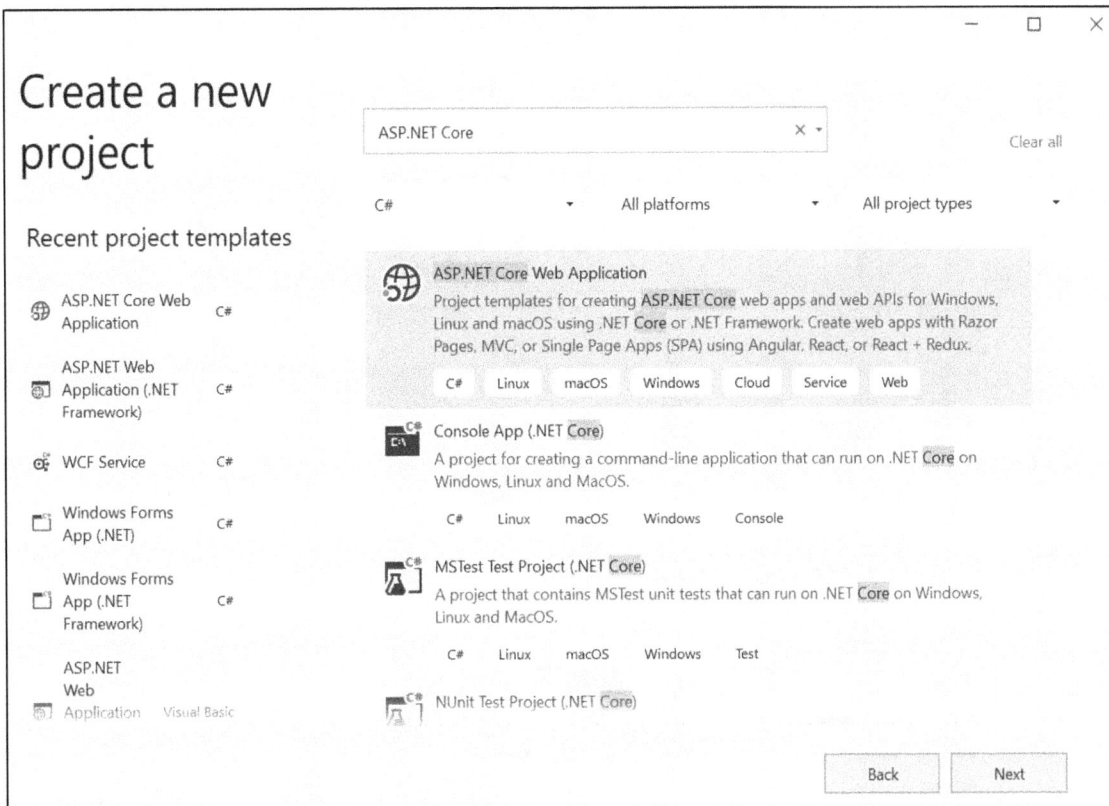

Figure 16-3: The *Create a New Project* window

You can search for templates by entering key words like "ASP.NET Core" and "C#" in the search box, as shown.

Select the *ASP.NET Core Web Application* (**not** *ASP.NET web application*) that includes the key word "C#" in the bottom line.

Click **Next**. You will see the *Configure your new project* window, shown in Figure 16-4, except for the input data.

Enter *Theater-RP* for the project name.

Select *WebProjects/Razor-pages* for the location.

Click **Create**. You will see the *Create a new ASP.NET Core Web Application* window, which shows the different project templates that you can use (see Figure 16-5).

Select *Empty* and click **Create**.

You will see the project window, similar to the one shown in Figure 16-6.

Configure your new project

ASP.NET Core Web Application   C#   Windows   Linux   macOS   Web

Project name

Theater-RP

Location

C:\WebProjects\Razor-pages\

Solution name ⓘ

Theater-RP

☐ Place solution and project in the same directory

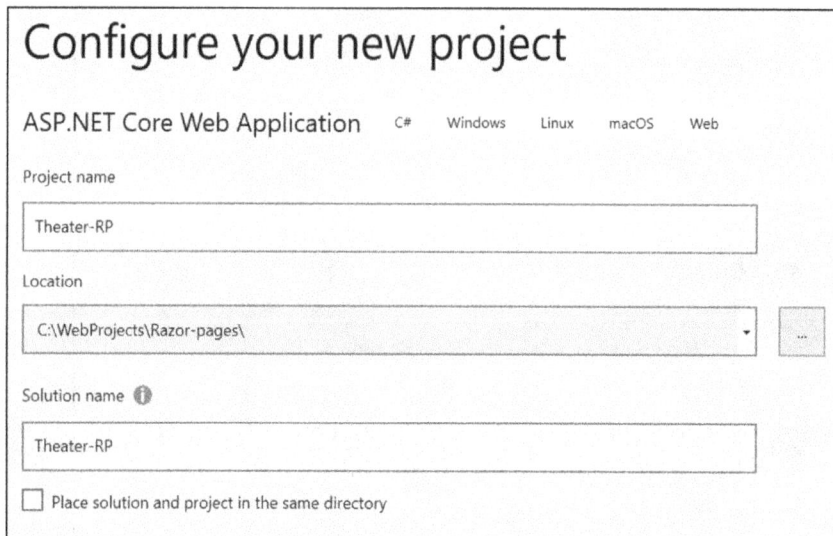

Figure 16-4: The *Configure your new project* window

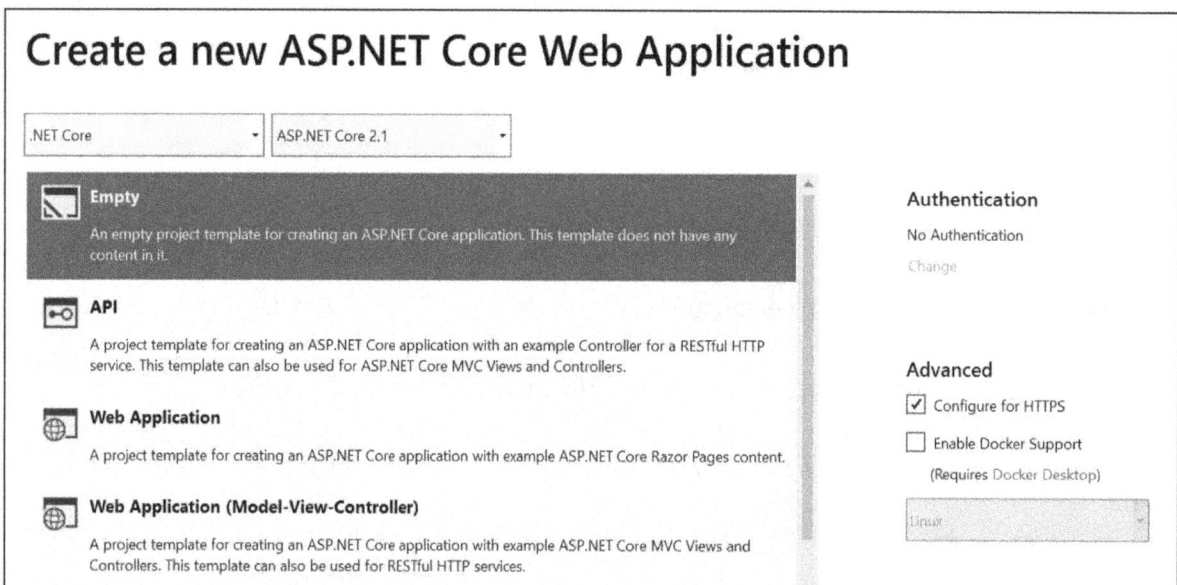

Create a new ASP.NET Core Web Application

.NET Core ▾   ASP.NET Core 2.1 ▾

Empty
An empty project template for creating an ASP.NET Core application. This template does not have any content in it.

API
A project template for creating an ASP.NET Core application with an example Controller for a RESTful HTTP service. This template can also be used for ASP.NET Core MVC Views and Controllers.

Web Application
A project template for creating an ASP.NET Core application with example ASP.NET Core Razor Pages content.

Web Application (Model-View-Controller)
A project template for creating an ASP.NET Core application with example ASP.NET Core MVC Views and Controllers. This template can also be used for RESTful HTTP services.

Authentication

No Authentication

Change

Advanced

☑ Configure for HTTPS

☐ Enable Docker Support

(Requires Docker Desktop)

Linux

Figure 16-5: The *Create a new ASP.NET Core Web Application* window

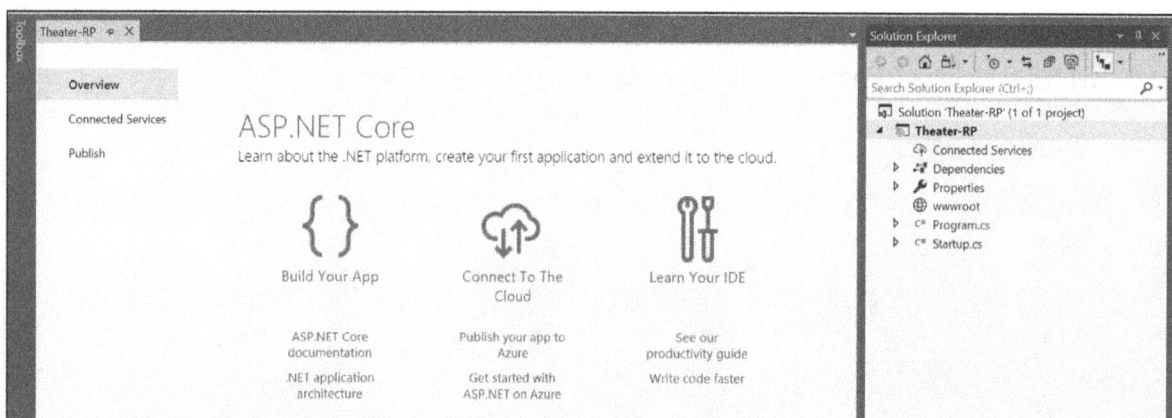

ASP.NET Core

Learn about the .NET platform, create your first application and extend it to the cloud.

Overview
Connected Services
Publish

Build Your App
ASP.NET Core documentation
.NET application architecture

Connect To The Cloud
Publish your app to Azure
Get started with ASP.NET on Azure

Learn Your IDE
See our productivity guide
Write code faster

Solution Explorer

Search Solution Explorer (Ctrl+;)

Solution 'Theater-RP' (1 of 1 project)
  Theater-RP
    Connected Services
    Dependencies
    Properties
    wwwroot
    Program.cs
    Startup.cs

Figure 16-6: The project window

Before creating a Razor page, you will create some commonly used folders within the project folder, add a database file, and configure the project to use Razor Pages.

Creating Folders

First, you will create the following folders:

1. The **Pages** folder to store the Razor Pages
2. The **App_Data** folder to store the database
3. The **Content** folder to store CSS and JavaScript code
4. The **Models** folder to store classes

Step 2: Create Pages, App_Data, Content, and Models folders within the Project folder.

Right-click the project name *Theater-RP* and select *Add > New Folder*.

Change the name of the folder to *Pages*.

Repeat the process to create *App_Data*, *Content*, and *Models* folders.

Adding a Database File

Next, you will add the database file named *Theater.mdf* to the project.

Step 3: Copy the Theater.mdf database from the Tutorial-starts\Data-files folder to the App_Data folder.

Right-click the *App_Data* folder, and select *Add > Existing Item*.

Select *Theater.mdf* from the *Tutorial-starts\Data-files* folder. Click *Add*.

Configuring the Project to Use Razor Pages

Lastly, you will configure the project to use Razor Pages, which involves setting it up to use MVC. Enabling the project to use MVC also enables Razor Pages.

Step 4: Configure the project to use Razor Pages.

Double-click the *Startup.cs* file in the Solution Explorer to open it.

You will see the startup class with two methods:

ConfigureServices(…)

Configure(…)

To work with ASP.NET Core 3.1, and to enable session state (described later), modify the ConfigureServices() and Configure() methods to include all the statements shown in Figure 16-7.

```
1  public void ConfigureServices(IServiceCollection services)
2  {
3      services.AddRazorPages(); //For v3.0
4      services.AddDistributedMemoryCache();
5      services.AddSession(); //enable Session State
6      services.AddMvc().AddRazorPagesOptions(o => //add Razor Pages conventions
7      {
8          o.Conventions.ConfigureFilter(new IgnoreAntiforgeryTokenAttribute());
9          o.Conventions.AddPageRoute("/Reservation", "");
10     });
11 }
12
13   // This method gets called by the runtime. Use this method to
         configure the HTTP request pipeline.
14 public void Configure(IApplicationBuilder app, IWebHostEnvironment env)
15 {
16     if (env.IsDevelopment())
17     {
18         app.UseDeveloperExceptionPage();
19     }
20     app.UseSession();//to enable Session State
21     app.UseRouting();
22
23     app.UseEndpoints(endpoints =>
24     {
25         endpoints.MapRazorPages();
26         //endpoints.MapGet("/", async context =>
27         //{
28         //    await context.Response.WriteAsync("Hello World!");
29         //});
30     });
31 }
```

Figure 16-7: The Startup.cs file

16.2.2 Adding a Razor Page

You will add Razor Pages to the Pages folder, which is the default location for all Razor Pages.

Step 5: Add a Razor page named Reservation.

Right-click the *Pages* folder in the Solution Explorer.

Select *Add > Razor Page*. Click *Add*

Select *Razor Page - Empty* from the *Add New Scaffold Item* window, as shown in Figure 16-8.

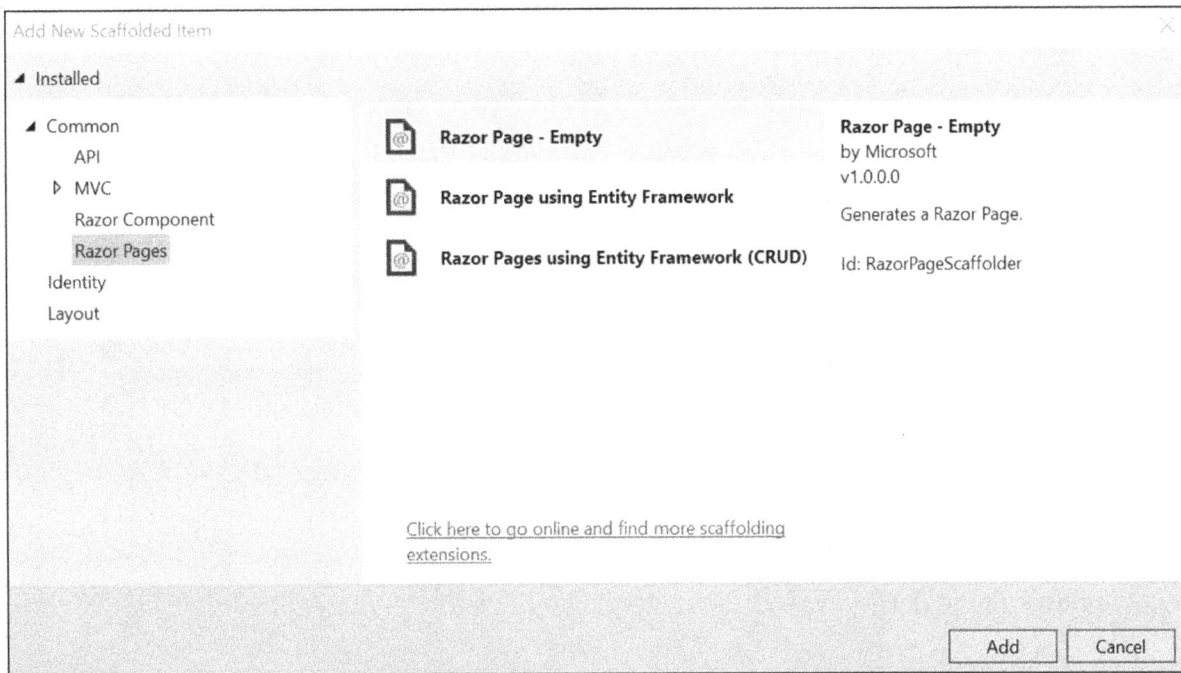

Figure 16-8: The Add New Scaffold Item window

Click **Add**. You will see the **Add New Item** window (Figure 16-9).

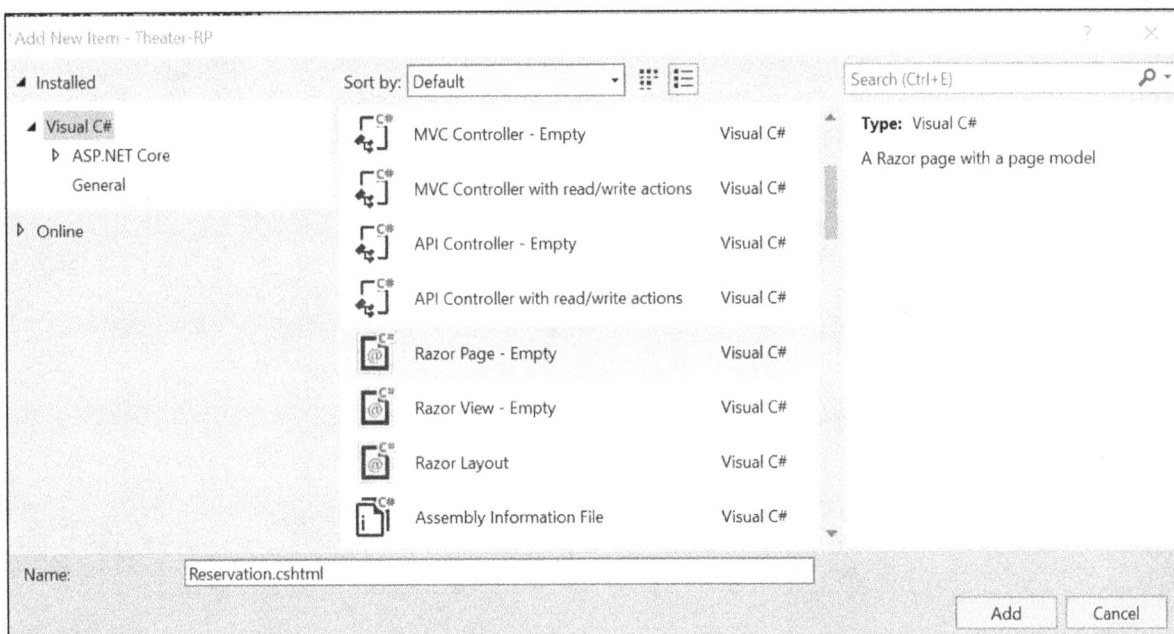

Figure 16-9: The Add Razor Page window

Select *Razor Page – Empty and enter Reservation* for the page name, as shown.

Click **Add**. The Solution Explorer will show two new files: **Reservation.cshtml** and **Reservation.cshtml.cs**.

16.2.3 Razor Page Files

A Razor page has two related files: the **Reservation.cshtml** file and the **Reservation.cshtml.cs** file, both stored in the **Pages** folder.

The **Reservation.cshtml** file, which represents the view (the user interface), stores the HTML code that structures the contents of the page and additional code like JavaScript and CSS. It is commonly referred to as the **view file** or the **content file**. This file also may contain embedded C# code.

Figure 16-10 shows the initial code in the content file. The *@page* directive at the beginning of the page, which is unique to Razor Pages, specifies that it is a Razor page.

Line 2 specifies that *ReservationModel* is the class that contains the C# code. This class is located within the Reservation.cshtml.cs file. The name of the model is formed by adding the word *Model* to the name of the page.

```
1   @page
2   @model Theater_RP.ReservationModel
3   @{
4   }
```

Figure 16-10: The .cshtml file

The **Reservation.cshtml.cs** file (code-behind file), shown in Figure 16-11, contains the ReservationModel class that is inherited from the **PageModel** class. This class contains the model (business logic and data), separating it from the view specified in the .cshtml file.

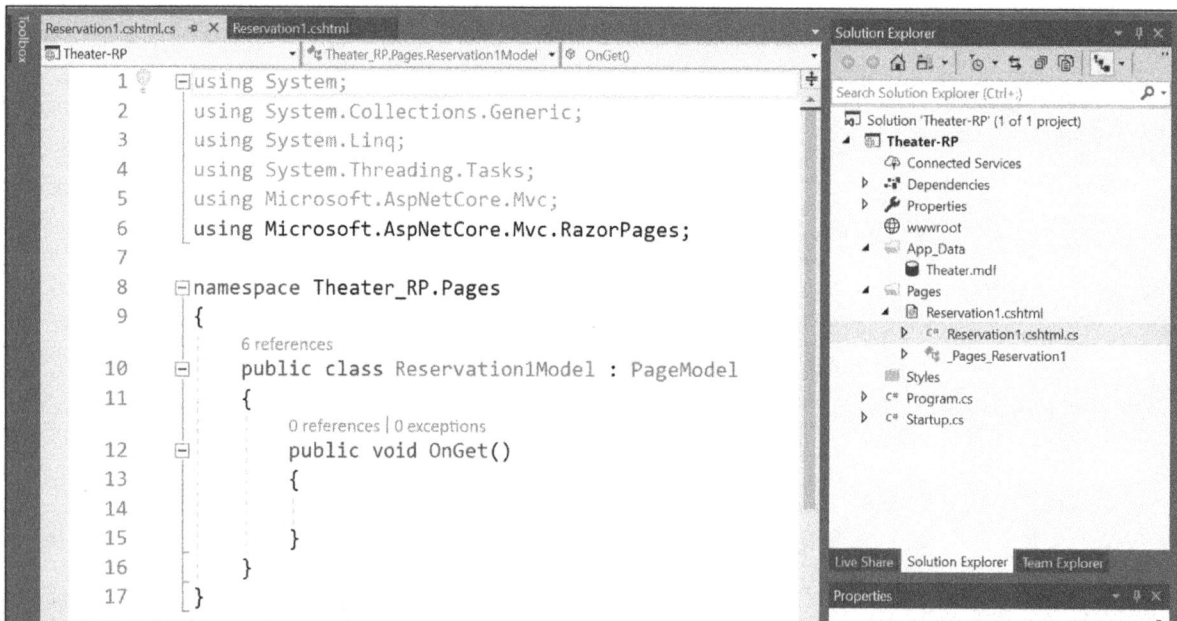

Figure 16-11: The .cshtml.cs file

Initially, the ReservationModel class has only the OnGet() event handler, as shown above. The OnGet() event handler is automatically invoked in response to a *GET* request (<form method="get">) for the page. *Another commonly used event handler is OnPost(), which is fired when the form with method="post" is submitted.* As you learned in Chapter 4, the GET request is used to fetch information from the server, and the POST request (<form method="post">) is used to send data to the server to create or update data. *You will use the OnGet() and OnPost() event handlers later in this chapter.*

First, you will develop the content (view) file to create the user interface.

HTML Code to Display the User Interface

The following HTML elements are used to develop the user interface shown earlier in Figure 16-1.

Table 16-2: HTML elements to build the user interface

Task	HTML element
Select performance	<select> -- (dropdown list)
Enter date	<input type="date">
Select zone	<input type="radio"> -- (radio buttons)
Specify member status	<input type="checkbox"> -- (checkbox)
Enter number of tickets	<input type="number">

Figure 16-12 presents the HTML code that creates the user interface. Except for the dropdown list, it is essentially the same code presented in the Reservation-1.html form in Chapter 4. The performance names in the dropdown list will be added from the database using server-side code, as described shortly.

```
1   @page
2   @model Theater_RP.Pages.ReservationModel
3   @{
4       Layout = null;
5   }
6   <!DOCTYPE html>
7   <html>
8   <head>
9     <meta name="viewport" content="width=device-width" />
10    <title>Reservation</title>
11    <!--Specify CSS for the field element -->
12      <style> fieldset {width: 26em;} </style>
13  </head >
14  <body>
15      <h1>Reservation</h1>
16      <form method="post" action="/Confirmation"> <br />
17        <fieldset>
18            <!--Dropdown list for performance will be added here-->
19
20            <!-- Input field for performance date -->
21            <label for="performDate">Date:</label>
22            <input type="date" name="performDate" id="performDate" ><br><br>
23
24            <!--Radio buttons to select a zone -->
25            <input type="radio" name="zone" id="suite" value="suite" />
26            <label for="suite">Suite</label>
27            <input type="radio" name="zone" id="premium" value="premium" />
28            <label for="premium">Premium</label>
29            <input type="radio" name="zone" id="circle" value="circle" />
30            <label for="circle">Circle</label>
```

Figure 16-12: HTML code for Reservation.cshtml (*continues*)

```
31              <input type="radio" name="zone" id="balcony" value="balcony"/>
32              <label for="balcony">Balcony</label>  <br> <br>
33
34              <!-- Checkbox to specify the member status -->
35              <input type="checkbox" name="memberStatus" id="memberStatus"
36                   value="true">
37              <label for="memberStatus">Are you a member?</label> <br> <br>
38
39              <!-- Input field to enter the number of tickets -->
40              <label for="tickets"># of tickets:</label>
41              <input type="number" name="tickets" id="tickets" min="1" max="10"
42                   step="1" value="0" /> br><br>
43
44              <!--Button to submit the form -->
45              <button type="submit" id="submitBtn" name="submitBtn"
46                   value="continue">Continue</button>
47           </fieldset>
48           </form>
49      </body>
50 </html>
```

Figure 16-12: HTML code for Reservation.cshtml (*continued*)

It should be noted that line 16 sets the method attribute of the form to "post" and the action attribute to "/Confirmation" to submit the form to the Confirmation page. The POST request invokes the OnPost event handler of the Confirmation page where the prices are computed.

Another important detail is that line 36 sets the value property of the checkbox to "true." When you submit the form, this value ("true") will be sent to the server if the checkbox is checked. If the checkbox is not checked, no value will be sent. This becomes important later when you bind the checkbox to a Boolean variable, because if you don't specify the value property, a value of "on" will be sent if the checkbox is checked.

Lines 11 and 12 specify the CSS.

Step 6: Add HTML code to the Reservation.cshtml to create the user interface.

Open Reservation.cshtml and replace the existing code with the code from Figure 16-12. You may copy the code from Tutorial-starts\Data-files\Ch16-Reservation-initial-html.txt.

Display the page in the Browser by clicking the run button:

▶ IIS Express ▾

The controls will be displayed in the browser, similar to how it is shown in Figure 16-1.

HTML Helpers and Tag Helpers

Instead of typing in the standard HTML markup for the elements in a form, you may generate them using HTML Helpers or Tag Helpers that are more efficient shortcuts.

Here is an example of an **HTML Helper:**

```
First Name: @Html.TextBox("fName")
```

The above HTML Helper would render the following HTML element:

```
First Name: <input id="fName" name="fName" type="text" value="" />
```

Tag Helpers also generate HTML elements, but they are more powerful than HTML Helpers. You can generate the entire set of elements on a form from the properties of a class. To do this, create a class that has properties corresponding to each element on the form. Assume that you have created a class named Reservation with a property of DateTime type named PerformDate. Consider the following <input> element with the *asp-for* tag helper.

```
<input asp-for="Reservation.PerformDate" />
```

The above tag helper would generate the attributes shown in the following HTML element:

```
<input type="datetime-local" data-val="true" data-val-required="The PerformDate field is required."
                    id="Reservation_PerformDate" name=" Reservation.PerformDate " value="" />
```

You will use Tag Helpers in the next chapter.

Next, we will look at the Razor syntax to add server-side code to create the dropdown list in the Razor.cshtml page.

Review Questions

1. How does ASP.NET differ from ASP.NET Core?

2. How is Razor Pages different from ASP.NET web forms?

3. What is the basic difference between the Razor Pages and MVC frameworks?

4. What is the difference between the contents of the .cshtml file and the .cshtml.cs file?

5. What does the following statement on the Reservation.cshtml file mean?

```
@model Theater_RP.ReservationModel
```

16.3 The Razor Syntax

A key aspect of Razor syntax is that you can embed C# server-side code within the HTML using the @ symbol. The server-based **Razor view engine** executes such code on the server and inserts the results (HTML markup or other content) into the HTML document before sending the page to the browser. The Razor view engine is not unique to the Razor Pages framework. As stated earlier, other frameworks like MVC also use the Razor view engine. Razor syntax was introduced in ASP.NET and adopted in ASP.NET MVC. However, the Razor Pages framework is supported only in ASP.NET Core.

As an example of Razor syntax, you can use C# code to display the current date on the Reservation page by inserting the code shown in line 3 in the body section.

```
1   <body>
2       <h3>Reservation</h3>
3       <h3>@DateTime.Now.ToLongDateString()</h3>
```

The server would execute the code and insert the date within the <h3> element before sending the page to the browser, shown as follows:

```
1   <body>
2       <h3>Reservation</h3>
3       <h3>Friday, May 1, 2020</h3>
```

The page would appear in the browser, like this:

Reservation

Friday, May 1, 2020

TRY IT

Insert the C# code within the body, and display the page. Make sure you delete the inserted code.

To insert a code block with multiple statements, you enclose the code block in curly brackets, as in Figure 16-13, which displays "Good Morning" or "Good Afternoon," based on the time of the day.

```
1   <body>
2       <h3>Reservation</h3>
3       @{
4           string message;
5           if (DateTime.Now <= DateTime.Parse("12:00 PM"))
6               {message = "Good Morning";}
7           else
8               {message = "Good Afternoon";}
9       }
10      <h3>@message, John</h3>
```

Figure 16-13: Code block

Note that, even though the *if* and *else* clauses contain only a single statement, they must each be enclosed in curly brackets. The server will execute the code from lines 4–8 and store the appropriate greeting in the variable, *message*. When the server executes the code, "@message" will be replaced by the value of the variable, *message*, changing the <h3> element to

```
<h3>Good Afternoon, John</h3>
```

This is what the page would look like in the browser:

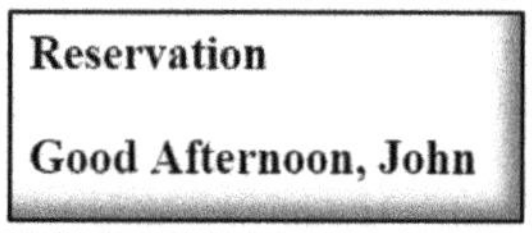

Reservation

Good Afternoon, John

TRY IT

Insert the code in Figure 16-13 into the Reservation page and display it in the browser. Don't delete the inserted code just yet.

The C# code in lines 3–9 of Figure 16-13, and C# code in general, also can be placed outside the HTML document, as shown in Figure 16-14.

```
1   @page
2   @model Theater_RP.Pages.ReservationModel
3   @{
4       Layout = null;
5   }
6   @{
7       string message;
8       if (DateTime.Now <= DateTime.Parse("12:00 PM"))
9       { message = "Good Morning"; }
10      else
11      { message = "Good Afternoon"; }
12  }
13  <!DOCTYPE html>
14  <html>
```

Figure 16-14: C# code outside the HTML document

The variable, *message,* can be accessed from within the body section of the HTML document to display its value, as shown here:

```
1    <body>
2        <h3>Reservation</h3>
3        <h3>@message, John</h3>
```

TRY IT

Move the code outside the HTML document as shown above and display the page.

You also may insert HTML code within a C# code block. For example, the HTML code to display the greeting (<h3>@message, John</h3>) may be placed within the C# code block, as shown in Figure 16-15.

```
1    <body>
2        <h3>Reservation</h3>
3        @{
4            string message;
5            if (DateTime.Now <= DateTime.Parse("12:00 PM"))
6                { message = "Good Morning"; }
7            else
8                { message = "Good Afternoon"; }
9            <h3>@message, John</h3>
10       }
```

Figure 16-15: HTML code within a code block

Because the code includes the HTML element <h3>, the entire code must be placed in the body of the document.

TRY IT

Insert the above code into the body and display the page.

Make sure you delete the inserted code that computes and displays the message.

16.3.1 Using Razor Syntax to Dynamically Create a Dropdown List

Let's use the Razor syntax to add performance names and base prices from the Performance table in the Theater.mdf database to the dropdown list on the Reservation form.

Visual Studio Community includes a database server, SQL Server Express LocalDB, that is a limited version of SQL Server. As stated earlier, if you are using Visual Studio for Mac, you may use a SQLite database file.

Figure 16-16 shows the performance table with sample records.

PerformanceId	PerformanceName	BasePrice	Description	Presenter	image
11	Young Irelanders	35	Each and every year the Irish Cultu...	ICA	*NULL*
12	Justin Hayward	30	Justin David Hayward is an English...	ACG	*NULL*
13	TedxOshkosh	59	We believe in the power of a great...	Tedx	*NULL*
14	Orchestral Presents	20	Join the Oshkosh Symphony Orch...	OSO	0xFFD8F...
15	Ladies of Laughter	30	Since 2012 the Ladies of Laughter...	*NULL*	*NULL*
16	Steve Earl & The Dukes	25	If you ever had any doubt about ...	ACG	*NULL*
17	Harmonious Wail	35	Join us for a night of Wail n' good...	Wail	*NULL*

Figure 16-16: Performance table

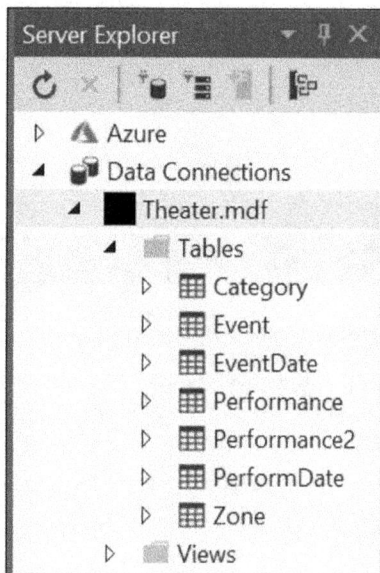

Figure 16-17: The Server Explorer window

You already added the Theater.mdf database file to the App_data folder in the project. You may view the database tables, columns, and data using *Server Explorer* or *SQL Server Object Explorer*. To view the database using Server Explorer, right-click the Theater.mdf file in the App_Data folder and select *Open*. You will see the Server Explorer window, as shown in Figure 16-17.

To view the data in the performance table, right-click *Performance* and select *Show Table Data*.

To view the database using the SQL Server Object Explorer, select *View > SQL Server Object Explorer*. You will see the SQL Server Object Explorer window shown in Figure 16-18.

Choose the database from the Databases tab, and view the tables and columns, as shown in Figure 16-19.

You may right-click the table name and select *View Designer* to view the Designer window, or select *View Data* to view the records within the table.

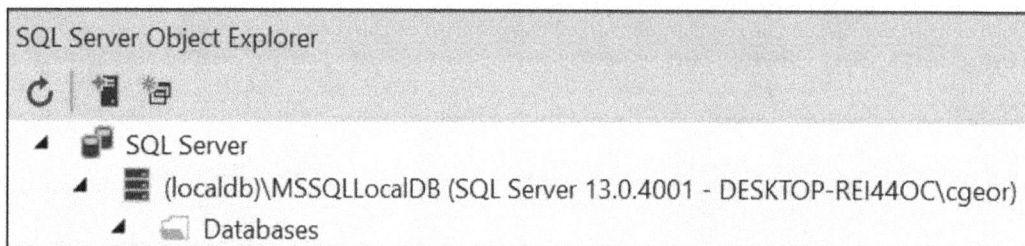

Figure 16-18: The SQL Server Object Explorer

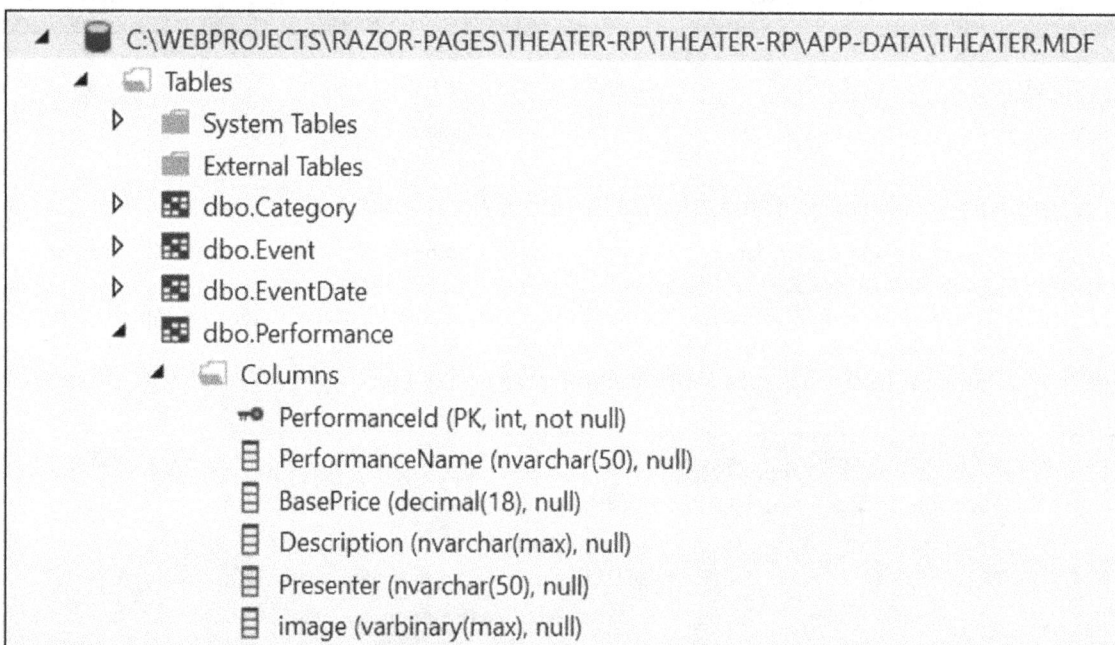

Figure 16-19: The Database tab

Accessing Database Tables from Razor Pages

There are two approaches to getting the performance names and base prices (or any other data) from the database. A traditional and general approach is to write C# code to create and use database access objects like SqlCommand and SqlDataReader to retrieve raw data.

An alternative is to use the DbContext class from the Entity Framework Core technology that lets you retrieve data from database tables as objects, with minimal code. However, it involves installing additional packages, generating classes like DbContext, and learning new concepts, which are outside the scope of this chapter. We will use the DbContext class in the next chapter on MVC. Here we will use the SqlCommand object to get the performance data.

Reading Data from a Database Using SqlCommand and SqlDataReader

The process of getting the performance names and base prices from the database table using SqlCommand and adding them to the dropdown list involves two primary steps:

1. Use the SqlCommand object to execute an SQL statement that reads performance names and base prices from the database and stores them in an SqlDataReader object.
2. Read the data from the SqlDataReader, and add performance names and base prices to the dropdown list.

Let's look at the first step. To execute an SQL statement that reads performance names and base prices, the Sql-Command object uses the SqlConnection object to make a connection to an SQL Server database. You can use the ConnectionString property of the SqlConnection object to specify the information needed to access the database, like the database name, server name, path, user name, and password. Figure 16-20 shows the method to implement this step. This method reads the performance data from the database and returns the DataReader that contains the data.

The *using System.Data.SqlClient* directive, which specifies the namespace containing the Command and DataReader objects, should be added to the page. In addition, you need to install the *NuGet* package, *System.Data. Sqlclient*, as described shortly.

```
1   public SqlDataReader PerformanceReader()
2   {
3       // Read performance data into SqlDataReader.
4       // Create an SqlCommand object
5       SqlCommand cmdPerformance = new SqlCommand();
6       // Create a Connection object and assign it to the Connection property:
7       cmdPerformance.Connection = new SqlConnection();
8       // Specify the connection string for the Connection object
9       cmdPerformance.Connection.ConnectionString =
            @"Data Source = (LocalDB)\MSSQLLocalDB;
10          AttachDbFilename =
            C:\WebProjects\Razor-pages\Theater-RP\Theater-RP\App_Data\Theater.mdf";
11      // Specify the SQL for the Command object
12      cmdPerformance.CommandText="SELECT PerformanceName,BasePrice FROM Performance";
13      cmdPerformance.Connection.Open();  // Open the connection
14      // Use the ExecuteReader method to get the data, and store it in a DataReader
15      SqlDataReader drPerformance = cmdPerformance.ExecuteReader();
16      return drPerformance;
17  }
```

Figure 16-20: Code to Read database data into the SqlDataReader

Lines 5–12 create the SqlCommand object named cmdPerformance. Please see the internal comments on individual statements.

If your database is not located in the place indicated by the string in line 10, you can right-click the database file, select Properties, and select *Database* property. The *Full Path* property will show the path/filename that can be copied into the code.

Line 13 opens the connection to the database.

Line 15 uses the ExecuteReader method of the Command object to execute the SQL and get the result data set, which is stored in the DataReader object named drPerformance.

Line 16 returns the DataReader object.
You will add the PerformanceReader() method to the PageModel class.

Invoking a PageModel Method from the Content Page

The PerformanceReader() method in Figure 16-20 gets the Performance names and base prices from the database and returns a DataReader object that contains these data items. You will invoke this method from the content page (.cshtml file) to get the names and prices and add them to the dropdown list.

To invoke a method in the PageModel class from the content page, use the following syntax:

```
Model.methodname
```

For example, you can call the PerformanceReader() method from the .cshtml file as follows:

```
Model.PerformanceReader()
```

Here, Model represents the class specified in the @model directive on the content page. The @model directive, shown here, specifies that the ReservationModel is the class:

```
@model Theater_RP.Pages.ReservationModel
```

Thus Model.PerformanceReader() means ReservationModel.PerformanceReader().
Let's install the NeGet package and create the PerformanceReader() method.

Installing the System.Data.Sqlclient NuGet Package

Step 7: Install the NuGet package, System.Data.Sqlclient.

From the Visual Studio menu, select *Tools > NuGet Package Manager > Manage NuGet Package for Solution*. The *NuGet – Solution* window opens.

Select the Browse tab and type in System.Data.Sqlclient in the search box to display available packages, as shown here.

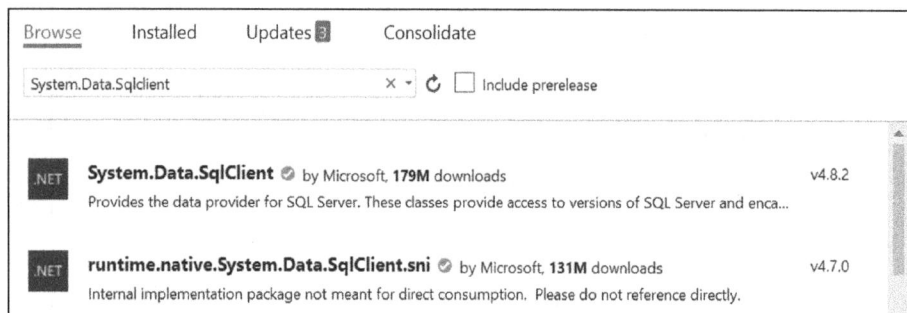

Browse	Installed	Updates 8	Consolidate	
System.Data.Sqlclient		× ▾ Ċ ☐ Include prerelease		
.NET	**System.Data.SqlClient** ✔ by Microsoft, **179M** downloads			v4.8.2
	Provides the data provider for SQL Server. These classes provide access to versions of SQL Server and enca...			
.NET	**runtime.native.System.Data.SqlClient.sni** ✔ by Microsoft, **131M** downloads			v4.7.0
	Internal implementation package not meant for direct consumption. Please do not reference directly.			

Select *System.Data.Sqlclient* and check the checkbox for *Theater-MVC*.

Click *Install*. The Package Manager installs the package.

When you are done, close the NuGet window.

Step 8: Get the Performance names and base prices from the database and store them in a DataReader.

Add the following directive to the Reservation.cshtml.cs file:

```
using System.Data.SqlClient;
```

Insert the PerformanceReader() method from Figure 16-20 into the **cshtml.cs** file immediately after the OnGet() function.

Make sure the AttachDbFilename property points to the full path of your database file.

Adding Performance Names and Base Prices from the DataReader to the Dropdown List

The second step is to create the dropdown list and read each row of the DataReader, and add the performance name and base price to the dropdown list. Data stored in a DataReader can be read sequentially using the Read() method of the DataReader. The Read() method reads the current row and advances the Reader to the next row. If there are no more rows to read, the Read() method returns the value *false*.

Because this method needs to access the dropdown list, the corresponding code is placed in the body section of the HTML document. Lines 6–19 in Figure 16-21 show the corresponding code.

```
1   <body>
2       <h3>Reservation</h3>
3       <form method="post"> action="/Confirmation"<br/>
4           <fieldset>
5               <!-- Create a dropdown list to select performance -->
6               <label for="performance">Performance:</label>
7               <select name="performance" id="performance">
8                   <option hidden value="prompt">Select a performance</option>
9                   <!-- Use the Read() method to read one row at a time-->
10                  @{
11                      SqlDataReader drPerformance = Model.PerformanceReader();
12                      while (drPerformance.Read() == true)
13                      {   // Get performance name & base price from the current row
14                          string performName =
                                drPerformance["PerformanceName"].ToString();
15                          string basePrice =
                                drPerformance["BasePrice"].ToString();
16                          // <option> element to add performance to the dropdown list
17                          <option value="@performName">@performName - $@basePrice</option>
18                      }
19                  }
20          </select> <br> <br>
```

Figure 16-21: Code to read each row from the DataReader and populate the dropdown list

Lines 6 and 7 create an empty dropdown list (with no items).

Line 8 adds the prompt "Select a performance" as the first item of the list.

Lines 10–19 presents a server-side C# code block, as indicated by the "@" symbol in line 10. This C# code is executed on the server when the page is rendered before the page is sent to the browser.

In line 11, Model.PerformanceReader() invokes the PerformanceReader() method. The SqlDataReader object returned by the method is stored in the variable drPerformance.

Line 12 uses the Read() method of the DataReader object to read the current row and advance the reader to the next row. The *while* loop processes the Read() method repeatedly until the Read() method returns the value false, indicating that there are no more rows to read.

You access a specific column of the current row by specifying the name of the column using the following pattern:

```
DataReader["columnname"]
```

Lines 14 and 15 get the PerformanceName and BasePrice fields.

Line 17 uses the <option> element to add the performance name to the dropdown list. Note that the @ symbol is needed to specify the variables performName and basePrice within the HTML element. The value attribute is set to the value of PerformName, which holds the performance name from the current row.

Step 9: Create the dropdown list and add performance names and base prices to the dropdown list.

Add the following directive to Reservation.cshtml file before the <!DOCTYPE html> element:

```
@using System.Data.SqlClient;
```

Add lines 5–20 from Figure 16-21 within the <form> element.

Display the page in the browser. The form should display a dropdown list that contains the performance names and base prices.

You can also apply the above method to dynamically generate the radio buttons for zone selection, instead of using static code to create each button, as specified in Exercise 16-1.

The current version of the Reservation form does not display the discount on the page when the user selects the member status. The discount will be computed on the server and displayed on the Confirmation page. For the benefit of the user, you could use a JavaScript function, just like you did in the HTML form in Chapters 4 and 5, to display the discount when the user selects member status. But because it is computed using server-side code, we will not repeat the JavaScript code here.

Next, you will develop the server-side code to look up the base price from the database and display it on the Confirmation page.

It's time to practice! Do Exercise 16-1 (see the end of the chapter).

Review Questions

6. Describe the Razor syntax and how it is used.

7. Is the following statement processed on the server or the client? Explain.

```
<h3>@DateTime.Now.ToLongDateString()</h3>
```

8. Consider the following statement within the HTML document:

```
@{SqlDataReader drPerformance = Model.PerformanceReader(); }
```

What does the word *Model* represent in the above statement?

16.4 Server-Side Processing of Input Data

This section describes how to access input data from the server and process it using server-side code.

16.4.1 Client-Side versus Server-Side Code

The HTML form that you developed in Chapters 4 and 5 used client-side code to compute and display the prices and discount based on the selections made by the user. These prices were displayed on the reservation page to help the user make selections. They were also used to compute the total cost to be billed to the customer.

However, compared to server-side code, client-side code is more vulnerable to tampering by users. Therefore the prices used for charging the customer should be computed using server-side code. Similarly, even though you can do data validation using the attributes of the HTML elements (e.g., `required`; `type="date"`; `max=10`) and JavaScript, it is recommended that you also do server-side validation because the clients cannot be trusted because of their vulnerability to malicious or unintentional but harmful actions by users. We will compute all prices on the server when the reservation form is submitted to the Confirmation page. You will learn server-side and client-side validation using *Data Annotation Attributes* in Chapter 17.

16.4.2 Server-Side Access of Input Data

A key aspect of processing data on the server is using server-side code to access input data. If you use ASP.NET web forms, the server-side code can directly access ASP.NET controls (server controls) on the form. However, the HTML controls on Razor Pages cannot be directly accessed from the server.

You can access the input data by retrieving the values sent by the browser to the server in a POST or GET HTTP request. For the Reservation form, which sends a POST request to the Confirmation page (`<form method="post" action="/Confirmation" >`), you can retrieve the data and process it from the OnPost event handler of the Confirmation form.

It is important to note that only values of elements that accept user input, like `<input>` and `<select>`, are sent with the HTTP request. Data displayed in `<output>` elements, if any, is not included in the request.

We will look at some of the commonly used methods to retrieve the form data.

Approaches to retrieve values of input fields from the HTTP request include

- using the `Request.Form` collection;
- model binding to parameters of the `OnPost()` or `OnGet()` method; and
- model binding to public properties on the `PageModel`.

Using the Request.Form Collection

`Request.Form` is a dictionary that contains the form data and is held in the body of the HTTP request. You access an element by specifying its **name**. For example, within the `OnPost()` method, you can get the number of tickets that the user entered into the `<input name="tickets" ...>` element as follows:

```
int numberOfTickets = int.Parse(Request.Form["tickets"]);
```

The `Request.Form` returns all values as strings. So, if the input data is not a string, the string returned by the `Request.Form` needs to be converted to its type. Another drawback is that you do not have the benefit of IntelliSense to specify element names. Further, when multiple data items need to be retrieved, the use of the Request.Form collection requires you to have a separate statement for each input value.

Just like you use the `Request.Form` collection to retrieve values of form fields, you can use the **`Request.Query`** collection to retrieve the values of query strings.

Model Binding Using Parameters

You may bind values in the HTTP request to parameters of the `OnPost()` or `OnGet` method. To do so, the parameters should have the same name as the elements whose values they bind to. Model binding is not case-sensitive. This is how you would specify the parameters to bind the values of form elements for selected performance, zone, date, and number of tickets.

```
OnPost(string performance, string zone, DateTime performDate, int tickets)
```

When the Reservation form is submitted, the value of each form element will be automatically stored in the corresponding parameter. Thus the value of the `<select name="performance">` element will be stored in the *performance* parameter. All you have to do to access the value of the selected performance within the `OnPost()` method is to use the *performance* parameter.

The value of the *zone* parameter will be the value of the selected zone (e.g., balcony) because all input elements for the radio buttons have the name, *zone*. Similarly, the parameters `performDate` and `tickets` will have the values of the input elements with matching names.

If an action method parameter name doesn't match the name of a value in the HTTP request (or query string), the parameter will be assigned the default value for its data type, like zero for *int* type. That means you won't get an error if you misspell an attribute name. However, if an input value cannot be cast to the data type of the corresponding parameter, an error message will be added to the `ModelState` property of the controller. Displaying such messages is discussed in Chapter 17, in the section on validation.

It should be noted that values of the `OnPost` and `OnGet` parameters are available only within the method when the method is executed. You may use "containers" like session state, `ViewBag`, or `ViewData` dictionary (explained later) to store values of parameters so that they can be used in content pages. For example, you may store the performance name in a `ViewBag`, as follows:

```
ViewBag.PerformanceName = performance;
```

`ViewBag` is essentially a container for data to be passed from the `PageModel` to the content page. It lasts only for the lifetime of the HTTP request until the view is rendered.

Instead of using simple variables as parameters, you may create a class (e.g., `Reservation`) that has public properties whose names match the names of the form elements. Such a class is presented later in Figure 16-22. You may specify an instance of the class as the parameter, as follows:

```
OnPost(Reservation reservation).
```

You may access an individual value, like the number of tickets, using the corresponding property of the `reservation` object using a statement like this:

```
int numberOfTickets = reservation.Tickets
```

When the `OnPost()` method is invoked by submitting the form, the `Reservation` object will be automatically created and initialized with the input values from the HTTP request.

You also may use parameters to bind to query strings. If a parameter doesn't have a matching form field, then the parameter will be bound to a matching query string, if any.

Model Binding to Properties

Model binding to public properties within the `PageModel` (.cshtml.cs file) provides another method for the server to get the values of input fields. A public property is created in the `PageModel` class using the syntax

```
Public datatype propertyname { get; set; }
```

The following statement creates a property named *Performance*:

```
public string Performance { get; set; }
```

Just like binding a parameter, you can bind a property to the value of an input field on the form, by giving it a name that matches the name of the input field. In addition, you must add the **[BindProperties]** attribute to the class.

If a property is bound, the value of the matching input field will be automatically stored in the property when the form is submitted. For example, a property named *Performance* will have the value of the input field named *performance*.

Similar to model binding using parameters, you can bind the entire set of values of input fields on a form to a single class. To do this, give each property of the class the same name as the input field whose value you want to bind it to. This is how you create a property named `ReservationData` that is of the type `Reservation`:

```
[BindProperty]
public Reservation ReservationData { get; set; }
```

You access an individual value, like the number of tickets, using the corresponding property:

```
int numberOfTickets = ReservationData.Tickets
```

Similar to model binding using parameters, you can use model binding using properties to bind a parameter to the value of a matching query string.

Server-Side Access of Non-input Data Using Hidden Fields

All three methods discussed above let the server access only input data because only values of elements that accept user input, like <input> and <select> elements, are posted to the server when a form is submitted. How can the server access non-input data? For example, suppose the reservation form displays the discount in an <output> element and the server needs to use the discount. You can store the discount in a hidden input element like <input type="hidden" id="hdnDiscount">, using the following statement:

```
document.getElementById("hdnDiscount").value = discount;
```

To get the discount from the hidden field, you could use another parameter with a name like `hdnDiscount`.

Displaying Reservation Data on the Confirmation Page

In this project, you will use model binding using a property to display the reservation data on the Confirmation page because properties also can be conveniently accessed from the content page to display their values. In Chapter 17, you will learn how to use model binding with parameters.

16.4.2.1 Creating a Class to Bind the Input Elements

The first step in model binding using properties is to create a class named *Reservation* with properties whose names match those of the input fields.

Figure 16-22 shows a class that has five properties (lines 4–8) corresponding to the five input data items (performance, performance date, zone, member status, and number of tickets). Note that the names of properties match the names of the corresponding elements in the Reservation form. Therefore, the value of each input element will be automatically stored in the matching property when the form is submitted. For example, the value of the performance element (the dropdown list) will be automatically stored in the Performance property of the Reservation.

In addition, there are four methods, BasePrice(), ZonePrice(), Discount(), and TotalCost(), as shown in lines 10–60 in Figure 16-22. You will use these methods later to compute the prices.

```
1  public class Reservation
2  {
3  // Properties to hold the input data
4  public string Performance { get; set; }
5  public DateTime PerformDate { get; set; }
6  public string Zone { get; set; }
7  public bool MemberStatus { get; set; }
8  public int Tickets { get; set; }
9  // Methods to compute prices
10 public decimal BasePrice()
11 {
12 // Create an SqlCommand object to get the base price.
13 decimal basePrice = 0;
14 SqlCommand cmdPerformance = new SqlCommand();
15 cmdPerformance.Connection = new SqlConnection();
16     cmdPerformance.Connection.ConnectionString = @"Data Source =
                (LocalDB)\MSSQLLocalDB;
17 AttachDbFilename = C:\WebProjects\Razor-pages\Theater-RP\Theater-RP\
                App_Data\Theater.mdf";
18 cmdPerformance.CommandText = "SELECT BasePrice FROM Performance WHERE
                PerformanceName=" + "'" + Performance + "'";
19 cmdPerformance.Connection.Open();
20 // Use ExecuteScalar method to get base price.
21 if (cmdPerformance.ExecuteScalar() != null)
22 basePrice = (decimal)cmdPerformance.ExecuteScalar();
23 return basePrice;
24 }
25 public decimal ZonePrice()
26 {
27 decimal zonePrice = 0;
28 switch (Zone)
29 {
30 case "suite":
31 zonePrice = 40;
32 break;
33 case "premium":
34 zonePrice = 25;
35 break;
36 case "circle":
37 zonePrice = 10;
38 break;
39 case "balcony":
40 zonePrice = 0;
```

Figure 16-22: Reservation class

```
41 break;
42 default:
43 break;
44 }
45 return zonePrice;
46 }
47 public decimal Discount()
48 {
49 decimal discount = 0;
50 if (MemberStatus)
51    discount = 10;
52 else
53    discount = 0;
54 return discount;
55 }
56 public decimal TotalCost()
57 {
58 return (BasePrice() + ZonePrice() - Discount()) * Tickets;
59 }
60 }
```

Figure 16-22: Reservation class

The `BasePrice()` method in lines 10–24 looks up the base price using the performance name stored in the `Performance` property of `Reservation`. It creates the `SqlCommand` object, similar to the one presented in Figure 16-20, except that it uses a `WHERE` clause (line 18) to find the base price, as shown here:

```
"SELECT BasePrice FROM Performance WHERE PerformanceName=" + "'" + Performance + "'"
```

In the `WHERE` clause, *BasePrice* and `PerformanceName` are the names of database fields, and `Performance` is the property in the `Reservation` class, which contains the name of the selected performance.

Line 22 uses the `ExecuteScalar()` method of the `SqlCommand` object to execute the SQL statement and get the base price.

16.4.2.2 Creating a Property in the ConfirmationModel

The second step is to create a property that is an instance of the `Reservation` class. You will create such a property, named `ReservationData`, in the `ConfirmationModel` (in the .cshtml.cs file), as shown in line 4 in Figure 16-23. Because the `ReservationData` property is an instance of the `Reservation` class that is bound to the input data, the value of an input field like *Tickets* can be retrieved from the corresponding property, `ReservationData.Tickets`.

```
1  public class ConfirmationModel : PageModel
2  {
3       [BindProperty]
4       public Reservation ReservationData { get; set; }
5
6       public void OnPost(){
7       // code to store reservation data in session state will be added here
8       }
9  }
```

Figure 16-23: ReservationData property

Step 10: Create a class named Reservation in the Models folder.

Right-click the *Models* folder and select *Add > New Item > Class*.

Change the name of the class file to *Reservation.cs*.

Add this using statement:

```
using System.Data.SqlClient;
```

Open Reservation.cs and add the properties and methods as shown in Figure 16-22.

You may copy the code from *Tutorial-starts\Data-files\Ch16-Reservation.txt*.

> **Step 11: Add a property named ReservationData to the PageMofel in Confirmation. cshtml.cs file.**

Add a Razor page named Confirmation, similar to how you created the Reservation page.

Add the code from lines 3 and 4 in Figure 16-23 to create the property in the Confirmation.cshtml.cs file.

Add a using statement for the Models:

```
using Theater_RP.Models;
```

16.4.2.3 Using the ReservationData Property on the Confirmation Page

Now you can use the `ReservationData` property within the `OnPost()` event handler and within the Confirmation view (.cshtml file). You will use the property in the `OnPost()` method to save the input data in session state for later use to restore the Reservation form. You also will use the property in the Confirmation view to display the data.

Note that when you click the submit button on the Reservation form, the reservation data will be submitted to the Confirmation page, as specified in the form element in the Reservation view:

```
<form method="post" action="Confirmation">
```

Because the form element has the method set to "post," the `OnPost()` method of the Confirmation page is invoked when the form is submitted.

Within the `OnPost()` method, you can get an individual data item like the selected zone from the `ReservationData.Zone` property. However, in the current application, you don't need to get individual data items in the `OnPost()` method. Instead, you will save the entire Reservation object in session state, to be used for restoring reservation fields after a page reload. The code to do this will be added later.

Individual properties of the `Reservation` object will be accessed from the content page to display them.

Figure 16-24 shows the steps involved in saving the reservation data and displaying it on the Confirmation page. The number of tickets entered into the `<input>` element is used as an example.

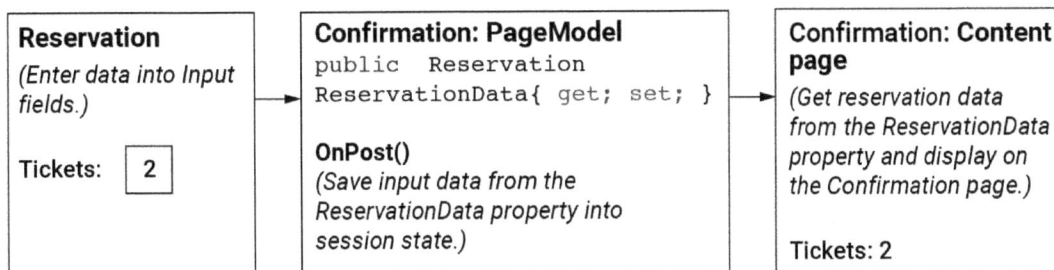

| Reservation
(Enter data into Input fields.)

Tickets: 2 | Confirmation: PageModel
`public Reservation`
`ReservationData{ get; set; }`

OnPost()
(Save input data from the ReservationData property into session state.) | Confirmation: Content page
(Get reservation data from the ReservationData property and display on the Confirmation page.)

Tickets: 2 |

Figure 16-24: Steps to save reservation data and display it on the confirmation page

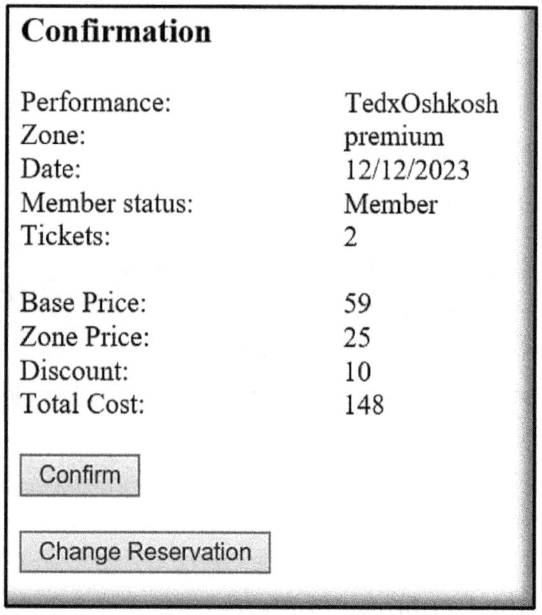

Confirmation

Performance:	TedxOshkosh
Zone:	premium
Date:	12/12/2023
Member status:	Member
Tickets:	2
Base Price:	59
Zone Price:	25
Discount:	10
Total Cost:	148

Confirm

Change Reservation

Figure 16-25: The Confirmation page

Displaying Reservation Data on the Confirmation Page

In this section, we will look at how to display the reservation data on the Confirmation page as shown in Figure 16-25.

A property of the `PageModel` class, like the `ReservationData` property, can be accessed from the view using the key word **Model**, just like you accessed `Performance-Reader()` method from the Reservation view by specifying `@Model.PerformanceReader()`. Thus, within the Confirmation view, you can access the ReservationData property by specifying `@Model.ReservationData`.

To get the value of the `Performance` property, specify `@Model.ReservationData.Performance`. Figure 16-26 shows the Confirmation view (.cshtml file). The above Razor syntax is used to get all reservation data items.

```
1   @page
2   @model Theater_RP.Pages.ConfirmationModel;
3   <!DOCTYPE> html
4   <html>
5   <head>
6       <meta name="viewport" content="width=device-width" />
7       <title>Confirmation</title>
8       <style>
9           .output {
10              position: fixed;
11              left: 12em;
12          }
13      </style>
14  </head>
15  <body>
16      <h3>Confirmation</h3>
17      <!--Get input data from ReservationData and display them in <label> elements-->
18      Performance: <label class="output"> @Model.ReservationData.Performance
                    </label> <br />
19      Zone:<label class="output">        @Model.ReservationData.Zone </label> <br />
20      Date:<label class="output">        @Model.ReservationData.PerformDate.
                            ToShortDateString()    </label> <br />
21      Member status:<label id="memberStatus" class="output"> </label> <br />
22      <script>
23          // Display member status as "Member" or "Non-member" //
24          var status = "";
25          var memberStatus;
26          memberStatus = "@Model.ReservationData.MemberStatus";
27          if (memberStatus=="True") //JS variables store bool "true" as string, "True"
28              status = "Member";
29          else
30              status = "Non-member";
31          document.getElementById("memberStatus").innerHTML = status;
32      </script>
33      Tickets:<label class="output"> @Model.ReservationData.Tickets.ToString()
                </label> <br /><br />
34
```

Figure 16-26: Confirmation view (*continues*)

```
35      <!-- Display prices in <label> elements -->
36      Base Price:<label Id="basePrice" class="output">
                    @Model.ReservationData.BasePrice().ToString("C") </label> <br />
37      Zone Price: <label Id="zonePrice" class="output">
                    @Model.ReservationData.ZonePrice().ToString("C") </label> <br />
38      Discount: <label Id="discount" class="output">
                    @Model.ReservationData.Discount().ToString("C") </label> <br />
39      Total Cost: <label Id="totalCost" class="output">
                @Model.ReservationData.TotalCost().ToString("C") </label> <br /><br />
40
41      <!-- Button to go back to reservation page -->
42      <a href="Reservation"><input type="button" value="Change Reservation"/></a><br />
43      <!-- Button to confirm reservation -->
44      <button type="button">Confirm</button>
45
46 </body>
47 </html>
```

Figure 16-26: Confirmation view (*continued*)

For example, lines 18–20 get the performance, zone, and date from the corresponding properties of the `Reser-vationData` property of the `PageModel`.

Lines 37–40 invoke the methods in the `Reservation` to compute the prices.

Line 43 adds a link to the `Reservation` page using the anchor element. Note that the anchor element's content is a button, instead of text, so the link looks like a button.

Line 45 creates a button to confirm the reservation. There is no code attached to this button to take any action.

Note that lines 8–13 specify the CSS to align the data.

Step 12: Add code into the Confirmation.cshtml file to display the reservation data.

You may copy the code from the *Ch16-Confirmation-cshtml.txt* file into the *Tutorial-starts\Data-files* folder.

To test the code, display the Reservation page, enter the data, and click the *Continue* button to display the Confirmation page. Make sure that the data and prices are displayed as shown in Figure 16-25.

It's time to practice! Do Exercise 16-2 (see the end of the chapter).

Review Questions

9. Why is it more desirable to use server-side code than client-side code when computing the cost of an order?

10. Describe the method of model binding using parameters.

11. Describe model binding to properties. What is meant by the term *bound property*?

12. Describe the Request.Form collection and how it is used.

13. Under what conditions is the OnGet() event handler of a page invoked?

16.5 Restoring Field Values after a Page Reload

Unlike in a web form that uses server controls, the HTML controls on the Razor page don't retain their values after a page reload. Let's test this.

Step 13: Test whether the Reservation page retains the input values when it is reloaded.

To test the code, display the reservation form, enter the reservation data, and press the *Continue* button to display the Confirmation page.

Click the *Change Reservation* button to go back to the reservation page.

The reservation page should be displayed in its initial state with no data displayed on it.

To restore the values of reservation fields, you can store the reservation data in session state within the `OnPost()` method on the Confirmation page, as shown in Figure 16-24. Then, you can retrieve it when the Reservation page is reloaded. If you reviewed Chapter 9, you should be familiar with session state. If not, it is described in the next subsection.

16.5.1 Using Session State to Share Data between Pages

Session state (`HttpSessionState`) lets you store data on the server, making data more secure than storing it on the client using methods like `sessionStorage` and `localStorage`, which were discussed in Chapter 5. Like `sessionStorage`, session state stores data as key/value pairs for an entire session. Unlike `sessionStorage`, data stored in session state can be accessed from any page. Use the **HttpContext.Session** property to store data in session state.

This is how you store the value of the variable, *performance*, in a session variable (key) named *Performance*:

```
HttpContext.Session.SetString("Performance", performance);
```

The methods `SetString()` and `SetInt32()` store `string` and `int` type data. The `Session` object has a third method, `Set()`, that stores byte arrays.

Corresponding to the three methods to set values, there are three methods to get values from session state: `GetString()`, `GetInt32()`, and `Get()`.

For example, this is how you retrieve the values stored in the session variables *Performance* and *Tickets*:

```
HttpContext.Session.GetString("Performance")
```

To use session state, make sure the file has the following use directive:

```
using Microsoft.AspNetCore.Http;      // to use session state
```

Storing Objects in Session State

Though you can store each data item in a separate session variable and share them individually, it is more convenient to store an object that contains all data items.

Because session state doesn't let you directly store an object in it, you need to convert objects to a string by a process called **serializing**. Serializing an object, in general, means converting an object to a stream of bytes so that it can be stored or transmitted and later reconstructed. Here's how you serialize the `Reservation` object:

```
string strReservation = JsonConvert.SerializeObject(reservation);
```

The `SerializeObject()` method of the `JsonConvert` class converts the reservation object to a string value in JSON format text. JSON (JavaScript Object Notation) is a human-readable data-interchange text format that is commonly used in web applications.

To use the `JsonConvert` class, you need to install the NuGet package *Newtonsoft.Json*, as specified in the next step, and add the following directive to use JSON:

```
using Newtonsoft.Json;
```

You can store the serialized object (`strReservation`) in a session variable named `StrReservation`, as follows:

```
HttpContext.Session.SetString("StrReservation", strReservation);
```

Here, `StrReservation` is the key and `strReservation` is the value.

The serialized object stored in a session variable can be retrieved as a string (JSON text), as shown in line 1 of the following code. Line 2 de-serializes the string using the `DeserializeObject()` method and stores it as an object in the reference variable, reservation. Unlike serialization, for deserialization, you need to specify the type you want to deserialize the string data to, which is specified by the class name inside angle brackets:

```
1   string strReservation = HttpContext.Session.GetString("StrReservation");
2   Reservation reservation = JsonConvert.DeserializeObject<Reservation>(strReservation);
```

Storing a Reservation Object in Session State and Retrieving It

The code to serialize the `ReservationData` property and store it in session state is shown in lines 3 and 4 in Figure 16-27.

```
1   public void OnPost(){
2       // Serialize and store reservation object in session state
3       string strReservation = JsonConvert.SerializeObject(ReservationData);
4       HttpContext.Session.SetString("StrReservation", strReservation);
5   }
```

Figure 16-27: OnPost() method of the Confirmation page

Line 3 serializes the object and stores the resulting string in the variable, `strReservation`.

Line 4 stores the string in the Session variable, `StrReservation`.

Data stored in session state can be accessed from the `PageModel` (Reservation.cshrml.cs) or from the content page (Reservation.cshtml). However, it is easier to do the retrieving and de-serialization in the `PageModel`. The retrieved object containing the reservation data can be stored in a property of type `Reservation` within the `Reservation-Model` (Reservation.cshtml.cs). So, you will create a `ReservationData` property in the `ReservationModel` and access it from the `Reservation` view, just like the `ReservationData` property in the `ConfirmationModel` is accessed from the Confirmation view.

Alternatively, you can store the object in a `ViewBag` and access it from the view. However, it is more convenient to use the `ReservationData` property with its IntelliSense feature.

Where do you place the code to de-serialize the object and store it in the `ReservationData` property? Because GET is the default method for page requests, the `OnGet()` event handler is invoked when you click the *Change Reservation* link on the Confirmation page. Therefore, you will place the code in the `OnGet()` method of the Reservation page, as shown in Figure 16-28.

```
1   public class ReservationModel : PageModel
2   {
3       [BindProperty(SupportsGet = true)]
4       public Models.Reservation ReservationData { get; set; }
5
6       public void OnGet() {
7           // Get reservation data to restore input fields.
8           string strReservation = HttpContext.Session.GetString("StrReservation"); // Get
                                    serialized object
9           if (strReservation != null)
10          { // Deserialize the object and store in ReservationData property
11           ReservationData = JsonConvert.DeserializeObject<Models.Reservation>(strReservation);
12          }
13          return;
14      }
```

Figure 16-28: OnGet() method of Reservation page.

Line 8 gets the serialized object. Line 11 deserializes the string and stores the object in the ReservationData property specified in line 4.

Step 14: Install the NuGet package, Newtonsoft.Json.

Select *Tools > NuGet Package Manager > Manage NuGet Package for Solution*. The *NuGet – Solution* window will open.

Select the *Browse* tab and type in *Newtonsoft.Json*.

Select *Newtonsoft.Json* and check the checkbox for *Theater-RP*.

Click *Install*. The *Package Manager* installs the package.

When you are done, close the *NuGet* window.

Step 15: Serialize and store the Reservation object in session state.

Add the following directives to the Confirmation.cshtml.cs page:

```
using Microsoft.AspNetCore.Http;
using Newtonsoft.Json;
```

Add the `OnPost()` method from Figure 16-27 to Confirmation.cshtml.cs.

Step 16: Add the ReservationData property to the Reservation PageModel

Add the code from lines 3 and 4 in Figure 16-28 to Reservation.cshtml.cs.

Step 17: De-serialize the Reservation object and save it in the ReservationData property.

Add the following directive to Reservation.cshtml.cs:

```
using Microsoft.AspNetCore.Http;
using Newtonsoft.Json;
```

Add the code from lines 7–12 in Figure 16-28 to the `OnGet()` method in Reservation.cshtml.

16.5.2 Restoring Reservation Data

Now you can get the data from the `ReservationData` property and display it in the elements in the content page. In a simple element like `<input type="number">` (for number of tickets), you can set the value attribute to `Model.ReservationData.Tickets`, as follows:

```
<input type="number" name="tickets" id="tickets" min="1" …
value="@Model.ReservationData.Tickets" />
```

However, for a group of radio buttons, for example, you need to use code to restore the selected button. You also need code to restore the checked state of a checkbox. So, rather than mixing such code with the HTML, you will keep the code to set the values of all elements in a separate `<script>` element, at the end of the Reservation.cshtml file, as shown in Figure 16-29. The code, with the comments, should be easy to follow.

```
1        </form>
2        <script>
3                    // Restore values of input fields
4                    document.getElementById("performance").value = "@Model.ReservationData.Performance";
5                    document.getElementById("performDate").value =
                        "@Model.ReservationData.PerformDate.ToString("yyyy-MM-dd"))"; //default date format
6                    // Restore zone selection
7                    zone = "@Model.ReservationData.Zone";
8                    switch (zone) {  // mark the selected zone
9                        case "suite":
10                           document.getElementById("suite").checked = true;
11                           break;
12                       case "premium":
13                           document.getElementById("premium").checked = true;
14                           break;
15                       case "circle":
16                           document.getElementById("circle").checked = true;
17                           break;
18                       case "balcony":
19                           document.getElementById("balcony").checked = true;
20                           break;
21                   }
22                   var memberStatus = "@Model.ReservationData.MemberStatus"
23                   if (memberStatus = = "True") {   // if the checkbox is checked
24                       document.getElementById("memberStatus").checked = true;
25                   }
26                   document.getElementById("tickets").value = "@Model.ReservationData.Tickets";
27       </script>
28   </body>
```

Figure 16-29: JavaScript to restore Reservation fields

An alternative to keeping the code in the `<script>` element is to keep the code in a JavaScript function named `RestoreReservation` within the Content/Scripts.js file, and call the function from the `<script>` element. In this case, you need to specify the Scripts.js file, within the `<head>` element as follows: `<script src="~/Content/Scripts.js"></script>`.

In Figure 16-29, line 4 sets the value attribute of the `<select>` element to the value of the `ReservationData.Performance` property that holds the performance selected by the user. This will cause the dropdown list to display the selected performance. The word "`Model`" in line 4 represents the *Theater_RP.ReservationModel* class specified on the content page:

```
@model Theater_RP.ReservationModel
```

Line 7 gets the name of the selected `zone` from the `Zone` property of Reservation. Lines 8–21 use a switch statement to check the radio button whose name matches the value of the `Zone` property.

Line 22 gets the value of the checkbox from the `MemberStatus` property. If the checkbox is checked, its value property will have the default value specified in the element, which is *true*. If the checkbox is not checked, the value property will be null.

Step 18: Restore the Reservation fields using values from the ReservationData property.

Add the script element and its contents as shown in Figure 16-29 to Reservation.cshtml. You may copy the code from Tutorial-starts\Date-files\Ch16-RestoreReservation.txt.

Step 19: Test the code.

To test the code, display the reservation page.

The dropdown list and the date field will look different from how they looked before adding the code to restore values. The dropdown list won't show the prompt (the first item of the element), and the date field will show the default date, "01/01/0001," as shown here:

Performance:	▼
Date:	01/01/0001

This is because even when the page is loaded for the first time, the following code that restores the field values is executed:

```
1    document.getElementById("performance").value="@Model.ReservationData.Performance";
2    document.getElementById("performDate").value =
          "@Model.ReservationData.PerformDate.ToString("yyyy-MM-dd")";
```

Because the default value of `Reservation.Performance` is null, the statement in line 1 results in showing no item in the dropdown list. Similarly, the date field shows the default value of 01/01/0001, which makes it take longer to select a date from the calendar.

You can address the above problem by providing appropriate default values for the `Performance` and `PerformDate` properties of the `Reservation` class, as follows:

```
1    public string Performance { get; set; } = "prompt";
2    public DateTime PerformDate { get; set; } = DateTime.Today;
```

In line 1, `"prompt"` is the value of the first `<option>` element in the dropdown list. Setting the `Performance` property to `"prompt"` causes the dropdown list to show the first item, which is a prompt.

An alternative approach is to not execute the code that restores the fields, the first time the page is loaded. This can be done by setting a flag to *true* in the Confirmation page before returning to the Reservation page and making sure the fields are restored only when the flag has the value true. We will use the first approach, which sets default values for `Performance` and `PerformDate`.

Step 20: Set default values for the Performance and PerformDate properties of the Reservation class.

Set the default values for the `Performance` and `PerformDate` properties of the `Reservation` class (in the Reservation.cs file), as shown in the code presented above.

Step 21: Test the code.

Display the reservation page. The dropdown list should show the prompt, and the date field should show today's date.

Enter the reservation data, click the **Continue** button, and click the **Change Reservation** link. The Reservation page should show the data selected/entered by the user.

In the above approach, you used session state to share data between pages. You may wonder why we cannot store the reservation data in **sessionStorage** before moving to the Confirmation page and use them when the user returns to the Reservation page. That would work. However, recall from Chapter 5 that `sessionStorage` is available only on the client. So, it is not as secure as session state, which is stored on the server.

It's time to practice! Do Exercise 16-3 (see the end of the chapter).

Review Questions

14. Why is it necessary to restore field values after a Razor page is refreshed?

15. Describe two methods to restore the field values after a page is reloaded.

The Complete Code

Reservation Page

The final HTML code in the Reservation.cshtml file, which combines the code shown in Figure 16-12 and Figure 16-21, can be found in Tutorial-starts\Data-files\Ch16-Reservation-cshtml.txt.

The final code in the Reservation.cshtml.cs file can be found in Ch16-Reservation-cshtml-cs.txt.

Confirmation Page

Figure 16-26 shows the final HTML code in the Confirmation.cshtml file.

The C# methods in the Confirmation.cshtml.cs file is provided in Tutorial-starts\Data-files\Ch16-Confirmation-C#-methods.txt.

Exercises

Exercise 16-1: Generate Radio Buttons

This exercise modifies the Reservation page by dynamically generating the radio buttons using zone names from the zone table.

Comment out the code in lines 25–32 in Figure 16-12, which creates each radio button using separate code. Insert code that dynamically generates the radio buttons using zone names from the zone table. You may follow the approach that you used to add <option> elements to create the dropdown list. You can use a loop to get each zone record from a DataReader, and use the zone name to create a radio button.

Exercise 16-2: Checkbox for Snacks

Add a second checkbox to the Reservation form that the user can check to choose snacks during the event. Update the Confirmation page so that it displays the cost for snacks and the total cost includes the snack price. To do this, add a method to Reservation to compute the snack price and update the TotalCost() method to include the snack price.

Exercise 16-3: Restore Snack Selection

Update the script element at the end of the Reservation.cshtml file so that the value of the checkbox for snacks is restored when the user returns to the Reservation form from the Confirmation form.

Assignments

E-Commerce Assignment: Razor Pages

This assignment develops a Razor Pages version of the web forms assignment *E-Commerce: Using Databases*, from Chapter 8.

P&I Laptops is a small business that assembles and sells a limited set of laptop models online. In this assignment, you will develop two web pages that are part of an e-commerce system.

Information on the different models sold by P&I and the software bundled with the laptops are maintained in the database **PI-Laptops**, which can be found in the folder Tutorial-starts\DataFiles. The database contains the following two tables:

Model

ModelId	ModelName	Description	CPU	OS	Touch	StartingPrice
101	S330	15 inch lap top with essential perf...	A6-9225	*NULL*	*NULL*	295.00
102	S380	Durable, ready for school, 14 inch ...	i3-8130U	Windows...	*NULL*	450.00
103	S580	S580 is powered by Intel Core™ i...	i5-8250U	Windows...	True	595.00
104	G780	14 inch touch-screen gaming lapt...	i7-8750H	Windows...	True	950.00
105	B780	A light-weight 2-in-one 15-inch to...	C i7-875...	Windows...	True	995.00

Software

SoftwareId	SoftwareName	Price
11	Office Home	119.95
12	Office Professional	149.95
13	Office 365 Home	69.95
14	Office 365 Personal	69.95

The StartingPrice field in the Model table represents the price for the default configuration.

Create a web project named PI-eComm-Razor within the WebProjects folder and develop a web page named **laptop-selection**. The web page allows the user to specify the model, software, whether a warranty is needed, and the quantity, as shown below. Choose an appropriate layout. What is shown here is just an example.

The dropdown list shows the model name and the price from the ModelName and StartingPrice fields in the Models table. Similarly, the radio buttons show the software and its price from the SoftwareName and Price fields in the Software table.

Use HTML elements to create a form and the input fields. The elements that accept user input should be bound to the properties of a class named ModelsClass, so that the input data can be displayed on a second page, named Cart. You need to create a class named ModelsClass with property names that match the names of elements to be bound to.

The **Select button** should display the Cart page.

The **Cart page** should display the following input data with proper labels: model name, software, whether a warranty is needed, and quantity. In addition, the Cart should compute the total cost using a method(s) in the ModelsClass, and display the total cost:

(Starting price + Cost of software + Cost of warranty) × Quantity.

The cost of the warranty is $99.

Provide a link on the Cart page to let the user go back to the Model-Selection page to make changes, if any. When the user goes back to the Model-Selection page, it should show all input data/selections made by the user. To do this, you may use session state or another method to pass data from the Cart page to the Model-Selection page.

Auto Rental Assignment: Razor Pages

This assignment develops a Razor Pages version of the Web Forms assignment *Auto Rental: Using Databases*, from Chapter 8.

You will develop a web project for "Ace Auto Rentals," which is a small business that rents automobiles. Ace would like you to develop two web pages: The first web page lets the user select a vehicle and specify related information. The second page computes the cost, and displays the input data and the cost.

Ace has a database named AutoRental.mdf that contains two tables, AutoTypes and Discount. The database can be found in the folder Tutorial-starts\Data-files. A subset of the records in the AutoTypes table is shown here.

AutoTypes

AutoTypeId	AutoType	Capacity	DailyRate	WeeklyRate
11	Standard SUV	5	45	250
12	Full-size SUV	7	65	400
13	Ludury SUV	5	55	350
14	Standard Minivan	7	60	400
15	12 Passenger Van	12	75	500

AutoType represents the type of vehicle, such as a standard SUV, an economy car, or a small pickup. AutoTypeId is a unique number that identifies each record in the RentalRate table. Capacity represents the passenger (seating) capacity.

Here is the Discount table:

Discount

Id	Category	DiscountPercent
11	BestRate	5
12	Government	15
13	Business	10
14	Senior	20

Use the Razor Pages framework to develop a web page named **Auto-selection**, within a web project named **Auto-Rental-Razor.** You may create the project within the WebProjects folder. The web page should let the user specify an auto type, the discount category, whether or not insurance is needed, and the number of days, shown as follows. Choose an appropriate layout. What is shown here is just an example.

Note that the dropdown list displays the auto type and daily rate from the AutoType and DailyRate fields of the AutoTypes tables. Similarly, the radio buttons show combined data from the DiscountCategory and DiscountPercent fields of the Discount table.

Use HTML elements to create a form and the input fields. The elements that accept user input should be bound to the properties of a class named RentalsClass, so that the input data can be displayed on a second page, named Confirmation. You need to create a class named RentalClass with property names that match the names of elements to be bound to.

The **Select** button should display the Confirmation page.

The **Confirmation page** should display the following input data with proper labels: auto type, start date, discount category, whether insurance is needed, and number of days. In addition, the Confirmation page should compute the total cost using a method(s) in the RentalsClass and display the total cost:

(Daily rate – Discount amount + Cost of insurance, if any) × Number of days.

The cost of insurance is $10/day. You may use the daily rate, irrespective of the number of days.

Provide a link on the Confirmation page to let the user go back to the Auto-Selection page to make changes, if any. When the user goes back to the Auto-Selection page, it should show all input data/selections made by the user. To do this, you may use session state or another method to pass data from the Confirmation page to the Auto-Selection page.

Chapter 17

ASP.NET Core MVC: An Introduction

Chapters 16 presented the Razor Pages framework for developing web applications, with a focus on server-side processing. Razor Pages is relatively simple because each Razor page is a self-contained unit that includes a model and a view for a single page. The MVC (Model View Controller) framework, which is becoming increasingly popular, especially for larger applications, includes a third component called a *controller*. The controller handles requests for web pages and controls the model and the view, providing several benefits, including ease of testing and maintenance. This chapter presents key concepts of MVC and provides you with the skills to use the *model*, *view*, and *controller* to build simple applications. You will learn how to develop a Core MVC form and bind input fields to a model using tag helpers to help pass input data to the server for processing.

Learning Objectives

After studying this chapter, you should be able to

- Create an ASP.NET Core web application using the MVC framework.
- Develop the model, view, and controller.
- Use Entity Framework to access databases.
- Use Ajax to lookup data and update a form.
- Use tag helpers to bind form elements to a model.

17.1 Introduction to ASP.NET Core MVC

MVC (Model View Controller) is a popular framework for developing web applications and web APIs. This development pattern is integrated with ASP.NET Core, which is an open source cross-platform framework for developing web applications.

The MVC model divides an application into three components:

- Model: Represents the data and contains the business logic.
- View: The presentation of data through the user interface, which typically uses HTML, CSS, tag helpers, JavaScript, and Razor code.
- Controller: Controls the model and the view. The controller receives requests for web pages and displays the views. The controller uses the model to pass data to the appropriate view.

Figure 17-1 shows a highly simplified representation of the interactions/relationships between the components. You will learn about these components and use them in the rest of this chapter and in Chapter 18. Some key aspects of how the components interact with each other are presented here.

When the user interacts with a web page (view), the view sends the corresponding information to the controller to take appropriate actions. For example, if the user clicks a link on a view, the view sends the information about the URL to the controller. The controller, in turn, renders the corresponding view using the Razor view engine (Razor engine). It should be noted that both MVC and Razor Pages frameworks use the same Razor engine to render web pages.

If rendering the view requires data from the model, the controller gets the data from the model and passes it to the view. The controller also uses the model to do computations and updates of data.

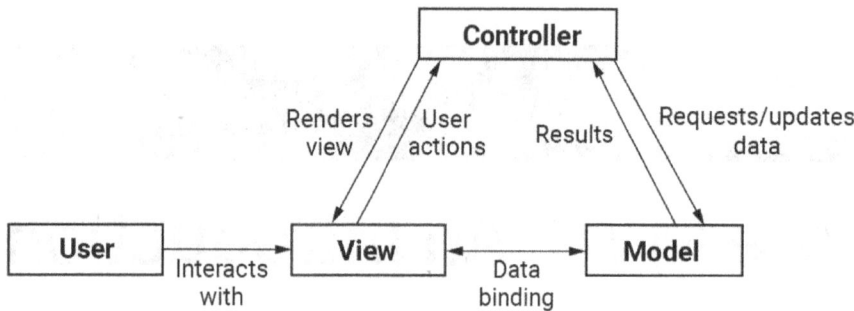

Figure 17-1: Components of MVC

When a view includes a form, the fields on the form can be bound to the fields of a model. Data binding makes it easier to generate input fields, validate them, and pass data to other views.

The process of separating the three components is called the *separation of concerns* of an application. A key advantage of this approach is that it makes the maintenance of applications easier, particularly for larger applications that involve significant processing. It also makes it easier for different members of a team to work on different components of an application. An additional benefit is that it enhances unit testing and the reuse of components. However, there is a higher learning curve, and it requires more effort to set up.

17.1.1 MVC vs. Razor Pages and ASP.NET Web Forms

In Razor Pages, the content page represents the view, and the page model combines certain functions of MVC's model and controller. Another key difference between Razor Pages and MVC is that each Razor page is a self-contained unit (page) that includes the model and the view, only for that page. With MVC, the grouping of the three components is project-wide. For example, the controller in the MVC framework may deal with multiple pages, which generally makes it more complex. Similarly, the views for multiple pages are grouped together, as are the models for the pages.

In an ASP.NET Web Forms application, the .aspx page contains the View, and the code-behind page with its event handling methods makes up the controller. The model in that system would be the C# classes that represent the problem domain, business logic, and data access. Thus, in both Razor Pages and .Net Web Forms applications, the three MVC components are not explicitly separated, making it easier to start developing an application; however, this makes it more difficult to cleanly separate the concerns of the application.

You will find that certain topics like model binding and data sharing are common to Razor Pages and MVC. To provide flexibility in choosing the sequence of chapters, some topics from Chapter 16 are briefly reviewed in this chapter.

17.1.2 ASP.NET vs. ASP.NET Core

As discussed in Chapter 16, ASP.NET and ASP.NET Core are platforms for developing web applications. ASP.NET Core, which is relatively new, is an open source cross-platform framework, whereas ASP.NET is Windows-based. ASP.NET Core supports multiple operating systems, including Windows, macOS, and Linux.

Both Razor Pages and ASP.NET Core MVC, which is the newer version of MVC, are integrated with the ASP. NET Core.

The older ASP.NET does not support Razor Pages or ASP.NET core MVC. But, it supports Windows Forms and an earlier implementation of MVC, called ASP.NET MVC, which is significantly different from the newer ASP. NET **Core** MVC. This earlier version of MVC is not supported in ASP.NET Core.

You may use Visual Studio Community (for Windows) or Visual Studio for Mac (for macOS), which provides an easy-to-use environment for developing ASP.NET Core MVC applications. The tutorials in this chapter are developed using ASP.NET Core 3.1 and Visual Studio 2019.

17.2 Project Structure and Components

This section uses a tutorial to help you understand the structure and components of an ASP.NET Core MVC project.

Tutorial: Reservation Page

This tutorial creates an MVC version of the reservation form that you developed in Chapters 4 and 5.

Creating an ASP.NET Core Web Application

First, you will create an ASP.NET Core project. An introduction to the Visual Studio environment is provided in Chapter 6.

Step 1: Create an ASP.NET Core web application.

Create a new folder named *MVC-1* within the folder *WebProjects,* which is the main folder that you created in Chapter 1 to store all projects.

Create a new ASP.NET Core web application within the folder *MVC-1*, as follows:

Open Visual Studio. You will see the start window with different options to get started, including *Open a project or solution* and *Create a new project*.

Select: *Create a new project.* You will see a *Create a new project* window similar to the one shown in Figure 17-2, except that the recent project templates may be different.

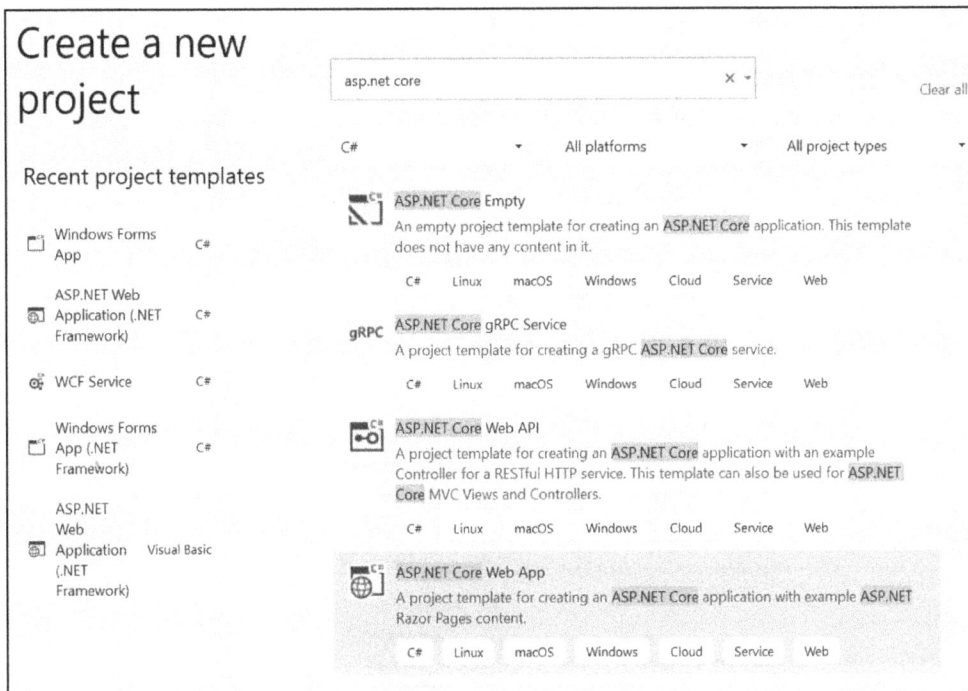

Figure 17-2: The *Create a new project* window

To make it easy to find the *ASP.NET Core Web App* framework for *C#*, enter *asp.net core* into the search box and select *C#* for the language.

Select the *ASP.NET **Core** Web App* (**not** *ASP.NET Web Application*) that includes the key word "C#" in the bottom line.

Click Next. A *Configure your new project* window, as shown in Figure 17-3, will be displayed.

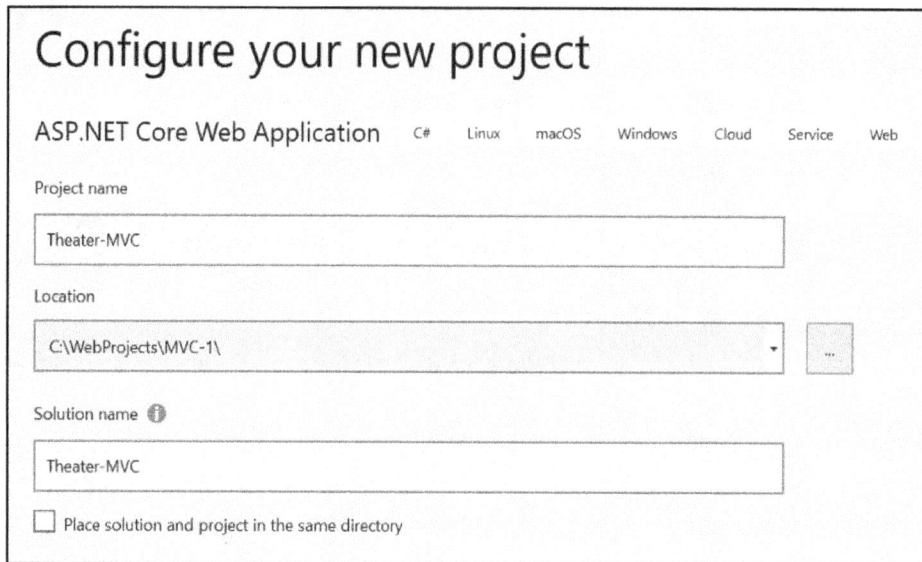

Figure 17-3: The *Configure your new project* window

Select *WebProjects\MVC-1* for the location.

Enter *Theater-MVC* for the project name.

Click *Create*. You will see a *Create a new ASP.NET Core Web Application* window, as shown in Figure 17-4, which displays the different project templates you can use.

Select *Web Application (Model-View-Controller)*, which sets up the app with certain initial folders and files.

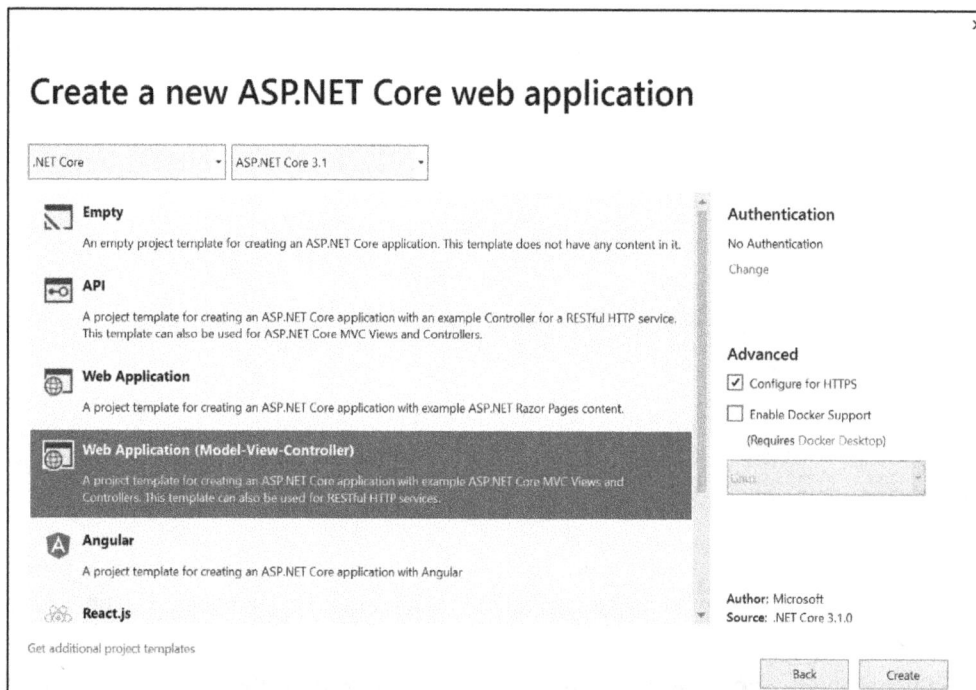

Figure 17-4: The *Create a new ASP.NET Core web application* window

Click *Create*.

You will see the project window, similar to the one shown in Figure 17-5.

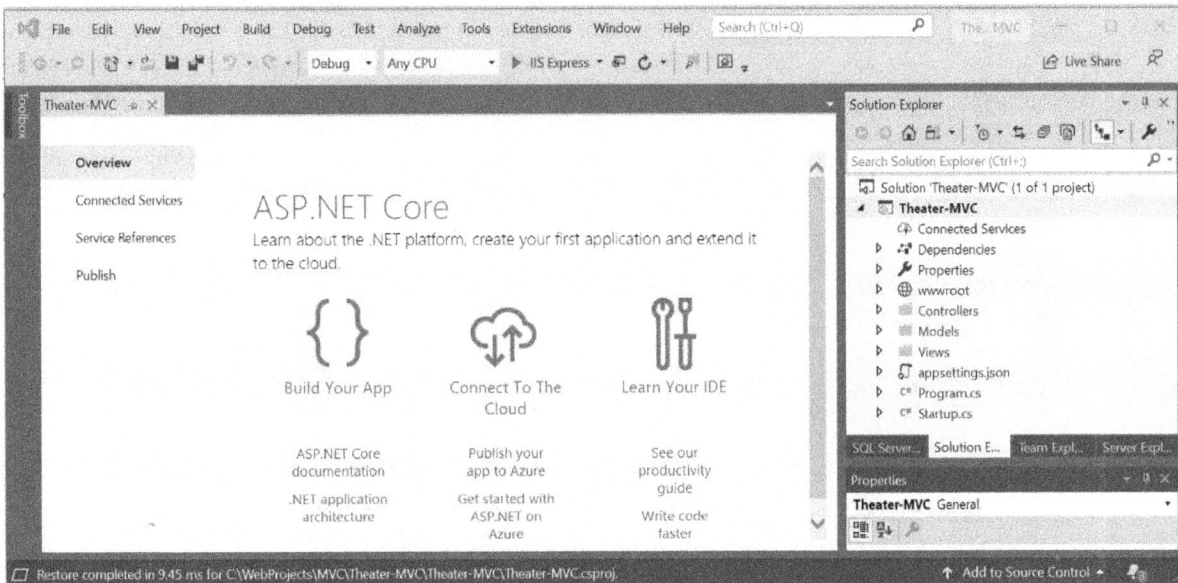

Figure 17-5: The project window

You may run the project to display the default Welcome page.

Step 2: Run the project.

Select *Debug > Start Without Debugging*.

If you get a warning about using SSL, you can click *Yes* to trust the self-signed IIS Express SSL Certificate.

You will see the *Welcome* page, as shown in Figure 17-6.

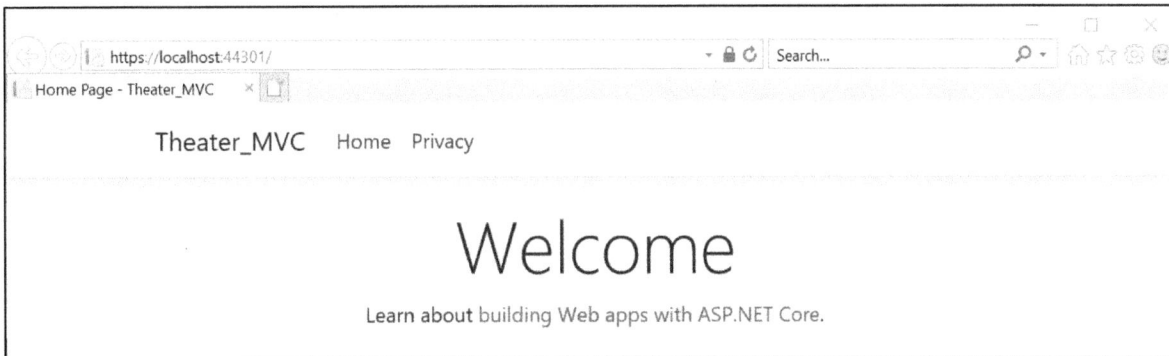

Figure 17-6: The Welcome page

Click the *Privacy* tab to display the privacy page. Click the *Home* tab to display the welcome page. Click the *Theater_MVC* tab, which also displays the Welcome page (called the Home page).

Close the browser.

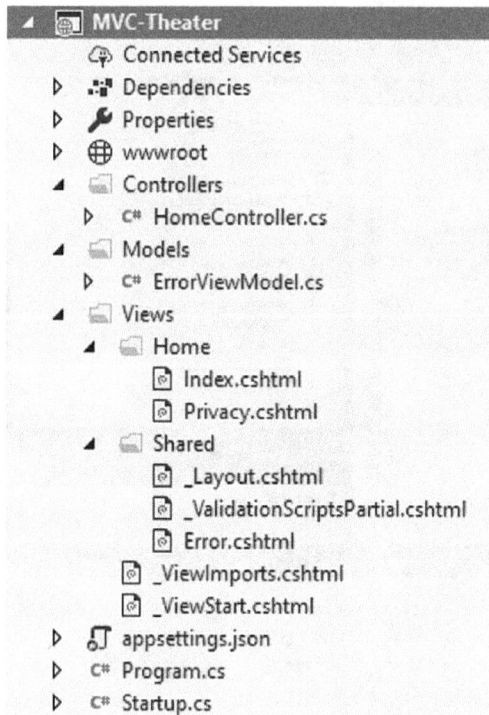

Figure 17-7: The Solution Explorer

If you are wondering where this page came from, this is one of two pages that are included in the project you created. Before creating your-own pages, let's look at the models, views, and controllers, that work behind the scenes to display the welcome page. These components are organized into corresponding folders: Models, Views, and Controllers.

17.2.1 The Views Folder and Views

A view consists of the user interface created using HTML, and it may contain Razor code, CSS, tag helpers, and JavaScript. A view is represented by a view file (.cshtml) that contains the HTML code. By convention, all view files should be stored in the Views folder.

As shown in Figure 17-7, the Views folder initially contains the *Home* and *Shared* folders.

The Home folder contains the Index.cshtml and Privacy.cshtml view files that display the Home page and the Privacy page, respectively. Note that the page displayed by the Index view is called the Home page.

In an MVC project, a controller controls one or more views. As you can see from the Solution Explorer, there is a controller named HomeController that controls the views within the Home folder. This brings up an important convention in MVC:

The name of the controller consists of the name of the folder that contains the views it controls, followed by the word *Controller*. Thus the HomeController controls the views in the Home folder (Index and Privacy).

Step 3: Open the files Index.cshtml and Privacy.cshtml, which display the Home and Privacy pages, respectively.

In the Solution Explorer, click the **Index.cshtml** file in the Views\Home folder to display the Home (welcome) page that matches Figure 17-8.

```
1  @{
2      ViewData["Title"] = "Home Page";
3  }
4
5  <div class="text-center">
6      <h1 class="display-4">Welcome</h1>
7      <p>Learn about <a href="https://docs.microsoft.com/aspnet/core">
                    building Web apps with ASP.NET Core </a> </p>
8  </div>
```

Figure 17-8: The Index.cshtml file

Note that this view doesn't have all the typical elements you see in an HTML document, like the header, footer, and menu. It also doesn't include the start and end tags for the HTML and body element. This is because there is a separate view called the **Layout** view (_Layout.cshtml) that contains such elements that are common to multiple pages. The controller will incorporate the Index view into the Layout view when the Home page is displayed. The Layout View functions similarly to the Master page used in .Net Web Forms.

You should note that this page uses the Razor syntax, signified by the @ symbol, in lines 1–3.

Line 2 stores the name "Home Page" in the ViewData dictionary, to share it with the Layout view.

Click the *Privacy.cshtml* file to display and view the privacy page, as shown in Figure 17-9.

```
1  @{
2      ViewData["Title"] = "Privacy Policy";
3  }
4  <h1>@ViewData["Title"]</h1>
5
6  <p>Use this page to detail your site's privacy policy.</p>
```

Figure 17-9: The Privacy page

You may close the two files.

The **Shared** folder, shown in Figure 17-10, contains the **Layout** and **Error** views specified by the _Layout. cshtml and Error.cshtml files.

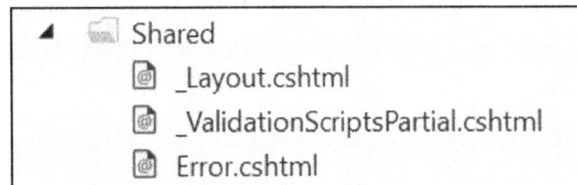

Figure 17-10: The Shared folder

The Layout View

As mentioned previously, the Layout view contains elements that are common to multiple pages. To display a specific view, the controller incorporates the elements from the specific view into the Layout view. Thus, to display the Privacy view, the controller incorporates the elements from the Privacy view into the Layout view. That is, the specific view is wrapped in the Layout view. The view (e.g., Privacy) that is incorporated into the Layout view is called a *child view*.

Step 4: Open the _Layout.cshtml file and view it.

A condensed version of the file is shown in Figure 17-11.

```
1   <!DOCTYPE html>
2   <html lang="en">
3   <head>
4       ...
5       <title>@ViewData["Title"] - Theater_MVC</title>
6       <link rel="stylesheet" href="~/lib/bootstrap/dist/css/bootstrap.min.css" />
7       <link rel="stylesheet" href="~/css/site.css" />
8   </head>
9   <body>
10      <header>
11          <nav class="navbar navbar-expand-sm navbar-toggleable-sm navbar-light
                       bg-white border-bottom box-shadow mb-3">
12              ...
13          <a class="navbar-brand" asp-area="" asp-controller="Home"
                       asp-action="Index" Theater_MVC</a>
14              ...
15          <a class="nav-link text-dark" asp-area="" asp-controller="Home"
                       asp-action="Index">Home</a>
16              ...
17          <a class="nav-link text-dark" asp-area="" asp-controller="Home"
                       asp-action="Privacy">Privacy</a>
18              ...
19          </nav>
20      </header>
21      <div class="container">
22          <main role="main" class="pb-3">
23              @RenderBody()
24          </main>
25      </div>
26
27      <footer class="border-top footer text-muted">
28          <div class="container">
29              © 2019 - Theater_MVC - <a asp-area="" asp-controller="Home"
                                   asp-action="Privacy">Privacy</a>
30          </div>
31      </footer>
32      <script src="~/lib/jquery/dist/jquery.min.js"></script>
33      <script src="~/lib/bootstrap/dist/js/bootstrap.bundle.min.js"></script>
34      <script src="~/js/site.js" asp-append-version="true"></script>
35      @RenderSection("Scripts", required: false)
36  </body>
37  </html>
```

Figure 17-11: The _Layout.cshtml file

The **RenderBody()** method that is called in line 23 renders the content of a child view like Index or Privacy, and incorporates it into the Layout view to develop the complete page. The method renders all content except those specified using @Section. A brief description of the different sections of code is as follows:

Line 5 displays the title stored by the child view into the ViewData dictionary.

Lines 6 and 7 specify the style sheets.

Lines 11–17 specify the navigation bar within the header section.

Line 29 displays the copy right within the footer.

Lines 32–34 specify the files that store the JavaScript code.

Line 35 calls the RenderSection() method to insert content that is specified in the *Scripts* section in the child view using *@Section Scripts{}*. The "required: false" parameter species that *Scripts* section is optional, so you won't get an error if the child view doesn't have a section named Scripts.

17.2.2 The Controllers Folder and Controllers

A controller handles the HTTP request for web pages. Handling a request typically includes getting the data needed for a page from models, processing the data, passing the data to the view, and displaying the view.

The *Controllers* folder has one controller named HomeController, which displays the views specified by the files in the Home folder (Index.cshtml and Privacy.cshtml). As stated earlier, by convention, the name of the controller should match the name of the folder that contains the views displayed by the controller. For example, the controller that controls the views in the Home folder should have the name HomeController.

Each controller is represented by a `Controller` class that is stored in a separate file (e.g., HomeController.cs).

```
▲  ▦ Controllers
    ▲  C# HomeController.cs
        ▷  ▓ HomeController
```

All controller classes should be stored in the Controllers folder.

Step 5: Open the HomeController.cs file and view it.

In the Solution Explorer, click the HomeController.cs file to open and view it. The HomeController.cs file appears as in Figure 17-12.

```
1   namespace Theater_MVC.Controllers
2   {
3       public class HomeController : Controller
4       {
5           private readonly ILogger<HomeController> _logger;
6
7           public HomeController(ILogger<HomeController> logger)
8           {
9               _logger = logger;
10          }
11
12          public IActionResult Index()
13          {
14              return View();
15          }
16
17          public IActionResult Privacy()
18          {
19              return View();
20          }
21
22          [ResponseCache(Duration = 0, Location = ResponseCacheLocation.None, NoStore = true)]
23          public IActionResult Error()
24          {
25              return View(new ErrorViewModel { RequestId = Activity.Current?.Id ??
                    HttpContext.TraceIdentifier });
26          }
27      }
28  }
```

Figure 17-12: The HomeController.cs file

As indicated in line 3, the `HomeController` is a class that inherits from the `Controller` class. The HomeController class contains four controller methods. We will look at two of them, **Index()** and **Privacy()**, which display the home page and the Privacy page, respectively.

Action Methods

The methods of a controller that runs in response to an HTTP request are called **action methods**. Thus both `Index()` and `Privacy()` are action methods because they are run in response to an HTTP request for the Home or Privacy pages. Typically, user interactions like clicking a link, submitting a form, or entering a URL cause a request to be sent to the server. MVC uses the information from the URL in the request to invoke the appropriate action method. It is important that you understand how these action methods work.

View() Method

Both action methods `Index()` and `Privacy()` contain the statement *return View()*, which calls the controller's **View**() method. Figure 17-13 shows what the `Index()` method looks like.

```
1   public IActionResult Index()
2       {
3               return View();
4       }
```

Figure 17-13: The `Index()` method

Though the `View()` method in line 3 doesn't specify which view should be returned (displayed), it is implied that it returns the view file that has the same name as the action method. Thus, by default, the **Index**() action method returns the **Index**.cshtml view file to display the Home (Welcome) page. Similarly, the `Privacy()` action method, as shown in Figure 17-14, returns the Privacy.cshtml view file to display the Privacy page.

```
1   public IActionResult Privacy()
2       {
3               return View();
4       }
```

Figure 17-14: The `Privacy()` action method

You can easily override the default to return a view file that has a name different from that of the action method. This is done by specifying the view name as an argument of the `View()` method, as shown here:

```
return View("viewname");
```

The `View()` method returns the view as a **ViewResult** object. However, the action methods have a return type of **IActionResult**, as in

```
public IActionResult Index
```

This is because the `ViewResult` implements the `IActionResult` interface, as does several other classes like `RedirectResult` and `JsonResult`. So, `IActionResult` works for all implementing classes and is commonly used when an action method may return different types of action results. In the current project, because the `Index()` method returns only `ViewResult` type, it is a better practice to use `ViewResult` as the return type, as in

```
public ViewResult Index
```

Later we will discuss the use of different return types. The `View()` method also has overloads that accept different arguments like the view name and model to be passed to the view. We will introduce some of these overloads later in the chapter.

How MVC Uses the URL

To understand the standard convention used in URLs, let's examine the URLs for the Home and Privacy pages. To display a page, the URL should specify the names of the controller and the action method. By convention, the name of the action method also should be the name of the view file for the page. Consider the following URL:

https://localhost:44301/Home/Privacy

MVC interprets *Privacy* to be the name of the `action` method and *Home* to be the name of the controller, as described in the next section on routing. So, it calls the Privacy() action method in the Home controller. The Privacy action, in turn, uses the `View()` method to display the Privacy.cshtml view that has the same name as the action method.

Similarly, to respond to the URL, https://localhost:44301/Home/Index, MVC calls the `Index()` action method in the Home controller, which displays the view specified in the Index.cshtml file.

Step 6: View the URL for the pages.

Run the project. Click the Privacy tab to display the privacy page.

Note that the URL of the privacy page is https://localhost:44301/Home/Privacy (your specific port number will likely be different).

Home is the name of the Controller (and also the name of the folder that contains the view). Privacy is the name of the method (action) that displays the view, and also the name of the view.

Click the Home tab to display the Home page. Note that the URL changes to https://localhost:44301/, though the complete URL for the Home page is https://localhost:44301/Home/Index.

This is because Home is the default controller (and also the folder) and Index is the default method (and also the view). So, it doesn't matter whether the URL is https://localhost:44301/, https://localhost:44301/Home, or https://localhost:44301/Home/Index.

17.2.3 Routing

The process of directing an HTTP request to an action method in a controller is called *routing*. An action method is also called an *end point*. ASP.NET Core MVC uses the Routing middleware for routing requests. In processing a request, MVC expects the URL to have the following structure:

Server:port/controller/action/id

For example, in the URL https://localhost:44301/Home/Index, *localhost* is the server, *44301* is the randomly assigned port number, *Home* is the controller, and *Index* is the action method. As described earlier, because Home is the default controller and Index is the default action method, you may omit them. The last part of the URL (*id*) represents a value for the optional *id* parameter of the action method. This value is passed to the action method when the method is invoked.

Each application configures a default routing template that can be found in the `Configure()` method in the Startup.cs file (Figure 17-15).

```
1   app.UseEndpoints(endpoints =>
2   {
3   endpoints.MapControllerRoute(
4   name: "default",
5   pattern: "{controller=Home}/{action=Index}/{id?}");
6   });
```

Figure 17-15: The Startup.cs file

Line 4 specifies that the name of the route is `"default,"` to indicate that it is the default route.

Line 5 defines Home as the default controller and `Index` as the default action, and also specifies that id is optional, as indicated by the "?" symbol.

You can change the default controller/action to any controller/action. For example, if you change line 5 to specify action="Privacy", the URLs https://localhost:44301/Home and https://localhost:44301/ will display the Privacy page. You also may specify additional routes, each with a different name. In this case, MVC will try the routes in the order they were added.

The above method of routing is called *conventional routing*. You should be aware that there is also another type of routing called *attribute routing* that places the route on the controller or the action. In this book, we will only cover conventional routing.

17.2.4 Other Folders and Files

There are a few other folders that you should be familiar with.

wwwroot Folder

The wwwroot folder is the root folder of the project. Folders for all static files, like JavaScript, CSS, Images, and library scripts, should be kept in the wwwroot folder. You can reference the files within the wwwroot folder using the base URL and folder/file name, as in http://localhost:<port>/js/scripts.js.

Startup.cs File

The Startup.cs file, as shown in Figure 17-16, contains code to configure the middleware for your application. Later in the tutorial, you will add code to this file. Two methods that are within the Startup.cs file are ConfigureServices() and Configure().

```
1  public class Startup
2  {
3      public Startup(IConfiguration configuration)
4      {
5          Configuration = configuration;
6      }
7      public IConfiguration Configuration { get; }
8      // This method gets called by the runtime. Use this method to add services to
       the container.
9      public void ConfigureServices(IServiceCollection services)
10     {
11         services.AddControllersWithViews();
12     }
13
14     // This method gets called by the runtime. Use this method to configure the HTTP
       request pipeline.
15     public void Configure(IApplicationBuilder app, IWebHostEnvironment env)
16     {
17         if (env.IsDevelopment())
18         {
19             app.UseDeveloperExceptionPage();
20         }
21         else
22         {
23             app.UseExceptionHandler("/Home/Error");
24             // The default HSTS value is 30 days. …
25             app.UseHsts();
26         }
27         app.UseHttpsRedirection();
28         app.UseStaticFiles();
29         app.UseRouting();
30         app.UseAuthorization();
31
32         app.UseEndpoints(endpoints =>
33         {
34             endpoints.MapControllerRoute(
35                 name: "default",
36                 pattern: "{controller=Home}/{action=Index}/{id?}");
37         });
38     }
39 }
```

Figure 17-16: The Startup.cs file

The **ConfigureServices()** method in lines 9–12 lets you configure services that need to be *injected* (passed) into the app. The design pattern that passes services to an app, rather than including the services within the app, is called **dependency injection (DI)**. For example, later you will add the following statement to this method to add *Session* to the DI container:

```
services.AddSession();
```

The **Configure()** method in lines 15–38 adds middleware to the HTTP request pipeline to be used by the app to process HTTP requests. For example, you will add the following statements to add Session and Routing middleware to the pipeline:

```
app.UseSession();
app.UseRouting();
```

The Configure() method also specifies how the controller and action method should be mapped to a URL (line 36).

appsettings.json File

appsettings.json is another configuration file. Typically, the connection string information needed to connect to a database is specified in this file.

Program.cs File

The Program.cs file, as shown in Figure 17-17, is the entry point for all applications. It gets executed first. This file contains the Main() method which builds the environment to host the application and starts the web application (line 5). The Program.cs file also specifies that Startup is the startup class (line 12).

```
1  public class Program
2  {
3      public static void Main(string[] args)
4      {
5          CreateHostBuilder(args).Build().Run();
6      }
7
8      public static IHostBuilder CreateHostBuilder(string[] args) =>
9          Host.CreateDefaultBuilder(args)
10             .ConfigureWebHostDefaults(webBuilder =>
11             {
12                 webBuilder.UseStartup<Startup>();
13             });
14 }
```

Figure 17-17: The Program.cs file

Review Questions

1. How does the MVC framework differ from the Razor Pages framework?
2. What are the benefits of separating an application into model, view, and controller?
3. How does ASP.NET differ from ASP.NET Core?
4. How is the name of a controller related to the name of the folder that contains the corresponding views?
5. How is the name of an action method related to the name of the view it displays?
6. What does the View() method do?
7. What is the function of the Routing middleware?
8. What is the significance of the Startup.cs file?

17.3 Creating a New Web Page

To help you better understand views, controllers, and models, here you will develop an MVC version of the Reservation page that was presented in Chapters 4 and 5. Figure 17-18 shows the page.

Compared to the Reservation page you developed in Chapters 4 and 5, the MVC version doesn't use JavaScript to compute and display the prices and discount on the Reservation form. These are computed using server-side code and displayed on the Confirmation page. If you are wondering why the base price is displayed both within the dropdown list and outside of it, this is done to demonstrate the use of Ajax. The display of the base price outside the dropdown list is not essential.

First, you will add the controller and the view for the page.

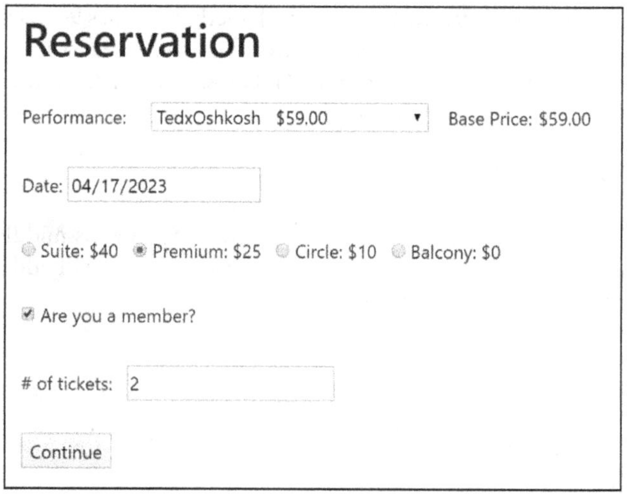

Figure 17-18: The Reservation page

17.3.1 Creating a Controller

A single controller can handle multiple pages. For example, the HomeController that has the Index and Privacy action methods handles both the Welcome (Home) and the Privacy pages that can be meaningfully grouped together. You will create a single controller named ReservationController that handles both the Reservation page and the Confirmation page.

> **Step 7: Add a controller named ReservationController.**

Right-click the Controllers folder in the Solution Explorer, and select *Add>controller*.

In the *Add New Scaffolded Item* dialog box, select *MVC Controller - Empty*, and click *Add*.

Change the controller name to **ReservationController**. Click *Add*.

You will see the ReservationController.cs file in the Controllers folder. Figure 17-19 shows what the file looks like.

```
1   using System;
2   using System.Collections.Generic;
3   using System.Linq;
4   using System.Threading.Tasks;
5   using Microsoft.AspNetCore.Mvc;
6
7   namespace Theater_MVC.Controllers
8   {
9       public class ReservationController : Controller
10      {
11          public IActionResult Index()
12          {
13              return View();
14          }
15      }
16  }
```

Figure 17-19: Index action method of ReservationController

The controller has one action method named `Index()`, similar to the `Index()` method in the Home controller. Next, you will create the view for the Reservation page.

17.3.2 Creating a View

As described earlier, a view contains the code for the user interface. View files are placed in the Views folder. To create a view that is displayed by the Index() action method in the ReservationController, the view should have the name *Index*. Further, because the name of the controller is ReservationController, the index view should be inside a folder named *Reservation*. You can make the process easier by having Visual Studio create the folder and view with the appropriate names. To do this, right-click the action method, Index(), and select *Add View*.

Step 8: Create the Index view for the Reservation page.

Right-click `public IActionResult Index()` (line 11 in Figure 17-19) and select **Add View**.

You will see the **Add MVC View** dialog box, as shown in Figure 17-20.

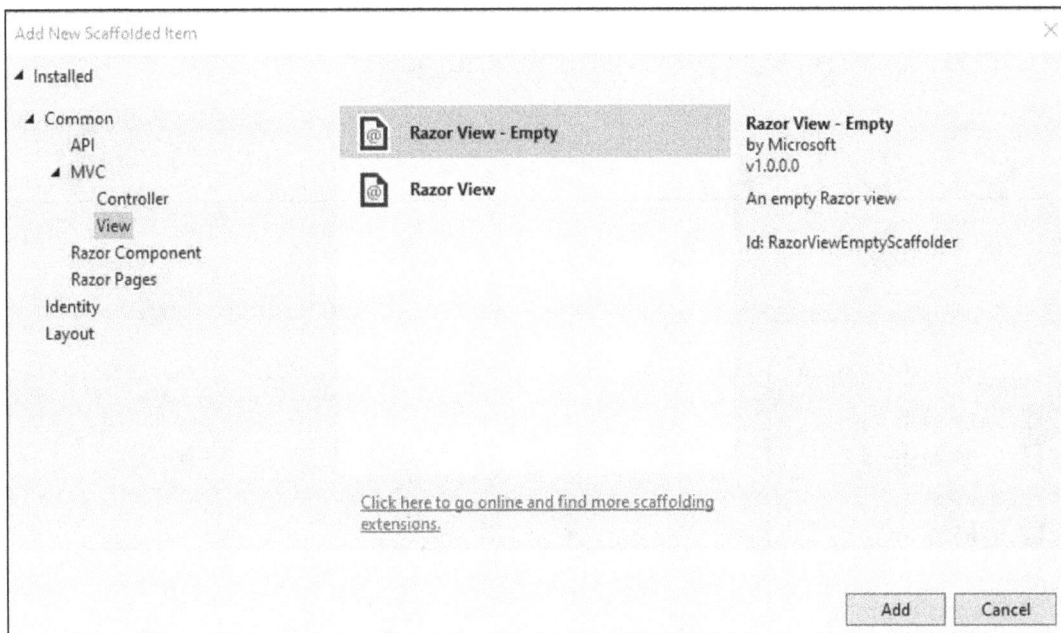

Figure 17-20: The Add MVC View dialog box

Click **Add**.

You will see the **Add New Item** window shown in Figure 17-21.

Select **Razor View - Empty** and click **Add**.

Visual Studio will create the Reservation folder within the Views folder and also create the Index.cshtml view file within the Reservation folder, as shown in Figure 17-22.

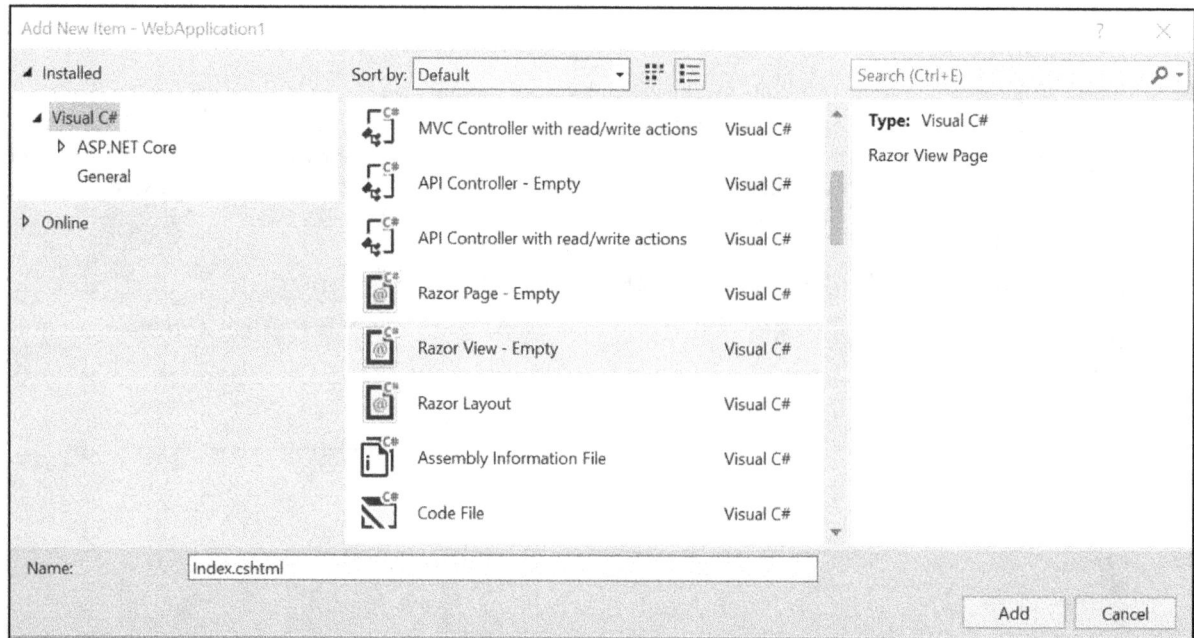

Figure 17-21: The Add New Item window

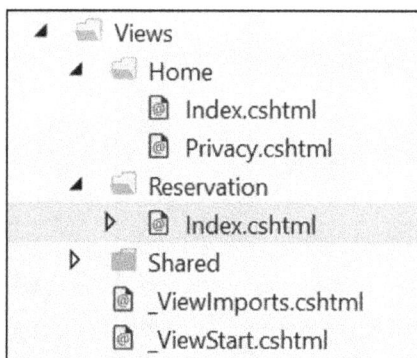

Figure 17-22: The Views folder

From this point forward, we will refer to the Reservation\Index view as the **Reservation view** or **Reservation page**.

Step 9: Open the Index.cshtml file and edit the heading.

Double-click Index.cshtml file in Solution Explorer to open the file. Replace the comment and initial code in the Index.cshtml file with the code shown in Figure 17-23.

```
1  @{
2      ViewData["Title"] = "Reservation Page";
3  }
4
5  <h1>Reservation</h1>
```

Figure 17-23: The Index.cshtml file

Line 2 stores the title in the ViewData dictionary so that it can be shared with the Layout view.

The above approach lets Visual Studio create the folder and the view from the controller. You also may create the folder and the view manually before creating the controller. To do this, follow these steps:

Right-click the Views folder, and create a new folder named Reservation.

Right-click the Reservation folder, and select *Add > New Item > Razor View*.

Adding a Link to the Navigation Bar

Next, you will edit the Layout view to add a link to the newly created Reservation page.

Step 10: Add a link to the Reservation page on the navigation bar in the Layout view.

Open the Layout view by clicking the file _Layout.cshtml in the Solution Explorer.

Add the code from lines 4–6 below the <li> element for Privacy, as shown in Figure 17-24.

```
1   <li class="nav-item">
2       <a class="nav-link text-dark" asp-area="" asp-controller="Home"
            asp-action= "Privacy">Privacy</a>
3   </li>
4   <li class="nav-item">
5       <a class="nav-link text-dark" asp-area="" asp-controller="Reservation"
            asp-action = "Index">Reservation</a>
6   </li>
```

Figure 17-24: _Layout.cshtml file

Line 5 specifies the controller (Reservation) and the action method (Index) to display the Reservation view, using the **tag helpers** asp-controller and asp-action. Tag helpers, as explained later in this chapter, provide an efficient way to render HTML elements on the server.

Click the green triangular button (▶) on the Toolbar to run the project. Check that the Reservation link is added to the navigation bar and that clicking the link displays the Reservation page, as shown in Figure 17-25.

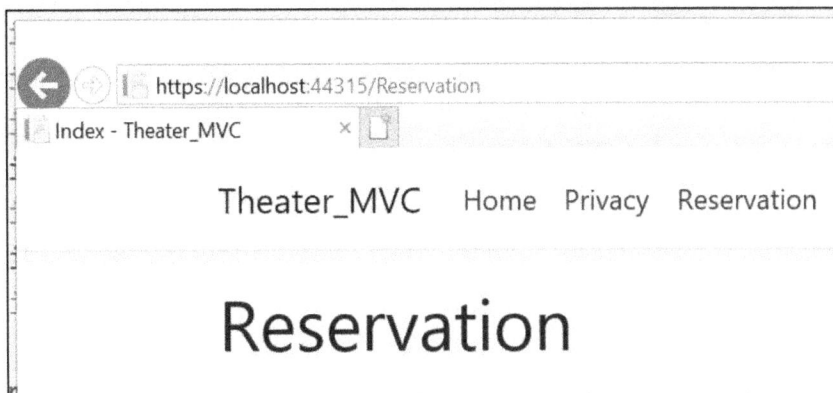

Figure 17-25: The Reservation page

Set the Reservation Page as the Default Page

Because you will be working with the Reservation page from now on, you may set it as the default page.

Step 11: Set the Reservation view as the default view.

Open the Startup.cs file. Change the name of the controller from *Home* to *Reservation*, as shown in line 5 below.

```
1   app.UseEndpoints(endpoints =>
2   {
3       endpoints.MapControllerRoute(
4           name: "default",
5           pattern: "{controller=Reservation}/{action=Index}/{id?}");
6   });
```

Run the project to display the Reservation page.

Adding User Interface Controls to the View

The Index.cshtml view file, shown in Figure 17-23, currently doesn't include any user interface controls, like a drop-down list or radio buttons.

Here, you will use standard HTML elements to create your user interface controls, as shown in Figure 17-26. Later, you will learn how to use tag helpers to generate the HTML and to bind the controls to a class (model) to make it easier to access the input data from the server.

As you might recognize, the elements in the view file, which create the user interface, are essentially the same as those in the HTML file in Chapter 4. They are also the same as the elements in the Reservation.cshtml file in the Razor Pages version of this project. However, the way you access the database to get the performance names for the dropdown list and to look up the base price is different. You will use Entity Framework Core technology to access the database, as explained shortly.

The segment of the code that adds performance names from the database to the dropdown list is incomplete, for now, as noted in line 11. The comments within the code describe the major sections (Figure 17-26).

```
1   @{
2       ViewData["Title"] = "Reservation Page";
3   }
4   <h1>Reservation</h1>
5   <form method="post" action="Reservation\Index">  <br />
6       <fieldset>
7           <label for="performance">Performance:</label>
8           <select name="performance" id="performance"
                    onchange="BasePrice(this.value)">
9               <option value="prompt">Select a performance</option>
10
11              <!-- code to add options to the dropdown list goes here -->
12
13          </select>    
14
15          <!--output element to display base price -->
16          <label for="basePrice" class="label"> Base Price: </label>
17          <output id="basePrice" class="output">0</output> <br><br>
18
19          <!-- Input field for performance date -->
20          <label for="performDate" required>Date:</label>
21          <input type="date" name="performDate" id="performDate" /> <br><br>
22
23          <!--Code to add radio buttons to let the user select a zone -->
24          <input type="radio" name="zone" id="suite" value="suite" />
25          <label for="suite">Suite: $40</label>
26          <input type="radio" name="zone" id="premium" value="premium" />
27          <label for="premium">Premium: $25</label>
28          <input type="radio" name="zone" id="circle" value="circle" />
29          <label for="circle">Circle: $10</label>
30          <input type="radio" name="zone" id="balcony" value="balcony" />
31          <label for="balcony">Balcony: $0</label>  <br> <br>
32
33          <!-- Checkbox to specify the member status -->
34          <input type="checkbox" name="memberStatus" id="memberStatus" value="true"/>
35          <label for="memberStatus">Are you a member?</label>  <br> <br>
36
37          <!-- Input field to enter the number of tickets -->
38          <label for="tickets"># of tickets:</label>
39          <input type="number" name="tickets" id="tickets" min="1" max="10"
                    step="1" value=0 /> <br> <br>
40
41          <!--Button to submit the form -->
42          <button type="submit" id="submitBtn" name="submitBtn"
                    value="continue">Continue</button>
43      </fieldset>
44  </form>
```

Figure 17-26: Initial HTML code for the Reservation view file

An important detail is that line 34 sets the value attribute of the checkbox to "true." When you submit the form, this value ("true") will be sent if and only if the checkbox is checked. If the checkbox is not checked, no value will be sent. If no value attribute is specified and the checkbox is checked, a value of "on" will be sent. This becomes important later, when you bind the checkbox to a Boolean variable to share the input data.

Line 5 posts the form to a second Index() action method of the Reservation controller, to be created in Chapter 18. This is not the Index() action method you already created, which, by default, is invoked in response to a GET request. The current form has method="post", so that the form data is submitted to an action method invoked by a POST request. So, you need an action method named Index() that is invoked in response to a POST request. Section 17.4, later in this chapter, and Chapter 18 describe how to use the HttpGet and HttpPost attributes to specify which type of request is accepted by an action method.

Note that the BasePrice() JavaScript method specified in line 8 to look up the base price has not yet been created.

Next, you will add the code from Figure 17-26 to the Reservation view file to create the user interface.

Step 12: Add elements to the Reservation view file to create the user interface.

Copy the code from the Ch17-Reservation-Index-html.txt file from the Tutorial-starts\Data-files folder and add it to the Reservation\Index.cshtml file. Note that the heading in the first line is already in the view file.

Step 13: Apply CSS to style the elements in the view.

Add the CSS shown in Figure 17-27 to the existing code in the file, wwwroot/CSS/site.css.

```
1    fieldset {width: 44em;}
2    .label {position: fixed; left: 36em;}
3    .output {position: fixed; left: 41em;}
```

Figure 17-27: CSS in the site.css file

Note that the _Layout.cshtml file already has a link to the site.css file:

```
<link rel="stylesheet" href="~/css/site.css" />
```

If you have a problem with the link, you may add the above CSS within a `<style>` element to the `<head>` section of the _Layout.cshtml file.

Step 14: Test the code.

Run the project and display the Reservation page. You should see the same familiar user interface from the Razor Pages version. The dropdown list is empty, except for the prompt, because you haven't added the code to display the performance names.

The next section explores how to access database tables using entity classes and the DbContext class to get the performance names and zone names. Here you will add the performance names to the dropdown list and use the zone names to dynamically generate the radio buttons.

17.4 Using Entity Framework Core to Access Databases

An important tool that makes it easy to work with databases is the **Entity Framework Core (EF Core)**. In this approach, you will use classes called **entity classes** (**model classes** or **models**) to hold the data and the logic. An entity class is just a standard class that is used to hold data from a database and the logic.

17.4.1 The Models Folder and Models

Consider the Performance table (from the Theater.mdf database), as shown in Figure 17-28, except for the field named *image*.

PerformanceId	PerformanceName	BasePrice	Description	Presenter
11	Young Irelanders	35	Each and every yea...	ICA
12	Justin Hayward	30	Justin David Haywa...	ACG
13	TedxOshkosh	59	We believe in the p...	Tedx
14	Oshkosh Orchestral	20	Join the Oshkosh S...	OSO
15	Ladies of Laughter	30	Since 2012 the Ladi...	*NULL*
16	Steve Earl & The Dukes	25	If you ever had an...	ACG
17	Harmonious Wail	35	Join us for a night ...	Wail

Figure 17-28: Performance table from the Theater.mdf database.

To hold the data from a row of this table, you will create a class (model) with public properties whose names match the names of the database fields. Thus the class will have properties named `PerformanceId`, `PerformanceName`, and so on. Class files are stored within the Models folder.

By default, the Models folder contains a class (model) named *ErrorViewModel* that can be used to display error messages. You add classes to the Models folder by right-clicking the Models folder and selecting *Add > class*.

Entity Framework Core

The EF Core provides a class named **DbContext** that allows you to connect to a database and query it without writing the traditional database access code. `DbContext` acts as an interface between the model classes and the database tables.

If you already have existing database tables, you can generate corresponding model classes and `DbContext` classes for the tables. Creating these classes for an existing database is called the **Database-First** approach. This is particularly suitable for when developers use databases created and maintained by others, like database administrators.

Alternatively, if you first create the entity and `DbContext` classes manually, you can create corresponding database tables from the classes using the **Code First** approach.

Here you will employ the Database First approach using the existing Performance table to generate an entity class to hold the performance data, and then use a `DbContext` class to access the data.

Generating Entity Class and DbContext Class from Existing Tables

Generating the entity and `DbContext` classes involves adding the database file to the project and installing two NuGet packages to enable the use of Entity Framework within the project.

> **Step 15: Add a folder named Data to the project folder and copy the database file Theater. mdf to the folder.**

Add a new folder named *Data* to the project folder.

Copy *Theater.mdf* from *Tutorial-starts\Data-files* to the *Data* folder.

> **Step 16: Install the packages Microsoft.EntityFrameworkCore.Tools and Microsoft. EntityFrameworkCore.SqlServer.**

From the Visual Studio menu, select *Tools > NuGet Package Manager > Manage NuGet Package for Solution*. The *NuGet−Solution* window opens.

Select the *Browse* tab and type in *Microsoft.EntityFrameworkCore* in the search box to display the packages shown in Figure 17-29.

Select *Microsoft.EntityFrameworkCore.Tools*, and check the checkbox for *Theater-MVC*.

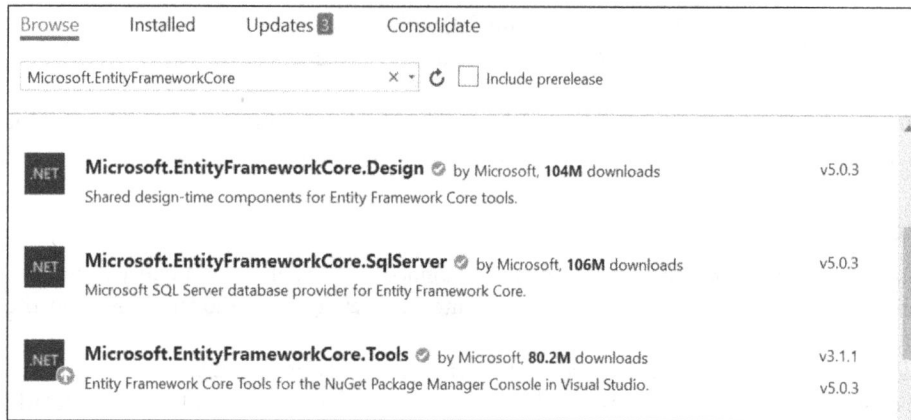

Figure 17-29: The NuGet-Solution window

Click *Install*. Click *OK* on the *Preview Changes* window, and click *I Accept* on the *License Acceptance* window. The Package Manager installs the package.

Repeat the process to install *Microsoft.EntityFrameworkCore.SqlServer*.

Close the *NuGet - Solution* window.

Generate Entity and DbContext Classes

Step 17: Generate the entity classes and the DbContext class from the Performance table and PerformDate table.

In the Visual Studio menu, select *Tools > NuGet Package Manager > Manage NuGet Console*. The console opens with the *PM>* prompt.

Run the following command from the *PM>* prompt. You may copy the command from *Tutorial-starts\ Data-files\Ch17-GenerateDbContext.txt*:

```
PM>Scaffold-DbContext "DataSource=(LocalDB)\MSSQLLocalDB;AttachDbFilename=C:\
WebProjects\MVC-1\Theater-MVC\Theater-MVC\Data\Theater.mdf;Integrated Security=True"
Microsoft.EntityFrameworkCore.SqlServer -OutputDir Models -Context
ReservationContext -Tables Performance, Zone -Force
```

Make sure that the path you specify (starting in the second line) is the path for the Theater.mdf file on your computer. You can find the path in the *Full Path* property of the file. In real-world applications, the database typically is stored on a server, which minimizes this problem.

ReservationContext (on the fifth line) is the name you give for the DbContext class to be created.

The fifth line specifies that DbContext class should be generated for the Performance and Zone tables.

After executing the command, the Models folder should show two entity class files named Performance. cs and Zone.cs, corresponding to the database tables, Performance and Zone, as well as a DbContext class file named ReservationContext.cs.

Step 18: View the Performance class.

Click the Performance.cs file in Solution Explorer to open it, and view the Performance class generated by EF Core.

Figure 17-30 shows the Performance class. The Performance class should have properties corresponding to the fields in the Performance table, as shown here. The question mark ("?") following the word *decimal* in line 7 indicates that *BasePrice* is an **optional property**, which means it can have null values. By default, numeric data types like int, double, and decimal cannot have null values.

```
1    namespace Theater_MVC.Models
2    {
3        public partial class Performance
4        {
5            public int PerformanceId { get; set; }
6            public string PerformanceName { get; set; }
7            public decimal? BasePrice { get; set; }
8            public string Description { get; set; }
9            public string Presenter { get; set; }
10           Public byte[ ] Image { get; set; }
11   }
```

Figure 17-30: The Performance class

Step 19: View the ReservationContext class.

Click the ReservationContext.cs file and view the ReservationContext class. Delete the Image property and save the class.

Figure 17-31 shows the ReservationContext class generated by EF Core. The ReservationContext class has two public properties named **Performances** (line 12) and **Zones** (line 13) that you will use to access the Performance and Zone tables, respectively. Note that the property names are plurals with an "s" at the end of the table names.

Make sure the folder for the database file (C:\\WebProjects\\MVC-1\\...) is correctly specified, as shown in line 18. The warning in line 17 indicates that it's not best practice to include a connection string in code. It is better to include it in a configuration file. In the next part of the tutorial, you will see how to move the connection string to the appsettings.json configuration file.

```
1    using System;
2    using Microsoft.EntityFrameworkCore;
3    using Microsoft.EntityFrameworkCore.Metadata;
4
5    namespace Theater_MVC.Models
6    {
7        public partial class ReservationContext : DbContext   {
8            public ReservationContext() {
9            }
10           public ReservationContext(DbContextOptions<ReservationContext>
                    options)  : base(options) {
11           }
12           public virtual DbSet<Performance> Performances { get; set; }
13           public virtual DbSet<Zone> Zones { get; set; }
14
15           protected override void OnConfiguring(DbContextOptionsBuilder
                optionsBuilder) {
16             if (!optionsBuilder.IsConfigured) {
17               //#warning …
18               optionsBuilder.UseSqlServer("Data Source=(LocalDB)\\MSSQLLocalDB;
                 AttachDbFilename=C:\\WebProjects\\MVC-1\\Theater-MVC\\
                 Theater-MVC\\Data\\Theater.mdf;Integrated Security=True");
19             }
20           }
21           ...
22           ...
23       }
24   }
```

Figure 17-31: The ReservationContext class

The `Performances` property of `ReservationContext` now can be used to get the Performance records.

17.4.2 Accessing Records from a Table Using DbContext

Before accessing the database, make sure that the connection string is specified in the appsettings.json file, as shown in Figure 17-32.

```
1   "AllowedHosts": "*",
2
3   "ConnectionStrings": {
4     "MvcReservationContext": "Data Source=(LocalDB)\\MSSQLLocalDB;AttachDbFilename=
          C:\\WebProjects\\MVC-1\\Theater-MVC\\Theater-MVC\\Data\\Theater.mdf;
          Integrated Security=True;MultipleActiveResultSets=true"
5   }
```

Figure 17-32: The appsettings.json file

Step 20: Specify the connection string in the appsettings.json file.

Specify the connection string in the appsettings.json file, as shown in lines 3 and 4 in Figure 17-32.

You may copy the string from the Ch-17-ConnString.txt file in the Project-starts\Data-files folder, and modify the path, if necessary

Note that you may also get the connection string from the Connection String property of your database in the Server Explorer window. To open the database, right-click the Theater.mdf file in Solution Explorer and select *Open*.

Make sure line 4 is in a **single line** in the appsettings file. Pasting the connection string may break it into multiple strings over multiple lines. You will need to fix it.

Next, you will use the `Performances` property of the `ReservationContext` to get the performance names from the Performance table and pass them to the Reservation view. To use the `Performances` property, you need to create a `ReservationContext` object within the `ReservationController`.

Figure 17-33 shows the code for creating a ReservationContext object named resContext.

```
1   public class ReservationController : Controller
2   {
3       // Create a ReservationContext object.
4       private ReservationContext resContext = new ReservationContext();
```

Figure 17-33: The ReservationController

You also need the using statement:

```
using Theater_MVC.Models;
```

Now the `Index` action method can pass data from the Performances property and pass to the view using `ViewBag`, as described next.

Passing Data from the Controller to the View Using ViewBag

You can pass data from the controller to the view using mechanisms like **model binding**, **ViewBag**, **ViewData**, and **TempData**.

The ViewBag is a property of the controller that lets you pass data from the controller to the view. The data only lasts for the lifetime of the HTTP request, until the view is rendered. You can store any type of data in a property of the ViewBag. Because ViewBag properties are dynamic, you don't need to explicitly cast the values stored in them, and ViewBag doesn't provide compile-time error checking.

ViewData is similar to ViewBag, except that it requires typecasting for getting data and checking for null values. TempData and model binding are described in Chapter 18, Section 18.1.

Here you will use ViewBag. The following is the general syntax to store a value in ViewBag:

```
ViewBag.Propertyname = value;
```

The property name can be any valid variable name. For example, you can store the performance data in the ViewBag as follows:

```
ViewBag.Performances = resContext.Performances;
```

Performances is the name of the ViewBag property.

Figure 17-34 shows the code in the Index() action method to get the performance data from the Performance property and store the data in the ViewBag.

```
1   using Microsoft.AspNetCore.Mvc;
2   using Theater_MVC.Models;
3   using Microsoft.EntityFrameworkCore;
4
5   namespace Theater_MVC.Controllers
6   {
7       public class ReservationController : Controller
8       {
9           // Create a ReservationContext object named resContext
10          private ReservationContext resContext = new ReservationContext();
11          [HttpGet]
12          public IActionResult Index()
13          {
14              // Get Performance records from the Performance property of resCotext
15              ViewBag.Performances = resContext.Performances;
16              return View();
17          }
18      }
19  }
```

Figure 17-34: Index() action of ReservationController

Line 10 creates the ReservationContext object named resContext.

Line 11 specifies the [HttpGet] attribute, which is the default, for the Index() method, to make it clear that this Index() method is invoked by a GET request.

Line 15 uses the Performances property of resContext to get the data and stores the data in the ViewBag.

Step 21: Modify the ReservationController to pass the Performance records to the view.

Add the using statements shown in lines 1–3. Insert the code from lines 9–11, 14, and 15 of Figure 17-34 into the ReservationController class.

Using the Data Passed by the Controller to a View

The ViewBag.Performances property can be accessed from the view to get the collection of records stored in it. You can use a *foreach* loop to access individual records, as shown in the code block (@{ }) in lines 13–18 of Figure 17-35.

```
1  @{
2      ViewData["Title"] = "Reservation Page";
3  }
4
5  <h1>Reservation</h1>
6  <form method="post" action="Reservation\Index">
7      <br />
8      <fieldset>
9          <label for="performance"> Performance: </label>
10         <select name="performance" id="performance" onchange="BasePrice(this.value)">
11             <option value="prompt">Select a performance</option>
12             <!-- Get each record  and add the performance names to the dropdown list -->
13             @{
14                 foreach (var performance in ViewBag.Performances)
15                 {
16                     <option value="@performance.PerformanceName">
17                       @performance.PerformanceName  
18                                          @performance.BasePrice.ToString("c") </option>
19                 }
20             }
21         </select>
```

Figure 17-35: Code to add performance names to the dropdown list

Line 14: The `foreach` loop iterates through each record in the `Performances` property and stores the record in the variable, `performance`.

Line 16: This line gets the performance name from the `PerformanceName` field of the current record. The value attribute of the `<option>` element, which specifies the value of the selected item, is set to the performance name.

Similarly, the content of the `<option>` element is set to the concatenation of performance name and base price. Thus each pass through the loop adds a performance name and the base price to the dropdown list. Note that the "@" symbol of Razor syntax is used to specify C# code within the HTML element.

Step 22: Insert the code in the Index.cshtml file to add performance names to the dropdown list.

Add the code from lines 12–18 in Figure 17-35 to the Reservation view file (Index.cshtml file).

To test the code, display the Reservation page and click the dropdown list.

The dropdown list should show all performance names and corresponding prices.

Note that selecting an item from the list will give an error because you haven't created the BasePrice() method specified in line 1.

Using Zone Data to Create Radio Buttons

Currently, the radio buttons are created using hard-coded values for zone names. The approach you used to dynamically add `<option>` elements to the `<select>` element also can be used to dynamically generate the radio buttons using zone names from the Zone table. You already generated an entity class named `Zone` (Zone.cs), as shown in Figure 17-36.

```
1  public partial class Zone
2  {
3      public int ZoneId { get; set; }
4      public string ZoneName { get; set; }
5      public decimal? ZonePrice { get; set; }
6  }
```

Figure 17-36: The Zone class

Within the ReservationController, you can use the Zones property of ReservationContext to get the Zone records and pass the data to the view using ViewBag. (This task is relegated to an exercise at the end of the chapter.)

It's time to practice! Do Exercise 17-1 (see the end of the chapter).

17.4.3 Looking Up Data from a Table Using Ajax

The Reservation page displays the base price whenever the user selects a performance from the dropdown list, as shown here:

Performance: TedxOshkosh $59.00   ▼   Base Price: $59.00

This is done by looking up the base price from the database. This feature is provided in the form primarily to demonstrate how Ajax can be used with MVC. Because the base price is displayed in the dropdown list, computing and displaying the base price outside the dropdown list is not essential to the project. If you are at the beginner level in programming, you may skip this section.

As you learned in Chapter 14, Ajax (Asynchronous JavaScript and XML) is a set of techniques available in the browser to do a partial update of the web page without having to submit an entire form. You also can use Ajax to send and receive data from the server in the background in different formats, like JSON (JavaScript Object Notation), HTML, XML, and text.

First, you will create a server-side method named BasePrice() to look up the base price. Then, you will create a JavaScript function to send the Ajax request to the server and get the result.

Lines 14–29 in Figure 17-37 show the C# method, BasePrice (that looks up the base price), within the ReservationController.

```
1   public class ReservationController : Controller
2   {
3       // Create a ReservationContext object and store in the variable, resContext
4       private ReservationContext resContext = new ReservationContext();
5
6       [HttpGet]
7       public ViewResult Index()
8       {
9           // Get Performance records from the Performance property of resCotext.
10          ViewBag.Performances = resContext.Performances;
11          return View();  // return the Reservation view.
12      }
13
14  public string BasePrice()
15  {
16      // Get the selected performance from the query string parameter, perform
17      string qryString = HttpContext.Request.QueryString.ToString();
18      string performName = HttpUtility.ParseQueryString(qryString).Get("perform");
19      decimal basePrice =0;
20      // Get each record from the Performance property of resContext object
21      foreach (var performance in resContext.Performances)
22      {
23          // Find the base price
24          if (performName == performance.PerformanceName)
25              basePrice = (decimal) performance.BasePrice;
26      }
27      string strBasePrice = basePrice.ToString("c");
28      return strBasePrice;
29  }
```

Figure 17-37: Server-side method to look up base price

The internal documentation within the code specifies the task performed in each section of the code.

Line 17 gets the entire query string using the Request property of HttpContext class. The HttpContext provides information about the current HTTP request.

Line 18 uses the `HttpUtility` class to get the query string item named "perform." The `HttpUtility`, found in the System.Web name space, provides methods that manipulate HTML strings.

Line 21 gets the current record from the `Performance` property of the `resContext` object created in line 4. The record is stored in the variable *performance*.

Line 24 gets the data from the `PerformanceName` field of the current record, and line 24 gets the data from the `BasePrice` field.

In line 25, the `BasePrice` property, which is of type *decimal?*, is cast to *decimal* because *decimal* is not the same as *decimal?* (which was briefly described in Section 17.4.1).

Line 28 returns the formatted base price as a string.

You need to add the following use directive to specify the name space for the `HttpUtility` class used in line 18:

```
using System.Web;
```

Ideally, a method like `BasePrice` that accesses the database to look up the base price should be in a class (e.g., Performance.cs) outside the controller to separate business logic from the controller function. However, the `Http-Context` class that you used in the `BasePrice` method to get the query string is not directly available in classes outside the controller. You need to use additional code to access the `HttpContext` class.

Step 23: Code the BasePrice() method.

Add the `BasePrice()` method, shown in Figure 17-37, to the ReservationController.cs file, and add the directive, using System.Web. The code for `BasePrice()` can be found in Tutorial-starts\Data-files\Ch17-BasePrice-C#.txt.

Add the following using statement:

```
using System.Web;
```

Sending an Ajax Request

Next, you will send an Ajax request (call) from the view to the server to invoke the `BasePrice()` method and get the base price. To send the request, create a JavaScript function, also named BasePrice(), that does the following:

- Create an **XMLHttpRequest** object (**XHR** object), provided by the browser, to send an Ajax request to the server and to receive the data returned by the server.
  - o Use the **open()** method of the object to specify the URL to send the request to.
  - o The open() method is used also to specify the query string to pass the performance name. The **send()** method sends the request.
- Specify what to do with the retrieved data, if the server processes the request successfully.

Figure 17-38 shows the JavaScript function, `BasePrice()`, that will be added to the Site.js file.

```
1          function BasePrice(performance) {
2              // Use Ajax to get base price
3              var xhttp = new XMLHttpRequest();        // Create an XMLHttpRequest object
4              // specify the function to be executed if readystate property changes
5              xhttp.onreadystatechange = function () {
6                if (this.readyState == 4 && this.status == 200) {  // if request was successful
7                    // Get the base price returned by the Ajax call and display in the output element
8                        document.getElementById("basePrice").innerHTML = this.responseText.toString();
9                    }
10             };
11             // specify the Ajax request
12             xhttp.open("GET", "/Reservation/BasePrice?perform=" + performance);
13             xhttp.send();   // send the request
14         }
```

Figure 17-38: The JavaScript method, BasePrice

Line 3 creates the XMLHttpRequest object, which is an object of the JavaScript environment of the browser.

Lines 5–10 specify what to do with the data sent by the server.

Line 5 specifies a function that will be called when the readyState property of the object changes – that is, when the status of the request changes.

Line 6 checks the value of the readyState and status properties. A readyState value of 4 indicates that the response is ready. The value ranges from 0 to 4. A status value of 200 indicates that the request was successful ("OK"). You may be familiar with the status value 404, which means that the page was not found.

Line 8 gets the base price returned by the Ajax call and displays it in the output element (<output id="base-Price" ...>) on the Reservation page. The data returned by the Ajax call is obtained from the responseText property of the XMLHttpRequest object.

Line 12 uses the open() method of the object to specify the Ajax request. The request specifies two parameters:

- the HTTP request method (GET)
- the URL to send the request to.

The URL in line 12 (*"/Reservation/BasePrice?perform="* + *performance*) specifies that the Ajax request should be sent to the *BasePrice* method in the *Reservation* controller, along with the query string parameter (perform) that specifies the performance name to be sent to the server.

Line 13 uses the send() method of the object to send the request.

Step 24: Add the client-side BasePrice method.

Add the JavaScript method, BasePrice(), shown in Figure 17-38, to the wwwroot\JS\Site.js file.

The code for BasePrice() can be found in Tutorial-starts\Data-files\Ch17-BasePrice-JS.txt.

Add the following statement within the <head> element of _Layout.cshtml file, and modify the path, if necessary:

```
<script src="C:\WebProjects\MVC\Theater-MVC\Theater-MVC\wwwroot\js\Sites.js"></script>
```

You also may add the BasePrice() function within a <script> element in the <head> section.

Step 25: Test the code.

Display the Reservation page and select a performance from the dropdown list. Make sure the base price for the selected performance is displayed in the output element to the right of the dropsown list.

17.4.4 Tag Helpers

Next, you will learn how you can enhance the development of web forms using tag helpers, which are popularly used in web development. Tag helpers provide an efficient and powerful way to generate HTML elements and to bind elements to properties of a model (class) to help access form data on the server.

As you will learn soon, to bind an element to a class property, you can give the same name for the element and the property. Using tag helpers to generate the elements makes it easier to specify the name and other attributes of an element and reduces the chances of error. The attribute name specified in a tag helper is automatically checked to make sure it matches a property in the class.

Tag helpers also allow you to enhance standard HTML elements by adding new server-side attributes to the elements. Further, they let you incorporate server-side code in rendering elements. Built-in tag helpers that come with the Core Framework are available for many commonly used HTML elements – but not for all. You also can create custom tag helpers.

Let's look at how to generate attributes of elements on a form and bind the elements to the properties of a class. The key steps are as follows:

1. Create a class that has a property corresponding to each input element on the form.

   You will create a class named Reservation, as shown in Figure 17-39. Later, you will add methods to this class to compute the base price, zone price, discount, and total cost.

```
1   using System;
2   using System.ComponentModel.DataAnnotations;
3   namespace Theater_MVC.Models
4   {
5   public class Reservation
6   {
7   public string Performance { get; set; }
8   [DataType(DataType.Date)]
9   public DateTime PerformDate { get; set; }  = DateTime.Today;
10  public string Zone { get; set; }
11  public bool MemberStatus { get; set; }
12  public int Tickets { get; set; }
13  }
14  }
```

Figure 17-39: The Reservation class

Line 8 is a **data annotation attribute** for the data type, DateTime. It specifies that the real data type of PerformDate is *Date*, which doesn't include the time. You use data annotation because you cannot directly specify the data type, *Date*, for the C# class property, PerformDate. Without the data attribute, the field will display both the date and the time.

2. Bind the model to the view using the @model directive in the view. You will add the following directive to the Reservation view (Index.cshtml file):

```
@model Reservation;
```

Specifying the @model directive lets the tag helpers access the properties of the Reservation class from within the view. A view that is bound to a model is called a **strongly-typed** view.

3. Use the names of the class properties in tag helpers to generate the elements, as explained next.

Some commonly used tag helpers are described in the following subsections.

asp-for Tag Helper

The asp-for tag helper generates attributes for common elements like <label>, <input>, and <select>.

<input> Element

The following is the pattern of the asp-for tag-helper in an <input> element to bind it to a property of the model object:

```
<input asp-for="propertyname">
```

For example, to generate an <input> element to enter the number of tickets, you would use the asp-for tag helper, as follows:

```
<input asp-for="Tickets" />
```

If you specify a value for asp-for that doesn't match the name of any property of the class, you will get an error.

You may wonder why there are no attributes like id, type, or value specified. For an <input> element, the asp-for attribute renders the id, name, type, and value attributes and additional attributes that specify data validation (explained later). The values of these attributes are determined by the name, data type, and value of the corresponding property of the Reservation class (Tickets). The above tag helper will generate the following HTML element, which can be viewed in the page source when you display the page in the browser.

```
<input type="number" data-val="true" data-val-required="The Tickets field is required."
       id="Tickets" name="Tickets" value="0" />
```

The values of the `id` and `name` attributes are determined by the value of the `asp-for` attribute (which is also the name of the corresponding property, `Tickets`, of the `Reservation` class).

The value property is set to zero, which is the default value of the `Tickets` property of the class.

The value of the type attribute is determined by the data type of the bound property of the class, as shown in Table 17-1.

Table 17-1: The values of the Type attribute

Data type of property	Type attribute for <input> element
`byte, sbyte, int, uint, short, ushort, long, ulong`	number
`float, double, decimal`	text
`string`	text
`bool`	checkbox
`DateTime`	datetime

You also may specify additional attributes for an element, along with tag helpers. For example, you may specify *min*, *max* and *step* attributes for an input element to validate its value, as follows:

```
<input asp-for="Tickets" min="1" max="10" step="1" />
```

This is how the resulting HTML element looks:

```
<input min="1" max="10" step="1" type="number" data-val="true"
    data-val-required="The Tickets field is required."  id="Tickets" name="Tickets" value="0" />
```

As shown in Table 17-1, when the data type of a property of the class is *bool*, asp-for renders the `<input>` element with the type attribute set to *checkbox*. In addition, the `value` attribute will be set to *true* so that the value *true* will be sent to the server if the checkbox is checked.

As an example, consider the tag helper

```
<input asp-for="MemberStatus" />
```

Because the MemberStatus property is of bool type, the above tag helper will render the type and value attributes as follows:

```
<input type="checkbox" data-val="true" data-val-required="The MemberStatus field is required."
    id="MemberStatus " name="MemberStatus" value="true" />
```

Table 17-1 also indicates that none of the data types renders a radio button element (`<input type="radio"`). Thus you will have to explicitly specify the type. You also will have to specify the value because radio buttons are used in groups, and the value of each button will be different from the id and name of the group. Here is an example of a radio button to let the user select a zone:

```
<input asp-for="Zone"  type="radio" value="suite" />
```

In this case, asp-for renders the id and name attributes:

```
<input type="radio" value="suite" id="Zone" name="Zone" />
```

<select> Element

For a `<select>` element, the asp-for tag helper generates the *id* and *name* attributes. Consider the following select element that uses the asp-for tag helper:

```
<select asp-for="Performance"></select>
```

This is what the rendered element looks like:

```
<select id="Performance" name="Performance"></select>
```

<label> Element

For the `<label>` element, the `asp-for` tag helper renders the *for* attribute. The `label` element

```
<label asp-for="Performance"> Performance: </label>
```

renders the following:

```
<label for="Performance"> Performance: </label>
```

asp-action and asp-controller Tag Helpers

The `asp-action` and `asp-controller` tag helpers are commonly used together in a **<form>** element to specify the action method and the controller for a form. Consider the `<form>` element for the Reservation form, which submits the data to the Confirmation page. The standard HTML code for the form is

```
<form  action="Reservation" method="post" >
```

In place of the action attribute, the tag helper would use the asp-action and asp-controller attributes to specify the action method and controller for the Confirmation view, as follows:

```
<form asp-action="Index" asp-controller="Reservation" method="post">
```

Building the Reservation Form Using Tag Helpers

Now that you have an understanding of tag helpers, you can use tag helpers to generate the elements in the Reservation form.

Figure 17-40 shows the Reservation form that uses tag helpers. Because the tag helpers used in this form were described previously, you should be able to understand the code without further explanation. (You can refer back to earlier sections of the chapter if any questions arise.)

```
1  <form asp-action="Index" asp-controller="Reservation" method="post">
2      <br />
3      <fieldset>
4          <label asp-for="Performance"> Performance: </label>
5          <select asp-for="Performance" onchange="BasePrice(this.value)">
6              <option hidden value="prompt">Select a performance</option>
7              <!-- Get each record and add performance names to the dropdown list -->
8              @{
9                  foreach (var performance in ViewBag.Performances)
10                 {
11                     <option value="@performance.PerformanceName">
                           @performance.PerformanceName  
                           @performance.BasePrice.ToString("c")</option>
12                 }
13             }
14         </select>     
15
16         <!--output element to display base price from BasePrice() function -->
17         <label for="basePrice" class="label"> Base Price: </label>
18         <output id="basePrice" class="output">0</output> <br /><br />
19
20         <!-- Input field for performance date -->
21         <label asp-for="PerformDate">Date:</label>
22         <input asp-for="PerformDate" class="Label" /><br /><br />
23
24         <!--Radio buttons to select a zone -->
25         <input asp-for="Zone" type="radio" value="suite" />
26         <label asp-for="Zone">Suite: $40</label>
27         <input asp-for="Zone" type="radio" value="premium" />
28         <label asp-for="Zone">Premium: $25</label>
29         <input asp-for="Zone" type="radio" value="circle" />
30         <label asp-for="Zone">Circle: $10</label>
31         <input asp-for="Zone" type="radio" value="balcony" />
32         <label asp-for="Zone">Balcony: $0</label> <br /><br />
33
34         <!-- Checkbox to specify the member status -->
35         <input asp-for="MemberStatus" />
36         <label asp-for="MemberStatus">Are you a member?</label><br /><br />
37
38         <!-- Input field to enter the number of tickets -->
39         <label asp-for="Tickets"># of tickets:</label>
40         <input asp-for="Tickets" /> <br /><br />
41
42         <!--Button to submit the form -->
43         <button type="submit" id="submitBtn" name="submitBtn"
                   value="continue">Continue</button><br>
44     </fieldset>
45 </form>
```

Figure 17-40: Reservation form with tag helpers

Lines 8–13 use the `foreach` loop to display the performance name and base price of each performance from the list stored in the `ViewBag`. Here we do not use the *asp-items* tag helper to display the options because each option displays values of two properties: `PerformanceName` and `BasePrice`. If you were to display only the `PerformanceNames` in the dropdown list, you could use the *asp-items* tag helper instead of the `foreach` loop, as follows:

```
1  <select asp-for="Performance" onchange="BasePrice(this.value)"
2  asp-tems="@(new SelectList(ViewBag.Performances, "PerformanceName","PerformanceName"))">
3      <option value="prompt">Select a performance</option>
4  </select>
```

Here, the tag helper uses the `SelectList()` method, which has the following pattern:

SelectList(list, value, text)

There is no tag helper specified for the `<output>` element in line 18 because tag helpers are not applicable to `<output>` elements. Also, note that the radio buttons are created using separate code for each button. You could use a foreach loop to generate these buttons using zone names from the Zone table, as specified in Exercise 17-1.

Now let's create the `Reservation` class, bind it to the view, and use tag helpers in the Reservation form, as explained previously.

Step 26: Create the Reservation class within the Models folder.

Right-click the *Models* folder and select *Add > class*.

Change the name of the class to *Reservation*. Click *Add*.

Add the properties shown in Figure 17-39. You may copy the code from the Ch17-Reservation-class-properties.txt file.

Make sure you add the *using* statement:

```
using System.ComponentModel.DataAnnotations;
```

Step 27: Bind the Reservation class to the Reservation view.

Add the following directive to the top of the Index.cshtml file (Reservation view):

```
@model Reservation;
```

Step 28: Use tag helpers to render elements in the Reservation view.

Comment out (or delete) the existing `<form>` element and all elements within the form using `@*` before the start tag and `*@` after the end tag, as follows:

```
@*
  <form method="post" action="Reservation">
    ...
  </form>
*@
```

Insert the code for the form and the elements from Figure 17-40 into the Index.cshtml file below the commented out code. Instead of typing in the code, you may either copy the commented out code and change it to use tag helpers, or copy the code from Tutorial-starts\Data-files\Ch17-Reservation-Index-Tag-Helpers.txt file.

The commented-out code is kept only for comparison purposes.

Step 29: Test the code.

Display the Reservation page. Make sure that the elements in the form behave the way they behaved before introducing tag helpers.

You may view the rendered HTML attributes by right-clicking the page in the browser and selecting *View page source*.

As stated earlier, the radio buttons are created using separate code for each button. If you haven't done **Exercise 17-1**, you may do it now. This exercise uses a foreach loop to generate these buttons using zone names from the Zone table.

Now that you have bound the form elements to the Reservation class, you can pass the form data to an action method using the Reservation object, as explained in the next section.

Review Questions

9. What is the role of DbContext class in accessing a database?

10. Describe ViewBag and what it is used for.

11. What are tag helpers? What are the benefits of using tag helpers?

It's time to practice! Do Exercise 17-2.

Exercises

Exercise 17-1: Generating Radio Buttons

This exercise generates radio buttons using data from the Zone table. Currently, the radio buttons are created using hard-coded values for zone names (lines 24–32 in Figure 17-40). Comment out these lines of code and replace them with code that dynamically generates the radio buttons using zone names and base prices from the Zone table.

Follow the approach that you used to dynamically add <option> elements to create the dropdown list. You already generated a class named Zone (Zone.cs). You can use the Zone property of ReservationContext, within the Reservation Controller, to get the Zone records and pass the data to the view using ViewBag (see Figure 17-34). Use a foreach loop to get each zone name and corresponding base price.

Currently, the value attribute of the <input> element is set to a fixed value like "suite" and the content of the <label> is "Suite", as shown here:

```
<input type="radio" name="zone" id="suite" value="suite" />
<label for="suite">Suite</label>
```

Instead, set the value attribute to the value of the ZoneName field in the Zone record. Similarly, use the zone name and base price from the Zone records to specify the content for the <label> element.

Exercise 17-2: Checkbox for Snacks

Add a second checkbox to the Reservation form that the user can check to choose snacks during the event. Add a property (e.g., IsSnackSelected) to the Reservation class to store the selection, and use tag helper to create the checkbox and bind it to the property.

Assignments

E-Commerce Assignment: MVC—Part 1

This assignment develops the Order page for essentially the same application described in the E-Commerce Assignment in Chapter 16, except that here you use the **MVC framework**.

P&I Laptops is a small business that assembles and sells a limited set of laptop models online. Information on the different models sold by P&I and the software bundled with the laptops are maintained in the database **PI-Laptops**, which can be found in the folder Tutorial-starts\DataFiles. The database contains the following two tables:

Model

ModelId	ModelName	Description	CPU	OS	Touch	StartingPrice
101	S330	15 inch lap top with essential perf...	A6-9225	*NULL*	*NULL*	295.00
102	S380	Durable, ready for school, 14 inch ...	i3-8130U	Windows...	*NULL*	450.00
103	S580	S580 is powered by Intel Core™ i...	i5-8250U	Windows...	True	595.00
104	G780	14 inch touch-screen gaming lapt...	i7-8750H	Windows...	True	950.00
105	B780	A light-weight 2-in-one 15-inch to...	C i7-875...	Windows...	True	995.00

The StartingPrice model is the price for the default configuration.

Software

SoftwareId	SoftwareName	Price
11	MS 365 Personal	65.00
12	MS 365 Family	99.00
13	Office Home & Student	149.00

Create a web project named PI-eComm-MVC and develop a web page named **Order.** You may develop the project within the WebProjects folder. The web page allows the user to select the model, software, whether a warranty is needed, and the quantity, as shown below. Choose an appropriate layout. What is shown here is an example:

Models: S380 - $450

Software:
  ○ MS 365 Personal - $65   ○ MS 365 Family - $99   ○ Office Home & Student - $149

  ✓ Ext. Warranty?

Quantity: 2

  Select

The dropdown list shows the laptop model name and the price from the ModelName and StartingPrice fields in the Models table. Similarly, the radio buttons show the software and their prices from the SoftwareName and Price fields in the Software table.

You should use tag helpers to create a form that contains the elements to accept the user input. The form should post the data to an action method named **Index.** The input elements should be bound to the properties of a class named **Order** to help display the input data on a second page, named **Cart.** Make sure the Order class has a property named OrderNumber that would serve as the primary for the table that will be generated from the class to save the order data in Part 2 of this assignment.

The **Select button** should be a submit button. You will develop the rest of the project, which involves validating, processing, and saving the data, in Part 2 of this assignment in Chapter 18.

Auto Rental Assignment: MVC—Part 1

This assignment develops the **auto-selection** page for the same application described in the Auto Rental Assignment in Chapter 16, except that here you use the MVC framework.

Ace would like you to develop two web pages: The first web page lets the user select a vehicle and specify related information. The second page computes the cost; displays the input data and the cost; and lets the user edit, cancel, or save the reservation. Part 1 develops the first web page, named *Rental*.

Ace has a database named AutoRental.mdf that contains two tables, AutoTypes and Discount. The database can be found in the folder Tutorial-starts\Data-files. A subset of the records in the AutoTypes table is shown here.

AutoTypes

AutoTypeId	AutoType	Capacity	DailyRate	WeeklyRate
11	Standard SUV	5	45	250
12	Full-size SUV	7	65	400
13	Ludury SUV	5	55	350
14	Standard Minivan	7	60	400
15	12 Passenger Van	12	75	500

AutoType represents the type of auto, such as a standard SUV, an economy car, or a small pickup. AutoTypeId is a unique number that identifies each record in the RentalRate table. Capacity represents the passenger (seating) capacity.
    Here is the Discount table:

Discount

Id	Category	DiscountPercent
11	BestRate	5
12	Government	15
13	Business	10
14	Senior	20

Use the **MVC** framework to develop a web page named **Rental** within the web project named **AutoRental-MVC.** You may develop the project within the WebProjects folder. The web page should let the user specify an auto type, the discount category, whether or not insurance is needed, and the number of days, as shown below. Choose an appropriate layout. What is shown here is just an example.

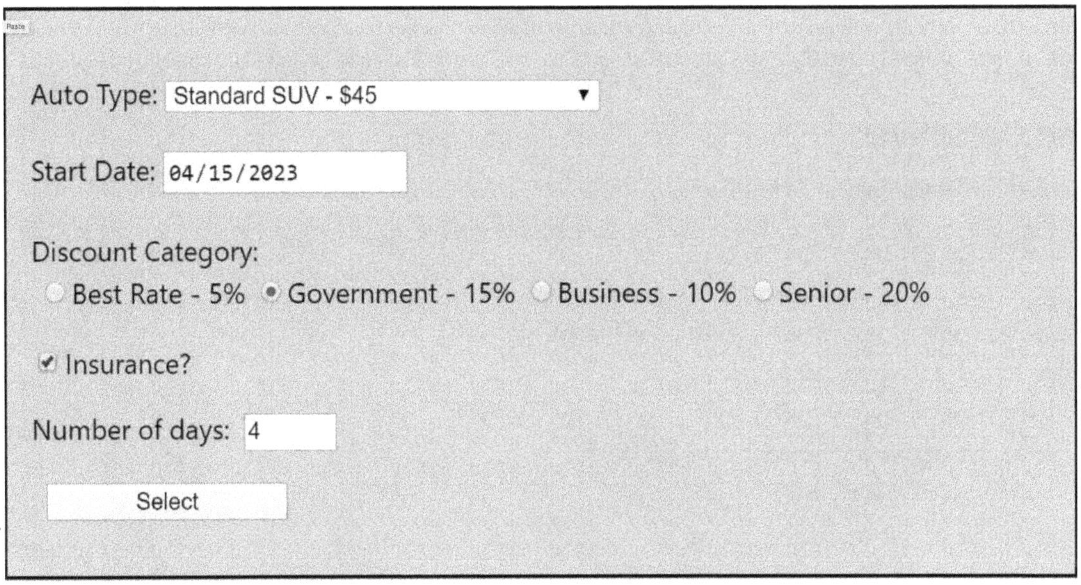

Note that the dropdown list displays the auto type and daily rate from the AutoType and DailyRate fields of the AutoTypes tables. Similarly, the radio buttons show combined data from the DiscountCategory and DiscountPercent fields of the Discount table.
    Use tag helpers to create a form and the elements that accept user input. The form should post the data to an action method named Index. The elements that accept user input should be bound to the properties of a class named RentalsClass, so that the input data can be displayed on a second page, named Confirmation. You need to create a class named RentalClass with property names that match the names of elements to be bound to.
    The **Select button** should be a submit button. You will develop the rest of the project, which involves validating, processing, and saving the data, in Part 2 of this assignment in Chapter 18.

ASP.NET Core MVC: Multipage Apps

Chapter 17 presented key concepts of MVC and provided you with the skills to use the *model*, *view*, and *controller* to create a Core MVC form using tag helpers. In this chapter, you will learn how to build more complex multipage applications that validate and process form data, share data between pages, present data in a view, and save data to a database. You will use models to bind and process data, migration files to generate database tables using the code-first approach, and session state to store data and a list of objects to work with multiple objects at a time.

Learning Objectives

After studying this chapter, you should be able to

- Use model binding to access input data.
- Process form data using models and share it between pages.
- Perform client-side and server-side validation of input data.
- Use view models to display data in a view.
- Restore the values of input elements after a page refresh.
- Use a code-first approach to generate database tables.
- Save data into a database table.
- Work with a list of objects

18.1 Processing Form Data and Sharing It

Web projects typically do server-side processing of user input from forms and share it with other pages. Unlike the server controls used in Web Forms, the standard HTML controls used in MVC cannot be directly accessed from the server. To help you learn how to access and process input data in an MVC app, you will retrieve the values of input fields from the Reservation form (Reservation\Index.cshtml file), compute prices, and display the input and total cost, as shown Figure 18-1, on a confirmation page (Reservation\Confirm.cshtml).

Confirmation							
Performance	**Date**	**Zone**	**Status**	**Tickets**	**Total Cost**		
Oshkosh Orchestral	11/29/2023	premium	Member	2	$70.00	Edit	Cancel
Confirm							

Figure 18-1: The Confirmation page

The *Edit* button lets the user go back to the Reservation view and edit the input data. The *Cancel* button lets the user go back to the Reservation view that displays the blank form, and the *Confirm* button saves the reservation to the database.

18.1.1 Components of the Web Project

Currently, the Reservation controller has one action method named Index. Since no http attribute (e.g., HttpGet or HttpPost) is specified for the `Index()` method, by default, this method is equivalent to an action method that has the HttpGet attribute specified, as shown here:

```
[HttpGet]
public ViewResult Index()
{
...
}
```

We will refer to this method as `HttpGet Index()`. The `HttpGet` attribute specifies that the method handles a GET request (`method="get"`). That is, this method is invoked when a GET request is sent to the URL to display the Reservation view (Reservation\Index.cshtml).

You will create a second `Index()` action method with an **HttpPost** attribute, which is invoked when a POST request is made by submitting the Reservation form. This method, which we will discuss shortly, will appear as follows:

```
[HttpPost]
public IActionResult Index(Reservation reservation)    {
...
}
```

We will refer to this method as the `HttpPost Index()` method. This method will get the input data using model binding and validate the data. If invalid, it will re-display the Reservation form. If valid, the method will store the data in session state and redirect the user to a third action method named `Confirm()`.

It should be noted that you cannot have two action methods with the same name, unless they accept different sets of arguments. This is true for any overloaded C# method. You can make them different by adding a dummy parameter to one method. This is not a problem in the current project because the HttpPost Index() method will have a parameter and the HttpGet Index() doesn't have any. Another option is to give a different name for this method. This option is examined in the section on validation.

Figure 18-2 shows the above-described components.

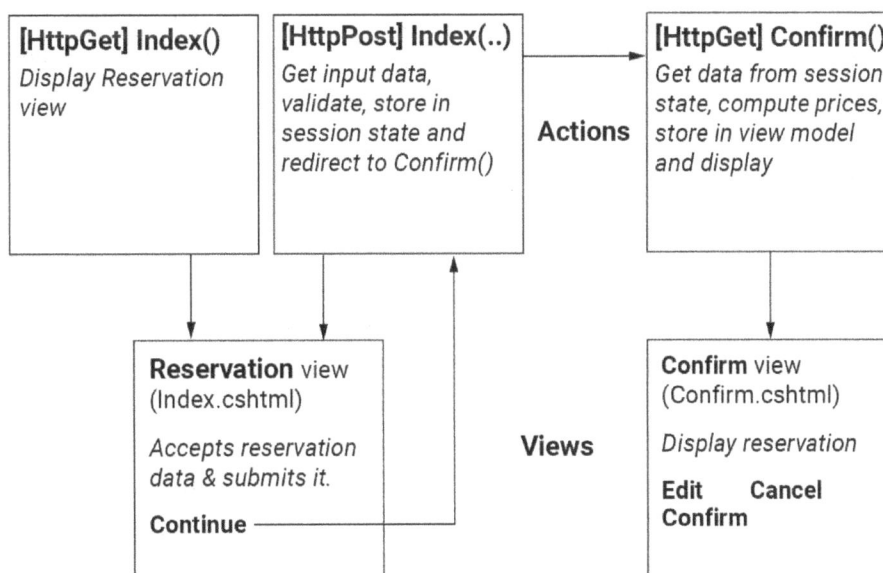

Figure 18-2: Components of the Theater app

The `Confirm()` method will get the data from the session state, computes prices, store the data in a view model class, and display the data on the Confirmation view (Reservation\Confirm.cshtml). The Confirmation view has buttons that let the user edit, cancel, or confirm (save) the reservation.

Thus there are two action methods for the Reservation page and one for the Confirmation page. All action methods are within the single controller, Reservation. There will be additional action methods for editing, canceling, and saving the reservation, which are discussed later.

In larger projects, typically there are multiple controllers. For example, all action methods that work with a certain database table may be grouped into one controller, or actions may be grouped together if they are related to a common task. You should be aware that it may not always be obvious which actions should be grouped into a controller.

Next, you will learn how the server-side code in the HttpPost Index() method can access the input data to validate and process it.

18.1.2 Model Binding in MVC

As you learned in Chapter 16 (Section 16.4.2), approaches to retrieve values of input fields from the HTTP request include

- using the Request.Form collection,
- model binding using parameters (parameter binding), and
- model binding using public properties.

You may review Section 16.4.2 (in Chapter 16) for a description of these approaches. These approaches work essentially the same way in MVC as in Razor Pages. The main difference is that in Razor Pages, the input values are accessed from the `OnPost()` or `OnGet()` methods in the `PageModel` class, whereas in MVC, they are accessed from the action methods like the `HttpPost Index()` in the controller class.

With Razor Pages, you used model binding using properties because of the convenience in accessing **properties** from the `PageModel` and the view. In this MVC version of the project, you will use **parameter binding** to help you learn this approach, which is popularly used in MVC.

Model binding is an exceptionally powerful feature. It is the process of retrieving data from various sources like input fields on a form, query string, or route data and providing the data to properties of a controller or parameters of action methods.

Let's look at the process of model binding using parameters in MVC. To access values of the Reservation form fields, you will bind them to a parameter of the `HttpPost Index()` action method. To bind multiple input values, the parameter must be of a complex type, like a class that has public properties whose names match the names of the input fields. You already created the `Reservation` class (to use in tag helpers) with such matching properties.

To access the input values, you specify a parameter of type `Reservation` in the `HttpPost Index()` action method, as shown here.

```
[HttpPost]
public IActionResult Index(Reservation reservation)    {
    . . .
}
```

When the `Index()` method is invoked by submitting the form, the input values are automatically collected from the HTTP request and used to populate the `Reservation` object. You may access an individual value, like the performance name, using the corresponding property of the reservation object, as follows:

```
string performance = reservation.Performance
```

As stated earlier, in order to invoke this Index() method when the Reservation form is submitted using the POST request, you need to add the HttpPost attribute immediately above the `Index()` method. Specifying the HTTP attribute is called **decorating** the action method with the attribute.

In addition to binding to posted values of form fields, model binding can be used to bind to route data and query string data. To bind a parameter, first MVC checks for matching posted values. If there is no matching value, it looks for matching route data. If there is none, it looks for query string values. The parameter will be bound to the first matching value.

Let's create the `HttpPost Index()` action method that displays the Reservation view, the `HttpGet Confirm()` method that displays the Confirm view and the Confirm view that displays the reservation data.

Step 1: Create a second action method, HttpPost Index(), within the Theater-MVC project.

Open the Theater-MVC project that you created in the WebProjects\MVC-1\ folder in Chapter 17. If you haven't created the project, your instructor has access to the completed project from Chapter 17 (Completed-Tutorials\Ch17-MVC-1\Theater-MVC).

Use the code in Figure 18-3 to add the `HttpPost Index()` action method within the Reservation controller. Again, this is different from the `HttpGet Index()` method you already created.

```
1   [HttpPost]
2   public IActionResult Index(Reservation reservation)
3   {
4       return View();
5   }
```

Figure 18-3: `HttpPost Index()` action method

Step 2: Create a third action method, Confirm(), to display the Confirm view.

Add the code in Figure 18-4 to create the `HttpGet Confirm()` action method within the Reservation controller to display the Confirm view.

```
1   [HttpGet]
2   public IActionResult Confirm()
3   {
4       return View();
5   }
```

Figure 18-4: `HttpGet Confirm()` action method

Step 3: Create a view named Confirm in the Views\Reservation folder.

Right-click *public IActionResult Confirm()*.

Select *Add View*. Choose *Razor View - Empty* and name it *Confirm*.

The Solution Explorer should show the file Confirm.cshtml within the Views\Reservation folder.

Add a link to the newly created Confirm view in the navigation bar as follows:

Open the Layout view (_Layout.cshtml) and add the following code:

```
<li class="nav-item">
    <a class="nav-link text-dark" asp-area="" asp-controller="Reservation"
    asp-action="Confirm">Confirmation</a>
</li>
```

We will refer to the Confirm view also as **Confirmation view** or **Confirmation page**.

Now you are ready to develop the code in the `HttpPost Index()` method to retrieve the input data, validate it, and share it with the `Confirm()` action, which computes the prices and displays the data in the Confirm view.

18.1.3 Processing Input Data and Displaying on the Confirmation Page

These are the steps needed to validate the input data, compute the total cost, and display the data in the Confirm view,

1. Add the necessary code within the `HttpPost Index()` method to do the following:
   a. Validate the data and return to the Reservation view, if any data item is invalid.
   b. Store the data in session state and redirect the user to the `Confirm()` action, if the data is valid.
2. Add methods to the `Reservation` class to compute prices.
3. Create a second class named `ReservationViewModel` to be used exclusively to pass the input data and computed prices from the `Confirm()` action method to the Confirm view. Such a class that is used to pass data to a view is called a *view model*, as explained shortly.
4. Develop the `Confirm()` action method that gets the data from the session state, computes the prices, stores the data in the view model, and passes the view model to the Confirm view.

Next, we will examine each step in more detail.

18.1.4 Data Validation

The `HttpPost Index()` method shown below has a parameter of type `Reservation`, which plays a key part in accessing the form data from the server.

```
[HttpPost]
public ViewResult Index(Reservation reservation)    {
    ...
}
```

As stated earlier, when this method is invoked by submitting the form, each property of the `Reservation` object is automatically populated with the corresponding input data item. In addition, if an input data item is not valid, MVC adds a message to the **ModelState** property of the controller and sets the **IsValid** property of Model-State to *false*. However, no validation message will be automatically displayed. You need to add elements to display the messages and add code within the `Index()` method to check the value of the `IsValid` property and redisplay the form if it is *false*.

Let's look at how to specify validation requirements and display validation messages.

Server-Side Validation

In Chapter 5, you learned that client-side validation can be done by specifying HTML attributes like `required`, `min`, `max`, and `step`. In Chapter 10, you used JavaScript to do client-side validation. Client-side validation is useful in giving feedback to users before a form is submitted, avoiding a round trip to the server. However, client-side validation cannot be fully trusted. For example, if the JavaScript is disabled on a client, the validation that uses JavaScript won't work. Further, it is easier to tamper with validation on the client. Server-side validation provides a more reliable method.

ASP.NET Core MVC provides a fairly easy and powerful way to do both server-side and client-side validation using **data attributes (data annotation attributes or validation attributes)** in the class that is bound to the input values.

Figure 18-5 shows examples of data attributes in the Reservation class that is bound to the input values.

```
1   public class Reservation
2   {
3       // Create a ReservationContext object
4       private ReservationContext resContext = new ReservationContext();
5       // Properties to hold reservation input data
6       public string Performance { get; set; }
7       [DataType(DataType.Date)]
8       public DateTime PerformDate { get; set; } = DateTime.Today;
9       [Required (ErrorMessage="The Zone field is required")]
10      public string Zone { get; set; }
11      public bool MemberStatus { get; set; }
12      [Range(1,10, ErrorMessage="Please enter a number between 1 and 10")]
13      public int Tickets { get; set; }
14  }
```

Figure 18-5: Reservation class with data attributes

Lines 9 and 12 show two different attributes that validate data. The optional *ErrorMessage* lets you override the default error message. Line 7 shows another attribute, discussed earlier, to specify the data type.

Required Attribute

Line 9 shows the **Required** attribute for the Zone property. This attribute is used to make sure that the value bound to a class property is not null. If the form is submitted with a null value for Zone (i.e., no zone is selected), MVC will do what it does with any invalid data: Add a validation message to the **ModelState** property of the controller and sets the **IsValid** property of ModelState to *false*. It is up to you to take action, if any, when the data is invalid.

Lines 4–8 in Figure 18-6 show how you can use the ModelState to check for any invalid value and to redisplay the reservation form, if any value is invalid. You also need to add the code to specify where validation messages should be displayed, as described shortly.

```
1   [HttpPost]
2   public IActionResult Index(Reservation reservation)
3   {
4       if (!ModelState.IsValid) // if input data is not valid
5       {
6           ViewBag.Performances = resContext.Performances; // store ReservationContext in ViewBag
7           return View(reservation);  // Re-display the Index view.
8       }
9       return RedirectToAction("Confirm", "Reservation");
10  }
```

Figure 18-6: HttpPost Index() method

Line 4 checks whether the IsValid property is false.

Line 6 stores the performance names from the ReservationContext object in the ViewBag to be used to display the performance names in the dropdown list. This is the same statement used in the HttpGet Index() method.

The *return* statement in line 7 redisplays the Reservation (Index) view. The View() method in line 7 passes the Reservation object to the view. However, it is not necessary to explicitly pass the Reservation object because the input values are bound to the Reservation object using tag helpers that will automatically redisplay the posted values. You could use just *return View()*.

Line 9 redirects the user to the Confirm page. This would only be executed if the ModelState is valid; otherwise, the return in line 7 would end the method execution.

The **RedirectToAction()** method causes MVC to process a different action method from within an action method. To redirect to a view, RedirectToActon() should specify the corresponding action method, and optionally, the controller name, route dictionary, and route values. To redirect to the Confirm view, line 9 specifies the Confirm()

action method and the Reservation controller. The `RedirectToAction()` method returns the view as a `Redirect-ToActionResult` object, which is a subtype of `ActionResult`.

Note that the `return` statement in line 7 provides an example of using an overload (a version) of the `View()` method that accepts the model name as an argument using the pattern

return View(*modelname*);

Property-Level Validation Message

You can display an individual validation message using the *Validation Message* tag helper, `asp-validation-for`, within a `<span>` element, shown as follows:

```
<span asp-validation-for="Zone" class="text-danger"></span><br /><br />
```

Setting `asp-validation-for` to `"Zone"` is the key part that tells MVC that this span element should display the validation message for the Zone field (radio button group), if any. The `"text-danger"` class causes the message to be displayed in red, as shown below on the far right:

Suite: $40   Premium: $25   Circle: $10   Balcony: $0   The Zone field is required.

The position of the `<span>` element determines the position of the message on the page.

Instead of displaying the default validation message, you may display a custom message using the `ErrorMessage` property, as follows:

```
[Required (ErrorMessage="Please select a zone")]
```

Step 4: Use the *Required* data attribute to make sure the user selects a zone.

Open the Reservation.cs file and add the Required attribute for `Zone` property, as shown in line 9, Figure 18-5.

Make sure the file includes the following directive:

```
using System.ComponentModel.DataAnnotations;
```

Step 5: Validate the input data.

Delete the `return` statement from the HttpPost Index() method in ReservationController.cs, and insert the code from Figure 18-6, lines 4–9, into the method.

Add a span element after the radio buttons in the Reservation view (Reservation\Index.cshtml file), as shown here, to display the validation message:

```
1    <label asp-for="Zone">Balcony: $0</label>
2    <span asp-validation-for="Zone" class="text-dange"></span> <br /><br />
```

To test the code, display the Reservation page and enter some data, but don't select a zone.

Click the **Continue** button. This results in submitting the reservation data to the `HttpPost Index()` method, which validates the data. Because the zone data is missing, the Reservation form will be re-displayed.

You should see the validation message next to the radio buttons.

Step 6: View the values of Reservation properties.

Put a break on the `return` statement in line 7 (click on the left margin) in the `HttpPost Index()` method in Figure 18-6.

Display the page and enter some data, but don't select a zone. Click the **Continue** button.

When the program breaks on line 7, display the Locals window (Debugg > Windows > Locals).

Expand the reservation object to view the properties and their values, including the *null* value for the Zone.

The data attribute, *Required*, applies only for properties that can have null values (nullable properties). Thus it can be used for properties like Zone and Performance, whose data type is *string*. It does not apply to numeric types like *int* and *decimal*, or the *datetime* type, which cannot have null values. Typically, this is not a problem, because for numeric types you want to check for a value of zero or values outside a range. If you want to apply a Required attribute to properties that cannot have null values (non-nullable properties like int and decimal), you can specify the type as nullable using the "?" symbol, as shown here for *int* type.

```
public int? Tickets { get; set; }
```

You should note that the Required attribute won't work for the dropdown list for performance. This is because even without the user selecting a performance, the field will have the value "prompt." As discussed later, you can use custom validation via code within the HttpPost Index() method to check the value of this field.

Range (min, max) Attribute

The Range (*min, max*) attribute is used to make sure a property value of numeric type is within the range specified by *min* and *max*.

Line 12 in Figure 18-5 specifies the Range attribute for the Tickets property, as shown here:

```
[Range(1,10)]
public int Tickets { get; set; }
```

You may display a custom message using the ErrorMessage property:

```
[Range(1,10, ErrorMessage="Please enter a number between 1 and 10")]
```

Step 7: Specify a range for the number of tickets.

Use the Range attribute to make sure that the number of tickets is between 1 and 10, and display a custom message if the number is outside the range. Use a <span> element to display the message. Run the project, enter a value that is outside the range for the number of tickets, and click the **Continue** button. You should see an error message.

Model-Level Validation Message

In addition to property-level messages or in place of them, you may display a single list of all validation messages or a single model-level message.

You can display a list of all validation messages using the *asp-validation-summary* tag helper within a <div> element, as shown below in line 1:

```
1  <div asp-validation-summary= "All" class="text-danger" ></div>
2  <form asp-action="Index" asp-controller="Reservation" method="post">
```

In this case, if a zone is not selected and the number of tickets is outside the specified range, a list of two validation messages will be displayed, as shown below, at the location where the <div> element is placed.

Reservation

- The Zone field is required.
- Please enter a number between 1 and 10

Performance: | TedxOshkosh $59.00 ▼ | Base Price:

If you display property-level messages, then also having a model-level list of those messages might look redundant. Instead, you may display a single model-level message by setting the asp-validation-summary to "Model-Only," as shown here:

```
<div asp-validation-summary= "ModelOnly"></div>
```

In this case, you need to specify a more general model-level message using code in the Index() method, as shown in Figure 18-7, line 6.

```
 1  public IActionResult Index(Reservation reservation)
 2  {
 3      if (!ModelState.IsValid) // if input data is not valid
 4      {
 5          ViewBag.Performances = resContext.Perfor-
mance; // store ReservationContext in ViewBag
 6          ModelState.AddModelError("", "Please fix the errors");
 7          return View(reservation);  // Re-display the Index view.  .
 8      }
 9      return RedirectToAction("Confirm", "Reservation");
10  }
```

Figure 18-7: Index() method with a model-level validation message

Line 6 adds the message to the ModelState using the **AddModelError()** method of ModelState, which has two parameters, as shown here:

AddModelError(*key*, *message*)

The parameter, *key*, represents the name of the property the message is associated with. If empty, the message will be added to ModelState as a model-level message.

When asp-validation-summary is set to "ModelOnly," the <div> element will display the single message:

Reservation

- Please fix the errors

Step 8: Display a single model-level message, if any value is invalid.

Add a <div> element in the Reservation view (index.cshtml) between the heading Reservation and the beginning tag of the <form> element, as shown here:

```
1  <h1>Reservation</h1>
2  <div asp-validation-summary="ModelOnly"></div>
3  <form asp-action="Index" asp-controller="Reservation" method="post"> <br />
```

Set the *asp-validation-summary* attribute of the div element to "ModelOnly."

Specify the model-level message in the HttpPost Index() method using the *ModelState.AddModelError()* method, as shown in line 6 in Figure 18-7.

Make sure you test the code.

MVC also adds error messages to the ModelState if an input value cannot be bound to the corresponding parameter. For example, trying to bind a text input value to an *int* type parameter will give a binding error. However, such errors are not likely to occur when you generate the input elements using tag helpers, which determine the type of the input element based on the data type of the corresponding class property.

Table 18-1 presents some additional data attributes.

Table 18-1: Data validation data attributes

Data annotation attribute	Description
Required	Makes sure that the property value is not null.
Range(min, max)	Makes sure that the property value of a numeric type is within the range specified by *min* and *max*.
StringLength(length) MaxLength(length)	Makes sure that the value of a string type property either doesn't exceed the specified length or it is null.
RegularExpression(pattern)	Make sure that the property value matches the specified pattern.
Display(Name = *string*)	Specifies how the name of the property should be displayed. Example: [Display (PerformDate = "Performance Date")]
DisplayFormat()	Specifies the format to display the property value. Example: [DisplayFormat(DataFormatString="{0:dd/MM/yy}")]
EmailAddress	Makes sure that the property value has an email format (text before and after an "@" symbol).
CreditCard	Makes sure that the property value has a credit card format.

Custom Server-Side Validation Using Code

There is no standard data attribute to do validations, like making sure that the performance date is not a past date. Such validations can be done using code or custom attributes. You can validate the PerformDate field using code within the HttpPost Index() method. To do this, you need to add a custom error message to the ModelState and associate it with the PerformDate property. Lines 4 and 5 in Figure 18-8 show the code.

```
1   [HttpPost]
2   public IActionResult Index(Reservation reservation)
3   {
4       if(reservation.PerformDate < DateTime.Today){    // if the date is a past date
5           ModelState.AddModelError("PerformDate", "Past dates are not valid"); // specify error message
6       }
7       if (!ModelState.IsValid) // if input data is not valid
8       {
9           ViewBag.Performances = resContext.Performance; // store ReservationContext in ViewBag
10          ModelState.AddModelError("", "Please fix the errors");
11          return View(reservation);  // Re-display the Index view
12      }
13      return RedirectToAction("Confirm", "Reservation");
14  }
```

Figure 18-8: Index() action with custom validation

Line 4 checks whether the date is a past date.

As you learned earlier, within the Index() method, you can get the value of any bound input field from the corresponding property of the Reservation object. When you use model binding, you have the benefit of using the IntelliSense feature to specify a certain property, like reservation.PerformDate.

Line 5 uses the AddModelError() to add an error message to the ModelState and also to associate the message with the PerformDate property by specifying *PerformDate* as the key. Recall that you used the AddModelError() method earlier to add a model-level message to the ModelState by providing the empty string as the key.

Step 9: Add the code to make sure the date selected by the user is not a past date.

Use the code from Figure 18-8 to display a property-level message if the date is invalid.

Add the using statement:

```
using System;
```

Make sure you test the code.

Currently, there is no validation to make sure that the user selects a performance. In Exercise 18-1, you will add the validation.

It's time to practice: Do Exercise 18-1 (see the end of the chapter).

Review Questions

1. Describe model binding in MVC.
2. Describe the benefits of server-side validation over client-side validation.

Client-Side Validation

Client-side validation, though not completely reliable, enhances user experience by providing immediate feedback without a roundtrip to the server. You can do client-side validation using three jQuery libraries that are added to your project when you use the MVC template: jQuery, jQuery validation, and jQuery unobtrusive validation.

You can view these libraries in the *lib* folder in the Solution Explorer. All you need to do to enable client-side validation is to add the following three <script> elements, in the order listed, to the head section of the _Layout. cshtml file:

```
1   <script src="~/lib/jquery/dist/jquery.min.js"></script>
2   <script src="~/lib/jquery-validation/dist/jquery.validate.min.js"></script>
3   <script src="~/lib/jquery-validation-unobtrusive/jquery.validate.unobtrusive.min.js"></script>
```

The jQuery Validate is a plug-in that makes it easy to do client-side validation without having to write JavaScript code. The jQuery Unobtrusive Validation helps you do client-side validation using the data attributes that you specified on the model, essentially copying the validation logic from the server to the client.

Client-side validation is performed by running JavaScript code when you submit the form. If there is invalid data, validation messages are displayed and the submission of the form is canceled. If not, the form is submitted. So, server-side validation won't take place until you fix all the errors identified by client-side validation. For example, if you enter a past date and don't select a zone, you will get a message for the zone when you submit the form, but the date won't be validated because it is done on the server.

Step 10: Add the <script> elements to enable client-side validation.

Use the code presented above to add the three script elements to the head section of the _Layout.cshtml file. You can use Intellisense to enter the paths, or you may copy the code from the file Ch17-jQuery-lib. txt in the Tutorial-starts\Data-files folder.

Step 11: Test the code.

Display the Reservation page and enter valid data for all fields, except that you enter an invalid number for tickets. Click the Continue button.

You will see a property-level message for tickets, but no model-level message is displayed. This is because model-level messages are added to the ModelState on the server.

Next, select a past date for the performance date, and enter an invalid number for tickets. Click the Continue button. You will get a property-level message for tickets, but there is no message for the date because the validation of the performance date is done on the server.

Enter a valid number for tickets and submit the form. Now you will get the property-level message for the date and the model-level message.

The user may find it illogical that not all error messages show up when doing client-side validation. To fix this, you may also create client-side JavaScript code that mirrors the custom server-side validations to make sure the user selects a performance from the dropdown list and that the performance date is not a past date.

Now that you have validated the input data, it can be shared with the Confirm() action method so that it can compute the prices and display the data on the Confirm view.

18.1.5 Action Method and View with Different Names

This section is a side note to discuss cases when the names of the action method and view are different. In the current tutorial, you have two action methods with the same name, Index: Http**Get** Index() and Http**Post** Index(). The HttpGet Index() method displays the Reservation view (Index.cshtml) and passes the Reservation object to the view using the statement:

```
return View(reservation);
```

The HttpPost Index() method also uses the same statement to display the Reservation view, if any data item is invalid. Note that the View() method doesn't specify the name of the view (Index) to be displayed because the view has the same name as the action method.

An action method may display a view with a name that is different from the name of the action. Instead of having two action methods with the same name, suppose you name the second action method Submit() instead of Index(), as follows:

```
[HttpPost]
public IActionResult Submit(Reservation reservation)
{
    ...
}
```

This method still needs to return the Index view to redisplay the reservation form if there is any invalid data. Thus the name of the view will be different from the name of the action.

If the Submit() method keeps the same return statement as the Index() method *(return View (reservation))*, you will get an error because MVC expects a view that has the same name, Submit, as the action method. But you don't have a view named Submit. When the view name is different from the action name, you need to specify the view name using the following pattern:

return View("*viewname*", *modelname*);

The second argument, *modelname*, is needed only if you want to pass a model to the view. The following statement within the Submit() action will display the Index view:

```
return View("Index", reservation);
```

Thus the same Index view can be used by HttpGet Index() and HttpPost Submit() methods.

In this case, you also will have to change the asp-action attribute of the <form> element to "Submit," as shown here, so that the form is submitted to the *Submit*() action:

```
<form asp-action="Submit" asp-controller="Reservation" method="post">
```

You will continue the tutorial without making the above changes. The name of the HttpPost Index() method will remain the same.

Next, we will look at how to pass the data to the Confirmation page.

18.1.6 Passing Data from the HttpPost Index() Action Method to the Confirm() Method

To display the reservation data on the Confirmation page, you will pass the reservation data from the HttpPost Index() method to the Confirm() method and, then, from the Confirm() method to the Confirm view. Two of the commonly used ways to pass data between action methods (within the same controller or different controllers) are **session state** and **TempData**.

Session State

Session state (HttpSessionState), as described in Chapter 16 (Section 16.5), lets you store data as key/value pairs on the server. Session state uses the **HttpContext.Session** property of the controller to store data that persists for an entire session, unless the app ends it, and can be accessed from any page.

By default, session state is not enabled in ASP.NET Core MVC. To enable session state, you need to add certain statements in the Startup.cs file, as shown in Figure 18-9. At the beginning of the ConfigureServices() method, add lines 3 and 4.

In addition, the *app.UseSession()* statement shown in line 15 should be added to the Configure() method above the app.UseEndpoints() statement.

```
1   public void ConfigureServices(IServiceCollection services)
2   {
3       services.AddDistributedMemoryCache();
4       services.AddSession();
5       services.AddControllersWithViews();
6   }
7
8   public void Configure(IApplicationBuilder app, IWebHostEnvironment env)
9   {
10      if (env.IsDevelopment())  {
11          ...
12      }
13      app.UseHttpsRedirection();
14      app.UseStaticFiles();
15      app.UseSession();
16          ...
17      app.UseEndpoints(endpoints =>  {
18          ...
19      });
20  }
21
```

Figure 18-9: Startup.cs file

Step 12: Enable the session state.

Open the Startup.cs file.

Add the code from lines 3 and 4 in Figure 18-9 to the ConfigureServices method.

Add the code from line 17 to the Configure method. Save and close the file.

Because session state doesn't let you directly store an object in it, you need to convert the Reservation object to a string by serializing it as follows:

```
string strReservation = JsonConvert.SerializeObject(reservation);
```

The following statement would store the serialized object (string) in a session state variable named StrReservation.

```
HttpContext.Session.SetString("StrReservation", strReservation);
```

Here, StrReservation is the key and strReservation is the value.

You can retrieve the string type data from the session state, de-serialize it, and store it in a Reservation object, as follows:

```
1   string strReservation = HttpContext.Session.GetString("StrReservation");
2   Reservation reservation = JsonConvert.DeserializeObject<Reservation>(strReservation);
```

Data stored in a session state variable can be accessed from an action method or a view. However, when you pass an object to a view using a session variable, you need to de-serialize the object, which is easier to do in the action method. From the action method, you can pass the de-serialized object to the view.

TempData

TempData, like session state and ViewBag, is a dictionary that stores data as key/value pairs, where the key is a string and the value is an object. It is a property of the controller. Whereas ViewBag is used to pass data between a controller and its view, TempData is used to pass data between action methods. You cannot use TempData to pass data to a view. Unlike the data stored in the session state, the data stored in TempData persists only until the data is read, unless you use the keep() method to keep the data longer. Data stored in ViewBag lasts only for the lifetime of the Http request, till the view is rendered.

Because the value stored in TempData is an object, you can directly store the Reservation object, as shown here.

```
TempData["ResData"] = reservation;
```

In this statement, ResData is the key, which could be any valid name.

You can get the Reservation object from TempData and cast it to Reservation type as follows:

```
1  if (TempData.Keys.Contains("ResData"))
2  Reservation reservation = (Reservation)TempData["ResData"];
```

So, should we use session state or TempData to share the Reservation object between controllers? Either can be used. Session state has broader application because it persists for an entire session, and you can pass data to an action method or a view. Later, you will need the reservation data in another action method named Edit that lets the user edit the data. Here you will use the session state that you are already familiar with from Chapter 16.

The code to serialize the Reservation object and store it in session state within the HttpPost Index() action is shown in lines 14 and 15 in Figure 18-10. The figure shows the complete code, including the code to validate data.

```
1   [HttpPost]
2   public IActionResult Index(Reservation reservation)
3   {
4       if(reservation.PerformDate < DateTime.Today){
5           ModelState.AddModelError("PerformDate", "Past dates are not valid");
6       }
7       if (!ModelState.IsValid) // if input data is not valid
8       {
9           ViewBag.Performances = resContext.Performanc-
es; // store ReservationContext in ViewBag
10          ModelState.AddModelError("", "Please fix the errors");
11          return View();  // Re-display the Index view.
12      }
13      // Serialize Reservation object and store in session state
14      string strReservation = JsonConvert.SerializeObject(reservation);
15      HttpContext.Session.SetString("StrReservation", strReservation);
16      return RedirectToAction("Confirm", "Reservation"); // Redirect to Confirm() action
17  }
```

Figure 18-10: The complete HttpPost Index() method.

Line 16 redirects MVC to the Confirm view. As described earlier, the **RedirectToAction()** method redirects to the Confirm view by specifying the Confirm() action method and the Reservation controller.

Step 13: Add code to the Index() action to store the Reservation object and redirect to the Confirm() action.

Add the code from lines 13–15 in Figure 18-10 to the HttpPost Index() method in ReservationController.cs.

You will also have to add the following directives to the ReservationController.cs file:

```
using Microsoft.AspNetCore.Http;
using Newtonsoft.Json;
```

18.1.7 Methods of Reservation Class to Compute Prices

Now you are ready to retrieve the data stored in the session state and compute prices in the Confirm() method. To compute prices, you will create methods in the Reservation class.

Figure 18-11 shows the modified Reservation class that includes four methods to compute prices. The Zone-Price(), Discount(), and TotalCost() methods should be easy to understand. The BasePrice() price method is similar to the BasePrice() method presented in Chapter 17, Section 17.4.3.

Line 6 creates an instance of the ReservationContext object that serves as an interface to the database.

Line 22 uses the Performance property of the ReservationContext object to get each record of the Performance table using a foreach loop. Lines 24 and 25 find the base price for the selected performance name by matching it with those in the table.

```
1   namespace Theater_MVC.Models
2   {
3       public class Reservation
4       {
5           // Create a ReservationContext object
6           private ReservationContext resContext = new ReservationContext();
7
8           // Properties to hold reservation input data
9           public string Performance { get; set; } = "";
10          [DataType(DataType.Date)]
11          public DateTime PerformDate { get; set; }
12           [Required (ErrorMessage="Please select a zone")]
13          public string Zone { get; set; }
14          public bool MemberStatus { get; set; }
15          [Range(1,10, ErrorMessage="Please enter a number between 1 and 10
16          public int Tickets { get; set; }
17
18          // Methods to compute prices
19          public decimal BasePrice()
20          {   // Look up base price
21              decimal basePrice = 0;
22              foreach (var performance in resContext.Performances)
23              {
24                  if (Performance == performance.PerformanceName)
25                      basePrice = (decimal)performance.BasePrice;
26              }
27              return basePrice;
28          }
29
30          public decimal ZonePrice()
31          {   // Compute zone price
32              decimal zonePrice = 0;
33              switch (Zone)
34              {
35                  case "suite":
36                      zonePrice = 40;
37                      break;
38                  case "premium":
39                      zonePrice = 25;
40                      break;
41                  case "circle":
42                      zonePrice = 10;
43                      break;
44                  case "balcony":
45                      zonePrice = 0;
46                      break;
```

Figure 18-11 Reservation class with the methods (*continues*)

```
47                  default:
48                      break;
49              }
50              return zonePrice;
51          }
52
53          public decimal Discount()
54          {   // Compute discount
55              decimal discount = 0;
56              if (MemberStatus)
57                  discount = 10;
58              else
59                  discount = 0;
60              return discount;
61          }
62
63          public decimal TotalCost()
64          {   // Compute total cost
65              return (BasePrice() + ZonePrice() - Discount()) * Tickets;
66          }
67      }
68  }
```

Figure 18-11 Reservation class with the methods (*continues*)

Step 14: Add methods to the Reservation class to compute prices.

Open the Reservation.cs file.

Add the code from lines 5 and 6 in Figure 18-11 to create the `ReservationContext` object.

Add the four methods shown in Figure 18-11.

You may copy the code from the file Ch18-Reservation-class-methods.txt in Tutorial-starts\Data-files folder.

Next, we will discuss how to pass the reservation data from the Confirm() method to the view.

18.1.8 Sharing Data between Action Method and View Using View Models

One way to pass the data from the Confirm() method to the Confirm view is to pass the Reservation object to the view. The view can then get the input data from the properties of the object and compute prices using the methods of the object.

However, it is recommended that the model class you use to pass data to a view should be a lightweight class that primarily contains data. It should not contain business logic (e.g., methods) that affects other parts of the project. Such a class that holds the data needed in the view is called a **view model**, whereas classes, like the Reservation class, that contain business logic are called the **domain model** or **business model**.

In addition to data, a view model may include formatting and validation of data specifically required for the view. For example, the MemberStatus property of the Reservation object is a Boolean that has true/false values. However, the Confirmation page needs to display the strings "Member" or "Non-member." You can provide a method in the view model class to do such a conversion, as described shortly.

If a view displays data from more than one model, you may combine the data from the two models into a single view model. By convention, view models that are kept within the project are stored within a folder named ViewModels. In larger projects, the view model may be kept outside the project as a service layer.

It is a common practice in MVC to attach the suffix "ViewModel" to the names of such view model classes. For example, use the name ReservationViewModel for the view model class that passes reservation data to the Confirmation view.

Figure 18-12 shows the view model that has properties to hold the input data and total cost for the current application.

```
1   namespace Theater_MVC.ViewModels
2   {
3       public class ReservationViewModel
4       {
5           public string Performance { get; set; }
6           public DateTime PerformDate { get; set; }
7           public string Zone { get; set; }
8           public bool MemberStatus { get; set; }
9           public int Tickets { get; set; }
10          public decimal TotalCost { get; set; }
11          public string Status(){
12              return MemberStatus ? "Member" : "Non-Member";
13          }
14      }
15  }
```

Figure 18-12: ReservationViewModel view model

The first five properties defined in lines 5–9 are the same as the properties in the Reservation class. These properties represent the input data. The last property (lines 10) will hold the total cost computed using the methods in the Reservation class. Lines 11–13 show the method, Status(), that returns the string "Member," if the Member-Status property has the value *true*, and returns "Non-member," if the value is *false*. You will use this method to display the member status.

Step 15: Create the ReservationViewModel view model.

Create a folder named ViewModels in the project folder, Theater-MVC.

Add a class named ReservationViewModel in the ViewModels folder.

Add the code shown in Figure 18-12 to the class. You may copy the code from the Ch18-ReservationViewModel-class.txt file in the WebProjects\Tutorial-starts\Data-files folder.

You will use this view model within the Confirm() method to pass all data items to the view.

18.1.9 Confirm() Action Method

Figure 18-13 shows the Confirm() action method, which gets the data from the session state, computes prices, stores the data and computed prices in the view model, and passes it to the view.

It's a good practice to avoid cluttering a controller with code representing business logic, like looking up base price and computing discount. Thus such computations are done using methods of the Reservation class.

```
1   [HttpGet]
2   public IActionResult Confirm()
3   {
4       // Create a ReservationViewModel
5       ReservationViewModel resViewModel = new ReservationViewModel();
6       // Get the serialized Reservation object from Session State, if it exists.
7       string strReservation = HttpContext.Session.GetString("StrReservation");
8       // Check whether the session state contains a Reservation object.
9       if (strReservation != null)
10      {
11          // De-serialize the Reservation object
12          Reservation reservation = new Reservation();
13          reservation = JsonConvert.DeserializeObject<Reservation>(strReservation);
14          // copy data from Reservation object to ReservationViewModel
15          resViewModel.Performance = reservation.Performance;
16          resViewModel.PerformDate = reservation.PerformDate;
17          resViewModel.Zone = reservation.Zone;
18          resViewModel.MemberStatus = reservation.MemberStatus;
19          resViewModel.Tickets = reservation.Tickets;
20          // Compute total cost and store in the view model object
21          resViewModel.TotalCost = reservation.TotalCost();
22      }
23      return View(resViewModel);  // pass the view model to Confirm view
24  }
```

Figure 18-13: `Confirm()` method of Reservation controller

Please see the internal comments for descriptions of each group of code.

Line 9 checks whether the session state contains the `Reservation` object, as indicated by a non-null value for `StrReservation`. Note that the session state won't have a `Reservation` object if the user displays the Confirmation view by clicking the Confirm button on the navigation bar. Clicking the Confirm button won't invoke the `HttpPost Index()` method where the session variable, `StrReservation`, is created. As you will see later, the session variable, `StrReservation`, will be removed when you cancel a reservation or save it.

Step 16: Add code to the Confirm() action method in the ReservationController.

Delete the `return` statement from the `Confirm()` method, and add the code shown in Figure 18-13 to the method. You may copy the code from the file Ch17-Confirm-method.txt in the Tutorial-starts\Data-files folder.

Add the following directives to the ReservationController.cs file:

```
using Theater_MVC.ViewModels;
```

18.1.10 Displaying Reservation Data in the Confirm View

The *Confirm* action method passes the view model (`ReservationViewModel` object) to the Confirm view using the following statement (Figure 18-13, line 23):

```
return View(ResViewModel);
```

To display this reservation data on the Confirm view, bind the Confirm view to the model using the following @model directive to specify the view model:

```
@model  ReservationViewModel;
```

Within the view file, use the @Model property to access the model object passed to the view. For example, to get the value of the `TotalCost` property, specify

```
@Model.TotalCost
```

Figure 18-14 shows the code for the Confirm view, which displays the data in a table row using the @Model property.

```
1   @{ViewData["Title"] = "Confirm";}
2   @using Theater_MVC.ViewModels;
3   @model ReservationViewModel;
4   <style>.output {position: fixed;left: 25em;}</style>
5
6   <h3>Confirmation</h3>
7   @if (@Model.Performance != null) // if view model contains reservation data
8   {
9       <table class="table table-bordered table-striped">
10          <thead>
11              <tr>
12                  <th>Performance</th>
13                  <th>Date</th>
14                  <th>Zone</th>
15                  <th>Status</th>
16                  <th>Tickets</th>
17                  <th>Total Cost</th>
18                  <th></th>
19                  <th></th>
20              </tr>
21          </thead>
22
23          <tbody>
24
25              <tr>
26                  <td>@Model.Performance</td>
27                  <td>@Model.PerformDate.ToShortDateString()</td>
28                  <td>@Model.Zone</td>
29                  <td>@Model.Status()</td>
30                  <td>@Model.Tickets</td>
31                  <td>@Model.TotalCost.ToString("C2")</td>
32                  <td><a asp-controller="Reservation" asp-action="Edit"
                            class="btn btn-primary"> Edit </a></td>
33                  <td><a asp-controller="Reservation" asp-action="Cancel"
                            class="btn btn-primary">Cancel</a></td>
34              </tr>
35          </tbody>
36      </table>
37      <br />
38      <!-- button to save the reservation -->
39      <a asp-controller="Reservation" asp-action="Save" class="btn btn-primary">Confirm</
    a>
40  }
41  else
42  {
43      <br /><h4> No current reservation </h4>
44  }
```

Figure 18-14: The Confirmation view

Line 7 checks whether the view model contains a value for the performance property of the view model. The performance property (and other properties) of the view model will have values only if the Confirm view is displayed by the Confirm() method, which stores the Reservation object in the session. Note that the Reservation object is removed from the session when the user saves or cancels a reservation. So, the properties won't have values if the Confirm view is displayed by other means, like clicking the Confirmation button on the navigation bar. In such cases, line 43 displays the message "No current reservation."

Lines 26–28 uses @Model property to get the performance, zone, and performance date from the corresponding properties of the view model. Similarly, lines 30 and 31 get the number of tickets and total cost.

Line 29, however, uses the Status() method of the view model class, rather than the MemberStatus property, to display "Member"/"Non-member" for member status.

Line 32 shows the *Edit* button that uses the `Edit()` action method, and line 33 shows the Cancel button that uses the `Cancel()` action method. Similarly, the Confirm button in line 39 uses the `Save()` action method. You will develop the `Edit()`, `Cancel()`, and `Save()` methods shortly.

The CSS specified in line 4 is specific to this page, and it overrules the styles specified in the _layout page and site.css file, which apply to all pages.

Step 17: Add the HTML code to the Confirmation view to display the reservation data.

Open Confirmation view (Confirm.cshtml) and add the code shown in Figure 18-14. You may copy the code from the file Ch17-Confirm-view.txt.

Add the following directives to the Confirm.cshtml file:

```
using Theater_MVC.ViewModels;
```

To test the code, display the Reservation page, enter the reservation data, and click the Continue button. The confirmation page should display the reservation data as shown in Figure 18-1.

Next, we will discuss the Edit, Cancel, and Confirm buttons, along with the corresponding action methods.

It's time to practice! Do Exercise 18-2 (see the end of the chapter).

Review Questions

3. What is a view model? How does it differ from domain/business models?

4. How does the `TempData` differ from `ViewBag`?

5. How does the session state differ from `TempData`?

18.2 Re-displaying Reservation Page and Restoring Field Values

The Edit, Cancel, and Confirm buttons on the Confirmation view provide links to go back to the reservation view. The Edit button also should restore the values of the input fields when the Reservation view is displayed. The Cancel and Confirm buttons should display the Reservation view in its original state with no input values displayed in the fields. This section discusses ways to go back to the Reservation view and how to pass data from the Confirmation view to the Reservation view.

18.2.1 Displaying a Page from Another Page

There are multiple ways to display one view from another view. A simple option is to use the JavaScript method *window.history.back()*. For example, you can create the following button on the Confirmation view to go back to the Reservation view:

```
<button type="button" onclick="history.back()">Back</button>
```

As discussed in Chapter 5, clicking this button has the same effect as clicking the browser's back button. Therefore, the Reservation form fields will retain all data previously entered/selected by the user, allowing the user to make changes to the data. However, this won't work if the user has the option to open other pages from the Confirmation page and come back to it. In such a scenario, `history.back()` will display whatever page was the previous page.

A second option, which is the one used in the Confirmation view, is to provide a link using the anchor element. To go to the Reservation page, we used an anchor element in the navigation bar (in _Layout.cshtml), as shown here.

```
<a class="nav-link text-dark" asp-area="" asp-controller="Reservation"
        asp-action="Index">Reservation</a>
```

The asp-controller and asp-action tag helpers in the anchor element specify the controller (Reservation) and action method (Index) that display the view.

Clicking the above link would display the Reservation form. However, the input data that you entered won't be displayed on the Reservation form when it is reloaded. Unlike a web form that uses server controls, HTML controls don't retain their values after a page reload.

Step 18: Test the Reservation button.

Display the Reservation page in the browser, enter reservation data, and click the Continue button to display the Confirmation page.

Click the Reservation button on the navigation bar. You should see the Reservation view in its original state with no input data displayed in it. Stop the project.

A third option to display the Reservation page is to use the button element that uses the `Url.Action` method to specify the url. For example, you can create the Reservation button as follows:

```
<button type="button" onclick="location.href='@Url.Action("Index", "Reservation")'">
                        Reservation</button>
```

In the above code, the location.href property is set to the URL of the page specified using the `Url.Action()` method. This works essentially the same as the anchor element.

The anchor and button elements shown above provide links directly to the Reservation page. In certain situations, it may be necessary to execute additional code before displaying another view. For example, to let the user edit a reservation, you need to retrieve the reservation data and pass it to the Reservation view. To execute the code for such actions, you can create an action method that will do the tasks and display the page.

18.2.2 The Edit() Action Method to Restore Field Values

Let's look at the `Edit()` action method that is invoked when the user clicks the *Edit* button. This method needs to retrieve the reservation data from the session state, pass it to the Reservation view, and display the view. Recall that the `HttpPost Index()` method stores the `Reservation` object, which contains the input data, in the session state. The `Edit()` method needs to get the Reservation object from the session state, and pass it to the Reservation view. Figure 18-15 shows the `Edit()` method.

```
 1  public ViewResult Edit()
 2  {   // Get Reservation object from sesssion state and pass it to Reservation view
 3      Reservation reservation = new Reservation();  // Create Reservation object
 4      //Get Serialized Reservation object from session state and de-serialize it
 5      string strReservation = HttpContext.Session.GetString("StrReservation");
 6      reservation = JsonConvert.DeserializeObject<Reservation>(strReservation);
 7      // Store data on performances in ViewBag
 8      ViewBag.Performances = resContext.Performances;
 9      return View("Index", reservation);//Pass Reservation object to Reservation view
10  }
```

Figure 18-15: The `Edit()` method

Lines 5 retrieves the serialized `Reservation` object from the session state, and line 6 de-serializes it—the same way you did it in the `Confirm()` method.

Line 8 gets the performance records from the Performance property of the `ReservationContext` object and stores them in the `ViewBag`. You will use the `ViewBag` to display the performance names in the dropdown list.

Line 9 uses the `View()` method to pass the Reservation object and to display the Reservation view (Index. cshtml). Because the elements on the Reservation form are bound to the properties of the `Reservation` object, the form will automatically display the reservation information from the object. You don't have to use the `@Model` property as you did in the Confirmation view, where the output fields are not bound to the Reservation class.

Note that if you display the Reservation view without passing the `Reservation` object (e.g., `return View("Index")`), the input data won't be displayed in the view.

You may wonder why the reservation data is not passed to the view using a view model class, like it is passed to the Confirmation view. Ideally, you should pass data to views using view models rather than the models that are used by the rest of the application. However, in the current situation, the input fields in the Reservation view already are bound to the Reservation model, and therefore they will automatically display the reservation data from the object without having to display the data from a view model using the @Model property.

Step 19: Create the Edit() action method.

Use the code presented above to create the `Edit()` action method within the `ReservationController` class.

Test the code: Enter reservation data, display the Confirmation view, and click the **Edit** button.

The Reservation page should show all input data, except the base price in the `<output>` element, which is not bound to the `Reservation` object. Restoring the `<output>` element requires additional code to store it in session state and displaying it from the session state. Because displaying base price is not an essential part of the Reservation form, you will not add the code to restore the base price.

Change an input data item, and display the Confirmation view. The Confirmation view should show the updated data.

View() vs. RedirectToAction()

Typically, to display a view, you invoke the action method that has the same name as the view. For example, you can display the Reservation view (Index.cshtml) by invoking the `HttpGet Index()` action method shown in Figure 18-16.

```
1  [HttpGet]
2  public ViewResult Index()
3  {
4      // Get Performance records from the Performance property of resCotext.
5      ViewBag.Performances= resContext.Performance;
6      Reservation reservation = new Reservation();  // Create Reservation object.
7      return View(reservation); // Return Reservation view with Reservation object.
8  }
```

Figure 18-16: `HttpGet Index()` action method

You can invoke the above `Index()` method from any other action method using *RedirectToAction()* method, as follows, to display the Reservation view.

```
return RedirectToAction("Reservation", "Index");
```

If you use this method to display the Reservation view from the `Edit()` action method, the view will be displayed without showing the input data for editing because the Reservation object passed from the above `Index()` method to the view doesn't contain any data.

Instead, the `Edit()` method uses the *View("Index", reservation)* method in line 9 of Figure 18-15 to display the Reservation view (Index.cshtml) directly (without invoking the `HttpGet Index()` action method). This is represented in Figure 18-17 by the arrow from the `Edit()` method to the Reservation view. The *View("Index", reservation)* method also passes the Reservation object to the Reservation view.

The next section presents the `Cancel` and `Save()` action methods that use the `RedirectToAction()` method to invoke the `HttpGet Index()` method to display the Reservation view in its original state with blank fields.

The `Edit()`, `Cancel()`, and `Save()` action methods, and their relationship to other action methods and views, are shown in Figure 18-17. Note that the `Edit()` action directly displays the Reservation view, whereas the `Cancel()` and `Save()` methods invoke the `HttpGet Index()` action methods to display the Reservation view.

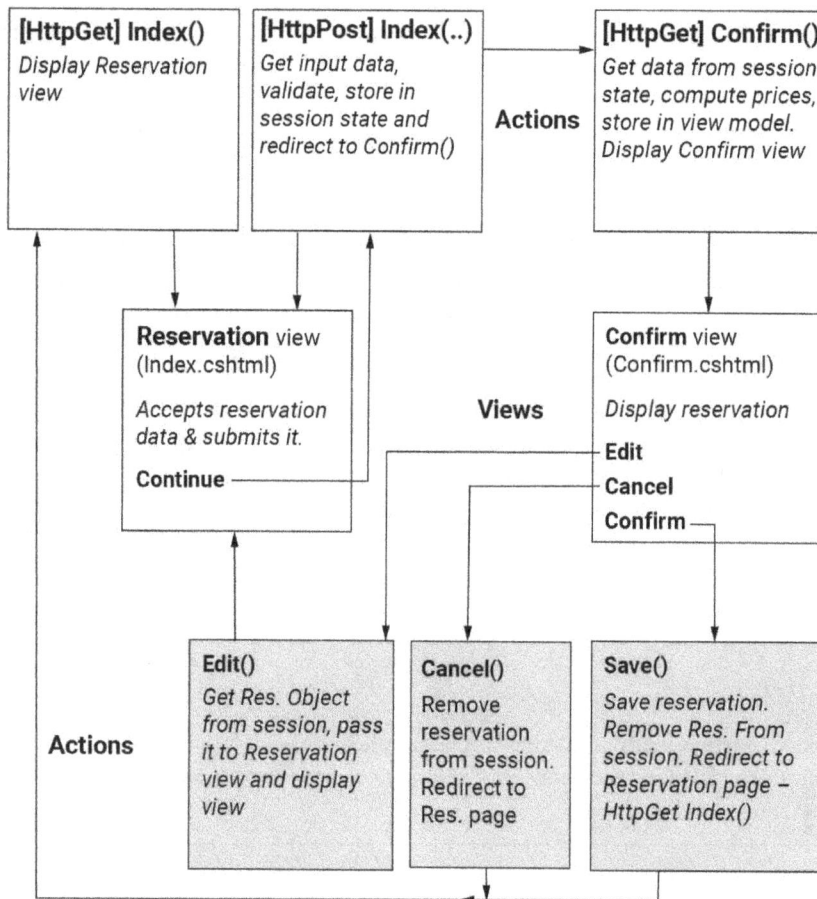

Figure 18-17: Action methods and views

18.2.3 The Cancel() Action Method

As shown in Figure 18-17, the `Cancel()` action method is invoked when the user clicks the Cancel button. This method is shown in Figure 18-18. First, it removes the reservation object from the session state so that re-displaying the Confirmation page using the navigation bar won't display the canceled reservation. Then, it displays the Reservation view in its original state with no input data displayed.

```
1   public IActionResult Cancel()
2   {
3       // Remove current reservation from session state
4       HttpContext.Session.Remove("StrReservation");
5       return RedirectToAction("Index");
6   }
```

Figure 18-18: The `Cancel()` method

Step 20: Create the Cancel() action method.

Use the code shown above to create the `Cancel()` action method.

To test the code, display the Reservation view, enter the reservation data, and click the Continue button to display the Confirmation view.

It should display the reservation data.

Click the Cancel button to return to the Reservation page.

Click the Confirmation button on the navigation bar.

The Confirmation view should be displayed with the message, "No current reservation."

Review Questions

6. What is the drawback of using a button that invokes the `history.back()` JavaScript function to go back to page, like going back to the Reservation page from the Confirmation page?

7. What is the difference between displaying a page using the `View()` method and `RedirectToAction()` method?

18.3 Saving a Reservation to the Database

The Confirmation view has a *Confirm* button to let the user save the reservation. You will create an action method named *Save* to save the data to a database table named *Reservation* that you will create next.

18.3.1 Code-First Approach to Generate the Reservation Table

Rather than manually creating the Reservation table, you will generate the Reservation table using the existing `Reservation` class and `DbContext` class. As you learned earlier, the method of generating a database and tables by first coding a `DbContext` class and model classes corresponding to the tables is called the Code First approach. In the current application, you didn't have to code the `DbContext` class. You were able to use the Database First approach to generate the `DbContext` class from the existing database (Section 17.3). Now you will use the `DbContext` class and the `Reservation` class to create the new table, Reservation, within the Theater.mdb database.

The `Reservation` class needs to be modified to include a property named `ReservationId` that will generate the primary key in the database table, and a second property named `CustomerId` to relate the reservation to a customer. Lines 6 and 7 in Figure 18-19 show the properties that will be added to the `Reservation` class.

```
1   public class Reservation
2   {
3       // Create a ReservationContext object
4       private ReservationContext resContext = new ReservationContext();
5       // Properties to hold reservation input data
6       public int ReservationId { get; set; }
7       public string CustomerId { get; set; }
8       public string Performance { get; set; }
9       [ ... ]
```

Figure 18-19: `ReservationId` and `CustomerId` properties of `Reservation` class

By convention, a property named *Id* or the class name followed by *Id* (e.g., `ReservationId`) is treated as the primary key of the table that is created from the class.

Step 21: Add the ReservationId and CustomerId properties to the Reservation class.

Open the Reservation.cs file and add the `ReservationId` and `CustomerId` properties as shown in lines 6 and 7 in Figure 18-19.

Save and close the file.

To generate a table using a class, EF Core requires the `DbContext` class to have a corresponding property of the `DbSet<Entity>` type. Thus, to generate the Reservation table using the `Reservation` class, you need to add a `Reservations` property of the type `DbSet<Reservation>` to the `DbContext` class. Here `<Reservation>` specifies the name of the entity class. Information from the properties of the `Reservation` class will be used to generate the Reservation table. As you will see soon, the `DbSet` class will also serve as the interface to add records to the `Reservation` table, as well as update and remove Reservation records.

The ReservationContext class already has two properties of the DbSet<Entity> type: Performances and Zones as shown in lines 8 and 9 in Figure 18-20. You will add the Reservations property, as shown in line 10.

```
1   public partial class ReservationContext : DbContext
2   {
3       public ReservationContext()  {
4       }
5       public ReservationContext(DbContextOptions<ReservationContext> options)
6           : base(options)  {
7       }
8       public virtual DbSet<Performance> Performances { get; set; }
9       public virtual DbSet<Zone> Zones { get; set; }
10      public virtual DbSet<Reservation> Reservations { get; set; }
```

Figure 18-20: ReservationContext class with Reservations property

Step 22: Add the property Reservations to the ReservationContext class.

Open the ReservationContext.cs file and add the Reservation property, as shown in line 10.

Save and close the file.

The Migration File

To create tables within a database, you will generate migration files that contain the necessary C# code to create the tables using the properties of the DbContext class and the model classes that specify the database fields. The migration files are easily created by executing the Add-Migration command from the Package Manager Console window.

Next, you will generate a *migration file* named *Initial* that contains the C# code. To generate this migration file, execute the Add-Migration command in the Package Manager Console window, as shown here.

```
PM> Add-Migration Initial
```

Step 23: Create the migration file named Initial.

Open the Package Manager Console:

> Select *Tools* > *NuGet Package Manager* > *Package Manager Console*.

The Package Manager Console opens with the prompt *PM>*.

Type the command, ***Add-Migration Initial***, and press the Enter key to execute it.

Executing the *Add-Migration Initial* command will create a folder named Migrations and the two classes within the folder, as shown below:

▲ ⬛ Migrations
    ▷ C# 20200626033225_Initial.cs
    ▷ C# ReservationContextModelSnapshot.cs

The file, Initial.cs, will be opened automatically. Figure 18-21 shows relevant parts of the *Initial* class that includes two methods, Up() and Down(). The Up() method contains code to generate the tables and the Down() method includes code to delete tables.

Because the ReservationContext class has the properties, Performances, Reservations, and Zones, the migration file has the code to create all three tables. The figure shows parts of the code to create the Performance table (lines 5–8) and the Zone table (lines 28–31). You already have the Performance and Zone tables. Therefore, you will comment out the two code segments that create these two tables.

Step 24: Comment out the code that creates the Performance and Zone tables.

Comment out the two CreateTable() methods shown in lines 5–8 and 28–31 in the _Initial.cs file, shown in Figure 18-21.

```
1   public partial class Initial : Migration
2   {
3       protected override void Up(MigrationBuilder migrationBuilder)
4       {
5           migrationBuilder.CreateTable(
6               name: "Performance",
7               ...
8               );
9
10          migrationBuilder.CreateTable(
11              name: "Reservations",
12              columns: table => new
13              {
14                  ReservationId = table.Column<int>(nullable: false)
15                      .Annotation("SqlServer:Identity", "1, 1"),
16                  CustomerId = table.Column<string>(nullable: true),
17                  Performance = table.Column<string>(nullable: true),
18                  PerformDate = table.Column<DateTime>(nullable: false),
19                  Zone = table.Column<string>(nullable: false),
20                  MemberStatus = table.Column<bool>(nullable: false),
21                  Tickets = table.Column<int>(nullable: false)
22              },
23              constraints: table =>
24              {
25                  table.PrimaryKey("PK_Reservation", x => x.ReservationId);
26              });
27
28          migrationBuilder.CreateTable(
29              name: "Zone",
30              ...
31              );
32      }
33
34      protected override void Down(MigrationBuilder migrationBuilder)
35      {
36          migrationBuilder.DropTable(
37              name: "Performance");
38          migrationBuilder.DropTable(
39              name: "Reservations");
40          migrationBuilder.DropTable(
41              name: "Zone");
42      }
43  }
```

Figure 18-21: The Initial.cs class

To create the Reservation table, you will execute the *Update-Database* command at the PM> prompt.

Step 25: Generate the Reservation table.

Enter the command, *Update-Database*, as shown here, and press the Enter key:

PM>Update-Database

Open the database and make sure the database contains the Reservation table.

To open the database, right-click *Data\Theater.mdf* in the Solutions Explorer, and select *Open*.

Expand the *Data Connnections, Theater.mdf*, and *Tables* nodes to view the list of tables.

It should be noted that the name of the newly generated table is *Reservations*, not *Reservation*, though the commonly used convention is to use singular names for database tables. This is because EF Core 2 uses the name of the DbSet property in the DbContext for the table. You can generate a table that has the name *Reservation* by using the name *Reservation* for the DbSet property. However, then you will be violating the convention of using plurals for DbSet property names. So, we will leave the table name as *Reservations*.

18.3.2 The Save() Action Method

To save data from an object into a database table, first, you will add the object to the DbSet collection. This is done using the Add() method of the DbSet<Entity> type property of DbContext, as follows:

DbContext.Property.Add(*object*);

To save the reservation data from the *reservation* object to the Reservation table, you will use the Add() method of the Reservation property of resContext, as shown here:

resContext.Reservations.Add(reservation);

After you add the object to the DbSet collection, save the data into the database table, using the SaveChanges() method of DbContext, as follows:

resContext.SaveChanges()

Figure 18-22 shows the Save() action method.

```
1   public IActionResult Save()
2   {
3       Reservation reservation = new Reservation();  // Create Reservation object
4       //Get Serialized Reservation object from the session state
5       string strReservation = HttpContext.Session.GetString("StrReservation");
6       // de-serialize the Reservation object
7       reservation = JsonConvert.DeserializeObject<Reservation>(strReservation);
8       // Add reservation to the Reservation table
9       resContext.Reservations.Add(reservation);
10      resContext.SaveChanges();
11      // Remove Reservation object form session state
12      HttpContext.Session.Remove("StrReservation");
13      return RedirectToAction("Index"); // Display Reservation view
14  }
```

Figure 18-22: The Save() action method

The comments within the method describe the code.

Note that the RedirectToAction() method in line 13 invokes the Index() action method to display the Reservation view in its original state with no data displayed in it.

Step 26: Create the Save() action method.

Use the code from Figure 18-22 to create the Save() action method.

Step 27: Test the code.

To test the code, display the Reservation view, enter reservation data, and click the Continue button to display the Confirmation view.

Click the Confirm button to save the data to the Reservation table and return to the blank Reservation form.

Make sure the Reservation table contains a new record with the reservation data that you entered.

To open the database, right-click Data\Theater.mdf in the Solutions Explorer, and select **Open.**

Expand the **Data Connnections, Theater.mdf,** and **Tables** nodes to display the list of tables.

Right-click the Reservation table and select **Show Table Data** to view the newly added record.

It's time to practice! Do Exercise 18-3 (see the end of the chapter).

18.4 Working with Multiple Reservations

The previous sections discussed how to create, edit, and save a single reservation at a time. Typical web applications would let users work with data for multiple items at a time. For example, an e-commerce site typically lets the user place an order that includes multiple products. In such cases, it would be convenient to work with a list consisting of multiple objects, with each object representing a different item.

You can use this approach in the current application to create a list of reservations that a user makes and share the entire list between action methods and views. You can let the user select a reservation from the list for editing or removal from the list. Because the reservation view model contains the input data and the total cost, you will create the list using the view model. You will use the list of view models to present the reservation information on the Confirmation page, as shown in Figure 18-23.

The *Edit* button allows the user to edit the selected reservation, and the *Remove* button removes it from the list. The *Add New* button lets the user add a new reservation to the list, and the *Confirm* button saves all reservations in the list to the database table.

Updating the project to let users work with multiple reservations requires changes in the existing methods and adding new ones. To preserve the methods in their current state for your future reference, you will copy the Theater-MVC project into a new folder named **MVC-2** and make changes in the copy.

Before discussing the code to add, edit, remove, and save reservations, we will examine how to create the list and use it to display multiple reservations on the Confirm view.

Figure 18-23: List of reservations on the Confirmation view

18.4.1 Creating and Maintaining a List of ReservationViewModel Objects

The reservation information is passed to the Confirm view from the `Confirm()` action method using the `ReservationViewModel` object. To store multiple view model objects, you will create the list

```
List<ReservationViewModel>
```

Each reservation made by a user will be added to the list until the list is saved in the session state.
Lines 9–49 in Figure 18-24 show the code needed in the `Confirm()` method to store the reservations in a list.

```
1   [HttpGet]
2   public IActionResult Confirm()
3   {
4       // Get the List of view model objects from session state, if it exists
5       List<ReservationViewModel> resViewModels = new List<ReservationViewModel>();
6       string strResViewModels = HttpContext.Session.GetString("StrResViewModels");
7       if (strResViewModels != null) // check whether the session variable exists
8       {   // de-serialize List of objects
9           resViewModels = JsonConvert.DeserializeObject<List<ReservationViewModel>>
                            (strResViewModels);
10      }
11      // Get serialized Reservation object from session state, if it exists.
12      Reservation reservation = new Reservation();
13      string strReservation = HttpContext.Session.GetString("StrReservation");
14
15      // Add the view model to the list, if there is a current reservation
           (i.e., if a reservation is being added or edited; not removed or opened
           from nav bar)
16      if (strReservation != null)
17      {
18          // de-serialize Reservation object
19          reservation = JsonConvert.DeserializeObject<Reservation>(strReservation);
20          // Create a ReservationViewModel object using the Reservation object
21          ReservationViewModel resViewModel = new ReservationViewModel();
22          // Copy input data from Reservation object to ReservationViewModel object
23          resViewModel.Performance = reservation.Performance;
24          resViewModel.PerformDate = reservation.PerformDate;
25          resViewModel.Zone = reservation.Zone;
26          resViewModel.MemberStatus = reservation.MemberStatus;
27          resViewModel.Tickets = reservation.Tickets;
28          // Compute total cost using TotalCost() methods and store in view model.
29          resViewModel.TotalCost = reservation.TotalCost();
30
31          // If it is an edit, remove the old view model object from the list
32          if (HttpContext.Session.GetString("Action") == "Edit")
33          {
34              // Get the index of the reservation being edited
35              int resIndex = (int)HttpContext.Session.GetInt32("ResIndex");
36              // Remove the old ViewModel
37              resViewModels.RemoveAt(resIndex);
38              // Set session state variable, Action, to empty string.
39              HttpContext.Session.SetString("Action", "");
40          }
41          // Add the view model object to the list of view models
42          resViewModels.Add(resViewModel);
43
44          // Serialize the list and store it in session state
45          strResViewModels = JsonConvert.SerializeObject(resViewModels);
46          HttpContext.Session.SetString("StrResViewModels", strResViewModels);
47          // Remove the current reservation from session state
48          HttpContext.Session.Remove("StrReservation");
49      }
50      // Pass the list to the Confirm view
51      return View(resViewModels);
52  }
```

Figure 18-24: The `Confirm()` method with the list

Line 5 creates an object of the type, `List<ReservationViewModels>`, which is a list of view models.

Line 6 gets the serialized list of view models from the session variable, *StrResViewModels*, if it exists. The variable won't exist if it is the first reservation (the session variable is created in line 45). As you learned earlier, the session state cannot store complex objects. So, the `List<>` object is serialized and stored as a string.

Lines 7–10 check whether the list exists, indicated by a non-null value for `StrResViewModels`, and de-serialize the list if it exists.

Line 13 gets the serialized `Reservation` object from the session variable, `StrReservation`, if the object exists.

Lines 16–48 create a view model, store the reservation data in it, and add it to the list, if the `Reservation` object exists. Note that the session state won't contain a `Reservation` object if the user displays the Confirmation view by clicking the Confirm button on the navigation bar. Clicking the Confirm button won't invoke the `HttpPost Index()` method where the session variable, `StrReservation`, is created. Similarly, if the Confirmation view is re-displayed after removing an item from the list, the session variable, `StrReservation`, won't exist because the `HttpPost Index()` method is not invoked.

Lines 19–27 copy the reservation data from the `Reservation` object into the `ReservationViewModel` object. Line 29 computes the total cost and stores it in the `TotalCost` property of the view model.

Lines 32–40 delete the old view model, if the current view model is an edit.

Line 42 adds the current view model to the list.

Lines 45 and 46 store the updated list in the session state, and line 48 removes the session variable, `StrReservation`.

Step 28: Make a copy of Theater-MVC project in a new folder named MVC-2.

Create a folder named MVC-2 in the WebProjets folder.

Copy the Theater-MVC folder from MVC-1 into the folder MVC-2.

Note that the name of the project, Theater-MVC, remains the same. You are creating a project with the same name in a different folder, MVC-2.

Before you make changes in the project to work with multiple reservations, you need to update the location of the database from WebProjects\MVC-1 to WebProjects\MVC-2 in the ReservationContext.cs file and appsettings.json file.

Step 29: Update the database location.

Open the Theater-MVC project from the MVC-2 folder.

Change WebProjects\MVC-1 to WebProjects\MVC-2 in the *OnConfiguring()* method in the ReservationContext.cs file.

Open the appsettings.json file in the SolutionExplorer and change WebProjets\\MVC-1 to WebProjects\\MVC-2 in the connection string.

Step 30: Delete the action methods that require significant changes in the updated version of the project.

Open the ReservationController.cs file.

Delete the following action methods completely, including the headers: **Cancel()**, **Confirm()**, **Edit()**, and **Save()**.

Step 31: Add the Confirm() method that adds the reservation to the list.

Use the code from Figure 18-24 to create the `Confirm()` method. You may copy the code from Tutorial-starts\Data-files\Ch18-Confirm-method.txt.

Add this using statement:

```
using System.Collections.Generic;
```

18.4.2 Displaying Items from a List

The previous version of the Confirm view (in the Theater-MVC project in the MVC-1 folder) displays the data from a single `ReservationViewModel` object. The new version (in the Theater-MVC project in the MVC-2 folder) should display all reservations from the list you created in the previous section.

The key differences of this version are:

1. The @model directive specifies the list, instead of ReservationViewModel class, as follows:

```
@model List<ReservationViewModel
```

2. Uses a *foreach* loop to get each view model object from the list and displays the data items from the view model in a row of the table.

3. Editing, removing, adding, and saving of reservations involves working with a list of reservations rather than a single reservation. Note that when adding new reservations, we don't check whether any of them already exists in the table.

4. As shown in Figure 18-23, the user interface includes a *Remove* button, in place of the Cancel button, to remove a reservation from the list, and an *Add New* button to add a new reservation to the list.

Figure 18-25 shows the Confirm.cshtml file for the current version.

```
1   @{ViewData["Title"] = "Confirm";}
2   @using Theater_MVC.ViewModels;
3   @model List<ReservationViewModel>;
4   <style>.output {position: fixed; left: 25em;}</style>
5
6   <h3>Confirmation</h3>
7   <table class="table table-bordered table-striped">
8       <thead>
9           <tr>
10              <th>Performance</th>
11              <th>Date</th>
12              <th>Zone</th>
13              <th>Status</th>
14              <th>Tickets</th>
15              <th>Total Cost</th>
16              <th></th>
17              <th></th>
18          </tr>
19      </thead>
20      <tbody>
21          @{int resIndex = 0;}
22          @foreach (var reservation in @Model)
23          {
24              <tr>
25                  <td>@reservation.Performance</td>
26                  <td>@reservation.PerformDate.ToShortDateString()</td>
27                  <td>@reservation.Zone</td>
28                  <td>@reservation.Status()</td>
29                  <td>@reservation.Tickets</td>
30                  <td>@reservation.TotalCost.ToString("C2")</td>
31
32                  <td><a asp-controller="Reservation" asp-action="Edit"
                          asp-route-index="@resIndex"
                          class="btn btn-primary"> Edit </a></td>
33                  <td><a asp-controller="Reservation" asp-action="Remove"
                          asp-route-index="@resIndex"
                          class="btn btn-primary">Remove</a></td>
34              </tr>
35              resIndex = resIndex + 1;
36          }
37      </tbody>
38  </table><br />
39  <!-- buttons to add and save reservation -->
40  <a asp-controller="Reservation" asp-action="Index" class="btn btn-primary">Add New</a>
41  <a asp-controller="Reservation" asp-action="Save" class="btn btn-primary">Confirm</a>
```

Figure 18-25: Confirmation view that displays multiple reservations

Line 3 specifies that the model is a list.

Lines 21–36 use a `foreach` loop to iterate over each reservation in the list and display it in a row of the table.

Line 21 initializes and line 35 increments the variable, `resIndex`, that represents the index of the reservation displayed in the current line.

Lines 32 and 33 shows the Edit and Remove buttons that invoke the action methods, `Edit()` and `Remove()`, to be developed. These buttons also pass the value of `resIndex` to the `Edit()` and `Remove()` action methods using the *asp-route* tag helper, as follows:

```
asp-route-index="@resIndex"
```

This index will allow you to pass the position in the List of a selected `Reservation` so the correct reservation is either edited or removed. The tag helper is explained in more detail, in the next section.

Line 40 shows the *Add New* button that invokes the action method `HttpGet Index()` to display the Reservation view. Line 41 shows the Confirm button that invokes the `Save()` action method, to be developed.

The asp-route-parametername Tag Helper

You can pass a data item from a view to an action method by specifying a parameter name in the `asp-route` tag helper in an anchor element using the pattern

asp-route-*routeparametername*

Here is an example from the code for the Edit button:

```
<a asp-controller="Reservation" asp-action="Edit" asp-route-index="@resIndex"> Edit </a>
```

Here, `index` is the name of the route parameter. If there is a web page with the url /Reservation/Edit/index, the above anchor will create a link to that page. If not, the value of index (`resIndex`) will be passed to the `Edit()` action method, as in the current project that doesn't have a web page with the URL, /Reservation/Edit/index. You may pass multiple data items using multiple asp-route tag helpers.

To get the value of the route parameter, `index`, you specify index as a parameter in the `Edit()` method, as follows:

```
Edit(int index)
```

You will create the `Edit()` method shortly.

The Add New Button

The Add New button shown in line 40 provides a link to the Reservation page by invoking the `HttpGet Index()` method. Clicking this button has the same effect as clicking the Reservation button on the navigation bar, which also invokes the `HttpGet Index()` method.

Step 32: Modify the Confirm view (Confirm.cshtml) to display the reservations.

Open the Confirm.cshtml file and modify the existing code as shown in Figure 18-25.

Step 33: Test the code in the Confirm() action method and Confirm view.

Display the Reservation form, enter reservation data, and click *Continue*.

The reservation data should be displayed in the Confirmation view.

Click the *Add New* button to display the Reservation form in its original state.

Enter data for another reservation and click *Continue*.

You should see both reservations listed on the Confirmation view.

Click the Reservation button on the navigation bar to display the Reservation form in its original state.

Click the Confirmation button on the navigation bar.

You should see the Confirmation view with the two reservations listed in it.

Next, you will develop the code to edit, remove, and save reservations.

18.5 Action Methods to Edit, Remove, and Save Reservations

In this section, you will develop the actions methods to edit, remove, and save reservations.

Figure 18-26 shows the relationship between the views and the action methods, including Edit(), Remove(), and Save(). A similar diagram is presented in Figure 18-17, which shows the views and actions to let the user make and save a single reservation at a time. The main difference of the current system is the addition of the HttpGet Remove() and HttpPost Remove() action methods and the Remove view, as shown on the right side of the figure. Further, there is no Cancel() action method. We will look at these action methods in more detail.

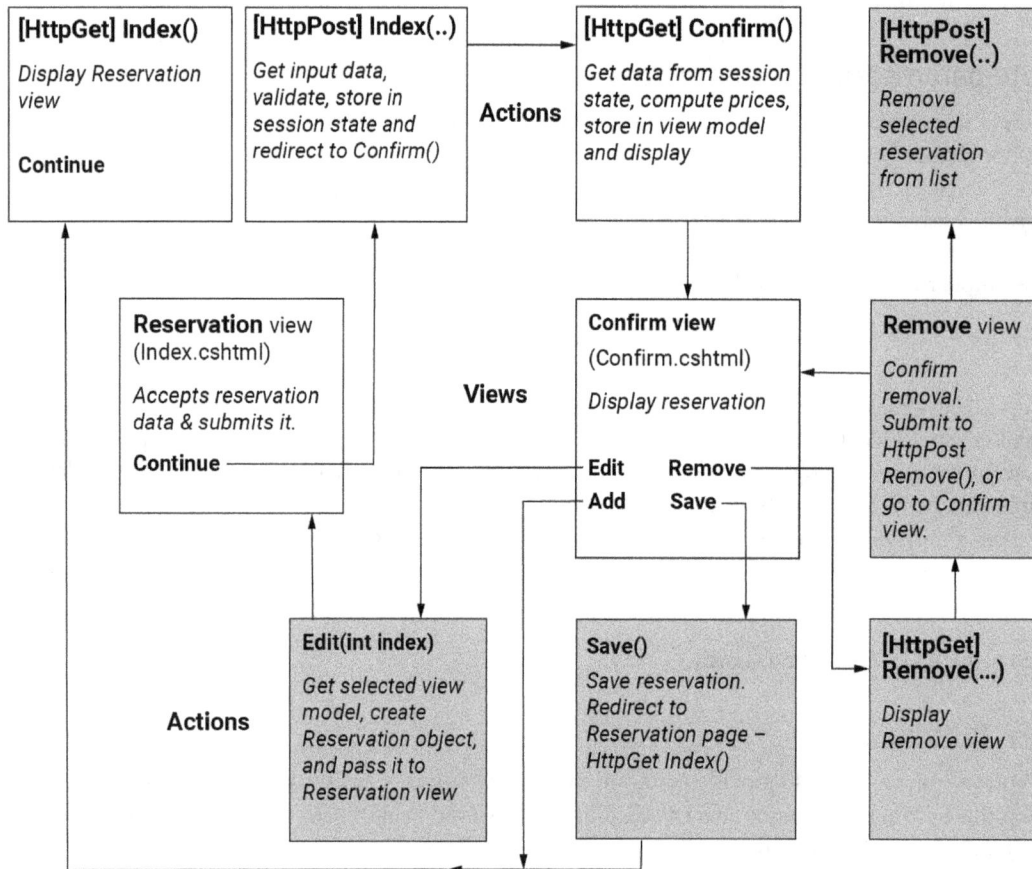

Figure 18-26: Actions methods and views

18.5.1 Editing the Selected Reservation: The Edit() Method

Clicking the *Edit* button in the Confirm view invokes the `Edit()` method and passes the index of the selected item to the method. To display the reservation data for editing, the `Edit()` method performs the following key tasks:

1. Get the list of reservation view models from the session state.

2. Get the selected view model from the list, using the index of the selected item.

3. Create a `Reservation` object from the view model.

4. Pass the `Reservation` object to the Reservation view.

Figure 18-27 shows the `Edit()` method.

```
1   public IActionResult Edit(int index) // Display selected reservation for editing
2   {
3       //Get the List of ReservationViewModel objects from session state
4       List<ReservationViewModel> resViewModels = new List<ReservationViewModel>();
5       string strResViewModels = HttpContext.Session.GetString("StrResViewModels");
6       resViewModels = JsonConvert.DeserializeObject<List<ReservationViewModel>>
                          (strResViewModels);
7       // Get the selected view model using its index
8       ReservationViewModel resViewModel = new ReservationViewModel();
9       resViewModel = resViewModels[index];
10      // Create a Reservation object from the view model
11      Reservation reservation = new Reservation();
12      reservation.Performance = resViewModel.Performance;
13      reservation.PerformDate = resViewModel.PerformDate;
14      reservation.Zone = resViewModel.Zone;
15      reservation.MemberStatus = resViewModel.MemberStatus;
16      reservation.Tickets = resViewModel.Tickets;
17
18      HttpContext.Session.SetString("Action", "Edit");//Store action type in session
19      HttpContext.Session.SetInt32("ResIndex", index);//Store the index in session
20      ViewBag.Performances = resContext.Performance;  //Store performances in ViewBag
21
22      return View("Index", reservation); // Pass Reservation object to Res. view
23  }
```

Figure 18-27: The Edit() method

Line 1 shows that the `Edit (int index)` method has a parameter named *index*, which gets the index of the selected reservation that is passed to the method by the Edit button using the `asp-route` tag helper:

```
asp-route-index="@resIndex"
```

It is important to note that the name of the parameter in the `Edit()` method must be the same as the name of the route parameter in the `asp-route` tag helper.

With the help of the comments within the code, you should be able to understand the tasks performed in each section of the code.

Note that the `Reservation` object that is passed to the Reservation view gets its values from the selected reservation view model (lines 12–16). Because the `HttpPost Index()` method stores each reservation object in the same session variable (`StrReservation`), only the last reservation is available from the session variable. An alternative is to add the `Reservation` objects (instead of the view models) to a separate list, similar to the list of view models, and use that list to get the selected reservation. Such a list can be used also in the `Save()` method to save all reservations.

Step 34: Create the Edit() action method.

Use the code from Figure 18-27 to create the `Edit()` action method within the `ReservationController` class. You may copy the code from Tutorial-starts\Data-files\Ch18-Edit-method.txt.

To test the code, add two reservations and display them on the Confirmation view.

Click the Edit button for one of the two reservations in the list. You should see the Reservation view that displays the data for the selected reservation.

Change one or more data items and click *Continue*. You should see the Confirmation view that displays the list with the updated data for the selected reservation.

You may repeat the test by editing the other reservation.

18.5.2 Removing the Selected Reservation: The Remove() Method

Clicking the Remove button on the Confirm view invokes the `HttpGet` `Remove()` method, which displays the Remove view, as shown in Figure 18-28.

This view displays the performance name and date for the selected reservation and asks the user to confirm the removal of the selected reservation. Clicking the Remove button invokes the `HttpPost Remove()` method that removes the selected reservation from the list and displays the Confirmation view with the updated list. The Cancel button re-displays the Confirmation page with no change.

Let's look at the code for the action methods and the view to remove a reservation.

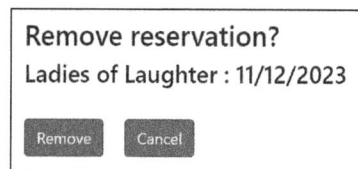

Figure 18-28: The Remove page in the browser

HttpGet Remove() Action Method

Similar to the Edit button, the Remove button (on the Remove view) that invokes the `HttpGet Remove()` method passes the index of the selected reservation to this method using the route parameter, `index`, as shown here:

```
<a asp-controller="Reservation" asp-action="Remove" asp-route-index="@resIndex" >Remove</a>
```

The `HttpGet Remove()` method, shown in Figure 18-29, gets the `index` using the route parameter (line 2), and performs the following tasks:

1. Stores the `index` in the session state (line 4) to pass it to the `HttpPost Remove()` method.

2. Gets the list of reservations from the session state (lines 9–11) and gets the selected reservation from the list (line 13).

3. Passes the selected reservation to the Remove view (line 14).

```
1   [HttpGet]
2   public IActionResult Remove(int index)
3   {
4       HttpContext.Session.SetInt32("ResIndex", index); //Store index in session state
5       // Get the List of reservation view model objects from session state
6       List<ReservationViewModel> resViewModels = new List<ReservationViewModel>();
7       string strResViewModels = HttpContext.Session.GetString("StrResViewModels");
8       resViewModels = JsonConvert.DeserializeObject<List<ReservationViewModel>>
                        (strResViewModels);
9       // Get the selected reservation from the list
10      ReservationViewModel resViewModel = resViewModels[index];
11      return View(resViewModel);
12  }
```

Figure 18-29: The `HttpGet Remove()` method

The Remove View

Figure 18-30 shows the code for the Remove view.

```
1   @using Theater_MVC.ViewModels;
2   @model ReservationViewModel;
3   @{ViewData["Title"] = "Delete";}
4
5   <h3>Remove reservation?</h3>
6   <h4>@Model.Performance  : @Model.PerformDate.ToShortDateString()</h4>
7   <form asp-action="Remove" method="post">
8       <button type="submit" class="btn btn-primary">Remove</button>
9       <button type="button" class="btn btnprimary" onclick="history.back()" >
        Cancel</button>
10  </form>
```

Figure 18-30: The Remove view

Line 2 specifies the model that is passed to the view (from the HttpGet Remove() method).

Line 6 displays the performance name and date.

The Remove button shown in line 8 submits the form to the HttpPost Remove() action method, as specified in the <form> element in line 7. It should be noted that, typically, when a form is submitted, some input data is sent to the server. However, here, the purpose of the form is not to submit any values. We use the form here to invoke the HttpPost Remove() action method when the user clicks the Remove button.

The Cancel button shown in line 9 uses the history.back() function to go back to the previous page, which is the Confirmation page. Be aware that older versions of browsers may have problems with history.back(). Creating a JavaScript function and calling it addresses this issue in many cases.

The HttpPost Remove() Method

The HttpPost Remove() method is invoked when the user clicks the Remove button on the Remove view.

This method removes the selected reservation by performing the following steps:

1. Get the list of reservation view models.

2. Get the index of the selected reservation from the session state and remove the reservation from the list.

3. Store the updated list in the session state.

4. Pass the list to the Confirmation view.

Figure 18-31 shows the code to implement these steps.

```
1   [HttpPost]
2   public IActionResult Remove()
3   {
4       // Get the List of reservation view model objects from session state
5       List<ReservationViewModel> resViewModels = new List<ReservationViewModel>();
6       string strResViewModels = HttpContext.Session.GetString("StrResViewModels");
7       resViewModels = JsonConvert.DeserializeObject<List<ReservationViewModel>>
8                   (strResViewModels);
9       // Remove the selected view model from the list
10      int resIndex =  (int) HttpContext.Session.GetInt32("ResIndex");
11      resViewModels.RemoveAt(resIndex);
12      // Store the updated list in session state
13      strResViewModels = JsonConvert.SerializeObject(resViewModels);
14      HttpContext.Session.SetString("StrResViewModels", strResViewModels);
15      // Pass the updated list to the view
16      return View("Confirm", resViewModels);
17  }
```

Figure 18-31: The HttpPost Remove() method

You should be able to follow the code with the help of the comments within the code.

Step 35: Create the HttpGet Remove() method, Remove view, and the HttpPostRemove() method.

Use the code from Figure 18-29 to create the `HttpGet Remove()` action method that displays the remove confirmation.

Add a view named Remove within the Views\Reservation folder and add the code from Figure 18-30 to the view (Remove.cshtml file).

Use the code from Figure 18-31 to create the `HttpPost Remove()` method that is called after the user confirms the removal and then actually removes the reservation. You may copy the code from Tutorial-starts\Data-files\Ch18-HttpPost-Remove-method.txt.

Step 36: Test the code for removing a reservation.

Add two reservations and display them on the Confirmation view.

Click the Remove button for a reservation, and click the Remove button on the Remove view.

You should see the Confirmation view with the selected reservation removed from the list.

Add another reservation, and click the Remove button for a reservation.

Click Cancel on the Remove view. The Confirmation view should be re-displayed with no change.

18.5.3 Saving Multiple Reservation: The Save() Action Method

The `Save()` action method is invoked when the user clicks the Confirm button. This method saves all reservations in the list to the Reservations table by performing the following steps:

1. Get the list of reservation view model objects from the session state.
2. For each view model in the list, create a Reservation object and add it to the table.
3. Remove the list from the session state and display the Reservation view.

Figure 18-32 shows the `Save()` method that implements these steps.

```
1   public IActionResult Save()
2   {
3       //Get the List of ReservationViewModel objects from session state
4       List<ReservationViewModel> resViewModels = new List<ReservationViewModel>();
5       string strResViewModels = HttpContext.Session.GetString("StrResViewModels");
6       resViewModels = JsonConvert.DeserializeObject<List<ReservationViewModel>>(strResViewModels);
7       // For each view model in the list, create a Reservation object and add it to the table
8       foreach (ReservationViewModel resViewModel in resViewModels)
9       {
10          Reservation reservation = new Reservation();
11          reservation.Performance = resViewModel.Performance;
12          reservation.PerformDate = resViewModel.PerformDate;
13          reservation.Zone = resViewModel.Zone;
14          reservation.MemberStatus = resViewModel.MemberStatus;
15          reservation.Tickets = resViewModel.Tickets;
16          // Add reservation to the Reservation table
17          resContext.Reservations.Add(reservation);
18          resContext.SaveChanges();
19      }
20      HttpContext.Session.Remove("StrResViewModels"); // Remove the list from session state
21      return RedirectToAction("Index"); // Display Reservation view
22  }
```

Figure 18-32: The `Save()` action method

Note that each time through the loop in lines 8–19, the reservation object gets its values from a view model object in the list. Recall that we don't keep the Reservation objects in a list; we keep only view model objects in the list. As discussed in the section on editing, an alternative is to maintain a parallel list of Reservation objects, similar to the list of view model objects. If you do that, you can get the Reservation objects from that list, and you won't need the code shown in lines 11–15.

Step 37: Create the Save() action method.

Use the code shown in Figure 18-32 to create the method. You may copy the code from the file *Tutorial-starts\Data-files\Ch18-Save-method.txt*.

Step 38: Test the Save() method.

To test the code, display the Reservation view, enter the reservation data, and click the Continue button to display the Confirmation view.

Click the **Add New** button and repeat the process to add another reservation.

The Confirmation page should show both sets of data.

Click the **Confirm** button to save the two reservations to the Reservation table and to return to the blank Reservation form.

Make sure the Reservation table contains two new records with the reservation data that you entered.

To open the database, right-click **Data\Theater.mdf** in the Solutions Explorer, and select **Open**.

Now the user can make multiple reservations, keep them in a list, add/remove reservations from the list, and save all reservations from the list to the Reservations table.

Exercises

Exercise 18-1: Validate Performance Name

Add the necessary server-side code to make sure that the user selects a performance before the reservation data is processed and displayed on the Confirmation page. Display a property-level message if the user didn't select a performance—that is, if the value of the performance field is "prompt."

Exercise 18-2: Checkbox for Snacks

If you haven't done Exercise 17-2, you should do it before trying this exercise.

Update the Confirmation page so that the price of snacks ($5), if any, is included in the total cost. To do this, add a method to the Reservation class to compute the snack price, and update the TotalCost() method to include the snack price in the total cost.

Exercise 18-3: Use of Two Controllers

Currently, a single controller controls the Reservation page and the Confirmation page. If you expand this project to include additional pages, then providing different controllers for different groups of pages may enhance maintenance. This exercise will help you get a basic understanding of working with two different controllers by performing the following tasks.

Create a second controller named Confirmation.

Make a copy of the HttpPost Confirm() action method and add it to the ConfirmationController class.

Within the Views folder, create a folder named *Confirmation* and copy the Confirm.cshtml file into the Confirmation folder.

Make sure you change the RedirectToAction() method in the HttpPost Index() method so that it specifies the Confirmation controller.

To test the code, temporarily comment out the Confirm() action method in the Reservation controller. Display the Reservation page, enter the data, and display the Confirmation page. Make sure the Confirmation page displays the reservation data, as before.

To change back to using a single controller (Reservation controller), uncomment the `Confirm()` action method in the Reservation controller. Change the `RedirectToAction()` method in the `HttpPost Index()` method so that it specifies the Reservation controller.

Assignments

E-Commerce Assignment: MVC—Part 2

This assignment develops essentially the same application described in the E-Commerce Assignment in Chapter 16, except that you use the **MVC framework**. This application has two web pages: *Order* and *Cart*. If you did Part 1 of this assignment in Chapter 17, you should have the Order page already developed. If not, do Part 1 (specified below) first, and then do Part 2.

Part 1

P&I Laptops is a small business that assembles and sells a limited set of laptop models online. In Part 1, you will develop a web page named Order within a web application named e-Comm-MVC, which is part of an e-commerce system.

Information on the different models sold by P&I and the software bundled with the laptops are maintained in the database **PI-Laptops** that can be found in the folder Tutorial-starts\DataFiles. The database contains the following two tables:

Model

ModelId	ModelName	Description	CPU	OS	Touch	StartingPrice
101	S330	15 inch lap top with essential perf...	A6-9225	*NULL*	*NULL*	295.00
102	S380	Durable, ready for school, 14 inch ...	i3-8130U	Windows...	*NULL*	450.00
103	S580	S580 is powered by Intel Core™ i...	i5-8250U	Windows...	True	595.00
104	G780	14 inch touch-screen gaming lapt...	i7-8750H	Windows...	True	950.00
105	B780	A light-weight 2-in-one 15-inch to...	C i7-875...	Windows...	True	995.00

StartingPrice is the price for the default configuration.

Software

SoftwareId	SoftwareName	Price
11	MS 365 Personal	65.00
12	MS 365 Family	99.00
13	Office Home & Student	149.00

Create a web project named PI-eComm and develop a web page named **Order.** The web page allows the user to place an order by specifying the model, software, whether a warranty is needed, and the quantity, as shown below. Choose an appropriate layout. Here is an example:

The dropdown list shows the laptop model name and the price from the Models table. Similarly, the radio buttons show the software and their prices from the Software table.

You should use tag helpers to create a form that contains the elements to accept the user input. You should post the form data to an action method named Index. The input elements should be bound to the properties of a class named **Order** to help display the input data on a second page, named **Cart.** Make sure the Order class has a property named OrderNumber that would serve as the primary key for the **Order** table, which will be generated from the class to save the order data, in Part 2.

Part 2

This part validates, processes, and displays the data on a page named Confirmation that lets the user edit, cancel, or save the order.

Part 2.1

When the user clicks the **Select** button, the input data should be validated as specified below, and if the data is valid, it should be displayed on the Cart page,.

Validation: Provide server-side validation to make sure that the user selects a model and specifies a positive integer for quantity. In addition, provide client-side validation for quantity. Optionally, you may use JavaScript for client-side validation to ensure the user selects a model.

Part 2.2

Display the following order data for a **single** laptop model on the Cart page, similar to how you displayed the theater reservation on the Confirmation page:

Order number, model name, software, whether a warranty is needed, quantity, and total cost.
Total cost: Quantity × (Starting price + Cost of software + Cost of warranty).
The cost of the warranty is $99.

Provide buttons on the Cart page to let the user edit or cancel an order. Provide a third button to save the order into a table named *Order* within the database PI-Laptops. Make sure you generate the Order table from the Order class. You may adopt the approach you used to generate the Reservation table from the Reservation class in the tutorial.

Part 2.3

If you did Part 2.2, you may copy the project file from Part 2.2 into a separate folder and modify the views and action methods to do this part. If not, start with the project developed in Part 2.1, and build on it.

Display the following data for **multiple** orders on the Cart page, similar to how multiple reservations are displayed on the Confirmation page:

Order number, model name, software, whether a warranty is needed, quantity, and total cost.
Total cost: Quantity × (Starting price + Cost of software + Cost of warranty).
The cost of the warranty is $99.

Provide buttons on the Cart page to let the user edit or remove any order displayed on the Cart page (see the Confirmation page from the tutorial). In addition, provide a third button to add a new order to the Cart and a fourth button to save all orders to a database table named *Order* within the database PI-Laptops. Make sure you generate the Order table from the Order class.

Auto Rental Assignment: MVC—Part 2

This assignment develops the same application described in the Auto Rental Assignment in Chapter 16, except that you use the MVC framework. This application has two web pages, *Rental* and *Confirmation*, within the web application named AutoRental-MVC. If you did Part 1 of this assignment from Chapter 17, you should have the Rental page already developed. If not, do Part 1 (copied here) first, and then do Part 2. The requirements for Part 1 are included here for your convenience.

Part 1

The assignment develops a web project for "Ace Auto Rentals," which is a small business that rents automobiles. Ace would like you to develop two web pages: (1) the **Rental** page that lets the user select a vehicle and specify related information, and (2) the **Confirmation** page that displays the rental information, including the cost, and lets the user add, edit, remove, and save the rental information. Part 1 develops the Rental page.

Ace has a database named AutoRental.mdf that contains two tables, AutoTypes and Discount. The database can be found in the folder Tutorial-starts\Data-files. A subset of the records in the AutoTypes table is shown here.

AutoTypes

AutoTypeId	AutoType	Capacity	DailyRate	WeeklyRate
11	Standard SUV	5	45	250
12	Full-size SUV	7	65	400
13	Ludury SUV	5	55	350
14	Standard Minivan	7	60	400
15	12 Passenger Van	12	75	500

AutoType represents the type of auto, such as standard SUV, economy car, and small pickup. AutoTypeId is a unique number that identifies each record in the RentalRate table. Capacity represents the passenger (seating) capacity.

Here is the Discount table:

Id	Category	DiscountPercent
11	BestRate	5
12	Government	15
13	Business	10
14	Senior	20

Use the **MVC** framework to develop a web page named **Rental**, within the web project named **AutoRental-MVC**. The web page should let the user specify an auto type, the discount category, whether or not insurance is needed, and the number of days, as shown below. Choose an appropriate layout. What is shown here is just an example.

Note that the dropdown list displays the auto type and daily rate from the AutoType and DailyRate fields of the AutoTypes tables. Similarly, the radio buttons show combined data from the DiscountCategory and DiscountPercent fields of the Discount table.

Use tag helpers to create a form and the elements that accept user input. The form should post the data to an action method named Index. The elements that accept user input should be bound to the properties of a class named **Rental**, so that the input data can be displayed on a second page, named **Confirmation**. Make sure the Rental class has a property named RentalId that would serve as the primary key for the Rental table, which will be generated from the class to save the rental data in Part 2.

Part 2

This part validates, processes, and displays the data on the Confirmation page that lets the user edit, cancel, or save the order.

Part 2.1

The **Select button** should let the user validate the input data as specified below and display the Confirmation page, if the data is valid.

Validation: Provide server-side validation to make sure that the user selects an auto type, specifies a start date that is not a past date, and enters a positive integer for the number of days. In addition, provide client-side validation for the number of days. Optionally, you may use JavaScript for client-side validation to ensure the user selects an auto type.

Part 2.2

Display the following order data for a **single** rental on the Confirmation page, similar to how you displayed the theater reservation on the Confirmation page:

Auto type, start date, discount category, whether insurance is needed, the number of days, and the total cost.

Total cost: Number of days × (Daily rate − Discount amount + Cost of insurance, if any).

The cost of insurance is $10/day. You may use the daily rate, irrespective of the number of days.

Provide buttons on the Confirmation page to let the user edit or cancel a rental. Provide a third button to save the rental data into a table named *Rental* within the database. Make sure you generate the Rental table from the Rental class. You may adopt the approach you used to generate the Reservation table from the Reservation class in the tutorial.

Part 2.3

If you did Part 2.2, you may copy the project file from Part 2.2 into a separate folder and modify the views and action methods to do this part. If not, start with the project developed in Part 2.1 and build on it.

Display the following data for **multiple** rentals on the Confirmation page, similar to how you displayed multiple theater reservations on the Confirmation page:

Auto type, start date, discount category, whether insurance is needed, the number of days, and the total cost.

Total cost: Number of days × (Daily rate − Discount amount + Cost of insurance, if any).

The cost of insurance is $10/day. You may use the daily rate, irrespective of the number of days.

Provide buttons on the Confirmation page to let the user edit or remove any rental displayed on the Confirmation page (see the Confirmation page from the tutorial). In addition, provide a third button to add a new rental to the Confirmation page and a fourth button to save all rentals to a database table, named *Rental* within the database. Make sure you generate the Rental table from the Rental class.

Note that the appendices are included in the e-textbook only. For those using the paperback version, PDFs of the appendices can be freely downloaded from the title's website at https://www.prospectpressvt.com/textbooks/philip-web-development. (Scroll to the red horizontal menu bar, then click on "Student Resources.")

Index

www.ingramcontent.com/pod-product-compliance
Lightning Source LLC
Chambersburg PA
CBHW060945210326
41598CB00031B/4724